Jürgen Appell, Józef Banaś, Nelson Merentes
Bounded Variation and Around

De Gruyter Studies in Nonlinear Analysis and Applications

Volume 17

Jürgen Appell, Józef Banaś, Nelson Merentes

Bounded Variation and Around

DE GRUYTER

Mathematics Subject Classification 2010
Primary: 26A45, 47H30; Secondary: 26A15, 26A16, 26A30, 26A42, 26A46, 45G05, 45G10, 46B25

Authors
Prof. Dr. Jürgen Appell
University of Würzburg
Department of Mathematics
Emil-Fischer-Str. 30
97074 Würzburg
Germany
jurgen@dmuw.de

Prof. Dr. Józef Banaś
Rzeszów University of Technology
Department of Mathematics
al. Powstańców Warszawy 8
35-959 Rzeszów
Poland
jbanas@prz.edu.pl

Prof. Dr. Nelson Merentes
Central University of Venezuela
School of Mathematics
Paseo Los Ilustres
Caracas 1020
Venezuela
nmerucv@gmail.com

ISBN 978-3-11-026507-1
e-ISBN 978-3-11-026511-8
Set-ISBN 978-3-11-026624-5
ISSN 0941-813X

Library of Congress Cataloging-in-Publication Data
A CIP catalog record for this book has been applied for at the Library of Congress.

Bibliographic information published by the Deutsche Nationalbibliothek
The Deutsche Nationalbibliothek lists this publication in the Deutsche Nationalbibliografie; detailed bibliographic data are available in the Internet at http://dnb.dnb.de.

Typesetting: le-tex publishing services GmbH, Leipzig
Printing and binding: Hubert & Co. GmbH & Co. KG, Göttingen
♾Printed on acid-free paper
Printed in Germany

MIX
Papier aus verantwortungsvollen Quellen
FSC® C016439

www.degruyter.com

Preface

Functions of bounded variation are fascinating mathematical objects which, since their definition by C. Jordan more than a century ago, have found much attention in real analysis, functional analysis, measure theory, integration theory, operator theory, Fourier analysis, nonlinear analysis, and even some fields of mathematical physics. Also, Jordan's original definition has been extended to more general classes of functions which are often motivated by (and in fact, lead to) convergence criteria for Fourier series or existence results for Riemann–Stieltjes integrals. Such general classes are of independent interest in the study of dual spaces in linear functional analysis and operator theory. For example, the classical Riemann–Stieltjes integral provides a natural one-to-one correspondence between the dual of the space $C([a,b])$ of continuous functions and the space $NBV([a,b])$ of (normalized) functions of bounded variation.

The basic facts about functions of bounded variation may be found in virtually any textbook on real or functional analysis. Among the application-oriented books which use functions of bounded variation, we mention the classical work of E. Giusti [122] and the recent monograph by L. Ambrosio, N. Fusco and D. Pallara [7]. However, to the best of our knowledge, there are no books which provide a thorough and self-contained account of functions of (generalized) bounded variation as an independent subject, the methods connected with their study, nor their applications to various analytical or geometrical problems. An exception is the French Lecture Notes *Variation totale d'une fonction* by M. Bruneau (1974, see [65] in the list of references) which, however, covers topics other than those treated here. Some kind of "rudimentary predecessor" of this book is the Spanish monograph *El operador de composición en espacios de funciones con algún tipo de variación acotada* by the third author and S. Rivas (1996, [226]). We remark that the monograph [226] does not contain full proofs, and only sketches very few applications to composition operators. Roughly speaking, the whole material treated in [226] is contained in Chapters 1, 2, 5, and 6 of the present book.

The purpose of this monograph is four-fold. Firstly, we want to collect the basic facts about functions of bounded variation and related functions, such as Lipschitz continuous or absolutely continuous functions, to present the main ideas which have been shown to be useful in studying their properties, and to provide a comparison of their importance and suitability for applications. Secondly, we want to study the (sometimes quite surprising and pathological) behavior of nonlinear composition and superposition operators in spaces of functions of bounded variation with a particular emphasis on continuity and boundedness in norm, or global and local Lipschitz conditions. (It is the second part of this book, treating nonlinear operator theory, which may be considered as complementary to the standard reference [271] and motivated us to submit it to the De Gruyter Series in Nonlinear Analysis and Applications.)

Third, some topics like Riemann–Stieltjes integrals and their role in the duality theory of Banach spaces which have not been considered in the above mentioned Spanish monograph will be treated in some detail. Finally, we will also discuss some newer developments which were still unknown when the monograph [226] appeared in 1996. In the last sections of nearly every chapter called "Comments," we discuss additional or peripheral topics of interest, alternate presentations, and historical commentary. Moreover, we try to take into account recent results and state several open problems; therefore, this book might also be a fruitful source of inspiration for young researchers and postgraduate students who are interested in the subject.

The reader will notice that apart from recent developments and open problems, the last section of each chapter contains a considerable number of exercises. This may be somewhat unusual for a research monograph; however, we are convinced that such exercises provide a deeper insight into the topic and might be useful for students during lectures and seminars on higher analysis. For a variety of reasons, we have deemed these exercises important for a full understanding of the material. Some of them include straightforward "computations," some are simply detail checking, and some unveil the seeds of ideas essential for later developments. Exercises which, in our opinion, are more technical are marked with an *, but this is sometimes a matter of taste.

Regarding the book as a whole, the choice of topics is necessarily far from comprehensive and has been made with a number of criteria in mind. The most obvious one is mathematical importance, but some additional topics were chosen because it is possible to discuss them in an easily accessible way, others because they have some unusual feature, and some because the authors felt that certain results should be treated separately in the comment section so as not to overburden the presentation in the main sections.

The only prerequisite for understanding this book is a modest background in real analysis, functional analysis, and operator theory. It addresses nonspecialists who want to get an idea of the development of the theory, methods, and applications of different notions of functions of bounded variation in the last 50 years as well as a glimpse into the diversity of the directions in which current research is moving. Apart from abstract results, we have tried to present many examples which give some indication as to the flavor of the subject. There are now several good books available with a particular emphasis on examples; we only mention the classical work *Counterexamples in Analysis* by B. R. Gelbaum and J. M. H. Olmstedt (1964, see [118]) and the more recent Russian book *Real Analysis in Exercises* by A. N. Bakhvalov et al. (2005, [39]) which is an almost inexhaustible source of beautiful examples and counterexamples.

The large number of examples is explained by the authors' conviction that monographs of this kind would not be much use if all they gave were formal definitions, abstract theorems, and technical proofs. To understand a concept, one needs to know what it means intuitively, why it is important, and why it was first introduced. Above all, if it is a fairly general concept, then one wants to know some good examples – ones that are not too simple nor too complicated. In fact, a good example is much easier to understand than a general definition, and more experienced readers will be able to work out a general definition by "abstracting" the important properties from an example, rather than finding a significant example by "concretizing" a general definition. Or, to put it the way one of our colleagues recently said: "To understand a theory, you needn't know all the theorems, but you have to know all the relevant examples!"

Some parts of the presentation may seem redundant, but we don't think that this is a flaw. For example, the spaces $\mathrm{WBV}_p([a,b])$ and $\mathrm{RBV}_p([a,b])$ of functions of bounded p-variation in the sense of Wiener and Riesz, respectively, are special cases of the more general spaces $WBV_\phi([a,b])$ and $RBV_\phi([a,b])$ of functions of bounded ϕ-variation in the sense of Young and Medvedev, respectively, which may be obtained by the special choice $\phi(u) = |u|^p$. However, instead of treating the general BV_ϕ-spaces right from the beginning, we start with the more special BV_p-spaces to make the concepts and results more transparent. Similarly, instead of presenting a fairly general characterization of locally Lipschitz continuous composition operators between the spaces $\mathrm{Lip}([a,b])$ and $\mathrm{BV}([a,b])$ (Theorem 5.10), we first prove the same characterization for operators from $\mathrm{BV}([a,b])$ into itself (Theorem 5.9), which has been not only the historical starting point of such results, but also the standard model of subsequent techniques for proving them. As for the order of the chapters themselves, the aim has been to make it the most natural one from a pedagogical point of view and to give the book some sense of direction.

The notation used in this book is standard. A detailed symbol index at the end may be helpful to find special notations, such as the numerous variations we consider in Chapters 1 and 2. The end of a definition is marked by ■, the end of a proof by □, and the end of an example by ♥.

This book could not have been realized without the possibility of meetings in Germany, Poland, and Venezuela, generously supported by the European Commission and the Ministry of Education and Research of Venezuela. In particular, the first author gratefully acknowledges hospitality of the Central University of Caracas (Venezuela) and of the Rzeszów University of Technology (Poland). A large debt of gratitude is also owed to our colleagues Daria Bugajewska, Dariusz Bugajewski, Janusz Matkowski, and Martin Väth for looking through some parts of an earlier version of the manuscript. Of course, the reader has to thank them for each passage that "works," and only blame us for those (hopefully, not too numerous) that don't.

Last, but by no means least, it is a great pleasure to thank Friederike Dittberner, Silke Hutt, and Anja Möbius from De Gruyter-Verlag Berlin, as well as le-tex Leipzig, for their a.i. (almost infinite) patience and constant kind support.

Würzburg, Rzeszów, Caracas, Summer 2013 The Authors

Contents

Introduction

In 1829, Johan Peter Gustav Lejeune Dirichlet proved that the Fourier series of a piecewise monotone real function is pointwise convergent. This result is now known as the *Dirichlet criterion* in the theory of Fourier series and seems to be the first mathematically rigorous proof of Fourier's conjecture which was raised in 1807 and published in his pioneering work *Théorie analytique de la chaleur* on the representability of functions by means of a trigonometric series. According to Béla Szőkefalvi–Nagy, the history of the Fourier series started with an exchange of letters between Jean–Baptiste d'Alembert, Leonhard Euler, and Daniel Bernoulli on the problem of the clamped vibrating string.

A further milestone was Camille Jordan's paper *Sur les séries de Fourier* (1881) in which he introduced the notion of *functions of bounded variation* and extended the Dirichlet criterion to this class of functions. In the same paper, he also proved that a function has bounded variation if and only if it may be represented as a difference of two monotonically increasing functions. In modern terminology, this means that the space of functions of bounded variation on $[a, b]$, usually denoted by $\mathrm{BV}([a, b])$, is the *linear hull of the set of all monotone functions* (which do not form a linear space on their own).

Subsequently, Jordan's definition has been generalized and extended in various directions, both from the viewpoint of the theory of real functions and from more application-oriented viewpoints. For example, in 1915, Charles De la Vallée–Poussin introduced the class of *functions of bounded second variation* and proved that every such function can be represented as a difference of two *convex functions*. In 1934, Mihael T. Popoviciu proved a parallel result for higher order variations, building on the notion of *functions of bounded k-th variation*. In case $k = 1$, one recovers Jordan's decomposition of classical BV-functions and in case $k = 2$, the De la Vallée–Poussin decomposition.

While these results seem to be more of theoretical interest, two other extensions have turned out to be extremely useful in applications to Fourier series, namely, Norbert Wiener's definition of *functions of bounded p-variation* (1924) and, more generally, Laurence C. Young's definition of *functions of bounded ϕ-variation* (1937), where ϕ is a suitable convex increasing "gauge function." A different notion of functions of bounded p-variation has been introduced by Frigyes Riesz (1910), and its generalization to bounded ϕ-variation by Yurij T. Medvedev (1953). Loosely speaking, it can be said that passing from p-variations to ϕ-variations is similar to passing from Lebesgue spaces to Orlicz spaces. The variations introduced by Riesz and Medvedev seem to be very natural from the functional analytic point of view. In fact, an important result (called the *Riesz lemma* by some people) states that a function f has bounded p-variation in Riesz's sense ($1 < p < \infty$) if and only if f is absolutely continuous and its derivative f' exists almost everywhere and belongs to the Lebesgue space L_p.

(This means that Riesz introduced Sobolev spaces, at least in the scalar case, several years prior to Sobolev.) A parallel result holds for functions f of bounded ϕ-variation in Medvedev's sense, where the L_p-condition on the derivative f' has to be replaced by some appropriate integrability condition for $\phi \circ f'$. Clearly, an analogous result for $p = 1$ is *not* true because functions in $RBV_1 = BV$ are generally not continuous, let alone absolutely continuous.

We notice that a "higher order variant" of Riesz's characterization leads to the following question: how can we characterize, by means of bounded variation techniques, all absolutely continuous functions f whose second derivative f'' exists almost everywhere and belongs to L_p? This problem was solved in 1991 by the third author who introduced the class of *functions of bounded* $(p, 2)$*-variation*, providing in this way a unified approach to the Riesz theory and the De la Vallée–Poussin theory. In the meantime, yet more general notions of bounded variation have been introduced and studied by Daniel Waterman (1976), Michael Schramm (1982), Boris Korenblum (1975), Hwa Jun Kim (2006), and others.

All of these notions and results, together with a large variety of applications, are scattered over many research papers which appeared during the last 50 years and are contained in the detailed list of references at the end of this book. However, as far as we know, they have not been collected in a single monograph, and thus the aim of this book is to fill the gap.

As far as applications are concerned, let us mention four fields in which functions of bounded variation turn out to be useful:
- They admit a decomposition into, hopefully, simpler functions.
- They are connected to geometric notions like curve length or surface area.
- They make it possible to define Riemann–Stieltjes type integrands.
- They provide (uniform) convergence results for Fourier series.

For the classical Jordan space $BV([a, b])$, this program has been accomplished quite successfully over the past 100 years:
- BV-functions may be represented as differences of increasing functions.
- BV-functions are precisely those whose graphs are rectifiable curves.
- BV-functions define Riemann–Stieltjes integrals for continuous integrands.
- BV-functions have, after a suitable normalization, convergent Fourier series.

As we will see, for the other spaces of functions of (generalized) variation, this ambitious program can be fulfilled only in part.

This monograph consists of 8 chapters. In Chapter 0, which is introductory, we collect everything on measure and integration, function spaces, and functional analysis which will be needed in the following chapters. Chapter 1 starts with the definition and discussion of the classical space $BV([a, b])$ of real *functions of bounded variation* on $[a, b]$, including Jordan's decomposition of functions in $BV([a, b])$ as differences of two increasing functions on $[a, b]$. We also consider the spaces $WBV_p([a, b])$ $(1 \leq p < \infty)$ of *functions of bounded p-variation* (in Wiener's sense) and discuss their basic properties and connections with other function spaces, with a particular emphasis on spaces of *continuous, Lipschitz continuous, Hölder continuous*, and *absolutely continuous functions*. In this chapter, we will also briefly discuss *functions of several* (in particular, two) *variables*, mainly due to Vyacheslav V. Chistyakov. The corresponding results are rather technical, but seem to have important applications.

As mentioned before, the space $\mathrm{BV}([a,b])$ has been generalized in various directions. Norbert Wiener and Laurence C. Young distorted the measurement of intervals in the range of functions by considering p-th powers or, more generally, continuous increasing *gauge functions* ϕ, introducing the classes $\mathrm{WBV}_p([a,b])$ and $\mathrm{WBV}_\phi([a,b])$. Subsequently, Daniel Waterman and Michael Schramm admitted countable families of such gauge functions in order to generalize the concept of variation, which leads to the classes $\Lambda\,\mathrm{BV}([a,b])$ and $\Phi\,\mathrm{BV}([a,b])$. As mentioned before, however, the most interesting generalization has been introduced by Frigyes Riesz and extended by Yurij T. Medvedev to the setting of gauge functions; we denote the corresponding classes by $\mathrm{RBV}_p([a,b])$ and $\mathrm{RBV}_\phi([a,b])$, respectively. All of these generalizations of the concept of bounded variation will be discussed in Chapter 2 which is, by far, the largest chapter in the book.

A flaw of all the extensions described above is the loss of an *effective decomposition* of a function from the corresponding classes into, hopefully, simpler functions, such as for Jordan's classical space $\mathrm{BV}([a,b])$. In 1975, Boris Korenblum considered a new kind of variation, called *κ-variation*, introducing a function κ for distorting the length $|t_j - t_{j-1}|$ of the subinterval $[t_{j-1}, t_j]$ in a partition $\{t_0, t_1, \ldots, t_m\}$ of the underlying interval, rather than the expression $|f(t_j) - f(t_{j-1})|$ in the range. One advantage of this alternate approach is that a *function of bounded κ-variation* may be decomposed into the difference of two simpler functions called *κ-decreasing functions* (the precise definition will be given in Section 2.5 below).

Absolutely continuous functions are intimately related to functions of bounded variation in several respects. First of all, absolute continuity is equivalent to the combination of three properties, namely, continuity, bounded variation, and invariance of (Lebesgue) nullsets; this is the assertion of the famous *Vitali–Banach–Zaretskij theorem*. Second, as mentioned above, the absolutely continuous functions on $[a,b]$ are precisely those functions $f \in \mathrm{BV}([a,b])$ whose derivatives f' (which exist almost everywhere, by Lebesgue's differentiation theorem) belong to $\mathrm{L}_1([a,b])$; moreover, the fundamental theorem of calculus (in the Lebesgue integral version) holds in this case which makes it possible to recover an absolutely continuous function, up to an additive constant, from its derivative. If we replace the condition $f \in \mathrm{AC}([a,b])$ with the stronger condition $f \in \mathrm{RBV}_p([a,b])$ $(1 < p < \infty)$ in this statement, we get precisely the condition $f' \in \mathrm{L}_p([a,b])$; this will be discussed in further detail, together with some generalization due to Yurij T. Medvedev, in Chapter 3.

In Chapter 4, we study *Riemann–Stieltjes integrals*. It is well known that by means of the classical Riemann–Stieltjes integral, we can construct a surjective isometry between the dual space of the Chebyshev space $C([a,b])$ and a subspace of (suitably normalized) functions in $\mathrm{BV}([a,b])$. Similarly, an analogous isometry makes it possible to identify the dual space of the Lebesgue space $L_p([a,b])$ $(1 < p < \infty)$ with the space $\mathrm{RBV}_{p/(p-1)}([a,b])$, where $\mathrm{RBV}_p([a,b])$ denotes the space of functions of bounded p-variation in Riesz's sense introduced in Chapter 2. This one-to-one correspondence is different from the usual identification of the dual space of $L_p([a,b])$ with $L_{p/(p-1)}([a,b])$; moreover, it does not require regularization here since all functions of bounded Riesz p-variation are continuous for $p > 1$. We may also consider the same isometry for the case $p = 1$ which leads to the space $\mathrm{RBV}_\infty([a,b]) = \mathrm{Lip}([a,b])$ of Lipschitz continuous functions on $[a,b]$, but not for the case $p = \infty$, which requires a different approach.

Given a function $h : \mathbb{R} \to \mathbb{R}$, in Chapter 5, we study the (autonomous) *nonlinear composition operator* C_h defined by

$$C_h f(x) := h(f(x)) \quad (a \le x \le b) \tag{1}$$

for $f : [a,b] \to \mathbb{R}$ belonging to several spaces of functions of (generalized) bounded variation and related spaces. It turns out that a typical condition, both necessary and sufficient, for C_h to map such a space into itself is a *local Lipschitz condition* on h on the real line, i.e. $|h(u) - h(v)| \le k(r)|u - v|$ for $|u|, |v| \le r$. Afterwards, we will briefly consider sufficient conditions on h under which the corresponding operator C_h is bounded and/or continuous in norm, or even uniformly continuous on bounded sets or on the whole space. In view of the applications, Lipschitz conditions for the operator C_h are of particular interest. In the second part of Chapter 5, we will show that a *global Lipschitz condition* for C_h in norm often leads to a strong *degeneracy* for h, namely, to affine functions of the form $h(u) = \alpha + \beta u$. Roughly speaking, this means that whenever a global Lipschitz condition is imposed on the operator C_h, the underlying problem is necessarily *linear*, and thus of very limited interest. On the other hand, a *local Lipschitz condition* for C_h is fulfilled for sufficiently large classes of nonlinear functions h (typically, those whose derivative h' exists and is locally Lipschitz). This emphasizes the need of imposing local Lipschitz conditions, rather than global conditions, if we want to apply classical *fixed point principles* like Banach's contraction mapping theorem.

In Chapter 6, we try to develop a parallel theory for the (nonautonomous) *nonlinear superposition operator* S_h defined by

$$S_h f(x) = h(x, f(x)) \quad (a \le x \le b)\,, \tag{2}$$

where h is now defined on the product $[a,b] \times \mathbb{R}$. It turns out that many results which hold in the autonomous case $h : \mathbb{R} \to \mathbb{R}$ are *not* true in the nonautonomous case, or simply unknown. Loosely speaking, the "terra incognita" is much larger here, and many problems are open. For example, conditions which are both necessary and sufficient for the function h to guarantee that the corresponding operator (2) maps a function space into itself are known only in a few exceptional cases.

Here is an example of what we mean. It is not hard to show that the autonomous operator (1) maps the space $C([a,b])$ into itself (i.e. $x \mapsto h(f(x))$ is continuous whenever $x \mapsto f(x)$ is continuous) if and only if h is continuous on the real line. Similarly, the operator (1) maps the space $C^1([a,b])$ into itself (i.e. $x \mapsto h(f(x))$ is continuously differentiable whenever $x \mapsto f(x)$ is continuously differentiable) if and only if h is continuously differentiable on the real line. The first result carries over without change to the nonautonomous case: the operator (2) maps the space $C([a,b])$ into itself (i.e. $x \mapsto h(x, f(x))$ is continuous whenever $x \mapsto f(x)$ is continuous) if and only if h is continuous on $[a,b] \times \mathbb{R}$. In contrast to this, however, the operator (2) may map the space $C^1([a,b])$ into itself (i.e. $x \mapsto h(x, f(x))$ is continuously differentiable whenever $x \mapsto f(x)$ is continuously differentiable) even if h is *discontinuous* somewhere on $[a,b] \times \mathbb{R}$! A corresponding sophisticated counterexample, due to Janusz Matkowski, is given in Section 6.2.

The final Chapter 7 is concerned with a few selected applications. Historically, the most important applications of the space BV and its various generalizations refer to the *Fourier series*. In the first two sections of Chapter 7, we will outline some of these applications, with a particular emphasis on the Waterman space ΛBV. Building on the existence and uniqueness results for nonlinear problems obtained in Chapters 5 and 6, we will also consider applications to *integral equations*. We remark, however, that the main purpose of this monograph is to outline the theory, so little attention is given to the applications themselves. Needless to say, the few applications discussed in the last chapter are by no means exhaustive; we hope that the material contained in this book will be a source for further research.

As pointed out in the Preface, we consider examples, counterexamples, and open problems of fundamental importance to get a deeper insight into the subject. Surprisingly enough, some of these open problems are sometimes easily formulated, but apparently hard to solve.

Here are two typical examples. The first question is related to admissible "inner functions" (i.e. changes of variables) which preserve bounded variation:

- Give conditions on a map $\tau : [a,b] \to [a,b]$, possibly both necessary and sufficient, such that $f \circ \tau \in \mathrm{BV}([a,b])$ for all $f \in \mathrm{BV}([a,b])$.

It is easy to see that monotonicity of τ is sufficient, but not necessary. Less obvious is the fact that bounded variation of τ is not sufficient. As we will see in Proposition 1.17, the correct condition on τ is *pseudomonotonicity*, a property which is intermediate between monotonicity and bounded variation.

The second question is related to admissible "outer functions" which preserve bounded variation and may be formulated as follows. Let the space $\mathrm{BV}([a,b])$ be equipped with the natural norm

$$\|f\|_{BV} = |f(a)| + \mathrm{Var}\,(f; [a,b])\,, \tag{3}$$

where $\mathrm{Var}\,(f; [a, b])$ denotes the total variation of f on $[a, b]$. Now, given $h : \mathbb{R} \to \mathbb{R}$, consider the (autonomous) composition operator C_h defined by (1), i.e. $C_h f := h \circ f$. Then the following three natural problems arise:

- Give conditions on h, possibly both necessary and sufficient, under which $h \circ f \in BV([a, b])$ for all $f \in BV([a, b])$, which means that the operator C_h maps the space BV into itself.
- Give conditions on h, possibly both necessary and sufficient, under which the operator C_h is bounded in the norm (3).
- Give conditions on h, possibly both necessary and sufficient, under which the operator C_h is continuous in the norm (3).

The first problem was solved by Michael Josephy in 1981 who showed that C_h maps BV into itself if and only if h is locally Lipschitz on the real line. The answer to the second question is almost trivial: a straightforward calculation shows that the operator C_h is *always* bounded in the norm (3) whenever it maps BV into itself. On the other hand, the answer to the third question is *unknown*: one does not know necessary and sufficient conditions for the continuity of the operator C_h in the norm (3). (Of course, conditions which are just sufficient are easily found.) In particular, we do not know whether continuity of C_h in norm follows from local Lipschitz continuity of h on $\mathbb{R}$, and we believe that a counterexample, if there is any, will presumably be rather complicated. Even worse, if we replace the autonomous operator (1) by the nonautonomous operator (2), all three questions stated above are open.

0 Prerequisites

This chapter is introductory and collects some of the material which will be needed in subsequent chapters. First, we recall basic notions and facts about (Lebesgue) integrable functions, with a particular emphasis on L_p spaces. Afterwards, we discuss some concepts from functional analysis, including the Hahn–Banach theorem and its consequences, and introduce several function spaces which are related to functions of bounded variation. Since some of our results are standard and beyond the scope of this monograph, we state them without proofs. The proofs of such results may be found in any textbook on measure theory or functional analysis. We remark that, although we restrict ourselves to subsets of the real line, many results carry over without change to measurable subsets of the Euclidean space $\mathbb{R}^n$.

0.1 The Lebesgue integral

We assume that the reader is familiar with the construction of the Lebesgue measure and Lebesgue integral on the real line. We will denote the Lebesgue measure of a measurable set $M \subseteq \mathbb{R}$ by $\lambda(M)$. As usual, properties are supposed to hold a.e. (almost everywhere), i.e. outside a Lebesgue nullset.

We start with a result which shows that convergence a.e. of a sequence of measurable functions means, roughly speaking, uniform convergence up to a set of arbitrarily small measure.

Theorem 0.1 (Egorov). *Let $M \subseteq \mathbb{R}$ be a measurable set with $\lambda(M) < \infty$, and let $(f_n)_n$ be a sequence of measurable functions $f_n : M \to \mathbb{R}$ satisfying*

$$f_n(x) \to f(x) \quad (n \to \infty) \tag{0.1}$$

a.e. on M. Then for each $\varepsilon > 0$, there exists a measurable subset $M_\varepsilon \subseteq M$ such that $\lambda(M \setminus M_\varepsilon) < \varepsilon$ and $(f_n)_n$ converges uniformly on M_ε to f.

The following result shows that measurability of a function means continuity up to a set of arbitrarily small measure.

Theorem 0.2 (Luzin). *Let $M \subseteq \mathbb{R}$ be a measurable set and $f : M \to \mathbb{R}$ a function. Then f is measurable on M if and only if for each $\varepsilon > 0$, we can find a closed subset $M_\varepsilon \subseteq M$ such that $\lambda(M \setminus M_\varepsilon) < \varepsilon$ and the restriction $f|_{M_\varepsilon} : M_\varepsilon \to \mathbb{R}$ of f to the set M_ε is continuous.*

The following Theorems 0.3–0.6 refer to sequences of Lebesgue integrable functions; the proofs may be found in any textbook on measure theory.

Theorem 0.3 (Levi). *Let $M \subseteq \mathbb{R}$ be a measurable set, and let $(f_n)_n$ be an increasing sequence of nonnegative measurable functions $f_n : M \to \mathbb{R}$ satisfying* (0.1) *everywhere on M. Then*

$$\int_M f(x)\,dx = \lim_{n\to\infty} \int_M f_n(x)\,dx\,, \tag{0.2}$$

where the functions and integrals are allowed to be infinite.

Theorem 0.4 (Lebesgue). *Let $M \subseteq \mathbb{R}$ be a measurable set, and let $(f_n)_n$ be a sequence of measurable functions $f_n : M \to \mathbb{R}$ satisfying* (0.1) *a.e. on M. Suppose that there exists an integrable function $F : M \to \mathbb{R}$ such that*

$$|f_n(x)| \le F(x) \quad (n \in \mathbb{N},\ x \in M)\,. \tag{0.3}$$

Then f is also integrable and satisfies (0.2).

Theorem 0.5 (Fatou). *Let $M \subseteq \mathbb{R}$ be a measurable set, and let $(f_n)_n$ be a sequence of nonnegative measurable functions $f_n : M \to \mathbb{R}$ satisfying* (0.1) *a.e. on M. Then*

$$\int_M f(x)\,dx \le \liminf_{n\to\infty} \int_M f_n(x)\,dx\,. \tag{0.4}$$

Theorem 0.6 (absolute continuity of the integral). *Let $M \subseteq \mathbb{R}$ be a measurable set, and let $f : M \to \mathbb{R}$ be integrable. Then for each $\varepsilon > 0$, there exists a $\delta > 0$ such that for any measurable subset $N \subseteq M$ with $\lambda(N) \le \delta$, we have*

$$\int_N |f(x)|\,dx \le \varepsilon\,. \tag{0.5}$$

The following result refers to measurable functions defined on the Euclidean space $\mathbb{R}^n$ and gives a sufficient condition under which the "order of integration" on lower dimensional sections may be interchanged.

Theorem 0.7 (Fubini). *Suppose that $f : \mathbb{R}^p \times \mathbb{R}^q \to \mathbb{R}$ is integrable on $\mathbb{R}^n$, where $p+q = n$ and f is written in the form $z = f(x,y)$ $(x \in \mathbb{R}^p, y \in \mathbb{R}^q)$. Then the function $f(x,\cdot) : \mathbb{R}^q \to \mathbb{R}$ is integrable on $\mathbb{R}^q$ for almost all $x \in \mathbb{R}^p$, and the function $f(\cdot,y) : \mathbb{R}^p \to \mathbb{R}$ is integrable on $\mathbb{R}^p$ for almost all $y \in \mathbb{R}^q$. Moreover, the equality*

$$\int_{\mathbb{R}^p} \left\{ \int_{\mathbb{R}^q} f(x,y)\,dy \right\} dx = \int_{\mathbb{R}^n} f(x,y)\,d(x,y) = \int_{\mathbb{R}^q} \left\{ \int_{\mathbb{R}^p} f(x,y)\,dx \right\} dy \tag{0.6}$$

holds.

A certain converse of the Fubini theorem is given by the following:

Theorem 0.8 (Tonelli). *Suppose that $f : \mathbb{R}^p \times \mathbb{R}^q \to \mathbb{R}$ is measurable, where $p + q = n$ and f is written in the form $z = f(x, y)$ $(x \in \mathbb{R}^p, y \in \mathbb{R}^q)$. Suppose that the function $f(x, \cdot) : \mathbb{R}^q \to \mathbb{R}$ is integrable on $\mathbb{R}^q$ for almost all $x \in \mathbb{R}^p$, and the function $F : \mathbb{R}^p \to \mathbb{R}$ defined by*

$$F(x) := \int_{\mathbb{R}^q} |f(x, y)|\, dy \tag{0.7}$$

is integrable on $\mathbb{R}^p$. Then f is integrable on $\mathbb{R}^n$.

We remark that one can construct examples which show that one cannot drop the absolute value of $f(x, y)$ in (0.7).

A property of functions which is particularly important (and is here supposed to hold a.e.) is *boundedness*. The *essential supremum* and *essential infimum* of a measurable function $f : M \to \mathbb{R}$ on M are defined by

$$\operatorname{esssup}\{f(x) : x \in M\} := \inf_{\lambda(N)=0} \sup\{f(x) : x \in M \setminus N\} \tag{0.8}$$

and

$$\operatorname{essinf}\{f(x) : x \in M\} := \sup_{\lambda(N)=0} \inf\{f(x) : x \in M \setminus N\}, \tag{0.9}$$

respectively, where the infimum in (0.8) and the supremum in (0.9) are taken over all nullsets $N \subseteq M$. Functions f with finite essential supremum and essential infimum are called *essentially bounded.*[1]

Definition 0.9. For $1 \le p < \infty$, the *Lebesgue space* $L_p([a, b])$ consists of all measurable functions $f : [a, b] \to \mathbb{R}$, for which

$$\int_a^b |f(x)|^p\, dx < \infty. \tag{0.10}$$

For $p = \infty$, the Lebesgue space $L_\infty([a, b])$ consists of all measurable functions $f : [a, b] \to \mathbb{R}$ which are essentially bounded on $[a, b]$. ∎

We will consider the space $L_p([a, b])$ with the usual norm

$$\|f\|_{L_p} := \begin{cases} \left\{\int_a^b |f(x)|^p\, dx\right\}^{1/p} & \text{for } 1 \le p < \infty, \\ \operatorname{esssup}\{|f(x)| : a \le x \le b\} & \text{for } p = \infty. \end{cases} \tag{0.11}$$

To be precise, (0.11) is not a norm on $L_p([a, b])$ since $\|f\|_{L_p} = 0$ does not imply that $f(x) \equiv 0$ everywhere, but only *almost everywhere* on $[a, b]$. This flaw may be overcome in two different, but equivalent ways. Either we identify any two functions f and g if

1 The word "essential" suggests that nullsets are negligible, i.e. f is bounded outside some negligible set.

the set of all $x \in [a,b]$ such that $f(x) \neq g(x)$ is a nullset, i.e. f and g coincide almost everywhere. Or (what is essentially the same), we introduce an equivalence relation on $L_p([a,b])$ by calling two functions f and g *equivalent* (and writing $f \sim g$) if

$$f - g \in N_p([a,b]) := \{h \in L_p([a,b]) : h(x) \equiv 0 \text{ a.e. on } [a,b]\}. \tag{0.12}$$

In this setting, (0.11) then defines a norm on the *quotient space* $L_p([a,b])/N_p([a,b])$ for which we still write $L_p([a,b])$ to avoid clumsy notation.[2] It is not hard to prove that $(L_p([a,b]), \|\cdot\|_{L_p})$ is a Banach space for each p.

Throughout this book, we denote by p' the *conjugate index* to p defined by[3]

$$p' := \begin{cases} \infty & \text{if } p = 1, \\ \frac{p}{p-1} & \text{if } 1 < p < \infty, \\ 1 & \text{if } p = \infty. \end{cases} \tag{0.13}$$

In the following proposition, we recall two important properties of the Lebesgue spaces $L_p([a,b])$. The first property is called *Hölder inequality*, and the second property shows that $L_p([a,b])$ is strictly decreasing with respect to the index p. A certain converse of the Hölder inequality is given in Exercise 0.39.

Proposition 0.10. *Let $1 \leq p \leq \infty$ and let p' defined by* (0.13)*. Then the following holds.*
(a) *From $f \in L_p([a,b])$ and $g \in L_{p'}([a,b])$, it follows that $fg \in L_1([a,b])$ with*

$$\|fg\|_{L_1} \leq \|f\|_{L_p}\|g\|_{L_{p'}}. \tag{0.14}$$

(b) *The strict inclusions*

$$L_\infty([a,b]) \subset L_q([a,b]) \subset L_p([a,b]) \subset L_1([a,b]) \quad (1 < p < q < \infty) \tag{0.15}$$

are true.

Proof. To prove (a), suppose first that $p = \infty$, and hence $p' = 1$. Then $|f(x)g(x)| \leq \|f\|_{L_\infty}|g(x)|$ (a.e. on $[a,b]$) implies

$$\|fg\|_{L_1} = \int_a^b |f(x)g(x)|\,dx \leq \|f\|_{L_\infty}\int_a^b |g(x)|\,dx = \|f\|_{L_\infty}\|g\|_{L_1}$$

which is (0.14). The case $p = 1$ and $p' = \infty$ follows by symmetry. It remains to prove (a) for $1 < p < \infty$.

2 Therefore, to be rigorous, the linear space $L_p([a,b])$ consists of *equivalence classes* of measurable functions, rather than individual functions. When working in these spaces, however, one usually treats their elements like individual functions, keeping in mind that the equality of f and g is meant in the sense of (0.12). For example, the characteristic function of the rational numbers in $[a,b]$ then belongs to the set $N_p([a,b])$ for every p, and so is "equal" to the zero function.
3 Note that $p'' = p$ for all p, and $p' = p$ if and only if $p = 2$.

Consider the graph of the function $x \mapsto x^{p-1}$ for $x \geq 0$ in the (x, y)-plane which coincides with the graph of $y \mapsto y^{p'-1}$ for $y \geq 0$ in the (y, x)-plane. Fix $\xi, \eta > 0$, denote by A the area of the region between this graph, the x-axis, and the line $x = \xi$, and by B, the area of the region between this graph, the y-axis, and the line $y = \eta$. A simple geometric reasoning then shows that

$$\xi\eta \leq A + B = \int_0^\xi x^{p-1}\,dx + \int_0^\eta y^{p'-1}\,dy = \frac{\xi^p}{p} + \frac{\eta^{p'}}{p'} . \tag{0.16}$$

Now, given $f \in \mathrm{L}_p([a,b])$ and $g \in \mathrm{L}_{p'}([a,b])$, we put[4]

$$\hat{f}(x) := \frac{f(x)}{\|f\|_{\mathrm{L}_p}}, \quad \hat{g}(x) := \frac{g(x)}{\|g\|_{\mathrm{L}_{p'}}} .$$

Then $\|\hat{f}\|_{\mathrm{L}_p} = \|\hat{g}\|_{\mathrm{L}_{p'}} = 1$, and taking $\xi := |\hat{f}(x)|$ and $\eta := |\hat{g}(x)|$ in (0.16) yields

$$\|\hat{f}\hat{g}\|_{\mathrm{L}_1} = \int_a^b |\hat{f}(x)\hat{g}(x)|\,dx \leq \int_a^b \frac{|\hat{f}(x)|^p}{p}\,dx + \int_a^b \frac{|\hat{g}(x)|^{p'}}{p'}\,dx = \frac{1}{p} + \frac{1}{p'} = 1 .$$

Taking into account the definition of $\hat{f}$ and $\hat{g}$ in this estimate and multiplying by $\|f\|_{\mathrm{L}_p}\|g\|_{\mathrm{L}_{p'}}$ gives (0.14).

To prove (b), we apply (a) to the special choice $g(x) \equiv 1$. For $f \in \mathrm{L}_\infty([a,b])$, we then get

$$\begin{aligned} \|f\|_{\mathrm{L}_q} &= \left(\int_a^b |f(x)|^q\,dx\right)^{1/q} \\ &\leq (b-a)^{1/q} \operatorname{esssup}\{|f(x)| : a \leq x \leq b\} = (b-a)^{1/q}\|f\|_{\mathrm{L}_\infty} . \end{aligned}$$

Similarly, for $p < q$ and $f \in \mathrm{L}_q([a,b])$, we get

$$\begin{aligned} \|f\|_{\mathrm{L}_p} &= \left(\int_a^b |f(x)|^p\,dx\right)^{1/p} \\ &\leq (b-a)^{(q-p)/pq}\left(\int_a^b |f(x)|^q\,dx\right)^{1/q} = (b-a)^{(q-p)/pq}\|f\|_{\mathrm{L}_q} , \end{aligned}$$

4 To be precise, we have to ensure that neither f nor g is zero a.e.; however, in this case, both the right-hand and left-hand side of (0.14) vanish, and there is nothing to prove.

while for $p > 1$ and $f \in \mathrm{L}_p([a,b])$, we get

$$\|f\|_{\mathrm{L}_1} = \int_a^b |f(x)|\,dx$$

$$\leq (b-a)^{(p-1)/p} \left(\int_a^b |f(x)|^p\,dx \right)^{1/p} = (b-a)^{(p-1)/p} \|f\|_{\mathrm{L}_p}$$

which completes the proof. □

We still have to prove that the inclusions in (0.15) are strict. For finite q, the function

$$f(x) := \begin{cases} 0 & \text{for } x = a\,, \\ \frac{1}{(x-a)^\alpha} & \text{for } a < x \leq b\,, \end{cases}$$

where $\alpha \in (0, 1/q)$, may serve as an example of a function $f \in \mathrm{L}_q([a,b]) \setminus \mathrm{L}_\infty([a,b])$. For the other inclusions, we now construct some more sophisticated examples which build on the following well-known fact from a first year calculus course: *for $\alpha, \beta \in \mathbb{R}$, the series*

$$\zeta(\alpha,\beta) := \sum_{n=1}^{\infty} \frac{1}{n^\alpha \log^\beta(n+1)} \tag{0.17}$$

converges if and only if either $\alpha > 1$ and β is arbitrary (in particular, $\beta = 0$), or $\alpha = 1$ and $\beta > 1$. This may be easily proved by means of the classical condensation theorem for series with decreasing terms.

Example 0.11. For $1 \leq p < \infty$, we construct a function

$$f \in \mathrm{L}_p([0,1]) \setminus \left(\bigcup_{q>p} \mathrm{L}_q([0,1]) \right).$$

Let $f(0) := 0$ and

$$f(x) := \frac{n^{1/p}}{\log^2(n+1)} \qquad \left(\frac{1}{n+1} < x \leq \frac{1}{n} \right)$$

for $n \in \mathbb{N}$. Then

$$\int_0^1 |f(x)|^p\,dx = \sum_{n=1}^{\infty} \int_{1/(n+1)}^{1/n} \frac{n}{\log^{2p}(n+1)}\,dx \leq \zeta(1,2p) < \infty\,,$$

and so $f \in \mathrm{L}_p([0,1])$. On the other hand, for $q > p$, we have

$$\int_0^1 |f(x)|^q\,dx = \sum_{n=1}^{\infty} \int_{1/(n+1)}^{1/n} \frac{n^{q/p}}{\log^{2q}(n+1)}\,dx = \sum_{n=1}^{\infty} \frac{n^{(q-p)/p}}{(n+1)\log^{2q}(n+1)} = \infty\,,$$

which shows that $f \notin \mathrm{L}_q([0,1])$ for any $q > p$. ♥

Example 0.12. For $1 < p \leq \infty$, we construct a function

$$f \in \left(\bigcap_{q<p} L_q([0,1]) \right) \setminus L_p([0,1]) .$$

Suppose first that $1 < p < \infty$. Let $f(0) := 0$ and

$$f(x) := n^{1/p} \quad \left(\frac{1}{n+1} < x \leq \frac{1}{n} \right)$$

for $n \in \mathbb{N}$. Then

$$\int_0^1 |f(x)|^p \, dx = \sum_{n=1}^{\infty} \int_{1/(n+1)}^{1/n} n \, dx = \sum_{n=1}^{\infty} \frac{1}{n+1} = \infty ,$$

and so $f \notin L_p([0,1])$. On the other hand, for $q < p$, we have

$$\int_0^1 |f(x)|^q \, dx = \sum_{n=1}^{\infty} \int_{1/(n+1)}^{1/n} n^{q/p} \, dx \leq \zeta(2 - q/p, 0) < \infty ,$$

which shows that $f \in L_q([0,1])$ for every $q < p$.

If $p = \infty$, we let $f(0) := 0$ and

$$f(x) := \log(n+1) \quad \left(\frac{1}{n+1} < x \leq \frac{1}{n} \right)$$

for $n \in \mathbb{N}$. Then

$$\int_0^1 |f(x)|^q \, dx = \sum_{n=1}^{\infty} \int_{1/(n+1)}^{1/n} \log^q(n+1) \, dx = \sum_{n=1}^{\infty} \frac{\log^q(n+1)}{n(n+1)} \leq \zeta(2, -q) < \infty ,$$

which shows that $f \in L_q([0,1])$ for every finite q. Clearly, $f \notin L_\infty([0,1])$. ♥

The Lebesgue space $L_p(I)$ $(1 \leq p \leq \infty)$ may also be defined for unbounded intervals $I \subseteq \mathbb{R}$ (and this case is even more important in applications). However, inclusions between such spaces are much more delicate, as the following three examples show which rely, as the preceding examples, on piecewise constant functions.

The first example shows that, in contrast to (0.15), $L_p(I) \not\subseteq L_q(I)$ for $q < p$ if $I \subseteq \mathbb{R}$ is an unbounded interval.

Example 0.13. For $1 < p \leq \infty$, we construct a function

$$f \in L_p([1,\infty)) \setminus \left(\bigcup_{q<p} L_q([1,\infty)) \right) .$$

In case $p = \infty$, we simply take $f(x) \equiv 1$. For $1 < p < \infty$, we let $f(1) := 0$ and

$$f(x) := \frac{1}{n^{1/p} \log(n+1)} \quad (n < x \leq n+1)$$

for $n \in \mathbb{N}$. Then

$$\int_1^\infty |f(x)|^p\,dx = \sum_{n=1}^\infty \int_n^{n+1} \frac{1}{n\log^p(n+1)}\,dx = \zeta(1,p) < \infty\,,$$

which shows that $f \in L_p([1,\infty))$. On the other hand, for $1 \le q < p$, we have

$$\int_1^\infty |f(x)|^q\,dx = \sum_{n=1}^\infty \int_n^{n+1} \frac{1}{n^{q/p}\log^q(n+1)}\,dx = \zeta(q/p,q) = \infty\,,$$

and so $f \notin L_q([1,\infty))$. ♥

The next example shows that in case of an unbounded interval $I \subseteq \mathbb{R}$, we may even find a function which belongs to $L_p(I)$ for precisely one prescribed value of p.

Example 0.14. For $1 \le p \le \infty$, we construct a function

$$f \in L_p([0,\infty)) \setminus \left(\bigcup_{q \ne p} L_q([0,\infty)) \right).$$

In case $p = 1$, we may use the function f from Example 0.11, and in case $p = \infty$, the function $f(x) \equiv 1$. Now, let $1 < p < \infty$. By Example 0.11, we may find a function

$$g \in L_p([0,1]) \setminus \left(\bigcup_{q > p} L_q([0,1]) \right).$$

We extend g to the whole semiaxis $[0,\infty)$ by taking $g(x) \equiv 0$ for $x > 1$. Similarly, by Example 0.13, we may find a function

$$h \in L_p([1,\infty)) \setminus \left(\bigcup_{q < p} L_q([1,\infty)) \right)$$

which we may also extend to $[0,\infty)$ by taking $h(x) \equiv 0$ for $0 \le x < 1$. The function $f := g + h$ then has the required property. ♥

Finally, in the last example of this series, we show that in case of an unbounded interval $I \subseteq \mathbb{R}$, there exist functions which belong to $L_r(I)$ precisely for r running over a prescribed interval $[p,q] \subset [1,\infty]$.

Example 0.15. For $1 \le p < q \le \infty$, let[5] $[p,q] \subset [1,\infty]$. We construct a function

$$f \in \left(\bigcap_{r \in [p,q]} L_r([0,\infty)) \right) \setminus \left(\bigcup_{s \notin [p,q]} L_s([0,\infty)) \right).$$

5 As usual, the sign $\subset$ denotes *strict* inclusion, so $p > 1$ or $q < \infty$.

For simplicity, let us assume that $1 < p < q < \infty$. By Example 0.11, we may find a function

$$g \in L_q([0,1]) \setminus \left(\bigcup_{r>q} L_r([0,1]) \right).$$

We extend g to the whole semiaxis $[0,\infty)$ by taking $g(x) \equiv 0$ for $x > 1$. Similarly, by Example 0.13, we may find a function

$$h \in L_p([1,\infty)) \setminus \left(\bigcup_{s<p} L_s([1,\infty)) \right)$$

which we may also extend to $[0,\infty)$ by taking $h(x) \equiv 0$ for $0 \le x < 1$. The function $f := g + h$ then belongs to $L_p([0,\infty)) \cap L_q([0,\infty))$, and the assertion follows from Exercise 0.3. ♥

There is an important generalization of Lebesgue spaces which is usually referred to as Orlicz spaces. We briefly discuss such spaces and need some preliminary constructions. The following notion will be often used in subsequent chapters.

Definition 0.16. We call a function $\phi : [0,\infty) \to [0,\infty)$ a *Young function*[6] if ϕ is continuous, convex, and satisfies $\phi(0) = 0$, $\phi(t) > 0$ for $t > 0$, and $\phi(t) \to \infty$ as $t \to \infty$. Given a Young function $\phi : [0,\infty) \to [0,\infty)$, we write $f \in \mathcal{L}_\phi([a,b])$ if $f : [a,b] \to \mathbb{R}$ is measurable and

$$\int_a^b \phi(|f(x)|)\,dx < \infty. \tag{0.18}$$

The set $\mathcal{L}_\phi([a,b])$ is called the *Orlicz class* generated by ϕ. ■

Typical examples of Young functions are $\phi(t) = t^p$ for $1 \le p < \infty$, $\phi(t) = e^t - 1$, or $\phi(t) = (t+1)\log(t+1)$. For the choice $\phi(t) = t^p$ with $1 \le p < \infty$, the Orlicz class $\mathcal{L}_\phi([a,b])$ coincides, of course, with the Lebesgue space $L_p([a,b])$, as a comparison of (0.10) and (0.18) shows.

Unfortunately, for general Young functions ϕ, the Orlicz class need not be a linear space, as the following simple example shows.

Example 0.17. Let $\phi(t) = e^t - 1$, and let $f : [0,1] \to \mathbb{R}$ be defined by

$$f(x) := \begin{cases} -\frac{1}{2}\log x & \text{for } 0 < x \le 1, \\ 0 & \text{for } x = 0. \end{cases}$$

---

6 We define a Young function ϕ here on the half-axis $[0,\infty)$; therefore, we have to write $\phi(|f(x)|)$ if f is a real function as in (0.18). To avoid the absolute value in the argument, some authors consider Young functions ϕ on the whole real axis by defining them first on $[0,\infty)$ and extending them afterwards as even functions to $(-\infty,0]$. As we will show in Lemma 1.36 in the next chapter, a Young function is always increasing.

Then

$$\int_0^1 \phi(|f(x)|)\,dx = \int_0^1 \left[\frac{1}{\sqrt{x}} - 1\right] dx < \infty,$$

so $f \in \mathcal{L}_\phi([a,b])$, but

$$\int_0^1 \phi(2|f(x)|)\,dx = \int_0^1 \left[\frac{1}{x} - 1\right] dx = \infty,$$

so $2f \notin \mathcal{L}_\phi([a,b])$. ♥

The reason for the phenomenon described in Example 0.17 will become clear later (Proposition 0.20). One may associate to every Orlicz class $\mathcal{L}_\phi([a,b])$, a linear space, called *Orlicz space*, in the following way. First, we observe that for any Young function ϕ, the set

$$A(\phi) := \{f \in \mathcal{L}_\phi([a,b]) : \int_a^b \phi(|f(x)|)\,dx \le 1\} \tag{0.19}$$

is convex, symmetric, balanced, and absorbing.[7] Therefore, the *Minkowski functional* associated to $A(\phi)$, i.e.

$$\mu_{A(\phi)}(f) := \inf\,\{\lambda > 0 : f/\lambda \in A(\phi)\} \tag{0.20}$$

is a *norm* on the linear space of all measurable functions $f : [a,b] \to \mathbb{R}$ with the property that $f/\lambda \in \mathcal{L}_\phi([a,b])$ for some $\lambda > 0$. Moreover, the closed unit ball in this norm coincides with the set $A(\phi)$ given in (0.19).

Definition 0.18. We call this linear space the *Orlicz space* $\mathrm{L}_\phi([a,b])$ generated by the Young function ϕ. In the sequel, we write $\|f\|_{\mathrm{L}_\phi}$ for (0.20) and call $\|f\|_{\mathrm{L}_\phi}$ the *Luxemburg norm* of f. ■

The natural question arises for what Young functions we have $\mathcal{L}_\phi([a,b]) = \mathrm{L}_\phi([a,b])$, which means that the Orlicz class $\mathcal{L}_\phi([a,b])$ itself is a linear space. This question may be answered by means of the following notion.

Definition 0.19. A Young function $\phi : [0,\infty) \to [0,\infty)$ satisfies a Δ_2*-condition* if

$$\phi(2t) \le M\phi(t) \quad (t \ge T) \tag{0.21}$$

for suitable constants $M > 0$ and $T > 0$. In this case, we write $\phi \in \Delta_2$. ■

Proposition 0.20. *The Orlicz class $\mathcal{L}_\phi([a,b])$ is a linear space, and hence coincides with $\mathrm{L}_\phi([a,b])$, if and only if $\phi \in \Delta_2$.*

7 The definition of these notions may be found in many textbooks on Functional Analysis, e.g. [270].

We do not prove Proposition 0.20; the proof may be found, for example, in the book [169]. Instead, let us look at some examples. The Young function $\phi(t) = t^p$ $(1 \leq p < \infty)$ obviously satisfies the Δ_2-condition (0.21) for arbitrary $T > 0$ and $M := 2^p$, and so $\mathcal{L}_\phi([a,b]) = \mathrm{L}_\phi([a,b]) = \mathrm{L}_p([a,b])$ in this case. Moreover, here, the Minkowski functional (0.20) coincides with the L_p norm (0.11) because

$$\mu_{A(\phi)}(f) = \inf\left\{\lambda > 0 : \int_a^b \phi\left(\frac{f(x)}{\lambda}\right) dx \leq 1\right\}$$
$$= \inf\left\{\lambda > 0 : \int_a^b |f(x)|^p \, dx \leq \lambda^p\right\} = \|f\|_{\mathrm{L}_p}.$$

On the other hand, the Young function $\phi(t) = e^t - 1$ from Example 0.17 cannot satisfy a Δ_2-condition since the corresponding Orlicz class $\mathcal{L}_\phi([a,b])$ is not a linear space, as we have seen there. In fact,

$$\lim_{t\to\infty} \frac{\phi(2t)}{\phi(t)} = \lim_{t\to\infty} \frac{e^{2t}-1}{e^t-1} = \lim_{t\to\infty} \frac{e^t - e^{-t}}{1-e^{-t}} = \infty, \tag{0.22}$$

and so $\phi \notin \Delta_2$.

It is illuminating to calculate the norm $\|f\|_{\mathrm{L}_\phi}$ for f and ϕ from Example 0.17. For any $\lambda > 0$, we have

$$\int_0^1 \phi(|f(x)|/\lambda)\, dx = \int_0^1 \left[\frac{1}{x^{1/\lambda}} - 1\right] dx < \infty$$

if and only if $\lambda > 1$, and in this case, for the integral, we get

$$\int_0^1 \left[\frac{1}{x^{1/\lambda}} - 1\right] dx = \left[\frac{\lambda}{\lambda-1} x^{1-1/\lambda} - x\right]_0^1 = \frac{1}{\lambda - 1}.$$

The smallest value of λ for which the last expression is ≤ 1 is $\|f\|_{\mathrm{L}_\phi} = \lambda = 2$.

We have defined $\mathrm{L}_\phi([a,b])$ as the set (actually, linear space) of all measurable functions $f : [a,b] \to \mathbb{R}$ such that $f/\lambda \in \mathcal{L}_\phi([a,b])$ for *some* $\lambda > 0$. It is also interesting to consider the set $E_\phi([a,b])$ of all measurable functions $f : [a,b] \to \mathbb{R}$ such that $f/\lambda \in \mathcal{L}_\phi([a,b])$ for *all* $\lambda > 0$; this is usually called the *small Orlicz space*. Equipped with the norm (0.20), this is a closed linear subspace of $\mathrm{L}_\phi([a,b])$ which coincides with $\mathrm{L}_\phi([a,b])$ if and only if $\phi \in \Delta_2$. This again illustrates the importance of the Δ_2-condition (0.21). Further important properties of the space E_ϕ will be considered in Exercises 0.22–0.24.

In rather the same way as we defined in (0.13) the conjugate index of a number p, we may associate to each Young function some kind of conjugate Young function. Given a Young function $\phi : [0,\infty) \to [0,\infty)$, we define its *conjugate Young function* $\phi^* : [0,\infty) \to [0,\infty)$ by

$$\phi^*(t) := \sup\{st - \phi(s) : s \geq 0\}. \tag{0.23}$$

By construction, we then have $\phi^{**} = \phi$ and

$$ab \le \phi(a) + \phi^*(b) \quad (a, b \ge 0)\,; \tag{0.24}$$

inequality (0.24) is usually referred to as *Young's inequality*. For example, an easy calculation shows that the conjugate Young function to

$$\phi(t) := \frac{1}{p} t^p$$

for $1 < p < \infty$ is

$$\phi^*(t) := \frac{1}{p'} t^{p'}\,,$$

where p' is given by (0.13). In this sense, the conjugate Young function plays the same role for Orlicz spaces as the conjugate index for Lebesgue spaces. Young's inequality (0.24) may then be considered as an analogue to Hölder's inequality (0.14).

0.2 Some functional analysis

In the following chapters, we will need three concepts from the theory of Banach spaces: equivalence of norms, imbeddings between spaces, and duality. We start with the concept of duality which is of fundamental importance in functional analysis and operator theory.

Recall that the *dual space* X^* of a real normed linear space consists of all bounded linear functionals[8] $\ell : X \to \mathbb{R}$, equipped with the norm[9]

$$\|\ell\|_{X^*} := \sup \left\{ |\langle x, \ell \rangle| : \|x\|_X \le 1 \right\} = \sup \left\{ |\langle x, \ell \rangle| : \|x\|_X = 1 \right\}. \tag{0.25}$$

One of the most important results on dual spaces is the Hahn–Banach theorem which asserts that a bounded linear functional on a subspace may be extended to the whole space without "increasing its size." For measuring the size of a functional, we recall that a *sublinear functional* is a map $p : X \to \mathbb{R}$ with the property that

$$p(x + y) \le p(x) + p(y), \quad p(\lambda x) = \lambda p(x) \quad (x, y \in X; 0 \le \lambda < \infty). \tag{0.26}$$

A standard example is $p(x) = \|x\|$, but more sophisticated choices of p lead to interesting results. With this notion, the announced extension theorem reads as follows.

8 Since we only use real vector spaces in this book, all functionals are considered real-valued. For this reason, we only recall the real version of the Hahn–Banach theorem.

9 Here, we adopt the usual notation $\langle x, \ell \rangle$ instead of $\ell(x)$ to emphasize the idea of duality pairing borrowed from scalar products in Hilbert space. We remark that the supremum in (0.25) could also be taken over the open unit ball, see Exercise 0.40.

Theorem 0.21 (Hahn–Banach). *Let X be a real normed space and p be some sublinear functional on X. Let $U \subset X$ be a linear subspace, and suppose that a functional $\ell \in U^*$ satisfies $\langle u, \ell \rangle \le p(u)$ for all $u \in U$. Then there exists a functional $\hat{\ell} \in X^*$ such that $\hat{\ell}|_U = \ell$, i.e.*

$$\langle u, \hat{\ell} \rangle = \langle u, \ell \rangle \quad (u \in U)$$

and $\langle x, \hat{\ell} \rangle \le p(x)$ for all $x \in X$.

We do not prove Theorem 0.21 since it is beyond the scope of this book. Instead, we prove three interesting consequences of this theorem which we will need in Chapter 4.

Corollary 0.22. *Let X be a real normed space and $U \subset X$ be a linear subspace. Then the following is true.*
(a) *Every functional $\ell \in U^*$ may be extended to a functional $\hat{\ell} \in X^*$ such that $\hat{\ell}|_U = \ell$ and $\|\hat{\ell}\|_{X^*} = \|\ell\|_{U^*}$.*
(b) *Given any $x_* \in X \setminus \{0\}$, one may find a functional $\hat{\ell} \in X^*$ such that $\|\hat{\ell}\|_{X^*} = 1$ and $\langle x_*, \hat{\ell} \rangle = \|x_*\|_X$.*
(c) *The equality*

$$\|x\|_X = \sup \left\{ |\langle x, \ell \rangle| : \|\ell\|_{X^*} \le 1 \right\} = \sup \left\{ |\langle x, \ell \rangle| : \|\ell\|_{X^*} = 1 \right\}, \tag{0.27}$$

which is dual to (0.25), *holds true.*

Proof. (a) It follows from the properties of a norm[10] that the map $p(x) := \|\ell\|_{U^*} \|x\|_X$ is a sublinear functional on X. Moreover, (0.25) implies that $\langle u, \ell \rangle \le p(u)$ for all $u \in U$. So, from Theorem 0.21, we conclude that ℓ admits an extension $\hat{\ell} \in X^*$ satisfying

$$\langle x, \hat{\ell} \rangle \le p(x) = \|\ell\|_{U^*} \|x\|_X \quad (x \in X),$$

and hence $\|\hat{\ell}\|_{X^*} \le \|\ell\|_{U^*}$, again by (0.25). The reverse estimate $\|\hat{\ell}\|_{X^*} \ge \|\ell\|_{U^*}$ is obvious.

(b) On the one-dimensional subspace $U := \mathbb{R}x_* = \{\lambda x_* : \lambda \in \mathbb{R}\} \subset X$, define $\ell \in U^*$ by $\langle \lambda x_*, \ell \rangle := \lambda \|x_*\|_X$. Clearly, $\langle x_*, \ell \rangle = \|x_*\|_X$. We claim that $\|\ell\|_{U^*} = 1$. In fact, this follows from the equalities

$$\|\ell\|_{U^*} = \sup \left\{ |\langle \lambda x_*, \ell \rangle| : \|\lambda x_*\| \le 1 \right\} = \sup_{\lambda > 0} \frac{\lambda \|x_*\|_X}{\|\lambda x_*\|_X} = 1 .$$

By (a), we may now extend ℓ to a functional $\hat{\ell} \in X^*$ which still has norm $\|\hat{\ell}\|_{X^*} = 1$, and (b) is proved.

(c) From the obvious estimate $|\langle x, \ell \rangle| \le \|\ell\|_{X^*} \|x\|_X$ which holds for all $x \in X$ and $\ell \in X^*$, it follows that

$$\|x\|_X \ge \sup \left\{ |\langle x, \ell \rangle| : \|\ell\|_{X^*} = 1 \right\} . \tag{0.28}$$

10 In this proof, we equip every norm with a subscript since we have to consider norms in the four different spaces X, U, X^*, and U^*. Later, we will drop the subscript when the underlying space is clear.

To prove the converse estimate, fix $x^* \in X$ (without loss of generality, $\|x^*\|_X = 1$). By statement (b), we may then find a functional $\ell \in X^*$ such that $\|\ell\|_{X^*} = 1$ and $\langle x^*, \ell\rangle = \|x^*\|_X$. In other words, for this functional ℓ, the supremum in (0.28) is attained, and so we get equality in (0.28). □

An important problem in functional analysis is to identify the dual space X^* of a given space X. In some cases, this is very easy. For example, it is not hard to see that the dual space of $X = \mathbb{R}^n$ is again $X^* = \mathbb{R}^n$. More precisely, the map $\Phi : \mathbb{R}^n \to (\mathbb{R}^n)^*$ defined by $\Phi(y) := \ell_y$ with

$$\langle x, \ell_y\rangle := \sum_{k=1}^{n} x_k y_k \quad (x \in \mathbb{R}^n)$$

is a *linear surjective isometry*[11] where the term "isometry" means that $\|\ell_y\|_{X^*} = \|y\|_X$ with $\|\ell_y\|_{X^*}$ denoting the functional norm (0.25) on $\mathbb{R}^n$.

A more interesting (and very important) example is the Lebesgue space $X = \mathrm{L}_p([a, b])$ which we considered in the first section. Without going into details, we state the corresponding result in the following Theorem 0.23.

For $1 \le p < \infty$, let p' denote the conjugate index (0.13) to p (in particular, $p' = \infty$ for $p = 1$). For fixed $g \in \mathrm{L}_{p'}([a, b])$, we define a functional $\ell_g : \mathrm{L}_p([a, b]) \to \mathbb{R}$ by

$$\langle f, \ell_g\rangle := \int_a^b f(t)g(t)\, dt \quad (f \in \mathrm{L}_p([a, b]))\,. \tag{0.29}$$

From Hölder's inequality (0.14), it follows that $\ell_g \in \mathrm{L}_p^*$. The interesting point is that *all* elements $\ell \in \mathrm{L}_p^*$ have this form:

Theorem 0.23 (Riesz). *For $1 \le p < \infty$, the dual space L_p^* of the space L_p with norm* (0.11) *may be identified with the space $\mathrm{L}_{p'}$. More precisely, the map $\Phi : \mathrm{L}_{p'}([a, b]) \to \mathrm{L}_p([a, b])^*$ defined by $\Phi(g) := \ell_g$, with ℓ_g as in* (0.29) *for $f \in \mathrm{L}_p([a, b])$ and $g \in \mathrm{L}_{p'}([a, b])$, is a linear surjective isometry.*

As before, the term "isometry" means that $\|\ell_g\|_{\mathrm{L}_p^*} = \|g\|_{\mathrm{L}_{p'}}$, where the norm on the dual space is given by (0.25). In an explicit form, this means that

$$\left(\int_a^b |g(x)|^{p'}\, dx\right)^{1/p'} = \sup\left\{\int_a^b f(x)g(x)\, dx : \int_a^b |f(x)|^p\, dx \le 1\right\} \tag{0.30}$$

for $1 < p < \infty$, and

$$\operatorname{esssup}\{|g(x)| : a \le x \le b\} = \sup\left\{\int_a^b f(x)g(x)\, dx : \int_a^b |f(x)|\, dx \le 1\right\} \tag{0.31}$$

11 Here, it is important to consider both spaces X and X^* equipped with the Euclidean norm; if we equip X with another norm, then we have to choose another norm on X^* to make Φ an isometry, see Exercises 0.42 and 0.43.

in case $p = 1$. Interestingly, the map Φ is also an isometry between L_1 and L_∞^*, which means that

$$\int_a^b |g(x)|\,dx = \sup\left\{\int_a^b f(x)g(x)\,dx : \operatorname{esssup}\{|f(x)| : a \le x \le b\} \le 1\right\}. \tag{0.32}$$

However, Theorem 0.23 is *not* true in case $p = \infty$ which means that $L_\infty^* \neq L_1$. The reason is that there are bounded linear functionals on L_∞ which cannot be written in the form ℓ_g as in (0.29), and so $\Phi : L_1([a,b]) \to L_\infty^*([a,b])$ is *not surjective*.

It is also interesting to observe that the formulas (0.30)–(0.32) may be made more precise by constructing functions for which the supremum in these formulas is actually achieved (Exercises 0.31, 0.33, and 0.36).

Theorem 0.23 shows that $L_p^* = L_{p/(p-1)}$ for $1 \le p < \infty$, where the duality between L_p^* and $L_{p/(p-1)}$ is given by (0.29). One could ask whether or not (0.29) also gives the general form of a bounded linear functional on the space $C([a,b])$, where now, of course, both functions f and g have to be chosen continuous. The following example shows that the answer is negative:

Example 0.24. Let $\ell : C([0,1]) \to \mathbb{R}$ be the evaluation functional at 0, i.e. the map defined by $\langle f, \ell\rangle := f(0)$. Clearly, ℓ is both linear and continuous, so $\ell \in C^*$ with $\|\ell\|_{C^*} = 1$. However, there is no $g \in C([0,1])$ such that ℓ may be represented in the form (0.29).

To see this, suppose that there exists a continuous function $g : [0,1] \to \mathbb{R}$ satisfying

$$\int_0^1 f(t)g(t)\,dt = f(0) \quad (f \in C([0,1]))\,. \tag{0.33}$$

Consider the sequence $(f_n)_n$ of continuous peak functions defined by

$$f_n(x) := \max\{1 - nx, 0\} = \begin{cases} 1 - nx & \text{for } 0 \le x < \frac{1}{n}, \\ 0 & \text{for } \frac{1}{n} \le x \le 1. \end{cases}$$

Then $f_n(0) = 1$ for all n, and $f_n \to f := \chi_{\{0\}}$ pointwise on $[0,1]$. Moreover,[12]

$$\int_0^1 f_n(t)g(t)\,dt \le \|g\|_C \int_0^1 f_n(t)\,dt = \frac{\|g\|_C}{2n}\,.$$

Finally, the majorization (0.3) holds with f_n replaced by $f_n g$ and the majorant function $F(x) \equiv \|g\|_C$ on $M = [0,1]$. Thus, Theorem 0.4 implies that

$$1 = f(0) = \int_0^1 f(t)g(t)\,dt = \lim_{n\to\infty} \int_0^1 f_n(t)g(t)\,dt = 0\,,$$

a palpable contradiction. ♥

12 The norm $\|\cdot\|_C$ in the space of continuous functions is defined in (0.45) below.

In Chapter 4, we will give a precise description of the dual space C^* of C, and this is one of the places where functions of bounded variation play a very prominent role, see Theorem 4.31.

Apart from the description of the dual space X^* of a given normed space X, another important problem consists of finding properties which carry over from X to X^* or vice versa. As a sample result, we discuss separability.

Definition 0.25. A normed space X is called *separable* if it contains a dense countable subset, i.e. a set $M = \{x_1, x_2, x_3, \ldots\}$ satisfying $\overline{M} = X$. ■

Here are some examples of separable and nonseparable spaces. Clearly, $\mathbb{R}^N$ is separable with any norm (choose $M := \mathbb{Q}^N$). Moreover, $C([a,b])$ is separable by the classical Weierstrass approximation theorems.[13] It is much harder to prove that the space $L_p([a,b])$ is also separable for $1 \leq p < \infty$; a proof may be found, for example, in [61]. However, it is easy to see that $L_\infty([a,b])$ is *not* separable:

Example 0.26. For any $c \in (a,b)$, let $f_c := \chi_{[a,c]}$ denote the characteristic function of the interval $[a,c]$. Clearly, $\|f_c - f_d\|_{L_\infty} = 1$ for $a < c < d < b$. This shows that the open balls

$$B_c := \left\{f \in L_\infty([a,b]) : \|f - f_c\|_{L_\infty} < 1/2\right\} \quad (a < c < b)$$

are *mutually disjoint* for different values of c, and there are uncountably many of them. However, this immediately implies that $L_\infty([a,b])$ cannot be separable. ♥

Example 0.26 and Theorem 0.23 (for $p = 1$) show that the separability of X does *not* imply the separability of X^*. The converse, however, is true:

Theorem 0.27. *If X^* is separable, then X is also separable.*

Proof. Let $\{\tilde{\ell}_1, \tilde{\ell}_2, \tilde{\ell}_3, \ldots\}$ be dense in X^*. Normalizing[14] $\ell_n := \tilde{\ell}_n/\|\tilde{\ell}_n\|_{X^*}$, we then get a dense subset $\{\ell_1, \ell_2, \ell_3, \ldots\}$ of the unit sphere in X^*.

By definition of the functional norm (0.25), we can find a sequence $(x_n)_n$ in the unit sphere of X such that $|\ell_n(x_n)| > 1/2$ for all $n \in \mathbb{N}$. Denote by M the rational span of $\{x_1, x_2, x_3, \ldots\}$, i.e. the set of all linear combinations of the x_n's with rational coefficients. Clearly, M is countable; we claim that M is dense in X.

In fact, assuming the contrary, we can find an element $\ell \in X^*$ satisfying $\|\ell\|_{X^*} = 1$ and vanishing on $\overline{M}$. In particular, we then have $\ell(x_n) = 0$ for all n. Choose ℓ_m from the set $\{\ell_1, \ell_2, \ell_3, \ldots\}$ defined above such that $\|\ell_m - \ell\|_{X^*} < 1/2$, which is possible since

13 To be precise, we first approximate a fixed continuous function on $[a,b]$ uniformly by a sequence of polynomials, and then each of these polynomials by a polynomial of the same degree, but with rational coefficients. Thus, we choose as M here the (countable) set of all polynomials with rational coefficients.

14 Obviously, it is no loss of generality to assume that all functionals $\tilde{\ell}_n$ are different from zero.

this set is dense in the unit sphere in X^*. Then $\|x_m\|_X = 1$ and $\ell(x_m) = 0$ imply that

$$\frac{1}{2} < |\ell_m(x_m)| \leq |\ell_m(x_m) - \ell(x_m)| + |\ell(x_m)| \leq \|\ell_m - \ell\|_{X^*} < \frac{1}{2},$$

a contradiction. □

Now, we recall the definition of equivalent norms and continuous imbeddings. We start with the following

Definition 0.28. Two norms $\|\cdot\|_X$ and $|||\cdot|||_X$ on a linear space X are called *equivalent* if there exist constants $M, m > 0$ such that

$$m\|f\|_X \leq |||f|||_X \leq M\|f\|_X \tag{0.34}$$

for all $f \in X$. ■

Geometrically, the equivalence of two norms means that every ball with respect to one of these norms may be included, possibly after diminishing its radius into a ball with respect to the other norm. We will consider various examples of equivalent and nonequivalent norms below.

Since convergent sequences and Cauchy sequences are the same with respect to two equivalent norms, a Banach space remains complete when passing to an equivalent norm. There is a nontrivial converse of this statement which states that whenever a linear space X is complete with respect to two norms $\|\cdot\|_X$ and $|||\cdot|||_X$, and one of the estimates in (0.34) holds, then the other estimate also holds.[15]

Definition 0.29. We say that a normed linear space $(X, \|\cdot\|_X)$ is *imbedded*[16] into another normed linear space $(Y, \|\cdot\|_Y)$ if $X \subseteq Y$ and

$$\|f\|_Y \leq c\|f\|_X \quad (f \in X) \tag{0.35}$$

for some constant $c > 0$ independent of f. In this case, we write $X \hookrightarrow Y$ and call c in (0.35) an *imbedding constant*. Moreover, the smallest possible imbedding constant c, i.e.

$$c(X, Y) := \sup\{\|f\|_Y : \|f\|_X \leq 1\}, \tag{0.36}$$

will be called the *sharp imbedding constant* for $X \hookrightarrow Y$ in the sequel. ■

The importance (and usefulness) of the imbedding condition (0.35) consists of the fact that $f_n \to f$ in X implies $f_n \to f$ in Y. If merely $X \subseteq Y$, but with unrelated norms, this need not be true.

15 This result is a consequence of the well-known *closed graph theorem* (or *open mapping theorem*, or *inverse mapping theorem*) in functional analysis, applied to the identity operator between $(X, \|\cdot\|_X)$ and $(X, |||\cdot|||_X)$.

16 Some authors state more precisely that $(X, \|\cdot\|_X)$ is *continuously imbedded* into $(Y, \|\cdot\|_Y)$ since (0.35) means that the identity operator is continuous between these spaces.

Now, we recall some definitions and facts about Banach spaces whose elements can be multiplied.

Definition 0.30. A normed linear space $(X, \|\cdot\|_X)$ is called an *algebra* if the product of two elements $f, g \in X$ also belongs to X and satisfies

$$\|fg\|_X \leq c\|f\|_X\|g\|_X \quad (f, g \in X) \tag{0.37}$$

for some constant $c > 0$ independent of f and g. If $(X, \|\cdot\|_X)$ is complete, X is called a *Banach algebra*. ■

Sometimes, (0.37) may be sharpened to

$$\|fg\|_X \leq \|f\|_X\|g\|_X \quad (f, g \in X), \tag{0.38}$$

i.e. we may take $c = 1$. In this case, we call $(X, \|\cdot\|_X)$ a *normalized algebra*. An interesting question is if under the hypothesis (0.37), we can introduce an equivalent norm $|||\cdot|||$ such that the stronger condition (0.38) holds in the new norm. The following result shows how this may be achieved [192] for function spaces which are contained in the linear space $B([a, b])$ of all bounded functions $f : [a, b] \to \mathbb{R}$ with norm[17]

$$\|f\|_\infty := \sup_{a \leq x \leq b} |f(x)| . \tag{0.39}$$

Proposition 0.31. *Let $(X, \|\cdot\|_X)$ be a Banach algebra of functions $f : [a, b] \to \mathbb{R}$ which satisfies $X \subseteq B([a, b])$ and*

$$\|fg\|_X \leq \|f\|_\infty\|g\|_X + \|f\|_X\|g\|_\infty \quad (f, g \in X), \tag{0.40}$$

where $\|\cdot\|_\infty$ is defined by (0.39). Then the space X equipped with the norm

$$|||f|||_X := \|f\|_\infty + \|f\|_X \quad (f \in X) \tag{0.41}$$

is a normalized Banach algebra, i.e.

$$|||fg|||_X \leq |||f|||_X|||g|||_X \quad (f, g \in X) . \tag{0.42}$$

Moreover, in case $X \hookrightarrow B([a, b])$, both norms $\|\cdot\|_X$ and $|||\cdot|||_X$ are equivalent.

Proof. By the obvious estimate $\|fg\|_\infty \leq \|f\|_\infty\|g\|_\infty$, for $f, g \in X$, we get

$$\begin{aligned}
|||fg|||_X &= \|fg\|_X + \|fg\|_\infty \leq \|fg\|_X + \|f\|_\infty\|g\|_\infty \\
&\leq \|f\|_X\|g\|_\infty + \|f\|_\infty\|g\|_X + \|f\|_\infty\|g\|_\infty \\
&\leq \|f\|_X\|g\|_\infty + \|f\|_\infty\|g\|_X + \|f\|_\infty\|g\|_\infty + \|f\|_X\|g\|_X \\
&= (\|f\|_X + \|f\|_\infty)(\|g\|_X + \|g\|_\infty) = |||f|||_X|||g|||_X ,
\end{aligned}$$

17 The norm (0.39) has to be carefully distinguished from the L_∞-norm (0.11). In fact, we may have $\|f\|_{L_\infty} < \infty$, but $\|f\|_\infty = \infty$, if f is unbounded on a nullset.

which shows that $(X, \|\!|\cdot|\!\|_X)$ is a Banach algebra satisfying (0.42). If $X \hookrightarrow B([a, b])$ with sharp imbedding constant $c(X, B)$, then

$$\|\!|f|\!\|_X = \|f\|_X + \|f\|_\infty \le (1 + c(X, B))\|f\|_X ,$$

while the converse estimate $\|f\|_X \le \|\!|f|\!\|_X$ is of course trivial. □

An example of how to apply Proposition 0.31 will be given in the next section (Example 0.43).

The following is a useful method to generate from a given function space a whole sequence of new function spaces by considering derivatives.

Definition 0.32. Given a space $(X, \|\cdot\|_X)$ of functions $f : [a, b] \to \mathbb{R}$, we denote by X^n ($n \in \mathbb{N}$) the set of all functions $f \in X$ such that all derivatives $f', f'', \dots, f^{(n)}$ also belong to X. A natural norm on X^n is then given by[18]

$$\|f\|_{X^n} := \sum_{j=0}^{n-1} |f^{(j)}(a)| + \|f^{(n)}\|_X . \tag{0.43}$$

A particularly important special case is $n = 1$, i.e. the space of all f for which the norm

$$\|f\|_{X^1} := |f(a)| + \|f'\|_X \tag{0.44}$$

is finite. ■

We will consider many examples of spaces of this type in Section 5.3 in connection with nonlinear operators.

Let $(X, \|\cdot\|_X)$ and $(Y, \|\cdot\|_Y)$ be two function spaces such that $X \hookrightarrow Y$ with imbedding constant $c(X, Y)$. If we associate to both X and Y the corresponding spaces X^n and Y^n with norm (0.43), for each $f \in X^n$, we have the estimate

$$\|f\|_{Y^n} = \sum_{j=0}^{n-1} |f^{(j)}(a)| + \|f^{(n)}\|_Y \le \sum_{j=0}^{n-1} |f^{(j)}(a)| + c\|f^{(n)}\|_X \le \max\{c, 1\}\|f\|_{X^n}$$

which shows that also $X^n \hookrightarrow Y^n$. One may also show that the sharp imbedding constant $c(X^n, Y^n) = \max\{c(X, Y), 1\}$ is independent of n.

0.3 Basic function spaces

There are two important subspaces of the space $(B([a, b]), \|\cdot\|_\infty)$ which will play a prominent role throughout this book. The first one is the classical Chebyshev space $C([a, b])$ of all *continuous* functions $f : [a, b] \to \mathbb{R}$ with norm

$$\|f\|_C := \max_{a \le x \le b} |f(x)| . \tag{0.45}$$

18 As usual, $f^{(0)} := f$.

It is well known that $(C([a,b]), \|\cdot\|_C)$ is a Banach space, and the convergence with respect to the norm (0.45) coincides with the uniform convergence on $[a,b]$.

A typical problem in analysis and topology is that of extending a function from a small to a larger domain without "increasing its size." One of the most important results of this type in functional analysis is the Hahn–Banach theorem for bounded linear functionals which we stated by Theorem 0.21. Another important extension result for continuous functions is the famous *Tietze–Urysohn extension theorem* which, in its simplest form, reads as follows.

Theorem 0.33 (Tietze–Urysohn). *Let $M \subset \mathbb{R}$ be closed and $f : M \to \mathbb{R}$ be continuous on M. Then there exists a continuous function $\hat{f} : \mathbb{R} \to \mathbb{R}$ such that*

$$\sup\{|\hat{f}(x)| : x \in \mathbb{R}\} = \sup\{|f(x)| : x \in M\} \tag{0.46}$$

and $\hat{f}|_M = f$, i.e. $\hat{f}$ extends f.

We do not prove Theorem 0.33 as the proof of this result is quite technical and can be found in every topology textbook.[19] It is clear that we need closedness of M in Theorem 0.33: if M is not closed and x_0 is an accumulation point of M which does not belong to M, then the continuous function $f(x) := 1/(x - x_0)$ does not have a continuous extension to $\mathbb{R}$.

We point out that Theorem 0.33 has an interesting consequence which is closely related to Luzin's theorem (Theorem 0.2) for measurable functions. For a measurable set $M \subseteq \mathbb{R}$ and two functions $f, g : M \to \mathbb{R}$, we set

$$\Delta(f,g) = \Delta(f,g;M) := \{x \in M : f(x) \neq g(x)\}. \tag{0.47}$$

Then the following result follows by combining Luzin's theorem and the Tietze–Urysohn extension theorem.

Theorem 0.34 (Luzin). *Let $f : [a,b] \to \mathbb{R}$ be a measurable function. Then for each $\varepsilon > 0$, we can find a continuous function $f_\varepsilon : [a,b] \to \mathbb{R}$ such that*

$$\lambda(\Delta(f, f_\varepsilon)) < \varepsilon, \tag{0.48}$$

where $\Delta(f, f_\varepsilon)$ is defined by (0.47).

Apart from the space $C([a,b])$, the second space we will often need in the sequel is the space $R([a,b])$ of all *regular* functions $f \in B([a,b])$, i.e. bounded functions which have, at most, removable discontinuities or discontinuities of first kind (jumps). In

19 We point out that Theorem 0.33 holds not only on the real line, but in the much more general setting of metric spaces, and even for certain topological spaces, see, e.g. [320]. A certain variant of this theorem with a simple constructive proof may be found in Exercise 0.65.

what follows, for $f : [a,b] \to \mathbb{R}$, we use the notation

$$D(f) := \{x : a \le x \le b, f \text{ is discontinuous at } x\}, \tag{0.49}$$

$$D_0(f) := \{x : a \le x \le b, f \text{ has a removable discontinuity at } x\}, \tag{0.50}$$

and

$$D_1(f) := \{x : a \le x \le b,\ f \text{ has a jump at } x\}. \tag{0.51}$$

So, we have

$$D(f) = D_0(f) \cup D_1(f) \tag{0.52}$$

if f is regular, and even

$$D(f) = D_1(f), \quad D_0(f) = \emptyset \tag{0.53}$$

if f is monotone. Although a regular function may be very far from being monotone, it shares an important property with monotone functions:

Proposition 0.35. *The discontinuity set* (0.49) *of a regular function $f : [a,b] \to \mathbb{R}$ is at most countable.*

Proof. We adopt the usual notation

$$f(x_0-) := \lim_{\substack{x\to x_0\\ x<x_0}} f(x), \quad f(x_0+) := \lim_{\substack{x\to x_0\\ x>x_0}} f(x) \tag{0.54}$$

for the unilateral limits of a function f at some point x_0. So f belongs to $R([a,b])$ if and only if both limits (0.54) exist for all $x_0 \in (a,b)$, as well as the unilateral limits $f(a+)$ and $f(b-)$.

Given a regular function $f : [a,b] \to \mathbb{R}$, we consider the "average function" $\overline{f}$ of f defined by[20]

$$\overline{f}(x) := \begin{cases} \frac{1}{2}(f(x-) + f(x+)) & \text{for } x \in D_1(f), \\ f(x) & \text{otherwise}. \end{cases}$$

Clearly,

$$D(\overline{f}) = D(f), \quad D_0(\overline{f}) = D_0(f), \quad D_1(\overline{f}) = D_1(f), \tag{0.55}$$

so it suffices to show that $D(\overline{f})$ is countable.

Let $x_0 \in D_0(f) \cap (a,b)$, and hence $f(x_0-) = f(x_0+) \neq f(x_0)$. Assume, for example, that $f(x_0-) = f(x_0+) < f(x_0)$; then, $\varepsilon := \frac{1}{2}(f(x_0) - f(x_0+)) > 0$. Choose $\delta > 0$ such that $f(x) < f(x_0) - \varepsilon$ for $x \in (x_0 - \delta, x_0) \cup (x_0, x_0 + \delta)$. This implies that the disc in $\mathbb{R}^2$ centered at $(x_0, f(x_0))$ with radius $\min\{\delta, \varepsilon\}$ does not contain any other point of the graph of f (or $\overline{f}$) than $(x_0, f(x_0))$.

20 The average function $\overline{f}$ is particularly useful in the theory of Fourier series; other kinds of regularization will be used in Definitions 1.2 and 4.27.

Now, let $x_0 \in D_1(f) \cap (a,b)$,and hence $f(x_0-) \neq f(x_0+)$. Assume, for example, that $f(x_0-) < f(x_0+)$; then, $\varepsilon := \frac{1}{3}(f(x_0+) - f(x_0-)) > 0$. Choose $\delta > 0$ such that $f(x) < f(x_0) + \varepsilon$ for $x \in (x_0 - \delta, x_0)$, and $f(x_0 > f(x_0+) - \varepsilon$ for $x \in (x_0, x_0 + \delta)$. This implies that the disc in $\mathbb{R}^2$ centered at $(x_0, \overline{f}(x_0))$ with radius $\min\{\delta, \varepsilon\}$ does not contain any other point of the graph of $\overline{f}$ than $(x_0, \overline{f}(x_0))$.

Thus, we have shown that the set of all points on the graph of $\overline{f}$, which are centers of the discs described above, contains only isolated points. By standard reasoning (see, e.g. [291]), this implies that the set of these points, and thus also the set $D(\overline{f})$, is at most countable. □

There is a remarkable result which links regular functions with continuous functions and goes back to Sierpiński [290].

Theorem 0.36 (Sierpiński). *A function f belongs to $R([a,b])$ if and only if it can be represented as composition $f = g \circ \tau$, where $\tau : [a,b] \to [c,d]$ is strictly increasing and $g \in C([c,d])$.*

Proof. We follow the idea of the proof in [290]. Suppose first that $f : [a,b] \to \mathbb{R}$ is regular. By Proposition 0.35, we then know that its discontinuity set (0.49) is countable, i.e.

$$D(f) = \{x_1, x_2, x_3, \ldots\}.$$

We define a function $p : [a,b] \to \mathbb{R}$ by

$$p(x) := \sum_{x > x_n} \frac{1}{2^n}, \tag{0.56}$$

where the sum is taken over all indices n such that $x_n < x$, and we put $p(x) := 0$ if there are no such indices. Evidently, the function p is increasing and takes its values in $[0,1]$. Afterwards, we define $\tau : [a,b] \to \mathbb{R}$ by

$$\tau(x) := \begin{cases} x + p(x) & \text{if } x \notin D(f), \\ x + p(x) + \frac{1}{4^m} & \text{if } x = x_m \in D(f). \end{cases} \tag{0.57}$$

Clearly, the function τ is strictly increasing, takes its values in $[c,d] = [a, b+2]$, and satisfies $D(\tau) = D(f)$. Being strictly monotone, the function τ admits an inverse $\tau^{-1} : E \to [a,b]$ on its range $E := \tau([a,b])$. So, we may define a map $g : E \to \mathbb{R}$ by

$$g(t) := f(\tau^{-1}(t)) \quad (t \in E). \tag{0.58}$$

By construction, this map satisfies $f = g \circ \tau$ on $[a,b]$; we claim that g is continuous on the closure $\overline{E}$ of E.

First, we show that g is continuous on $E = \tau([a,b])$. Thus, fix $t_* = \tau(x_*) \in E$. If $x_* \in D(f)$, it is easy to see that t_* is an isolated point of E, and there is nothing to prove. Therefore, assume that $x_* \notin D(f)$, which means that f is continuous at x_*.

Given $\varepsilon > 0$, choose $\delta > 0$ such that $|x - x_*| < \delta$ implies $|f(x) - f(x_*)| < \varepsilon$. Putting $t_1 := \tau(x_* - \delta)$ and $t_2 := \tau(x_* + \delta)$, we see that $t_1 < t_* < t_2$ since τ is strictly increasing.

Now, let t be an arbitrary point in $E \cap (t_1, t_2)$, and let $x := \tau^{-1}(t)$. Then

$$\tau(x_* - \delta) < \tau(x) < \tau(x_* + \delta)\,,$$

and hence $x_* - \delta < x < x_* + \delta$, i.e. $|x - x_*| < \delta$, and so

$$|g(t) - g(t_*)| = |f(x) - f(x_*)| < \varepsilon\,. \tag{0.59}$$

We have shown that for each $\varepsilon > 0$, we can find t_1, t_2 such that $t_1 < t_2$ and (0.59) holds for any $t \in E \cap (t_1, t_2)$. However, this means that g is continuous at t_*, and also on the whole set E since $t_* \in E$ was arbitrary.

One must now prove that g is continuous at every accumulation point $t_* \in \overline{E} \setminus E$. Being an accumulation point of E, the point t_* may be approximated by sequences in E. However, we point out that we cannot find both an increasing sequence $(t_n)_n$ in E and a decreasing sequence $(t_n')_n$ in E, both converging to t_*. In fact, the elements $x_n := \tau^{-1}(t_n)$ and $x_n' := \tau^{-1}(t_n')$ would then satisfy $x_n < x_* < x_n'$, where $x_n \to x_*$ as $n \to \infty$, and hence $t_n < \tau(x_*) < t_n'$. Letting $n \to \infty$ in this estimate, we get $t_* = \tau(x_*)$, which means that $t_* \in E$, contradicting our choice of t_*.

Therefore, we know that there exists either an increasing sequence $(t_n)_n$ in E converging to t_*, or a decreasing sequence $(t_n')_n$ in E converging to t_*. Suppose that $(t_n)_n$ is increasing with $t_n \to t_*$ as $n \to \infty$. The sequence $(x_n)_n$ with $x_n := \tau^{-1}(t_n)$ is then also increasing; moreover, $(x_n)_n$ is bounded from above because

$$\lim_{x\to-\infty} \tau(x) = -\infty, \qquad \lim_{x\to\infty} \tau(x) = \infty\,.$$

Thus, $x_n \to x_*$ for some $x_* \in [a, b]$, and so

$$\lim_{n\to\infty} f(x_n) = \lim_{n\to\infty} g(\tau(x_n)) = \lim_{n\to\infty} g(t_n)\,.$$

Consequently, by putting

$$\lim_{n\to\infty} g(t_n) =: g(t_*),$$

we see that the function g is continuous, by construction, on the closure $\overline{E}$ of E.

Applying now Theorem 0.33 to $M := \overline{E}$, we get a continuous function $\hat{g} : \mathbb{R} \to \mathbb{R}$ which satisfies $f = \hat{g} \circ \tau$ on $[a, b]$. This proves the "only if" part of our assertion.

The proof of the "if part" is much simpler. Suppose that $f = g \circ \tau$, where $\tau : [a, b] \to [c, d]$ is strictly increasing, and $g : [c, d] \to \mathbb{R}$ is continuous. Given $x_0 \in [a, b]$, choose an increasing sequence $(x_n)_n$ in $[a, b]$ converging to x_0. Since τ is monotonically increasing, there exists $t_0 \in [c, d]$ such that $\tau(x_n) \to t_0$ as $n \to \infty$, and hence

$$f(x_n) = g(\tau(x_n)) \to g(t_0) \quad (n \to \infty)$$

by our continuity assumption on g. This shows that the left limit $f(x_0-)$ of f at x_0 exists. The existence of the right limit $f(x_0+)$ is proved similarly, and so we conclude that $f \in R([a, b])$. □

In what follows, we will refer to Theorem 0.36 as the *Sierpiński decomposition* of $f \in R([a,b])$. Other types of decomposition will be considered in Theorems 1.28 and 1.41 in the next chapter. We illustrate Theorem 0.36 by means of a very simple example.

Example 0.37. On $[a,b] = [0,2]$, consider the function $f := \chi_{\{1\}}$. Clearly, $D(f) = D_0(f) = \{1\}$. We apply Sierpiński's construction to this function to represent it in the form $f = g \circ \tau$ with g continuous and τ monotone.

The functions (0.56) and (0.57) become

$$p(x) = \begin{cases} 0 & \text{if } 0 \le x \le 1\,, \\ \frac{1}{2} & \text{if } 1 < x \le 2 \end{cases}$$

and

$$\tau(t) = \begin{cases} t & \text{if } 0 \le t < 1\,, \\ \frac{5}{4} & \text{if } t = 1\,, \\ t + \frac{1}{2} & \text{if } 1 < t \le 2\,. \end{cases}$$

Therefore, τ is a strictly increasing bijection between $[0,2]$ and $E = [0,1) \cup \{\frac{5}{4}\} \cup (\frac{3}{2}, \frac{5}{2}]$ with inverse

$$\tau^{-1}(s) = \begin{cases} s & \text{if } 0 \le s < 1\,, \\ 1 & \text{if } s = \frac{5}{4}\,, \\ s - \frac{1}{2} & \text{if } \frac{3}{2} < s \le \frac{5}{2}\,. \end{cases}$$

The function $g : E \to [0,2]$ given by

$$g(s) = f(\tau^{-1}(s)) = \begin{cases} 1 & \text{if } s = \frac{5}{4}\,, \\ 0 & \text{if } s \in [0,1) \cup (\frac{3}{2}, \frac{5}{2}] \end{cases}$$

may be extended continuously to the whole interval $[0, \frac{5}{2}]$ by putting

$$\hat{g}(s) := \begin{cases} 0 & \text{if } 0 \le s \le 1\,, \\ 4s - 4 & \text{if } 1 < s \le \frac{5}{2}\,, \\ 6 - 4s & \text{if } \frac{5}{4} \le s < \frac{3}{2}\,, \\ 0 & \text{if } \frac{3}{2} \le s \le \frac{5}{2}\,. \end{cases}$$

Then $\hat{g}$ is continuous, τ is increasing, and $f = \hat{g} \circ \tau$ as claimed. ♥

Note that we have used Proposition 0.35 in the proof of Theorem 0.36. Vice versa, we could have proved Proposition 0.35 by means of the statement of Theorem 0.36. In fact, if we represent $f \in R([a,b])$ in the form $f = g \circ \tau$ as in Sierpiński's theorem, then the continuity of g implies that $D(f) \subseteq D(\tau)$; but $D(\tau)$ is at most countable since τ is monotone. Of course, the inclusion $D(f) \subseteq D(\tau)$ does not give any information on the type of discontinuities of f; for instance, in Example 0.37, we have $D(f) = D(\tau) = \{1\}$, f has a removable discontinuity at 1, but τ has a jump at 1.

Now, we recall an important auxiliary characteristic which "counts" the number of solutions of the equation $f(x) = y$ for a continuous (or, more generally, regular) function f.

Definition 0.38. Given a continuous function $f : [a,b] \to \mathbb{R}$, we define a function $I_f : \mathbb{R} \to \mathbb{N}_0 \cup \{\infty\}$ by denoting $I_f(y)$ the number of elements of the set

$$f^{-1}(y) \cap [a,b] = \{x : a \leq x \leq b,\ f(x) = y\}. \tag{0.60}$$

This function is called the *Banach indicatrix* of f. More generally, we may generalize this definition to regular functions $f : [a,b] \to \mathbb{R}$ in the following way. Given $f \in R([a,b])$, we extend the graph of f by adjoining a vertical line connecting the unilateral limits $f(x_0-)$ and $f(x_0+)$ at each point of discontinuity of f. So, if f has a jump at $x_0 \in (a,b)$ and

$$f(x_0-) \leq y \leq f(x_0+),$$

the point x_0 is added to $f^{-1}(y) \cap [a,b]$ and counted for $I_f(y)$. ■

Obviously, denoting

$$m(f) = m(f; [a,b]) := \inf\{f(x) : a \leq x \leq b\} \tag{0.61}$$

and

$$M(f) = M(f; [a,b]) := \sup\{f(x) : a \leq x \leq b\}, \tag{0.62}$$

we may restrict the domain of the Banach indicatrix I_f to the interval $[m(f), M(f)]$. We remark that the Banach indicatrix I_f of a continuous function f is always a measurable function (see, e.g. [156]). Later (Proposition 1.27), we will establish a relation between the so-called *total variation* of a function f and the (Lebesgue) integrability of I_f over the interval $[m(f), M(f)]$. A very special version of this result may be found in Exercise 0.14.

Using the notation (0.61) and (0.62), we may pass from a given nonconstant function f to the function $F : [a,b] \to \mathbb{R}$ defined by

$$F(x) := \frac{f(x) - m(f)}{M(f) - m(f)} \qquad (a \leq x \leq b)$$

which often has the same properties as f, but, in addition, satisfies $0 \leq F(x) \leq 1$. This will be useful, for example, in the proof of Proposition 2.24 in the next chapter.

Apart from the spaces $B([a,b])$, $C([a,b])$, and $R([a,b])$, we now consider spaces of "more regular" functions. Applying the definition of the higher order spaces X^n from Section 0.2 to the space $X = C([a,b])$ gives the classical function spaces $C^n([a,b])$ with norm (0.43), that is,

$$\begin{aligned} \|f\|_{C^n} &:= \sum_{j=0}^{n-1} |f^{(j)}(a)| + \|f^{(n)}\|_C \\ &= |f(a)| + |f'(a)| + \ldots + |f^{(n-1)}(a)| + \max_{a\leq x\leq b} |f^{(n)}(x)|. \end{aligned} \tag{0.63}$$

Note that instead of (0.63), we could use the equivalent norm

$$\begin{aligned}|||f|||_{C^n} &:= \sum_{j=0}^{n} \|f^{(j)}\|_C \\ &= \max_{a\le x\le b} |f(x)| + \max_{a\le x\le b} |f'(x)| + \ldots + \max_{a\le x\le b} |f^{(n)}(x)|\,.\end{aligned} \tag{0.64}$$

In particular, the two norms

$$\|f\|_{C^1} = |f(a)| + \|f'\|_C, \qquad |||f|||_{C^1} = \|f\|_C + \|f'\|_C \tag{0.65}$$

are equivalent on the space $C^1([a,b])$; this is an immediate consequence of the mean value theorem for differentiable functions.

We come now to another important function class which is situated "between" $C^1([a,b])$ and $C([a,b])$.

Definition 0.39. A function $f : [a,b] \to \mathbb{R}$ is called *Lipschitz continuous* if there exists a constant $L > 0$ such that

$$|f(x) - f(y)| \le L|x - y| \quad (a \le x, y \le b)\,. \tag{0.66}$$

More generally, f is called *Hölder continuous* (or *α-Lipschitz continuous* for $0 < \alpha \le 1$) if there exists a constant $L > 0$ such that

$$|f(x) - f(y)| \le L|x - y|^\alpha \quad (a \le x, y \le b)\,. \tag{0.67}$$

We denote the set of all Lipschitz continuous functions on $[a,b]$ by $\mathrm{Lip}([a,b])$, and the set of all α-Lipschitz continuous functions on $[a,b]$ by $\mathrm{Lip}_\alpha([a,b])$. ■

Denoting by

$$\mathrm{lip}(f) = \mathrm{lip}(f; [a,b]) := \sup_{x\ne y} \frac{|f(x) - f(y)|}{|x - y|} \tag{0.68}$$

the minimal Lipschitz constant L in (0.66) and, for $0 < \alpha < 1$, by

$$\mathrm{lip}_\alpha(f) = \mathrm{lip}_\alpha(f; [a,b]) := \sup_{x\ne y} \frac{|f(x) - f(y)|}{|x - y|^\alpha} \tag{0.69}$$

the minimal Hölder constant L in (0.67), one may show that the spaces $\mathrm{Lip}([a,b])$ and $\mathrm{Lip}_\alpha([a,b])$, equipped with the norms[21]

$$\|f\|_{\mathrm{Lip}} := |f(a)| + \mathrm{lip}(f)\,, \tag{0.70}$$

and

$$\|f\|_{\mathrm{Lip}_\alpha} := |f(a)| + \mathrm{lip}_\alpha(f)\,, \tag{0.71}$$

respectively, are Banach spaces. Moreover, one may easily prove that

$$\mathrm{Lip}_\alpha([a,b]) \subseteq \mathrm{Lip}_\beta([a,b]) \quad (0 < \beta \le \alpha \le 1)\,, \tag{0.72}$$

21 Thus, in case $\alpha = 1$, we drop the subscript 1 and write lip and Lip rather than lip_1 and Lip_1.

where the inclusion is strict in case $\alpha > \beta$ (Exercise 0.45). The inclusions

$$C^1([a,b]) \subseteq \mathrm{Lip}([a,b]) \subseteq \mathrm{Lip}_\alpha([a,b]) \subseteq C([a,b]) \tag{0.73}$$

show that Lipschitz and Hölder continuity is some kind of "intermediate property" between continuity and continuous differentiability. The inclusions (0.72) and (0.73) are even continuous imbeddings:

Proposition 0.40. *For any $\alpha \in (0,1)$, the imbeddings*

$$C^1([a,b]) \hookrightarrow \mathrm{Lip}([a,b]) \hookrightarrow \mathrm{Lip}_\alpha([a,b]) \hookrightarrow C([a,b]) \tag{0.74}$$

hold.

Proof. Let $C^1([a,b])$ be equipped with the first norm in (0.65), and let $f \in C^1([a,b])$ and $x, y \in [a,b]$ be fixed. By the mean value theorem, we then have

$$|f(x) - f(y)| = |f'(\xi)|\,|x-y| \le \|f'\|_C|x-y|$$

for some ξ between x and y, and so $\mathrm{lip}(f) \le \|f'\|_C$. Comparing this with (0.68) yields

$$\|f\|_{\mathrm{Lip}} = |f(a)| + \mathrm{lip}(f) \le \|f\|_{C^1} \quad (f \in C^1),$$

which means that $C^1([a,b]) \hookrightarrow \mathrm{Lip}([a,b])$. Now, let $f \in \mathrm{Lip}([a,b])$ and $L > \mathrm{lip}(f)$. Then

$$\frac{|f(x)-f(y)|}{|x-y|^\alpha} = \frac{|f(x)-f(y)|}{|x-y|}|x-y|^{1-\alpha} \le L(b-a)^{1-\alpha}$$

for $x, y \in [a,b]$ and $\alpha \in (0,1)$. Since $L > \mathrm{lip}(f)$ was arbitrary, we conclude that $\mathrm{Lip}([a,b]) \hookrightarrow \mathrm{Lip}_\alpha([a,b])$. Finally, let $f \in \mathrm{Lip}_\alpha([a,b])$ and $L > \mathrm{lip}_\alpha(f)$. Then

$$|f(x)| \le |f(x) - f(a)| + |f(a)| \le L|x-a|^\alpha + |f(a)| \le L(b-a)^\alpha + |f(a)|\,.$$

Since $L > \mathrm{lip}_\alpha(f)$ was arbitrary, we conclude that $\mathrm{Lip}_\alpha([a,b]) \hookrightarrow C([a,b])$, and the proof is complete. □

The following simple example shows that all inclusions in (0.73) are strict for $\alpha < 1$.

Example 0.41. Let $[a,b] = [0,1]$. The function $f(x) := |x - x_0|$, where $x_0 \in (0,1)$ is fixed, clearly belongs to $\mathrm{Lip}([0,1])$ with $\mathrm{lip}(f) = 1$, but is not differentiable at x_0. The function $f(x) = x^\alpha$ belongs to $\mathrm{Lip}_\alpha([0,1])$, though in case $\alpha < 1$, not to $\mathrm{Lip}([0,1])$. Finally, let $f : [0,1] \to \mathbb{R}$ be defined by

$$f(x) := \begin{cases} \frac{1}{\log \frac{2}{x}} & \text{for } 0 < x \le 1\,, \\ 0 & \text{for } x = 0\,. \end{cases} \tag{0.75}$$

L'Hospital's rule shows that f is continuous at zero, and thus on the whole interval $[0,1]$. On the other hand, for any $\alpha > 0$, we have

$$\lim_{x\to 0+} \frac{|f(x) - f(0)|}{x^\alpha} = \lim_{x\to 0+} \frac{1}{x^\alpha \log \frac{2}{x}} = \infty\,,$$

and so $f \notin \mathrm{Lip}_\alpha([0,1])$ for *any* $\alpha \in (0,1]$. ♥

There is an analogous result[22] to Theorem 0.33 for Hölder continuous (in particular, Lipschitz continuous) functions; here, the proof is even constructive. This result is usually referred to as *McShane extension theorem* [210].

Theorem 0.42 (McShane). *Let $M \subset \mathbb{R}$ and $f : M \to \mathbb{R}$ be Hölder continuous on M with Hölder exponent $\alpha \in (0,1]$. Then there exists a Hölder continuous function $\hat{f} : \mathbb{R} \to \mathbb{R}$ such that $\hat{f}|_M = f$ and $\operatorname{lip}_\alpha(\hat{f}) = \operatorname{lip}_\alpha(f)$.*

Proof. Define $\hat{f} : \mathbb{R} \to \mathbb{R}$ by

$$\hat{f}(x) := \sup\{f(z) - \operatorname{lip}_\alpha(f)|x-z|^\alpha : z \in M\} \quad (x \in \mathbb{R}). \tag{0.76}$$

Since $\operatorname{lip}_\alpha(f)|x-z|^\alpha \geq 0$, with equality precisely for $x = z$, it is clear that $\hat{f}(x) = f(x)$ for $x \in M$. For general points $x, y \in \mathbb{R}$, we have

$$\begin{aligned}|\hat{f}(x) - \hat{f}(y)| &= \left|\sup_{z\in M}\,[f(z) - \operatorname{lip}_\alpha(f)|x-z|^\alpha] - \sup_{z\in M}\,[f(z) - \operatorname{lip}_\alpha(f)|y-z|^\alpha]\right| \\ &\leq \sup_{z\in M} \operatorname{lip}_\alpha(f)\,\big||x-z|^\alpha - |y-z|^\alpha\big| \leq \operatorname{lip}_\alpha(f)\,|x-y|^\alpha\,.\end{aligned}$$

This shows that $\hat{f} \in \operatorname{Lip}_\alpha(\mathbb{R})$ with $\operatorname{lip}_\alpha(\hat{f}) \leq \operatorname{lip}_\alpha(f)$. The converse inequality $\operatorname{lip}_\alpha(\hat{f}) \geq \operatorname{lip}_\alpha(f)$ is trivial since $\hat{f}$ extends f. □

The Hölder spaces $\operatorname{Lip}_\alpha([a,b])$ provide a good example for applying Proposition 0.31:

Example 0.43. For $0 < \alpha \leq 1$, let

$$X = \operatorname{Lip}_\alpha^o([a,b]) := \{f \in \operatorname{Lip}_\alpha([a,b]) : f(a) = 0\}$$

be equipped with the the norm $\|f\|_{\operatorname{Lip}_\alpha} = \operatorname{lip}_\alpha(f)$. Then $X \subseteq C([a,b]) \subseteq B([a,b])$ and, for $f, g \in X$ and $x, y \in [a,b]$,

$$\begin{aligned}\operatorname{lip}_\alpha(fg) &= \sup_{x\neq y} \frac{|(fg)(x) - (fg)(y)|}{|x-y|^\alpha} \\ &\leq \sup_{x\neq y} \frac{|f(x)g(x) - f(x)g(y)|}{|x-y|^\alpha} + \sup_{x\neq y} \frac{|f(x)g(y) - f(y)g(y)|}{|x-y|^\alpha} \\ &\leq \sup_{a\leq x\leq b} |f(x)| \sup_{x\neq y} \frac{|g(x) - g(y)|}{|x-y|^\alpha} + \sup_{a\leq y\leq b} |g(y)| \sup_{x\neq y} \frac{|f(x) - f(y)|}{|x-y|^\alpha} \\ &= \|f\|_\infty \operatorname{lip}_\alpha(g) + \|g\|_\infty \operatorname{lip}_\alpha(f)\,.\end{aligned}$$

This shows that (0.40) is satisfied since $\operatorname{lip}_\alpha(\cdot)$ is a norm on X. By Proposition 0.31, we know that the norm

$$|||f|||_{\operatorname{Lip}_\alpha} := \|f\|_C + \|f\|_X = \max_{a\leq x\leq b} |f(x)| + \operatorname{lip}_\alpha(f) \quad (f \in X) \tag{0.77}$$

22 As Theorem 0.33, the following Theorem 0.42 holds not only on the real line, but in the much more general setting of metric spaces.

is equivalent to the norm (0.71) on $\mathrm{Lip}_\alpha^o([a,b])$ and turns the space $(\mathrm{Lip}_\alpha^o, |||\cdot|||_{\mathrm{Lip}_\alpha})$ into a normalized Banach algebra. ♥

Of course, we may apply the construction of the higher order spaces X^n also to $X = \mathrm{Lip}([a,b])$ or, more generally, $X = \mathrm{Lip}_\alpha([a,b])$. This leads to the following

Definition 0.44. According to (0.43), we equip the linear space $\mathrm{Lip}^n([a,b])$ with the natural norm

$$\begin{aligned} \|f\|_{\mathrm{Lip}^n} &:= \sum_{j=0}^{n-1} |f^{(j)}(a)| + \|f^{(n)}\|_{\mathrm{Lip}} \\ &= |f(a)| + |f'(a)| + \ldots + |f^{(n)}(a)| + \mathrm{lip}(f^{(n)}) \end{aligned} \tag{0.78}$$

or the equivalent norm

$$\begin{aligned} |||f|||_{\mathrm{Lip}^n} &:= \sum_{j=0}^{n-1} \|f^{(j)}\|_C + \|f^{(n)}\|_{\mathrm{Lip}} \\ &= \max_{a\le x\le b} |f(x)| + \max_{a\le x\le b} |f'(x)| + \ldots + \max_{a\le x\le b} |f^{(n)}(x)| + \mathrm{lip}(f^{(n)}) . \end{aligned} \tag{0.79}$$

The analogous construction for Hölder spaces is left to the reader (Exercise 0.46). ■

In the study of both linear and nonlinear operators in function spaces, it is sometimes useful to know that restricting the discussion to a special domain (like $[a,b] = [0,1]$) does not affect the generality. To this end, we introduce some notation.

Definition 0.45. Given real numbers a, b, c, d with $a < b$ and $c < d$, consider the map $\ell : [c,d] \to [a,b]$ defined by

$$\ell(t) := \frac{b-a}{d-c}(t-c) + a \quad (c \le t \le d) . \tag{0.80}$$

Clearly, ℓ is an affine C^∞-diffeomorphism[23] between $[c,d]$ and $[a,b]$ with inverse

$$\ell^{-1}(s) = \frac{d-c}{b-a}(s-a) + c \quad (a \le s \le b) . \tag{0.81}$$

We call a function space X *shift-invariant* if there exist numbers $M, m > 0$ such that

$$m\|f\|_{X([a,b])} \le \|f \circ \ell\|_{X([c,d])} \le M\|f\|_{X([a,b])} . \tag{0.82}$$

Here, the interval $[a,b]$ in $X([a,b])$ denotes, of course, the domain of definition of functions from the space X. ■

---

23 To be precise, we should write $\ell_{a,b,c,d}$ because (0.80) depends of course on our choice of intervals. However, we drop the indices so as not to overburden the notation.

The two-sided estimate (0.82) means that the linear operators L and L^{-1} defined by

$$Lf = f \circ \ell, \quad L^{-1}g = g \circ \ell^{-1} \tag{0.83}$$

are bounded from $X([a,b])$ into $X([c,d])$ respectively from $X([c,d])$ into $X([a,b])$. In some cases, one may even choose $M = m = 1$ in (0.82), which means that the operator L in (0.83) is a linear surjective isometry. In this case, changing the interval does not increase or decrease the norm. We illustrate this by some simple examples.

Example 0.46. A straightforward calculation shows that (0.82) holds with $M = m = 1$ for $X = C$, $X = B$, $X = \mathrm{L}_\infty$, or $X = R$. On the other hand, (0.82) holds for $X = \mathrm{L}_1$ with

$$m = M = \frac{d-c}{b-a},$$

and for $X = C^1$ with

$$m = \min\left\{\frac{b-a}{d-c}, 1\right\}, \quad M = \max\left\{\frac{b-a}{d-c}, 1\right\}. \tag{0.84}$$

To calculate the constants m and M for $X = \mathrm{Lip}_\alpha$, observe that

$$\mathrm{lip}_\alpha(f \circ \ell; [c,d]) = \left(\frac{b-a}{d-c}\right)^\alpha \mathrm{lip}_\alpha(f; [a,b]) \quad (0 < \alpha \le 1).$$

Thus, we conclude that (0.82) holds for $X = \mathrm{Lip}$ with m and M as in (0.84), and for $X = \mathrm{Lip}_\alpha$ with

$$m = \min\left\{\left(\frac{b-a}{d-c}\right)^\alpha, 1\right\}, \quad M = \max\left\{\left(\frac{b-a}{d-c}\right)^\alpha, 1\right\}. \tag{0.85}$$

Further calculations of m and M may be found in Exercises 0.57–0.60. ♥

At this point, we introduce two interesting families of parameter-dependent functions which we will consider over and over in what follows to illustrate our abstract results.

Definition 0.47. Given $\alpha, \beta \in \mathbb{R}$, consider the function $f_{\alpha,\beta} : [0,1] \to \mathbb{R}$ defined by

$$f_{\alpha,\beta}(x) := \begin{cases} x^\alpha \sin x^\beta & \text{for } 0 < x \le 1, \\ 0 & \text{for } x = 0. \end{cases} \tag{0.86}$$

Clearly, this function is harmless for $\alpha > 0$ and $\beta > 0$, but interesting for $\alpha < 0$ or $\beta < 0$. We will call (0.86) the *oscillatory function*[24] determined by (α, β) in what follows. ■

24 Of course, the function (0.86) is "oscillatory" only for $\beta < 0$; however, we use this name for all values of α and β.

It is very instructive to determine all values of $\alpha, \beta \in \mathbb{R}$ for which $f_{\alpha,\beta}$ belongs to one of the spaces introduced so far. As a sample result, we do this for the spaces C, Lip, and C^1 in the following Proposition 0.48. Further results in this direction may be found in Exercises 0.7, 0.8, 0.52, 0.54 and 0.55, and in many of the forthcoming chapters.

Proposition 0.48. *For $\alpha, \beta \in \mathbb{R}$, let $f_{\alpha,\beta} : [0,1] \to \mathbb{R}$ be defined by* (0.86). *Then the following holds.*
(a) $f_{\alpha,\beta} \in C([0,1])$ *if and only if $\alpha > 0$ and β is arbitrary, or $\alpha \le 0$ and $\beta > -\alpha$.*
(b) $f_{\alpha,\beta} \in \mathrm{Lip}([0,1])$ *if and only if α is arbitrary and $\beta \ge 1-\alpha$.*
(c) $f_{\alpha,\beta} \in C^1([0,1])$ *if and only if α is arbitrary and $\beta > 1-\alpha$.*

Proof. (a) The continuity of $f_{\alpha,\beta}$ in the case $\alpha > 0$ is clear since $|\sin x^\beta| \le 1$ for any $\beta \in \mathbb{R}$. For $\alpha = 0$, we get the function $f_{0,\beta}(x) = \sin x^\beta$ which is continuous at 0. Finally, in the case $\alpha < 0$, L'Hospital's rule shows that

$$\lim_{x\to 0+} \frac{\sin x^\beta}{x^{-\alpha}} = -\frac{\beta}{\alpha} \lim_{x\to 0+} x^{\alpha+\beta} \cos x^\beta . \tag{0.87}$$

So, in this case, $f_{\alpha,\beta}$ is continuous at 0 if and only if[25] $\beta > -\alpha$.
(b) Together with the function (0.86), we consider its "twin sister,"

$$g_{\alpha,\beta}(x) := \begin{cases} x^\alpha \cos x^\beta & \text{for } 0 < x \le 1 , \\ 0 & \text{for } x = 0 . \end{cases} \tag{0.88}$$

Clearly, a continuous function $f : [a,b] \to \mathbb{R}$ which is differentiable on (a,b) is Lipschitz continuous on $[a,b]$ if and only if its derivative is bounded on $[a,b]$. This means that we have to find all pairs (α, β) for which the derivative

$$f'_{\alpha,\beta}(x) = \alpha f_{\alpha-1,\beta}(x) + \beta g_{\alpha+\beta-1,\beta}(x) \quad (0 < x \le 1) \tag{0.89}$$

is bounded near zero. The second term in (0.89) is bounded near zero only if $\beta \ge 1-\alpha$. For $\alpha \ge 1$, the first term is also bounded. On the other hand, for $\alpha < 1$, the factor $x \mapsto x^{\alpha-1}$ is unbounded, but (0.89) shows that, nevertheless, the first term remains bounded if $\beta \ge 1-\alpha$.

(c) We have to show that the function (0.89) has the limit $f'_{\alpha,\beta}(0)$ for $x \to 0+$. In the case $\alpha > 1$ and $\beta > 1-\alpha$, we have

$$\lim_{x\to 0+} f'_{\alpha,\beta}(x) = \alpha \lim_{x\to 0+} f_{\alpha-1,\beta}(x) - \beta \lim_{x\to 0+} g_{\alpha+\beta-1,\beta}(x) = 0$$

since both exponents $\alpha - 1$ and $\alpha + \beta - 1$ are positive. Similarly, in the case $\alpha = 1$ and $\beta > 1-\alpha$, the limit is zero since we still have $\beta > 0$ in the first term. Finally, in the case

25 The limit (0.87) also exists in case $\beta = -\alpha$, but has the "wrong" value 1, so $f_{\alpha,-\alpha}$ has a removable discontinuity at zero.

$\alpha < 1$ and $\beta > 1 - \alpha$, L'Hospital's rule shows that again

$$\lim_{x\to 0+} f'_{\alpha,\beta}(x) = \alpha \lim_{x\to 0+} f_{\alpha-1,\beta}(x) - \beta \lim_{x\to 0+} g_{\alpha+\beta-1,\beta}(x)$$
$$= \left(\frac{\alpha\beta}{1-\alpha} - \beta\right) \lim_{x\to 0+} g_{\alpha+\beta-1,\beta}(x) = 0\,.$$

In all other cases, the limit does not exist. □

The second family of functions we will often use in the sequel is constructed over the interval $[0, 1]$ as follows.

Definition 0.49. Let $C = (c_n)_n$ and $D = (d_n)_n$ be two positive and decreasing real sequence converging to 0. In addition, $(c_n)_n$ is supposed to satisfy

$$\sum_{n=1}^{\infty} c_n = 1\,. \tag{0.90}$$

By means of such sequences, we construct a continuous function $Z_{C,D} : [0, 1] \to \mathbb{R}$ in the following way. We put $Z_{C,D}(0) = 0$ and let $Z_{C,D}$ increase linearly by d_1 on the interval $[0, c_1]$ so that $Z_{C,D}(c_1) = d_1$. Then we let $Z_{C,D}$ decrease linearly by d_2 on $[c_1, c_1 + c_2]$, increase linearly by d_3 on $[c_1 + c_2, c_1 + c_2 + c_3]$, decrease linearly by d_4 on $[c_1 + c_2 + c_3, c_1 + c_2 + c_3 + c_4]$, and so on. So, the explicit form of $Z_{C,D}$ reads

$$Z_{C,D}(x) = \begin{cases} 0 & \text{for } x = 0\,, \\ \sum_{k=1}^{n} (-1)^{k+1} d_k & \text{for } x = \sum_{k=1}^{n} c_k\,, \\ \text{linear} & \text{otherwise}\,. \end{cases} \tag{0.91}$$

We call (0.91) the (general) *zigzag function* determined by (C, D) in what follows. A particularly important special case is

$$c_n := \frac{1}{2^n}, \quad d_n := \frac{1}{n^\theta} \tag{0.92}$$

for some $\theta > 0$, i.e.

$$Z_\theta(x) = \begin{cases} 0 & \text{for } x = 0\,, \\ 1 - \frac{1}{2^\theta} + \frac{1}{3^\theta} - + \ldots + \frac{(-1)^{n+1}}{n^\theta} & \text{for } x = 1 - \frac{1}{2^n}\,, \\ \text{linear} & \text{otherwise}\,. \end{cases} \tag{0.93}$$

In this case, we call (0.93) a *special zigzag function.* ■

It follows from the construction and continuity of the zigzag function (0.91) that

$$Z_{C,D}(1) = \sum_{k=1}^{\infty} (-1)^{k+1} d_k, \quad Z_\theta(1) = \sum_{k=1}^{\infty} \frac{(-1)^{k+1}}{k^\theta}\,. \tag{0.94}$$

Again, it is illuminating to determine all sequences $C = (c_n)_n$ and $D = (d_n)_n$ for which the zigzag function (0.91) belongs to the function classes introduced so far. Of course, the function (0.91) is always continuous, by construction, but not differentiable at its peaks. In the following proposition, we characterize all sequences $C = (c_n)_n$ and $D = (d_n)_n$ for which (0.91) is Hölder (in particular, Lipschitz) continuous.

Proposition 0.50. *For sequences $C = (c_n)_n$ and $D = (d_n)_n$ as above, the general zigzag function* (0.91) *belongs to* $\mathrm{Lip}_\alpha([0,1])$ $(0 < \alpha \le 1)$ *if and only if*

$$\sup\{d_n c_n^{-\alpha} : n = 1, 2, 3, \ldots\} < \infty . \tag{0.95}$$

Proof. Denoting the supremum in (0.95) by $S_\alpha(C, D)$, we show that

$$\mathrm{lip}_\alpha(Z_{C,D}; [0,1]) \le S_\alpha(C, D) \le 4\ \mathrm{lip}_\alpha(Z_{C,D}; [0,1]) . \tag{0.96}$$

The left inequality in (0.96) follows from a simple calculation, taking into account the slope of $Z_{C,D}$ between two successive peaks. To prove the right inequality, observe that the alternating series in (0.91) converges to some real number d^*, say, since $(d_n)_n$ is monotonically decreasing and converges to zero. Now, if s and $t > s$ belong to the same interval between two adjacent peaks, we have

$$|Z_{C,D}(s) - Z_{C,D}(t)| \le S_\alpha(C, D)|s - t|^\alpha ,$$

by definition of $S_\alpha(C, D)$. If s and $t > s$ belong to two intervals with just one peak in between, say $s < c_1 + c_2 + \ldots + c_n =: C_n < t$, then

$$\begin{aligned}|Z_{C,D}(s) - Z_{C,D}(t)| &\le |Z_{C,D}(s) - Z_{C,D}(C_n)| + |Z_{C,D}(C_n) - Z_{C,D}(t)| \\ &\le S_\alpha(C, D)|s - C_n|^\alpha + S_\alpha(C, D)|C_n - t|^\alpha \le 2S_\alpha(C, D)|s - t|^\alpha .\end{aligned}$$

In general, let $\hat{s} > s$ be the smallest number at which $Z_{C,D}$ attains the value d^*, and let $\hat{t} < t$ be the largest number with this property.[26] Then

$$\begin{aligned}&|Z_{C,D}(s) - Z_{C,D}(t)| \\ &\quad\le |Z_{C,D}(s) - Z_{C,D}(\hat{s})| + |Z_{C,D}(\hat{s}) - Z_{C,D}(\hat{t})| + |Z_{C,D}(\hat{t}) - Z_{C,D}(t)| \\ &\quad= |Z_{C,D}(s) - Z_{C,D}(\hat{s})| + |Z_{C,D}(\hat{t}) - Z_{C,D}(t)| \\ &\quad\le 2S_\alpha(C, D)|s - \hat{s}|^\alpha + 2S_\alpha(C, D)|\hat{t} - t|^\alpha \le 4S_\alpha(C, D)|s - t|^\alpha .\end{aligned}$$

This proves the second estimate in (0.96), and we are done. □

In case of the special zigzag function (0.93), Proposition 0.50 implies the following somewhat surprising corollary.

Corollary 0.51. *The special zigzag function* (0.93) *does not belong to* $\mathrm{Lip}_\alpha([0,1])$ *for any* $\theta \in (0, 1]$ *and* $\alpha \in (0, 1]$.

---

26 From a geometric reasoning, we see that $\hat{s}$ belongs to the interval following s, and $\hat{t}$ belongs to the interval preceding t.

Proof. Choosing c_n and d_n as in (0.92), we obtain

$$\sup\{d_n c_n^{-\alpha} : n = 1, 2, 3, \ldots\} = \sup\{n^{-\theta} 2^{n\alpha} : n = 1, 2, 3, \ldots\} = \infty$$

since the exponential term $2^{n\alpha}$ grows essentially faster than the power type term n^{θ}. By (0.96), we see that $\mathrm{lip}_{\alpha}(Z_{\theta}; [0, 1]) = \infty$, and so $Z_{\theta} \notin \mathrm{Lip}_{\alpha}([0, 1])$ for any $\alpha \in (0, 1]$. □

The oscillation function (0.86) and the (special) zigzag function (0.93) are very useful for constructing counterexamples. For further reference, in the following Table 0.1, we collect the values of α, β, and θ, respectively, for which these functions belong to the function spaces considered so far. An essential extension of this table will be given in Table 2.4 in Chapter 2.

Table 0.1. Oscillation functions and zigzag functions.

	The function $f_{\alpha,\beta}$	*The function* Z_{θ}
belongs to $C([0, 1])$ *iff*	$\alpha > 0$ or $\alpha \leq 0$ and $\alpha + \beta > 0$	always
belongs to $C^1([0, 1])$ *iff*	see Exercise 0.54	never
belongs to $L_1([0, 1])$ *iff*	see Exercise 0.7	always
belongs to $\mathrm{Lip}([0, 1])$ *iff*	$\alpha + \beta \geq 1$	never
belongs to $\mathrm{Lip}_{\gamma}([0, 1])$ *iff*	see Exercise 0.52	never

It is clear that the zigzag function Z_{θ} never belongs to spaces of differentiable functions. On the other hand, being bounded on $[0, 1]$, Z_{θ} trivially belongs to $L_p([0, 1])$ for any p. Table 2.4 in Chapter 2 will contain other functions spaces in which the behavior of zigzag functions is more interesting.

The Hölder spaces Lip_{α} may be generalized in different ways; we consider one generalization which consists of replacing the map $t \mapsto t^{\alpha}$ by a so-called modulus of continuity:

Definition 0.52. An increasing function $\omega : [0, \infty) \to [0, \infty)$ is called *modulus of continuity* if $\omega(0) = 0$ and $\omega(t) > 0$ for $t > 0$, $\omega(s + t) \leq \omega(s) + \omega(t)$, and ω is continuous at 0. ■

A standard example of a modulus of continuity is of course $\omega(t) = t^{\alpha}$ for $0 < \alpha \leq 1$. Definition 0.52 is inspired by classical moduli of continuity which are defined for $f : [a, b] \to \mathbb{R}$ by

$$\omega_{\infty}(f; \delta) = \omega_{\infty}(f, [a, b]; \delta) := \sup_{0 \leq h \leq \delta} \{|f(x + h) - f(x)| : a \leq x \leq b - h\} \tag{0.97}$$

and

$$\omega_p(f; \delta) = \omega_p(f, [a, b]; \delta) := \sup_{0 \leq h \leq \delta} \left\{ \int_a^{b-h} |f(x + h) - f(x)|^p \, dx \right\}^{1/p} \tag{0.98}$$

for $1 \leq p < \infty$. In fact, the following is true.

Proposition 0.53. *The relation*

$$\lim_{\delta\to 0+} \omega_\infty(f;\delta) = 0 \tag{0.99}$$

holds for every function $f \in C([a,b])$, *while the relation*

$$\lim_{\delta\to 0+} \omega_p(f;\delta) = 0 \tag{0.100}$$

holds for every $f \in \mathrm{L}_p([a,b])$.

Proof. The validity of (0.99) for continuous f is clear. To prove that (0.100) holds for $f \in \mathrm{L}_p([a,b])$, we fix $\varepsilon > 0$ and choose a continuous function[27] $g : [a,b] \to \mathbb{R}$ such that $\|f - g\|_{\mathrm{L}_p} \le \varepsilon$. Since g is uniformly continuous on $[a,b]$, we find a $\delta > 0$ such that $|g(x) - g(y)| \le \varepsilon(b-a)^{-1/p}$ for $x, y \in [a,b]$ satisfying $|x-y| \le \delta$. Combining these two estimates, for $0 \le h \le \delta$, we get

$$\begin{aligned}
&\left\{\int_a^{b-h} |f(x+h) - f(x)|^p\,dx\right\}^{1/p} \\
&\le 2\|f-g\|_{\mathrm{L}_p} + \left\{\int_a^{b-h} |g(x+h) - g(x)|\,dx\right\}^{1/p} \\
&\le 2\varepsilon + \frac{\varepsilon}{(b-a)^{1/p}}(b-a)^{1/p} = 3\varepsilon,
\end{aligned}$$

proving the assertion. □

Proposition 0.53 states that

$$\omega_p(f;\delta) = o(1) \quad (\delta \to 0+)$$

for every $f \in \mathrm{L}_p$. One may show that the essentially stronger condition

$$\omega_p(f;\delta) = o(\delta) \quad (\delta \to 0+)$$

is satisfied only for constant functions, see Exercise 0.76.

By means of the moduli (0.97) and (0.98), we may define generalized Hölder spaces as follows.

Definition 0.54. Let $\omega : [0,\infty) \to [0,\infty)$ be an arbitrary modulus of continuity in the sense of Definition 0.52, and let $\omega_\infty(f;\cdot)$ and $\omega_p(f;\cdot)$ be defined by (0.97) and (0.98), respectively. For $f : [a,b] \to \mathbb{R}$, put

$$\mathrm{lip}_{\omega,p}(f) = \mathrm{lip}_{\omega,p}(f;[a,b]) := \sup_{\delta>0} \frac{\omega_p(f;\delta)}{\omega(\delta)} \quad (1 \le p \le \infty). \tag{0.101}$$

---

27 Here, we use the fact that the space $C([a,b])$ is *dense* in the space $\mathrm{L}_p([a,b])$ with respect to the norm (0.11).

We write $f \in \mathrm{Lip}_{\omega,p}([a,b])$ if $\mathrm{lip}_{\omega,p}(f) < \infty$ and call $\mathrm{Lip}_{\omega,p}([a,b])$ the *generalized Hölder space* defined by ω and p. We consider this space equipped with the norm

$$\|f\|_{\mathrm{Lip}_{\omega,p}} := \begin{cases} \|f\|_C + \mathrm{lip}_{\omega,\infty}(f) & \text{if } p = \infty\,, \\ \|f\|_{L_p} + \mathrm{lip}_{\omega,p}(f) & \text{if } p < \infty\,. \end{cases} \tag{0.102}$$

In case $\omega(t) = t^\alpha$ $(0 < \alpha < 1)$, we write $\mathrm{lip}_{\alpha,p}(f)$ instead of $\mathrm{lip}_{\omega,p}(f)$ and $\mathrm{Lip}_{\alpha,p}([a,b])$ instead of $\mathrm{Lip}_{\omega,p}([a,b])$. ■

Clearly, taking $p = \infty$ in Definition 0.54, we get the Hölder space $\mathrm{Lip}_{\alpha,\infty}([a,b]) = \mathrm{Lip}_\alpha([a,b])$; this follows from the fact that the condition $\omega_\infty(f,\delta) \le \omega(\delta)$ for all $\delta \ge 0$ is equivalent to the condition

$$|f(s) - f(t)| \le \omega(|s-t|) \quad (a \le s, t \le b)\,.$$

We remark that the spaces $\mathrm{Lip}_{\alpha,p}$ are related to the classical Hölder spaces Lip_β through the imbeddings

$$\mathrm{Lip}_\alpha([a,b]) \hookrightarrow \mathrm{Lip}_{\alpha,p}([a,b]) \hookrightarrow \mathrm{Lip}_{\alpha-1/p}([a,b])$$

(Exercise 0.71).

We will return to this class of spaces in Chapter 2, where we will compare them with imbeddings into spaces of functions of generalized bounded variation.

To conclude this section on function spaces, we briefly consider compact sets. In many applications, it is important to describe compact subsets of a Banach space by an "intrinsic" characterization. The simplest such description is known in the finite dimensional space $\mathbb{R}^n$, where a subset $K \subset \mathbb{R}^n$ is compact if and only if K is both closed and bounded.[28] On the other hand, in every infinite dimensional space, closedness and boundedness is still necessary for compactness, but never sufficient. In fact, in such spaces, one has to add a third property (called "compactness criterion") to get compactness.

We briefly recall two well-known compactness criteria, the first in the space $C([a,b])$, and the second in the space $L_p([a,b])$ for $1 \le p < \infty$. They both build on the moduli of continuity (0.97) and (0.98); we cite them without proof.

Proposition 0.55 (Arzelà–Ascoli). *Let $\omega_\infty(f;\delta)$ be defined by* (0.97). *Then a subset $K \subset C([a,b])$ is compact if and only if K is closed, bounded, and satisfies* (0.99) *uniformly with respect to $f \in K$, i.e.*

$$\lim_{\delta \to 0+} \sup\,\{\omega_\infty(f;\delta) : f \in K\} = 0\,. \tag{0.103}$$

28 This result is known as Heine–Borel compactness criterion.

Proposition 0.56 (Kolmogorov). *Let $\omega_p(f;\delta)$ be defined by* (0.99). *Then a subset $K \subset$ $L_p([a,b])$ is compact if and only if K is closed, bounded, and satisfies* (0.100) *uniformly with respect to $f \in K$, i.e.*

$$\lim_{\delta\to 0+} \sup\{\omega_p(f;\delta) : f \in K\} = 0\,. \tag{0.104}$$

The property expressed in relation (0.103) is usually called the *equicontinuity* of K. Therefore, if $(f_n)_n$ is a bounded sequence in an equicontinuous set $K \subset C([a,b])$, we may always find a subsequence which is uniformly convergent on $[a,b]$. A weak analogue to this in the space of functions of bounded variation will be proved in Theorem 1.11 in the first chapter.

0.4 Comments on Chapter 0

In any textbook on measure and integration theory, the Lebesgue spaces L_p constitute an important ingredient. In our examples and counterexamples in Section 0.1, we followed the book [39] which is an extremely valuable source. Classical monographs on Orlicz spaces are [169] or [263].

The general theorems on Banach spaces and bounded linear functionals treated in Section 0.2 may be found in every textbook on functional analysis, we mention [146, 171, 266, 270, 302, 319, 320]. A standard reference on function spaces is [172]. An updated version [249] of it just appeared in this book series.

All of function spaces we have discussed in this chapter are Banach spaces, and the same is true for other spaces we will introduce in the following chapters. An example of an incomplete function space is the linear space $P([a,b])$ of polynomials $p(x) = a_n x^n + a_2 x^2 + a_1 x + a_0$ (of any degree n) with the norm

$$\|p\|_C = \max_{a\le x\le b} |p(x)| \tag{0.105}$$

inherited from the larger space $C([a,b])$. Since $P([a,b])$ is not closed[29] in $C([a,b])$, the space $(P([a,b]), \|\cdot\|_C)$ is incomplete.

For $m(f)$ as in (0.61) and $M(f)$ as in (0.62), in 1925, Banach [40] introduced the indicatrix $I_f : [m(f), M(f)] \to [0,\infty]$ of a continuous function $f : [a,b] \to \mathbb{R}$ as

$$I_f(y) := \#\{f^{-1}(y) \cap [a,b]\}\,, \tag{0.106}$$

29 In fact, the classical Weierstrass approximation theorem shows that the closure of $P([a,b])$ in the norm (0.45) is the whole space $C([a,b])$. Even worse, since the monomials form a countably infinite (algebraic) basis in $P([a,b])$, it follows from the so-called *Baire category theorem* that there is *no norm* on $P([a,b])$ which makes this space complete, see also Exercise 0.27.

where $\#A$ denotes the cardinality of the set A. He also proved that f has bounded variation if and only if

$$\int_{m(f)}^{M(f)} I_f(y)\,dy < \infty\,, \tag{0.107}$$

see Proposition 1.27 in the next chapter. Later, Natanson [238] and others called N_f the *Banach indicatrix* of f. If f is merely regular, one may use the Sierpiński decomposition (Theorem 0.36) $f = g \circ \tau$ of f, with g being continuous und τ being strictly increasing, to extend Banach's definition to $f \in R([a,b])$ by putting $N_f := N_g$. This extension due to Lozinskij [186] is possible since a strictly increasing change of variables does not change the number of elements in $f^{-1}(y)$.

Some authors call the estimate (0.16) *Hölder inequality*. However, usually this term refers to the inequality

$$\sum_{k=1}^{n} \alpha_k\beta_k \le \left(\sum_{k=1}^{n} \alpha_k^p\right)^{1/p} \left(\sum_{k=1}^{n} \beta_k^{p'}\right)^{1/p'}\,, \tag{0.108}$$

where $\alpha_1, \dots, \alpha_n, \beta_1, \dots, \beta_n$ are positive real numbers and the exponents p and p' are related by (0.13).

Regular functions (also called *quasicontinuous functions* by some authors, e.g. [76]), are closely related to step functions. Recall that $f : [a,b] \to \mathbb{R}$ is called a *step function* if finitely many points $a = t_0 < t_1 < \dots < t_m = b$ exist such that f is constant[30] on each open interval (t_{j-1}, t_j) for $j = 1, 2, \dots, m$. We will write $S([a,b])$ for the linear space of all step functions on $[a,b]$. Obviously, the (strict) inclusion $S([a,b]) \subset R([a,b])$ holds, which means that step functions are regular. The following proposition which we will prove in view of its importance makes this more precise.

Proposition 0.57. *The equality*

$$\overline{S([a,b])} = R([a,b]) \tag{0.109}$$

holds, where the closure in (0.109) *is taken in the norm* (0.39).

Proof. First, suppose that $f \in \overline{S([a,b])}$, which means that for every $\varepsilon > 0$, there exists a step function g such that $\|f - g\|_\infty \le \varepsilon$. We have to show that f has unilateral limits at every point.

Fix $x_0 \in [a,b)$; we show that the right limit $f(x_0+)$ exists, see (0.54). Since g is a step function, we can find a $\delta > 0$ such that g is constant on $(x_0, x_0 + \delta)$. However, this implies that

$$|f(s) - f(t)| \le |f(s) - g(s)| + |g(s) - g(t)| + |g(t) - f(t)| \le 2\varepsilon$$

for $s, t \in (x_0, x_0 + \delta)$, and so $f(x_0+)$ exists. The proof for the left limit $f(x_0-)$ for $x_0 \in (a,b]$ is similar, and so we have shown that f is regular.

30 Here, f is allowed to take on any arbitrary value at the points $t_0, t_1, \dots, t_m$ themselves.

Conversely, suppose now that $f \in R([a,b])$, and let $\varepsilon > 0$. For each $x \in [a,b]$, we can find a $\delta = \delta(\varepsilon, x)$ such that $|f(s)-f(t)| \le \varepsilon$ for $s,t \in (x-\delta(\varepsilon,x),x)\cup(x,x+\delta(\varepsilon,x))$, by definition of the unilateral limits (0.54). Since the intervals $\{((x-\delta(\varepsilon,x),x+\delta(\varepsilon,x)) : a \le x \le b\}$ form an open cover and $[a,b]$ is compact, we only actually need finitely many of them to do the job, say

$$[a,b] \subseteq (x_0 - \delta(\varepsilon,x_0), x_0 + \delta(\varepsilon,x_0)) \cup \ldots \cup (x_n - \delta(\varepsilon,x_n), x_n + \delta(\varepsilon,x_n)) .$$

Choose points $a = t_0 < t_1 < \ldots < t_m = b$ such that each interval (t_{j-1}, t_j) is contained in some $(x_i - \delta(\varepsilon,x_i), x_i)$ or in some $(x_i, x_i + \delta(\varepsilon,x_i))$. Then $|f(s) - f(t)| \le \varepsilon$ whenever $s,t \in (t_{j-1}, t_j)$. Now, define $g : [a,b] \to \mathbb{R}$ by

$$g(x) := \begin{cases} f(t_j) & \text{for } x = t_j \ (j = 0,1,\ldots,m), \\ f\left(\frac{1}{2}(t_j + t_{j-1}\right) & \text{for } t_{j-1} < x < t_j \ (j = 1,2,\ldots,m). \end{cases}$$

Clearly, $g \in S([a,b])$ and $\|f-g\| \le \varepsilon$, so $f \in \overline{S([a,b])}$. □

The shift invariance of some spaces in the sense of Definition 0.45 is discussed in [16] in connection with nonlinear composition operators; we will come back to this in Chapter 5. The oscillation functions are studied in detail in [11], while the zigzag functions are discussed in [250] in connection with certain spaces of functions of bounded variation which we will study in Section 2.2.

Compactness criteria like those given in Propositions 0.55 and 0.56 are extremely useful for proving existence theorems in both linear and nonlinear functional analysis. The range of applications of such existence theorems may be enlarged by considering so-called *measures of noncompactness* which are closely related to the characteristics (0.103) and (0.104), see, e.g. [6, 10, 26, 41, 42].

The generalized Hölder spaces $\mathrm{Lip}_{\omega,p}$ and $\mathrm{Lip}_{\alpha,p}$ introduced in Definition 0.54 are quite useful in the theory of Fourier series which we will briefly discuss in Chapter 7. In particular, it is interesting to establish continuous imbeddings between such spaces. For example, the following two results [141, 142, 305] are known:

Proposition 0.58 (Hardy–Littlewood). *Let $\omega : [0,\infty) \to [0,\infty)$ be a modulus of continuity, and let $0 < \beta \le 1 \le p < q < \infty$. Then the imbedding*

$$\mathrm{Lip}_{\omega,p}([a,b]) \hookrightarrow \mathrm{Lip}_{\beta,q}([a,b])$$

holds if and only if

$$\lim_{\delta\to 0+} \frac{\omega(\delta)}{\delta^{\beta+1/p-1/q}} < \infty .$$

In particular, $\mathrm{Lip}_{\alpha,p}([a,b]) \hookrightarrow \mathrm{Lip}_{\beta,q}([a,b])$ if and only if

$$\alpha - \beta \ge \frac{1}{p} - \frac{1}{q}$$

for $\beta < \alpha \le 1$.

Proposition 0.59 (Ulyanov). *Let $\omega : [0,\infty) \to [0,\infty)$ be a modulus of continuity, and let $1 \le p < q < \infty$. Then the imbedding*

$$\mathrm{Lip}_{\omega,p}([a,b]) \hookrightarrow \mathrm{L}_q([a,b])$$

holds if and only if

$$\sum_{n=1}^{\infty} n^{q/p-2}\omega^q\left(\frac{1}{n}\right) < \infty .$$

In particular, $\mathrm{Lip}_{\alpha,p}([a,b]) \hookrightarrow \mathrm{L}_q([a,b])$ *if and only if*

$$\frac{p}{q} - \alpha q < 1$$

for $0 < \alpha \le 1$.

More imbedding theorems of this type, also in connection with functions of bounded variation, will be discussed in Sections 2.2 and 2.8.

0.5 Exercises to Chapter 0

We state some exercises on the topics covered in this chapter; exercises marked with an asterisk $*$ are more difficult.

Exercise 0.1. Let $M \subseteq \mathbb{R}$ be a measurable set and $f : M \to \mathbb{R}$ a measurable function. For $c > 0$, let

$$M_c(f) := \{x \in M : |f(x)| > c\} .$$

Prove that

$$\inf\{c > 0 : \lambda(M_c(f)) = 0\} = \inf_{f\sim g} \sup\{|g(x)| : x \in M\} ,$$

where the supremum is taken over all functions $g : M \to \mathbb{R}$ which are equivalent to f.

Exercise 0.2. From the proof of Proposition 0.10 (b), it follows that

$$\mathrm{L}_\infty([a,b]) \hookrightarrow \mathrm{L}_q([a,b]) \hookrightarrow \mathrm{L}_p([a,b]) \hookrightarrow \mathrm{L}_1([a,b]) \quad (1 < p < q < \infty) .$$

Calculate the sharp imbedding constants $c(\mathrm{L}_\infty, \mathrm{L}_q)$, $c(\mathrm{L}_q, \mathrm{L}_p)$, and $c(\mathrm{L}_p, \mathrm{L}_1)$ for these values of p and q.

Exercise 0.3. Let $1 \le p < q \le \infty$, and suppose that $f \in \mathrm{L}_p(I) \cap \mathrm{L}_q(I)$, where $I \subseteq \mathbb{R}$ is an unbounded interval. Show that then

$$f \in \bigcap_{p \le r \le q} \mathrm{L}_r(I) .$$

Why is this statement trivial for bounded intervals I?

Exercise 0.4. The *convolution* $f * g$ of two functions $f, g \in L_1(\mathbb{R})$ is defined by

$$(f * g)(x) := \int_{-\infty}^{\infty} f(x-t)g(t)\,dt\,.$$

Use Fubini's theorem (Theorem 0.7) for $p = q = 1$ to show that $(L_1(\mathbb{R}), *)$ is a Banach algebra such that

$$\|f * g\|_{L_1} \leq \|f\|_{L_1}\|g\|_{L_1}$$

for all $f, g \in L_1(\mathbb{R})$. Calculate the convolution $f * f$ of the characteristic function $f = \chi_{[a,b]}$ with itself.

Exercise 0.5. Let $\rho : \mathbb{R} \to \mathbb{R}$ be defined by

$$\rho(x) := \begin{cases} c\exp\left(-\frac{1}{1-|x|^2}\right) & \text{for } |x| < 1\,, \\ 0 & \text{for } |x| \geq 1\,, \end{cases}$$

where

$$c := \int_{-\infty}^{\infty} \rho(x)\,dx\,.$$

For $\varepsilon > 0$, define $\rho_\varepsilon : \mathbb{R} \to \mathbb{R}$ by

$$\rho_\varepsilon(x) := \frac{1}{\varepsilon}\rho\left(\frac{x}{\varepsilon}\right)\,.$$

The function ρ_ε is called the *mollifier*; this name is explained by the fact that $\rho_\varepsilon * f \in C^\infty(\mathbb{R})$ for every $f \in L_1(\mathbb{R})$, where $*$ denotes the convolution introduced in Exercise 0.4. Prove this and calculate the derivative of $\rho_\varepsilon * f$.

Exercise 0.6*. With ρ and ρ_ε as in Exercise 0.5, prove that

$$\|\rho_\varepsilon * f\|_{L_1} \leq \|f\|_{L_1} \quad (\varepsilon > 0)$$

and

$$\lim_{\varepsilon\to 0+} \|\rho_\varepsilon * f - f\|_{L_1} = 0$$

for every $f \in L_1(\mathbb{R})$. Conclude that $f \in L_1(\mathbb{R})$ is separable.

Exercise 0.7. Show that the function $f_{\alpha,\beta}$ from (0.86) belongs to $L_1([0,1])$ if and only if $\beta \geq 0$ and $\alpha + \beta > -1$, or $\beta < 0$ and $\alpha > -1$.

Exercise 0.8. Show that the function $f_{\alpha,\beta}$ from (0.86) belongs to $L_1([1,\infty))$ if and only if $\beta \leq 0$ and $\alpha + \beta < -1$, or $\beta > 0$ and $\alpha < -1$.

Exercise 0.9. For $\mu \in \mathbb{R}$, and let $f_\mu : \mathbb{R} \to \mathbb{R}$ be defined by

$$f_\mu(x) := \begin{cases} \frac{|\log x|^\mu}{x^{1/p}} & \text{for } x \neq 0\,, \\ 0 & \text{for } x = 0\,. \end{cases}$$

Prove that $f_\mu \in L_p([0,1])$ if and only if $\mu p < -1$.

Exercise 0.10. Determine all values of μ for which the functions f_μ from Exercise 0.9 belongs to $L_p([1,\infty))$.

Exercise 0.11. Let $M \subseteq \mathbb{R}$ be a measurable set. A sequence $(f_n)_n$ of measurable functions $f_n : M \to \mathbb{R}$ converges *in measure* to some measurable function $f : M \to \mathbb{R}$ if

$$\lim_{n\to\infty} \lambda(M_c(f_n - f)) = 0$$

for all $c > 0$, where $M_c(f)$ is defined as in Exercise 0.1. Show that convergence of $(f_n)_n$ to f a.e. on M implies convergence of $(f_n)_n$ in measure to f, though not vice versa.

Exercise 0.12. Let $\phi : [0,\infty) \to [0,\infty)$ be an increasing continuous function satisfying $\phi(0) = 0$ and $\phi(u) > 0$ for $u > 0$. Let $M \subseteq \mathbb{R}$ be a measurable set, and let $f_n : M \to \mathbb{R}$ and $f : M \to \mathbb{R}$ be measurable functions such that

$$\int_M \phi(|f_n(x) - f(x)|)\, dx \to 0 \quad (n \to \infty)\,.$$

Prove that the sequence $(f_n)_n$ converges in measure on M to f (Exercise 0.11).

Exercise 0.13. For a function $f \in L_1([a,b])$, prove the equality

$$\int_a^b |f(x)|\, dx = \int_0^\infty \lambda(M_t(f))\, dt\,,$$

where $M_c(f)$ is defined as in Exercise 0.1. More generally, prove that

$$\int_a^b |f(x)|^p\, dx = p \int_0^\infty t^{p-1}\lambda(M_t(f))\, dt$$

for $f \in L_p([a,b])$, $p \geq 1$.

Exercise 0.14. Let $f : [a,b] \to \mathbb{R}$ be monotonically increasing. Use Exercise 0.13 to prove that

$$\int_{-\infty}^\infty I_f(y)\, dy = \int_{f(a)}^{f(b)} I_f(y)\, dy = f(b) - f(a)\,,$$

where I_f denotes the Banach indicatrix of f introduced in Definition 0.38.

Exercise 0.15. Let $M \subset \mathbb{R}$ be a measurable set and $f : M \to \mathbb{R}$ a measurable function. With the notation of Exercise 0.1, let

$$]|f|[:= \inf\{c + \lambda(M_c(f)) : c > 0\}\,.$$

Prove that $]|f|[$ is always finite if M has finite measure, and construct an example which shows that $]|f|[$ may be infinite if M has infinite measure.

Exercise 0.16. Prove that the quantity $]|\cdot|[$ defined in Exercise 0.15 satisfies the triangle inequality

$$]|f+g|[\leq]|f|[+]|g|[\,.$$

Moreover, use Luzin's theorem (Theorem 0.2) to show that $]|f|[=0$ if and only if $f(x)=0$ a.e. on M.

Exercise 0.17. Show by means of an example that the quantity $]|\cdot|[$ defined in Exercise 0.15 is not absolutely homogeneous, i.e. it is *not* true that $]|\lambda f|[=|\lambda|\,]|f|[$ for all $\lambda\in\mathbb{R}$, and so $]|\cdot|[$ is not a norm.

Exercise 0.18. Given a measurable set $M\subset\mathbb{R}$ of finite measure, denote by $S(M)$ the linear space of all measurable functions $f:M\to\mathbb{R}$, where we identify functions which coincide a.e. on M. Prove that $d(f,g):=]|f-g|[$, with $]|\cdot|[$ as in Exercise 0.15, defines a metric on $S(M)$. Moreover, prove that the metric space $(S(M),d)$ is complete.

Exercise 0.19. With the notation of Exercise 0.18, prove that convergence of a sequence $(f_n)_n$ in the metric d coincides with convergence in measure (Exercise 0.11).

Exercise 0.20. Find estimates for the constants m and M from (0.82) in the Orlicz space $X=\mathrm{L}_\phi$ introduced in Definition 0.18.

Exercise 0.21. Let ϕ be some Young function and ϕ^* its conjugate Young function (0.23). On $\mathrm{L}_\phi([a,b])$, consider the *Orlicz–Amemiya norm*

$$|||f|||_{\mathrm{L}_\phi}:=\inf\left\{\frac{1}{\mu}\int_a^b\phi(\mu f(x))\,dx:\ \mu>0\right\}.$$

Show that this norm is equivalent to the Luxemburg norm (0.20) on $\mathrm{L}_\phi([a,b])$. Calculate this norm in case $\phi(t)=t^p$, i.e. $\mathrm{L}_\phi=\mathrm{L}_p$.

Exercise 0.22. Prove that the dual space to $(E_\phi([a,b]),\|\cdot\|_{\mathrm{L}_\phi})$ is the space $(\mathrm{L}_{\phi^*}([a,b]),|||\cdot|||_{\mathrm{L}_{\phi^*}})$, where E_ϕ is the small Orlicz space defined in Section 0.1, ϕ^* denotes the conjugate Young function (0.23), and $|||\cdot|||_{\mathrm{L}_\phi}$ is the norm introduced in Exercise 0.21. More specifically, show that the map $\Phi(g):=\ell_g$ with ℓ_g given by (0.29), defines a linear surjective isometry between L_{ϕ^*} and E_ϕ^*.

Exercise 0.23. Given $f\in\mathrm{L}_\phi([a,b])$, denote by f_n the *truncation at height* n of f defined by

$$f_n(x):=\begin{cases}f(x) & \text{if } |f(x)|\leq n,\\ 0 & \text{if } |f(x)|>n.\end{cases}$$

Prove that

$$\lim_{n\to\infty}\|f-f_n\|_{\mathrm{L}_\phi}=\operatorname{dist}(f,E_\phi)=\inf\{\|f-g\|:g\in E_\phi\},$$

where $E_\phi([a,b])$ is the small Orlicz space defined in Section 0.1. Deduce that $E_\phi([a,b])$ may be characterized as the closure of all essentially bounded functions on $[a,b]$ in the norm (0.11).

Exercise 0.24. Prove that $(L_\phi([a,b]), \|\cdot\|_{L_\phi})$ is separable if and only if $\phi \in \Delta_2$.

Exercise 0.25. For $f : [a,b] \to \mathbb{R}$, consider the following two statements:
(a) there exists a bounded function $g : [a,b] \to \mathbb{R}$ such that $f \sim g$;
(b) $f \in L_\infty([a,b])$, i.e. f is essentially bounded on $[a,b]$.
Do any of these statements imply the other one?

Exercise 0.26. For $f : [a,b] \to \mathbb{R}$, consider the following two statements:
(a) there exists a continuous function $g : [a,b] \to \mathbb{R}$ such that $f \sim g$;
(b) f is a.e. continuous on $[a,b]$.
Do any of these statements imply the other one?

Exercise 0.27. Prove that the linear space of all polynomials

$$p(x) = a_n x^n + a_{n-1} x^{n-1} + \ldots + a_2 x^2 + a_1 x + a_0$$

with either the norm

$$\|p\| := |a_0| + |a_1| + \ldots + |a_n|$$

or the norm

$$|||p||| := \max\{|a_0|, |a_1|, \ldots, |a_n|\}$$

is not a Banach space. Are these two norms equivalent? Is one of them equivalent to the norm (0.105)?

Exercise 0.28. Suppose that $(X, \|\cdot\|_X)$ satisfies the hypotheses of Proposition 0.31, and let $c(X,B)$ be the sharp imbedding constant of $X \hookrightarrow B([a,b])$. Prove that $(X, |||\cdot|||_X)$ equipped with the norm

$$|||f|||_X := 2c(X,B)\|f\|_X \quad (f \in X)$$

is then a Banach algebra.

Exercise 0.29. Apply the preceding Exercise 0.28 to $X = \mathrm{Lip}_\alpha^o([a,b])$ and compare with Example 0.43.

Exercise 0.30. Let $(X, \|\cdot\|_X)$ be a Banach algebra satisfying (0.37). Show that the norm $|||\cdot|||_X$ defined by[31]

$$|||f|||_X := \sup\{\|fg\|_X : \|g\|_X = 1\}$$

is equivalent to $\|\cdot\|_X$ and satisfies (0.42).

---

31 Clearly, this norm is nothing else but the usual operator norm of the linear multiplication operator $g \mapsto fg$ which, by (0.37), is bounded on X.

Exercise 0.31. Show that the supremum in (0.30) is attained in the following sense: for every function $g \in L_{p'}([a,b]) \setminus \{0\}$, there exists a function $f_0 \in L_p([a,b])$ such that $\|f_0\|_{L_p} = 1$ and

$$\|g\|_{L_{p'}} = \int_a^b f_0(x)g(x)\,dx\,.$$

Moreover, prove that the function f_0 is unique (up to equivalence).

Exercise 0.32. Prove the formula (0.31) which is the analogue to (0.30) for $p = 1$ and $p' = \infty$.

Exercise 0.33. Show that the supremum in formula (0.31) is attained in the following sense: for every function $g \in L_\infty([a,b]) \setminus \{0\}$, there exists a function $f_0 \in L_1([a,b])$ such that $\|f_0\|_{L_1} = 1$ and

$$\|g\|_{L_\infty} = \int_a^b f_0(x)g(x)\,dx\,.$$

Exercise 0.34. Show that the function $f_0 \in L_1([a,b])$ from Exercise 0.33 is, in general, *not* unique. More precisely, find functions $g \in L_\infty([0,1]) \setminus \{0\}$, $f_1 \in L_1([0,1])$, and $f_2 \in L_1([0,1])$ such that $\|f_1\|_{L_1} = \|f_2\|_{L_1} = 1$ and $f_1 \neq f_2$, but

$$\|g\|_{L_\infty} = \int_0^1 f_1(x)g(x)\,dx = \int_0^1 f_2(x)g(x)\,dx\,.$$

Exercise 0.35. Although Theorem 0.23 is not true for $p = \infty$, prove that the formula (0.32) holds true which is the analogue to (0.30) for $p = \infty$ and $p' = 1$.

Exercise 0.36. Show that the supremum in formula (0.32) is attained in the following sense: for every function $g \in L_1([a,b]) \setminus \{0\}$, there exists a function $f_0 \in L_\infty([a,b])$ such that $\|f_0\|_{L_\infty} = 1$ and

$$\|g\|_{L_1} = \int_a^b f_0(x)g(x)\,dx\,.$$

Exercise 0.37. Show that the function $f_0 \in L_\infty([a,b])$ from Exercise 0.36 is, in general, *not* unique. More precisely, find functions $g \in L_1([0,1]) \setminus \{0\}$, $f_1 \in L_\infty([0,1])$, and $f_2 \in L_\infty([0,1])$ such that $\|f_1\|_{L_\infty} = \|f_2\|_{L_\infty} = 1$ and $f_1 \neq f_2$, but

$$\|g\|_{L_1} = \int_0^1 f_1(x)g(x)\,dx = \int_0^1 f_2(x)g(x)\,dx\,.$$

Exercise 0.38. Show that the equality holds in the Hölder inequality (0.14) for $1 < p < \infty$ if and only if $\alpha|f(x)|^p = \beta|g(x)|^{p'}$ $(a \le x \le b)$ for some constants $\alpha, \beta \ge 0$ with $\alpha + \beta > 0$.

Exercise 0.39*. Prove the following converse of Hölder's inequality. Let $g : [a, b] \to \mathbb{R}$ be a measurable function, and suppose that $fg \in L_1([a, b])$ for every function $f \in L_p([a, b])$; then, $g \in L_{p'}([a, b])$.

Exercise 0.40. Show that the functional norm (0.25) may be equivalently defined by taking the supremum over $\|x\| < 1$.

Exercise 0.41. Denote by $\hat{\ell} : L_\infty([0, 1]) \to \mathbb{R}$ the Hahn–Banach extension of the bounded linear functional $\ell : C([0, 1]) \to \mathbb{R}$ in Example 0.24. Prove that there is no $g \in L_1([0, 1])$ such that $\hat{\ell}$ may be represented in the form (0.29).

Exercise 0.42. Show that the dual space to $(\mathbb{R}^n, \|\cdot\|_1)$, where $\|x\|_1 := |x_1| + \ldots + |x_n|$, is isometrically isomorphic to the space $(\mathbb{R}^n, \|\cdot\|_\infty)$, where $\|x\|_\infty := \max\{|x_1|, \ldots, |x_n|\}$.

Exercise 0.43. With the notation of Exercise 0.42, show that the dual space to $(\mathbb{R}^n, \|\cdot\|_\infty)$ is isometrically isomorphic to the space $(\mathbb{R}^n, \|\cdot\|_1)$.

Exercise 0.44. Show that the set $S([a, b])$ of all step functions is a subalgebra of the normalized algebra $(B([a, b]), \|\cdot\|_\infty)$.

Exercise 0.45. Considering the function $f_\beta(x) := |x|^\beta$, show that the inclusion (0.72) is strict in case $\alpha > \beta$. More precisely, prove that for each $\beta \in (0, 1)$, there exists a function

$$f \in \mathrm{Lip}_\beta([0, 1]) \setminus \left(\bigcup_{\alpha > \beta} \mathrm{Lip}_\alpha([0, 1]) \right).$$

Exercise 0.46. In analogy to Definition 0.44, define spaces $\mathrm{Lip}_\alpha^n([a, b])$ for $0 < \alpha \leq 1$ and $n \in \mathbb{N}$, and prove possible inclusions between them for varying n and α.

Exercise 0.47. Prove that the norms (0.63) and (0.64) are equivalent on $C^n([a, b])$.

Exercise 0.48. Prove that the norms (0.45) and (0.70) are *not* equivalent on $\mathrm{Lip}([a, b])$.

Exercise 0.49. Prove that the norms (0.45) and (0.65) are *not* equivalent on $C^1([a, b])$.

Exercise 0.50. Is the space $\mathrm{Lip}_\alpha([a, b])$ with one of the norms (0.71) or (0.77) a Banach algebra?

Exercise 0.51. Calculate the McShane extension (0.76) of the function $f \in \mathrm{Lip}_\alpha([0, 1])$ defined by $f(x) := x^\alpha$.

Exercise 0.52. Given $\gamma \in (0, 1)$, show that the function (0.86) belongs to $\mathrm{Lip}_\gamma([0, 1])$ if and only if α is arbitrary and $\beta \geq 1 - \alpha/\gamma$. Compare this with Proposition 0.48 (b).

Exercise 0.53. For $k \in \mathbb{N}$, calculate the norms (0.11), (0.45), (0.63), (0.64), (0.70), (0.71), (0.77), (0.78), and (0.79) of the function $f_k(x) := x^k$ over $[0, 1]$.

Exercise 0.54. For $\alpha, \beta \in \mathbb{R}$ with $\beta < 0$, let $f_{\alpha,\beta} : [0, 1] \to \mathbb{R}$ be defined by (0.86). Prove the following statements.
(a) The n-th derivative $f_{\alpha,\beta}^{(n)}$ exists on $[0, 1]$ if and only if $\alpha > 1 + (n-1)(1-\beta)$.
(b) The n-th derivative $f_{\alpha,\beta}^{(n)}$ exists and is bounded on $[0, 1]$ if and only if $\alpha \geq n(1-\beta)$.
(c) The n-th derivative $f_{\alpha,\beta}^{(n)}$ exists and is continuous on $[0, 1]$ if and only if $\alpha > n(1-\beta)$.

Exercise 0.55. Let $f_{\alpha,\beta} : [0, 1] \to \mathbb{R}$ be defined by (0.86). Find all pairs (α, β) for which $f_{\alpha,\beta}$ belongs to $\mathrm{Lip}^n([0, 1])$.

Exercise 0.56. Show that the inclusion (0.72) is a continuous imbedding, i.e.

$$\mathrm{Lip}_\alpha([a, b]) \hookrightarrow \mathrm{Lip}_\beta([a, b]) \quad (0 < \beta \leq \alpha \leq 1),$$

with sharp imbedding constant $c(\mathrm{Lip}_\alpha, \mathrm{Lip}_\beta) = \max\{(b-a)^{\alpha-\beta}, 1\}$.

Exercise 0.57. Show that the two norms (0.78) and (0.79) are equivalent on $\mathrm{Lip}^n([a, b])$. Is $\mathrm{Lip}^n([a, b])$ with one of these norms a Banach algebra?

Exercise 0.58. Calculate the constants m and M from (0.82) in the space $X = C^n$.

Exercise 0.59. Calculate the constants m and M from (0.82) in the space $X = \mathrm{Lip}^n$.

Exercise 0.60. Calculate the constants m and M from (0.82) in the space $X = \mathrm{Lip}_\alpha^n$, see Exercise 0.46.

Exercise 0.61. Calculate the constants m and M from (0.82) in the space $X = \mathrm{L}_p$ for $1 < p < \infty$.

Exercise 0.62. Construct a function space which is not shift-invariant in the sense of Definition 0.45.

Exercise 0.63. Show that the sharp imbedding constant for $\mathrm{Lip}([a, b]) \hookrightarrow \mathrm{Lip}_\alpha([a, b])$ is $c(\mathrm{Lip}, \mathrm{Lip}_\alpha) = \max\{(b-a)^{1-\alpha}, 1\}$. Compare with Exercise 0.56.

Exercise 0.64. Show that the sharp imbedding constant for $\mathrm{Lip}_\alpha([a, b]) \hookrightarrow C([a, b])$ is $c(\mathrm{Lip}_\alpha, C) = \max\{(b-a)^\alpha, 1\}$.

Exercise 0.65. Let $M \subset \mathbb{R}$ be closed and $f : M \to [1, 2]$ continuous on M. Define $\hat{f} : \mathbb{R} \to \mathbb{R}$ by

$$\hat{f}(x) := \begin{cases} f(x) & \text{for } x \in M, \\ \frac{\inf\{f(y)|x-y| : y \in M\}}{\inf\{|x-y| : y \in M\}} & \text{for } x \notin M. \end{cases}$$

Show that $\hat{f}$ is a continuous extension of f with $\hat{f}(\mathbb{R}) \subseteq [1, 2]$.

Exercise 0.66. Prove that the sequence $(f_n)_n$ defined by $f_n(x) := \frac{1}{2}\sin nx$ converges in every Hölder space $\mathrm{Lip}_\alpha([0, 1])$ $(0 < \alpha < 1)$ to zero, but not in the space $\mathrm{Lip}([0, 1])$.

Exercise 0.67. Prove that the sequence $(f_n)_n$ of C^∞-functions $f_n(x) := \left(x^2 + \frac{1}{n}\right)^{1/4}$ converges in the space $C([0,1])$ with norm (0.45) to $f(x) = \sqrt{x}$. Does it also converge in the Hölder space $\mathrm{Lip}_{1/2}([0,1])$ with norm (0.71)?

Exercise 0.68. Calculate the norms $\|Z_{C,D}\|_{L_1}$ and $\|Z_\theta\|_{L_1}$ of the general zigzag function (0.91) and the special zigzag function (0.93) in $L_1([0,1])$.

Exercise 0.69. If $f \in B([a,b])$ only has countably many points of discontinuity, does it follow that $f \in R([a,b])$?

Exercise 0.70. Let $f, g \in S([a,b])$, and let $(f \vee g)(x) := \max\{f(x), g(x)\}$ and $(f \wedge g)(x) := \min\{f(x), g(x)\}$. Does it follow that $f \vee g \in S([a,b])$ and $f \wedge g \in S([a,b])$?

Exercise 0.71. Let $f, g \in R([a,b])$, and let $f \vee g$ and $f \wedge g$ be defined as in Exercise 0.69. Does it follow that $f \vee g \in R([a,b])$ and $f \wedge g \in R([a,b])$?

Exercise 0.72. Show that the continuous imbeddings

$$\mathrm{Lip}_\alpha([a,b]) \hookrightarrow \mathrm{Lip}_{\alpha,p}([a,b]) \hookrightarrow \mathrm{Lip}_{\alpha-1/p}([a,b])$$

hold for $p\alpha > 1$, where $\mathrm{Lip}_{\alpha,p}$ denotes the generalized Hölder space introduced in Definition 0.54. Compare with Proposition 0.58.

Exercise 0.73. In the terminology of Exercise 0.72, calculate the sharp imbedding constants $c(\mathrm{Lip}_\alpha, \mathrm{Lip}_{\alpha,p})$ and $c(\mathrm{Lip}_{\alpha,p}, \mathrm{Lip}_{\alpha-1/p})$ for $p\alpha > 1$.

Exercise 0.74. Calculate the sharp imbedding constant $c(\mathrm{Lip}_{\alpha,p}, \mathrm{Lip}_{\beta,q})$, where α, β, p and q are as in Proposition 0.58.

Exercise 0.75. Prove the following generalization of Theorem 0.42: given a set $M \subset \mathbb{R}$ and a modulus of continuity $\omega : [0,\infty) \to [0,\infty)$, suppose that $f \in \mathrm{Lip}_{\omega,\infty}(M)$, see Definition 0.54. Then there exists a function $\hat{f} \in \mathrm{Lip}_{\omega,\infty}(\mathbb{R})$ such that $\hat{f}|_M = f$ and $\mathrm{lip}_{\omega,\infty}(\hat{f}) = \mathrm{lip}_{\omega,\infty}(f)$.

Exercise 0.76. Suppose that a function $f \in L_p([a,b])$ satisfies

$$\omega_p(f,[a,b];\delta) = o(\delta) \quad (\delta \to 0+),$$

where $\omega_p(f,[a,b];\delta)$ denotes the integral modulus of continuity (0.98). Show that f is constant a.e. on $[a,b]$.

1 Classical BV-spaces

In this chapter, we start our analysis of the classical function space $BV([a,b])$ and its generalization $WBV_p([a,b])$ ($1 \leq p < \infty$). We discuss the basic properties of these spaces and their connections with other function spaces, with a particular emphasis on spaces of continuous, Lipschitz continuous, Hölder continuous, and absolutely continuous functions. Moreover, we will study the Jordan and Federer decompositions of BV-functions, as well as Helly's selection principle for bounded sequences of BV-functions which, in a certain sense, is an analogue to the Arzelà–Ascoli compactness criterion for sequences of continuous functions. In the last section, we briefly study functions of two variables and introduce the corresponding space $BV([a,b] \times [c,d])$ and related spaces.

1.1 Functions of bounded variation

We start with the definition and properties of the classical function space $BV([a,b])$ which, as far as we know, goes back to Camille Jordan [153, 154]. Throughout the following, we denote by $\mathcal{P}([a,b])$ the family of all *partitions* of the interval $[a,b]$, i.e. all finite sets $P = \{t_0, t_1, \dots, t_m\}$ ($m \in \mathbb{N}$ variable) with

$$a = t_0 < t_1 < \cdots < t_{m-1} < t_m = b\,. \tag{1.1}$$

The number

$$\mu(P) := \max\{t_j - t_{j-1} : j = 1, 2, \dots, m\} \tag{1.2}$$

is called the *mesh size* of the partition P. If $t_j - t_{j-1}$ is independent of j, i.e.

$$t_1 - t_0 = t_2 - t_1 = \cdots = t_m - t_{m-1} = \frac{b-a}{m}\,,$$

the partition $P = \{t_0, t_1, \dots, t_m\}$ is called *equidistant*.

Definition 1.1. Given a partition $P = \{t_0, t_1, \dots, t_m\} \in \mathcal{P}([a,b])$ and a function $f : [a,b] \to \mathbb{R}$, the nonnegative real number

$$\mathrm{Var}(f, P) = \mathrm{Var}(f, P; [a,b]) := \sum_{j=1}^{m} |f(t_j) - f(t_{j-1})| \tag{1.3}$$

is called the *variation* (or *Jordan variation*) of f on $[a,b]$ with respect to P. Moreover, the (possibly infinite) number

$$\mathrm{Var}(f) = \mathrm{Var}(f; [a,b]) := \sup\left\{\mathrm{Var}(f, P; [a,b]) : P \in \mathcal{P}([a,b])\right\}\,, \tag{1.4}$$

where the supremum is taken over all partitions of $[a,b]$, is called the *total* (Jordan) *variation* of f on $[a,b]$. In case $\mathrm{Var}(f; [a,b]) < \infty$, we say that f is a *function of bounded variation* (or *function of bounded Jordan variation* on $[a,b]$ and write $f \in BV([a,b])$. ■

In this and subsequent chapters, we will need two important subsets of $BV([a,b])$ which we introduce right now.

Definition 1.2. We denote by $BV^o([a,b])$ the subset of all functions $f \in BV([a,b])$ satisfying $f(a) = 0$. Furthermore, we write $NBV([a,b])$ for the set of all functions $f \in BV^o([a,b])$ which are right continuous at any point $x_0 \in [a,b)$, i.e. satisfy

$$f(x_0+) = \lim_{x \to x_0+} f(x) = f(x_0) \quad (a \leq x_0 < b). \tag{1.5}$$

Functions in $NBV([a,b])$ will be called *normalized* in what follows. ∎

For further use, we collect in the following proposition some important properties of the quantities (1.3) and (1.4).

Proposition 1.3. *The quantities* (1.3) *and* (1.4) *have the following properties.*

(a) *The variation* (1.4) *is subadditive with respect to functions, i.e.*

$$\operatorname{Var}(f+g;[a,b]) \leq \operatorname{Var}(f;[a,b]) + \operatorname{Var}(g;[a,b]) \tag{1.6}$$

for $f, g : [a,b] \to \mathbb{R}$.

(b) *The variation* (1.4) *is homogeneous with respect to functions, i.e.*

$$\operatorname{Var}(\mu f;[a,b]) = |\mu| \operatorname{Var}(f;[a,b]) \tag{1.7}$$

for $\mu \in \mathbb{R}$.

(c) *The estimate*

$$|f(x) - f(y)| \leq \operatorname{Var}(f;[x,y]) \tag{1.8}$$

holds for $a \leq x < y \leq b$.

(d) *Every function* $f \in BV([a,b])$ *is bounded with*

$$\|f\|_\infty \leq |f(a)| + \operatorname{Var}(f;[a,b]), \tag{1.9}$$

where the norm $\|\cdot\|_\infty$ *is given by* (0.39).

(e) *Every monotone function* $f : [a,b] \to \mathbb{R}$ *belongs to* $BV([a,b])$ *with*

$$\operatorname{Var}(f;[a,b]) = |f(b) - f(a)|\,.$$

(f) *The variation* (1.3) *is monotone with respect to partitions, i.e.*

$$\operatorname{Var}(f,P;[a,b]) \leq \operatorname{Var}(f,Q;[a,b])$$

for $P, Q \in \mathcal{P}([a,b])$ *with* $P \subseteq Q$.

(g) *The variation* (1.4) *is additive with respect to intervals, i.e.*

$$\operatorname{Var}(f;[a,b]) = \operatorname{Var}(f;[a,c]) + \operatorname{Var}(f;[c,b]) \tag{1.10}$$

for $a < c < b$.

Proof. For two functions $f, g : [a, b] \to \mathbb{R}$, $\mu \in \mathbb{R}$, and any partition $P = \{t_0, t_1, \dots, t_m\} \in \mathcal{P}([a, b])$, we have

$$\begin{aligned}\operatorname{Var}(f + g, P) &= \sum_{j=1}^{m} |(f + g)(t_j) - (f + g)(t_{j-1})| \\ &\le \sum_{j=1}^{m} |f(t_j) - f(t_{j-1})| + \sum_{j=1}^{m} |g(t_j) - g(t_{j-1})| = \operatorname{Var}(f, P) + \operatorname{Var}(g, P)\end{aligned}$$

and

$$\operatorname{Var}(\mu f, P) = \sum_{j=1}^{m} |(\mu f)(t_j) - (\mu f)(t_{j-1})| = |\mu| \sum_{j=1}^{m} |f(t_j) - f(t_{j-1})| = |\mu| \operatorname{Var}(f, P),$$

which proves (a) and (b). To prove (c), it suffices to consider the special partition $P = \{x, y\} \in \mathcal{P}([x, y])$ and to observe that then $|f(y) - f(x)| = \operatorname{Var}(f, P; [x, y])$. Similarly, considering the partition $P_x = \{a, x, b\} \in \mathcal{P}([a, b])$, we obtain

$$|f(x) - f(a)| \le |f(b) - f(x)| + |f(x) - f(a)| = \operatorname{Var}(f, P_x; [a, b]),$$

and hence

$$|f(x)| \le |f(a)| + \operatorname{Var}(f, P_x; [a, b]) \le |f(a)| + \operatorname{Var}(f; [a, b]), \tag{1.11}$$

which proves (d) after taking the supremum over $x \in [a, b]$.

Now, let $P = \{t_0, t_1, \dots, t_m\} \in \mathcal{P}([a, b])$ and $f : [a, b] \to \mathbb{R}$ be increasing.[1] Then

$$\operatorname{Var}(f, P; [a, b]) = \sum_{j=1}^{m} |f(t_j) - f(t_{j-1})| = \sum_{j=1}^{m} [f(t_j) - f(t_{j-1})] = f(b) - f(a),$$

and hence $\operatorname{Var}(f; [a, b]) = f(b) - f(a)$ as well. In case of a decreasing function $f : [a, b] \to \mathbb{R}$, the same argument shows that $\operatorname{Var}(f; [a, b]) = f(a) - f(b)$, and so we have proved (e).

Clearly, the property (f) and the additivity formula in (g) follow directly from (1.3) and (1.4). □

It is clear that one cannot expect equality in (1.6); for example, for $f(t) = t$ and $g(t) = -t$, we have $\operatorname{Var}(f; [0, 1]) = \operatorname{Var}(g; [0, 1]) = 1$, but $\operatorname{Var}(f + g; [0, 1]) = 0$.

Part (e) of Proposition 1.3 provides a link between monotone functions and functions of bounded variation. One could ask if this could be in some sense inverted. The next example shows, however, that the converse of (e) is very far from being true: there exist functions of bounded variation which are not monotone on *any* interval.

1 We call a function f *increasing* if $x < y$ implies $f(x) \le f(y)$, and *decreasing* if $x < y$ implies $f(x) \ge f(y)$. So, in contrast to some authors, we do not require the strict inequalities $f(x) < f(y)$ and $f(x) > f(y)$, respectively. In that case, we call f *strictly increasing* and *strictly decreasing*, respectively.

Example 1.4. We arrange the rational numbers between 0 and 1 in a sequence, i.e. $(0,1)\cap\mathbb{Q}=\{r_1,r_2,r_3,\dots\}$, and define a function $f:[0,1]\to\mathbb{R}$ by

$$f(t):=\begin{cases}2^{-k} & \text{for } t=r_k\,,\\ 0 & \text{otherwise}\,.\end{cases}$$

Clearly, f is not monotone on any interval $[a,b]\subseteq[0,1]$. On the other hand, we claim that $\mathrm{Var}(f;[0,1])=2$, and hence $f\in BV([0,1])$.

To see this, we introduce some notation. For a bounded function $f:[a,b]\to\mathbb{R}$ and any subset $A\subseteq[a,b]$, we call

$$\mathrm{osc}(f;A):=\sup_{t\in A}f(t)-\inf_{t\in A}f(t) \tag{1.12}$$

the *oscillation* of f on A. Let $P=\{t_0,t_1,\dots,t_m\}$ be an arbitrary partition of $[0,1]$. For a fixed $j\in\{1,2,\dots,m\}$, denote

$$k_j:=\min\{i:r_i\in[t_{j-1},t_j]\}\,.$$

Observe that every number in the finite sequence $(k_1,k_2,\dots,k_m)$ may appear at most two times, namely, as the right or left endpoint of two adjacent intervals generated by P. This yields

$$\begin{aligned}\mathrm{Var}(f,P;[0,1])&=\sum_{j=1}^{m}|f(t_j)-f(t_{j-1})|\le\sum_{j=1}^{m}\mathrm{osc}(f;[t_{j-1},t_j])\\&=\sum_{j=1}^{m}2^{-k_j}\le 2\sum_{j=1}^{\infty}2^{-j}=2,\end{aligned}$$

and so $\mathrm{Var}(f;[0,1])\le 2$, and hence $f\in BV([0,1])$. To prove the equality $\mathrm{Var}(f;[0,1])=2$, we construct a special partition in the following way. For fixed $n\in\mathbb{N}$, we rearrange the first n rational points $r_1,r_2,\dots,r_n$ in increasing order, and hence $r_1<r_2<\dots<r_n$. Afterwards, we put

$$s_0:=0,\quad s_1:=r_1,\quad s_3:=r_2,\quad s_5:=r_3,\quad \dots\quad ,s_{2n-1}:=r_n,\quad s_{2n}:=1$$

and choose irrational points $s_2\in(s_1,s_3)$, $s_4\in(s_3,s_5)$, $s_6\in(s_5,s_7)$, …, $s_{2n-2}\in(s_{2n-3},s_{2n-1})$. For the corresponding partition $\tilde{P}=\{s_0,s_1,\dots,s_{2n}\}\in\mathcal{P}([0,1])$, we get then

$$\mathrm{Var}(f,\tilde{P};[0,1])=\sum_{j=1}^{2n}|f(s_j)-f(s_{j-1})|=2\sum_{j=1}^{2n}2^{-j}=\frac{1-\left(\frac{1}{2}\right)^{2n}}{1-\frac{1}{2}}=2\left(1-2^{-2n}\right),$$

which may be taken arbitrarily close to 2 by choosing n sufficiently large. We conclude that the total variation of f on $[0,1]$ is precisely 2. ♥

Now comes an important point. Although functions of bounded variation have no monotonicity behavior, there is a natural interconnection between bounded variation and monotonicity which is the statement of the classical *Jordan decomposition theorem* [153].

Theorem 1.5. *A function $f : [a,b] \to \mathbb{R}$ is of bounded variation if and only if it may be represented in the form $f = p_f - n_f$, where both p_f and n_f are monotonically increasing functions.*

Proof. The fact that the sum or difference of two monotone functions has bounded variation is an immediate consequence of Proposition 1.3 (a) and (e); the nontrivial part is the converse.

Given $f \in BV([a,b])$, consider the *variation function* $V_f : [a,b] \to \mathbb{R}$ defined by

$$V_f(x) := \mathrm{Var}(f; [a,x]) . \tag{1.13}$$

We take $p_f := V_f$. Clearly, p_f is increasing with $V_f(a) = 0$ and $V_f(b) = \mathrm{Var}(f; [a,b])$. So, the only point to show is that the function $n_f := V_f - f$ is increasing as well. However, for $a \le x < y \le b$, we have

$$f(y) - f(x) \le \mathrm{Var}(f; [x,y]) = V_f(y) - V_f(x) ,$$

by (1.8) and (1.10), and so

$$n_f(y) - n_f(x) = V_f(y) - f(y) - V_f(x) + f(x) \ge 0$$

as claimed. □

In what follows, we will refer to the representation $f = V_f - (V_f - f)$ constructed in Theorem 1.5 as the *Jordan decomposition* of $f \in BV([a,b])$. The following result is a slight modification of Theorem 1.5.

Theorem 1.6. *A function $f : [a,b] \to \mathbb{R}$ is of bounded variation if and only if it may be represented in the form $f = p_f - n_f + f(a)$, where both p_f and n_f are monotonically increasing nonnegative functions.*

Proof. We define $p_f, n_f : [a,b] \to \mathbb{R}$ by

$$p_f(x) := \frac{1}{2}\left(V_f(x) + f(x) - f(a)\right), \quad n_f(x) := \frac{1}{2}\left(V_f(x) - f(x) + f(a)\right)$$

with V_f given by (1.13). By construction, we then have

$$f(x) = p_f(x) - n_f(x) + f(a), \quad V_f(x) = p_f(x) + n_f(x) .$$

Moreover, for $a \le x < y \le b$, we get

$$p_f(y) - p_f(x) = \frac{1}{2}\left(V_f(y) + f(y) - V_f(x) - f(x)\right) \ge 0$$

and

$$n_f(y) - n_f(x) = \frac{1}{2}\left(V_f(y) - f(y) - V_f(x) + f(x)\right) \geq 0,$$

by Proposition 1.3 (e), which shows that p_f and n_f are in fact increasing. Since both $p_f(a) = 0$ and $n_f(a) = 0$, it follows that p_f and n_f are also nonnegative. □

The variation function (1.13) employed in Theorems 1.5 and 1.6 has many interesting properties on its own. Moreover, some properties of $f \in BV([a,b])$ are "reflected" in V_f, and vice versa (see Proposition 1.7 and Theorem 1.26 below). It may also be used to prove the following majorant principle: a function $f : [a,b] \to \mathbb{R}$ belongs to $BV([a,b])$ if and only if there exists an increasing function $g : [a,b] \to \mathbb{R}$ such that $|f(x)-f(y)| \leq g(y) - g(x)$ for any interval $[x,y] \subseteq [a,b]$. In fact, if such a function g exists, we have $\mathrm{Var}(f, P; [a,b]) \leq g(b)-g(a)$ for any $P \in \mathcal{P}([a,b])$, and hence $f \in BV([a,b])$. Conversely, in case $f \in BV([a,b])$, we may simply choose $g := V_f$, as (1.8) shows.

An important example of the "interaction" between f and V_f is given in the following

Proposition 1.7. *Suppose that $f \in BV([a,b])$ is continuous at some point $x_0 \in [a,b]$; then, the function V_f from* (1.13) *is also continuous at x_0. The converse statement is also true.*

Proof. Let $\varepsilon > 0$, and suppose first that f is continuous at x_0, where it is no loss of generality to assume that $x_0 < x < b$. Consider the difference $V_f(x) - V_f(x_0)$. Choose a partition $P = \{t_0, t_1, \ldots, t_m\} \in \mathcal{P}([x_0, b])$ such that

$$\mathrm{Var}(f; [x_0, b]) < \mathrm{Var}(f, P; [x_0, b]) + \varepsilon.$$

Afterwards, we choose $\delta \in (0, t_1 - x_0)$ such that $|f(x) - f(x_0)| < \varepsilon$ for $0 < x - x_0 < \delta$ which is possible by the continuity of f at x_0. Then for these x, we have

$$\begin{aligned}
V_f(x) - V_f(x_0) &= \mathrm{Var}(f; [x_0, x]) = \mathrm{Var}(f; [x_0, b]) - \mathrm{Var}(f; [x, b]) \\
&< \mathrm{Var}(f, P; [x_0, b]) + \varepsilon - \mathrm{Var}(f; [x, b]) \\
&\leq |f(x) - f(x_0)| + |f(x) - f(t_1)| + \sum_{j=2}^{m} |f(t_j) - f(t_{j-1})| - \mathrm{Var}(f; [x, b]) + \varepsilon \\
&\leq |f(x) - f(x_0)| + \varepsilon < 2\varepsilon.
\end{aligned}$$

This shows that $V_f(x_0+) = V_f(x_0)$, and so V_f is right continuous at x_0. The left continuity of V_f at x_0 may be proved in the same way.

To prove the converse statement, suppose now that V_f is continuous at $x_0 \in [a,b]$. For $x \geq x_0$, we then have

$$|f(x) - f(x_0)| \leq \mathrm{Var}(f; [x_0, x]) = V_f(x) - V_f(x_0) \to 0 \quad (x \to x_0+),$$

while for $x \leq x_0$, we have

$$|f(x) - f(x_0)| \leq \mathrm{Var}(f; [x, x_0]) = V_f(x_0) - V_f(x) \to 0 \quad (x \to x_0-).$$

This proves the second assertion. □

An alternative proof of Proposition 1.7 is contained in Exercise 1.6. In Theorem 1.26 below, we give a more systematic discussion of the “interplay” between the variation function V_f and its parent function f.

Theorems 1.5 and 1.6 explain why many “nice” properties of monotone functions (like Riemann integrability or differentiability a.e., see Exercises 1.28 and 1.29 carry over to functions of bounded variation. In particular, a function $f \in BV([a, b])$ has, at most, many countably points of discontinuity in $[a, b]$, all being of first kind (jumps) or removable. On the other hand, the following well-known example shows that a continuous function need not have bounded variation.

Example 1.8. Let $f: [0, 1] \to \mathbb{R}$ be defined by

$$f(x) := \begin{cases} x \sin \frac{1}{x} & \text{for } 0 < x \leq 1, \\ 0 & \text{for } x = 0. \end{cases} \tag{1.14}$$

Clearly, f is continuous on $[0, 1]$. To show that $f \notin BV([0, 1])$, we construct partitions with the property that the oscillation sums in (1.3) become as large as possible. This may be achieved by choosing $P = \{t_0, t_1, \dots, t_m\}$ in such a way that t_{j-1} and t_j give alternating maxima and minima of f, i.e.

$$t_0 := 0, \qquad t_1 := \frac{2}{(2m-1)\pi}, \qquad t_2 := \frac{2}{(2m-3)\pi}, \qquad \dots,$$

$$\dots, \qquad t_{m-2} := \frac{2}{5\pi}, \qquad t_{m-1} := \frac{2}{3\pi}, \qquad t_m := 1.$$

In fact, we then have $|f(t_j) - f(t_{j-1})| \geq 2t_{j-1}$ for $j = 2, 3, \dots, m-1$, and so the sum in (1.3) becomes arbitrarily large if we choose m sufficiently large.[2] ♥

A refinement of Example 1.8 will be discussed in Exercises 1.8 and 1.9. In the next section, we will discuss relations between bounded variation and continuity in more detail.

Now, we provide an important statement concerning the possibility of passing to the pointwise limit for a sequence of BV-functions.

Proposition 1.9. *Let $(f_n)_n$ be a sequence in $BV([a, b])$ which converges pointwise on $[a, b]$ to some function f. Then*

$$\operatorname{Var}(f; [a, b]) \leq \liminf_{n \to \infty} \operatorname{Var}(f_n; [a, b]). \tag{1.15}$$

Consequently, the pointwise limit of a sequence of functions with equibounded variations on the interval $[a, b]$ is a function of bounded variation on $[a, b]$.

Proof. If the right-hand side of (1.15) is infinite, there is nothing to prove. Thus, suppose that

$$\liminf_{n \to \infty} \operatorname{Var}(f_n; [a, b]) \leq L$$

2 The reason is, of course, the divergence of the harmonic series.

for some constant $L > 0$. Then, keeping in mind the properties of the limit inferior of a real sequence, we can choose a subsequence $(f_{n_k})_k$ of the sequence $(f_n)_n$ such that $\mathrm{Var}(f_{n_k}; [a,b]) \leq L$ for $k \in \mathbb{N}$. Fix an arbitrary partition $P = \{t_0, t_1, \dots, t_m\} \in \mathcal{P}([a,b])$. Then we obtain

$$\mathrm{Var}(f_{n_k}, P; [a,b]) = \sum_{j=1}^{m} |f_{n_k}(t_j) - f_{n_k}(t_{j-1})| \leq L\,,$$

and so, passing to the limit as $k \to \infty$,

$$\mathrm{Var}(f, P; [a,b]) = \sum_{j=1}^{m} |f(t_j) - f(t_{j-1})| \leq L\,.$$

Since L is independent of the partition P, this implies that $\mathrm{Var}(f; [a,b]) \leq L$, and the assertion follows. The last statement is an immediate consequence. □

In what follows, we shall describe the structure of the set $BV([a,b])$. First of all, let us observe that, by Proposition 1.3 (a) and (b), $BV([a,b])$ forms a real vector space with respect to the standard operations on functions. For an arbitrary function $f \in BV([a,b])$, let us put

$$\|f\|_{BV} := |f(a)| + \mathrm{Var}(f; [a,b])\,. \tag{1.16}$$

It is easy to check that (1.16) defines a norm on the linear space $BV([a,b])$. The following useful statement is concerned with the completeness of this space.

Proposition 1.10. *The space $BV([a,b])$ equipped with the norm* (1.16) *is a Banach space which is continuously imbedded into the space $B([a,b])$ of all bounded functions on $[a,b]$ with norm* (0.39). *Moreover, $BV([a,b])$ is an algebra with*

$$\mathrm{Var}(fg) \leq \|f\|_\infty \mathrm{Var}(g) + \|g\|_\infty \mathrm{Var}(f) \tag{1.17}$$

for all $f, g \in BV([a,b])$. Even better, we have

$$\|fg\|_{BV} \leq \|f\|_{BV}\|g\|_{BV} \quad (f, g \in BV([a,b])), \tag{1.18}$$

which means that $BV([a,b])$ is a normalized algebra.

Proof. Assume that $(f_n)_n$ is a Cauchy sequence in $BV([a,b])$ with respect to the norm (1.16). Fix $\varepsilon > 0$ and choose a natural number n_0 such that for all $m, n \in \mathbb{N}$ with $m, n \geq n_0$, we have

$$\|f_n - f_m\|_{BV} = |f_n(a) - f_m(a)| + \mathrm{Var}(f_n - f_m; [a,b]) \leq \varepsilon\,. \tag{1.19}$$

This inequality implies, in particular,

$$|f_n(a) - f_m(a)| \leq \varepsilon \tag{1.20}$$

for $m, n \geq n_0$, and so the real sequence $(f_n(a))_n$ converges to some real number which we denote by $f(a)$. Letting $m \to \infty$ in (1.20), we get

$$|f_n(a) - f(a)| \leq \varepsilon \tag{1.21}$$

for $n \geq n_0$. Further, from (1.19), we also obtain

$$\mathrm{Var}(f_n - f_m; [a, b]) \leq \varepsilon \quad (m, n \geq n_0). \tag{1.22}$$

Now, fix an arbitrary point $x \in (a, b]$. Then, in view of (1.22) and Proposition 1.3 (g), we deduce that $\mathrm{Var}(f_n - f_m; [a, x]) \leq \varepsilon$. So, taking the special partition $P_x := \{a, x\}$ of the interval $[a, x]$, we have

$$\begin{aligned} &\left|f_n(x) - f_m(x)\right| - \left|f_n(a) - f_m(a)\right| \\ &\quad \leq \left|[f_n(x) - f_m(x)] - [f_n(a) - f_m(a)]\right| \\ &\quad = \mathrm{Var}(f_n - f_m, P_x; [a, x]) \leq \varepsilon, \end{aligned}$$

which yields

$$\left|f_n(x) - f_m(x)\right| \leq \left|f_n(a) - f_m(a)\right| + \varepsilon \leq 2\varepsilon \quad (a \leq x \leq b). \tag{1.23}$$

However, this means that for any fixed $x \in [a, b]$, the sequence $(f_n(x))_n$ is a Cauchy sequence, and so it has a pointwise limit which we denote by $f(x)$.

Next, by fixing in (1.23) the index n and letting $m \to \infty$, we obtain $|f_n(x) - f(x)| \leq 2\varepsilon$ for $n \geq n_0$ and $x \in [a, b]$. This shows that the function sequence $(f_n)_n$ is even uniformly convergent on $[a, b]$ to the function f.

It remains to show that $(f_n)_n$ converges to f in variation, i.e. in the norm (1.16). To this end, we first notice that $(f_n)_n$ is bounded in $BV([a, b])$, i.e. there exists a constant $M > 0$ such that

$$\|f_n\|_{BV} \leq M \quad (n = 1, 2, \dots).$$

In particular, we have $\mathrm{Var}(f_n; [a, b]) \leq M$ for all $n \in \mathbb{N}$. Combining this with the last statement in Proposition 1.9 ensures that the limit function f is of bounded variation on the interval $[a, b]$. Further, observe that keeping n fixed in (1.22) and passing to the limit inferior as $m \to \infty$, by virtue of Proposition 1.9, we get

$$\mathrm{Var}(f_n - f; [a, b]) \leq \liminf_{m \to \infty} \mathrm{Var}(f_n - f_m; [a, b]) \leq \varepsilon$$

for $n \geq n_0$. This proves the completeness of $BV([a, b])$ in the norm (1.16).

The fact that $BV \hookrightarrow B$ (with sharp imbedding constant $c(BV, B) = 1$) is only a reformulation of (1.9).

Now, we prove that $BV([a, b])$ is an algebra satisfying (1.17). Fix $f, g \in BV([a, b])$, and let $P = \{t_0, t_1, \dots, t_m\}$ be an arbitrary partition of $[a, b]$. By Proposition 1.3 (d), both functions f and g are then bounded, and

$$\begin{aligned}\mathrm{Var}(fg, P) &= \sum_{j=1}^{m} |f(t_j)g(t_j) - f(t_{j-1})g(t_{j-1})| \\ &\leq \sum_{j=1}^{m} |f(t_j) - f(t_{j-1})|\,|g(t_j)| + \sum_{j=1}^{m} |g(t_j) - g(t_{j-1})|\,|f(t_{j-1})| \\ &\leq \mathrm{Var}(f, P)\,\|g\|_\infty + \mathrm{Var}(g, P)\,\|f\|_\infty,\end{aligned}$$

so by passing to the supremum with respect to $P \in \mathcal{P}([a, b])$, we obtain (1.17).

Finally, let us prove (1.18). Since $f \mapsto \mathrm{Var}(f; [a, b])$ is a norm on the set $BV^o([a, b])$ of all $f \in BV([a, b])$ satisfying $f(a) = 0$, from Proposition 0.31, we know that $BV^o([a, b])$ with norm

$$|||f|||_{BV} = \mathrm{Var}(f) + \|f\|_\infty \tag{1.24}$$

is a normalized Banach algebra. However, we can get rid of both the term $\|f\|_\infty$ in (1.24) and the condition $f(a) = 0$ by using the decomposition of f given in Theorem 1.6. In fact, considering for $f, g \in BV([a, b])$ the corresponding functions p_{fg} and n_{fg} of the product, we obtain

$$\|fg\|_{BV} = V_{fg}(b) + |f(a)g(a)| = p_{fg}(b) + n_{fg}(b) + |f(a)g(a)|\,.$$

Thus, using the subadditivity (1.6) of the total variation, we get

$$\begin{aligned}\|fg\|_{BV} &\leq p_f(b)p_g(b) + p_f(b)n_g(b) + |g(a)|p_f(b) + n_f(b)p_g(b) \\ &\quad + n_f(b)n_g(b) + |g(a)|n_f(b) + |f(a)|p_g(b) + |f(a)|n_g(b) + |f(a)g(a)| \\ &= (p_f(b) + n_f(b) + |f(a)|)(p_g(b) + n_g(b) + |g(a)|) = \|f\|_{BV}\|g\|_{BV}\end{aligned}$$

which proves (1.18). □

Proposition 1.9 shows that finite variation carries over from a sequence to its pointwise limit if this limit exists. Now, we prove a somewhat surprising statement about bounded sequences in $BV([a, b])$ which asserts the *existence* of such pointwise limits and is known as *Helly's selection principle.*

Theorem 1.11 (Helly). *Let $(f_n)_n$ be a bounded sequence in $BV([a, b])$ with respect to the norm* (1.16). *Then $(f_n)_n$ contains a subsequence which converges pointwise on $[a, b]$ to some $f \in BV([a, b])$.*

Proof. We use Theorem 1.5 to reduce the problem to monotone functions in the following way. Putting

$$g_n(x) := V_{f_n}(x) = \mathrm{Var}(f_n; [a, x]), \quad h_n(x) := g_n(x) - f_n(x)\,,$$

we know that g_n and h_n are increasing and bounded, and $f_n = g_n - h_n$ for all n. We claim that the sequences $(g_n)_n$ and $(h_n)_n$ contain subsequences which converge pointwise on $[a, b]$ to some increasing function g and h, respectively; then, the function $f := g - h$ has the required properties.

So, let $g_n : [a, b] \to \mathbb{R}$ be monotonically increasing with $|g_n(x)| \leq c < \infty$ for all $n \in \mathbb{N}$ and $a \leq x \leq b$. Let $E := \{r_1, r_2, r_3, \ldots\}$ be some countable dense subset of $[a, b]$, where $r_1 := a$ and $r_2 := b$. Since $|g_n(r_1)| \leq c$ for all n, by the classical Bolzano–Weierstrass theorem, we find a subsequence $(g_n^{(1)})_n$ of $(g_n)_n$ which converges at the point r_1. Similarly, since $|g_n^{(1)}(r_2)| \leq c$ for all n, we find a subsequence $(g_n^{(2)})_n$ of $(g_n^{(1)})_n$ which converges at both points r_1 and r_2. Having constructed $(g_n^{(k-1)})_n$ in this way, we choose a subsequence $(g_n^{(k)})_n$ of $(g_n^{(k-1)})_n$ which converges at all points $r_1, r_2, \ldots, r_k$. Thus, the diagonal sequence $(g_{n_k})_k$ defined by $g_{n_k}(x) := g_k^{(k)}(x)$ converges at every point $r_j \in E$. Now, we define $g : [a, b] \to \mathbb{R}$ by

$$g(x) := \begin{cases} \lim_{k\to\infty} g_{n_k}(x) & \text{if } x = r_j \in E\,, \\ \sup_{r_j < x} \lim_{k\to\infty} g_{n_k}(r_j) & \text{otherwise}\,, \end{cases} \tag{1.25}$$

where the supremum in (1.25) is taken over all elements $r_j \in E$ with $r_j < x$. Then g is bounded and monotonically increasing, by construction, and $g_{n_k}(x) \to g(x)$ pointwise on E as $k \to \infty$.

We know that the set D of discontinuity points of g is finite or at most countably infinite, and so is the set $D \cup E$. Fix $x \in [a, b] \setminus (D \cup E)$. Given $\varepsilon > 0$, we may choose $r_i, r_j \in E$ such that

$$r_i < x < r_j\,, \quad g(r_j) - g(r_i) < \varepsilon \tag{1.26}$$

since g is continuous at x. Moreover, by the pointwise convergence of $(g_{n_k})_k$ to g on E, we find some $k_0 \in \mathbb{N}$ such that

$$|g_{n_k}(r_i) - g(r_i)| < \varepsilon\,, \quad |g_{n_k}(r_j) - g(r_j)| < \varepsilon \tag{1.27}$$

for $k \geq k_0$. Combining (1.26) and (1.27) yields

$$\begin{aligned} g(x) - \varepsilon < g(r_i) &\leq g_{n_k}(r_i) + \varepsilon \leq g_{n_k}(x) + \varepsilon \\ &\leq g_{n_k}(r_j) + \varepsilon < g(r_j) + 2\varepsilon \leq g(x) + 2\varepsilon \end{aligned} \tag{1.28}$$

since $r_i < x < r_j$ and all functions occurring in (1.28) are increasing. This shows that $g_{n_k}(x) \to g(x)$ pointwise on $[a, b] \setminus D$ as $k \to \infty$. However, the set D is at most countable and the sequence $(g_{n_k})_k$ is uniformly bounded. Thus, we may again use the diagonal procedure described above and find another subsequence which converges pointwise on the whole interval $[a, b]$. Clearly, the limit function g obtained in this way is increasing (Exercise 1.17), and so we are done. □

Interestingly, Helly's selection principle may also be used to give an alternative proof of the completeness of the space $(BV([a, b]), \|\cdot\|_{BV})$. To see this, let $(f_n)_n$ be a Cauchy sequence with respect to the norm (1.16); without loss of generality, we assume that $f_n(a) = 0$ since otherwise, we pass from f_n to the function $f_n - f_n(a)$. Given $\varepsilon > 0$, choose $n_0 \in \mathbb{N}$ such that (1.22) holds. By Helly's selection principle (Theorem 1.11), we find a subsequence $(f_{n_k})_k$ of $(f_n)_n$ which converges pointwise on $[a, b]$ to some $f \in BV([a, b])$. For any partition $P \in \mathcal{P}([a, b])$ and all $m \geq n_0$, we then get

$$\mathrm{Var}(f_m - f, P; [a, b]) = \lim_{k \to \infty} \mathrm{Var}(f_m - f_{n_k}, P; [a, b]) \leq \varepsilon,$$

and hence $\|f_m - f\|_{BV} \leq \varepsilon$ for $m \geq n_0$, which proves the assertion.

Obviously, the monotone functions on a fixed interval $[a, b]$ do not form a linear space, as may easily be seen by considering $f(t) = t^2$ and $g(t) = 1 - t$ on $[a, b] = [0, 1]$, say (see also Exercise 1.11). Theorem 1.5 and Proposition 1.3 (a), (b) and (e) show that $BV([a, b])$ is the *linear hull* (or *span*) of the monotone functions, i.e. the smallest linear space which contains all monotone functions. This is another way of introducing functions of bounded variation.

We remark that the problem of whether or not the quotient f/g of two functions $f, g \in BV([a, b])$ belongs again to $BV([a, b])$ is more delicate, see Exercises 1.1 and 1.2. Moreover, it is illuminating to compare the two conditions $f \in BV([a, b])$ and $|f| \in BV([a, b])$, where $|f|$ is defined by $|f|(x) := |f(x)|$, see Exercises 1.3–1.5.

In Chapter 5 of this monograph, we will study the so-called *composition operator problem* in great detail for the space $BV([a, b])$ (and many similar function spaces), which consists of determining conditions, possibly both necessary and sufficient, on a function $g : \mathbb{R} \to \mathbb{R}$, under which the composition $g \circ f$ belongs to $BV([a, b])$ whenever f belongs to $BV([a, b])$. Here, we briefly discuss the dual *substitution operator problem* of characterizing all "admissible changes of variables," i.e. the problem of determining conditions on a transformation $\tau : [a, b] \to [c, d]$, under which $f \circ \tau$ belongs to $BV([a, b])$ for all $f \in BV([c, d])$. For example, we have the following necessary and sufficient condition in case of an increasing bijection τ.

Proposition 1.12. *Given a function $g : [c, d] \to \mathbb{R}$, let $\tau : [a, b] \to [c, d]$ be continuous and strictly increasing*[3] *with $\tau(a) = c$ and $\tau(b) = d$. Then $g \circ \tau \in BV([a, b])$ if and only if $g \in BV([c, d])$.*

Proof. Given a partition $P = \{t_0, t_1, \ldots, t_m\} \in \mathcal{P}([a, b])$, the set

$$\tau(P) = \{\tau(t_0), \tau(t_1), \ldots, \tau(t_m)\}$$

is then a partition of $[c, d]$, by our assumptions on τ. Therefore, in case $g \in BV([c, d])$, we have

$$\mathrm{Var}(g \circ \tau, P; [a, b]) = \mathrm{Var}(g, \tau(P); [c, d]),$$

3 The classical intermediate value theorem implies that under these hypotheses, τ is then a homeomorphism, i.e. a bijection with continuous inverse.

and hence

$$\mathrm{Var}(g \circ \tau; [a, b]) \leq \mathrm{Var}(g; [c, d]) .$$

Applying this reasoning to the function $\tau^{-1} : [c, d] \to [a, b]$ (which has the same properties as τ), we conclude that also

$$\mathrm{Var}(g; [c, d]) = \mathrm{Var}(g \circ \tau \circ \tau^{-1}; [c, d]) \leq \mathrm{Var}(g \circ \tau; [a, b]) .$$

This shows that $\mathrm{Var}(g; [c, d]) = \mathrm{Var}(g \circ \tau; [a, b])$, and so the assertion follows. □

Our proof shows that under the hypotheses of Proposition 1.12, the map $g \mapsto g \circ \tau$ is a (surjective) *isometry* between the two function spaces $(BV([a, b]), \|\cdot\|_{BV})$ and $(BV([c, d]), \|\cdot\|_{BV})$ since $g(\tau(a)) = g(c)$ and $g(\tau(b)) = g(d)$.

Observe that the "if" part in the statement of Proposition 1.12 is also valid if τ is merely monotone, but not necessarily continuous. In fact, if $g : [c, d] \to \mathbb{R}$ has bounded variation, we may decompose g, by Theorem 1.5, as difference $g = p_g - n_g$, where both p_g and n_g are increasing. If $\tau : [a, b] \to [c, d]$ is increasing (respectively decreasing), then $g \circ \tau = p_g \circ \tau - n_g \circ \tau$, where both $p_g \circ \tau$ and $n_g \circ \tau$ are increasing (respectively decreasing), and so $g \circ \tau : [a, b] \to \mathbb{R}$ has bounded variation as well. On the other hand, the following simple example shows that one cannot drop the continuity assumption on τ in the "only if" part of Proposition 1.12.

Example 1.13. Define $\tau : [0, 4] \to [0, 4]$ by $\tau(0) := 0$ and $\tau(t) := 3 + \frac{1}{4}t$ for $0 < t \leq 4$. Then τ is strictly increasing with $\tau(0) = 0$ and $\tau(4) = 4$, but discontinuous at $t = 0$. The function $g : [0, 4] \to \mathbb{R}$ defined by

$$g(x) := \begin{cases} 0 & \text{for } 0 \leq x \leq 1 , \\ \tan \frac{\pi}{2}(x - 1) & \text{for } 1 < x < 2 , \\ 0 & \text{for } 2 \leq x \leq 4 \end{cases}$$

does not belong to $BV([0, 4])$ since it is unbounded near $x = 2$. On the other hand, the function $(g \circ \tau)(t) \equiv 0$ trivially belongs to $BV([0, 4])$. ♥

Proposition 1.12 shows that, roughly speaking, homeomorphisms are suitable changes of variables which preserve bounded variation. Since being homeomorphic is a strong property, this is not surprising. So the question arises whether or not one can describe the *precise* class of "admissible changes of variables" which preserve bounded variation. This leads to a new class of maps which has been introduced, as far as we know, by Josephy [155].

Definition 1.14. For $n = 1, 2, 3, \ldots$, we denote by $\mathfrak{J}_n$ the family of sets $M \subseteq \mathbb{R}$ which may be represented as a union of n intervals.[4] Obviously, $\mathfrak{J}_n \subset \mathfrak{J}_{n+1}$ since each interval is a union of two nonempty subintervals.

4 Here, the intervals may be closed, open, half-open, or singletons.

By $J_n([a,b])$, we denote the class of all bounded functions $f : [a,b] \to \mathbb{R}$ such that $f^{-1}([\alpha,\beta]) \in \mathcal{J}_n$ for any interval[5] $[\alpha,\beta] \subset \mathbb{R}$. Moreover, we put

$$J([a,b]) := \bigcup_n J_n([a,b]) \tag{1.29}$$

and call functions $f \in J([a,b])$ *pseudomonotone* in what follows. ■

The following proposition shows that pseudomonotone functions are intermediate between monotone functions and functions of bounded variation.

Proposition 1.15. *Every monotone function is pseudomonotone, and every pseudomonotone function has bounded variation.*

Proof. The pseudomonotonicity of a monotone function f follows from the fact that $f^{-1}([\alpha,\beta])$ is always an interval (closed, open, or half-open) if f is monotonically increasing or decreasing,[6] and so $f^{-1}([\alpha,\beta]) \in \mathcal{J}_1$ for any interval $[\alpha,\beta] \subset \mathbb{R}$.

Now, let $f \in J_n([a,b])$ for some $n \in \mathbb{N}$; we show that

$$\mathrm{Var}(f;[a,b]) \le 4(n+1)\|f\|_\infty, \tag{1.30}$$

where $\|\cdot\|_\infty$ denotes the norm (0.39). If (1.30) is false, we find a partition $P = \{t_0, t_1, \dots, t_m\} \in \mathcal{P}([a,b])$ such that

$$\sum_{j=1}^{m} |f(t_j) - f(t_{j-1})| > 4(n+1)\|f\|_\infty. \tag{1.31}$$

Some nondegenerate interval $[A,B] \subseteq [-\|f\|_\infty, \|f\|_\infty]$ is covered more than $2(n+1)$ times by subintervals $[f(t_{j-1}), f(t_j)]$ or $[f(t_j), f(t_{j-1})]$ of our partition P in (1.31), and at least $n+1$ among them are of type $[f(t_{j-1}), f(t_j)]$. Consequently,

$$[A,B] \subseteq \bigcap_{k=1}^{n+1} [f(x_k), f(y_k)],$$

where $x_k = t_{j-1}$ and $y_k = t_j$ for some $j \in \{1,2,\dots,m\}$, and $x_1 < y_1 < \dots < x_{n+1} < y_{n+1}$, and so

$$f(x_k) \le A < B \le f(y_k) \quad (k = 1,2,\dots,n+1).$$

Now, choosing α and β such that $A < \alpha < B$ and $\beta > \max\{f(y_1),\dots,f(y_{n+1})\}$, we see that $f(x_k)$ belongs to $[\alpha,\beta]$ for all k, while $f(y_k)$ does not belong to $[\alpha,\beta]$ for any k. This means that $f^{-1}([\alpha,\beta]) \notin \mathcal{J}_n$, contradicting our assumption. □

---

5 Without loss of generality, we take into account only intervals $[\alpha,\beta]$ satisfying $f^{-1}([\alpha,\beta]) \cap [a,b] \neq \emptyset$.

6 The fact that $f^{-1}(I)$ is an interval for each interval I is even *equivalent* to the monotonicity of a function f, see Exercise 1.10. Of course, for *strictly* monotone functions, we simply have $f^{-1}([\alpha,\beta]) = [f^{-1}(\alpha), f^{-1}(\beta)]$ if f is increasing, and $f^{-1}([\alpha,\beta]) = [f^{-1}(\beta), f^{-1}(\alpha)]$ if f is decreasing. If f is not strictly monotone, however, the interval $f^{-1}([\alpha,\beta])$ need not be closed.

Finding a pseudomonotone function which is not monotone is easy. In the following example,[7] we give a function $f \in BV([0,1])$ which is not pseudomonotone.

Example 1.16. Let $f : [0,1] \to \mathbb{R}$ be defined by

$$f(x) := \begin{cases} x^2 \sin^2 \frac{1}{x} & \text{for } 0 < x \leq 1\,, \\ 0 & \text{for } x = 0\,. \end{cases} \tag{1.32}$$

A straightforward calculation shows[8] that $\operatorname{Var}(f;[0,1]) \leq 4$, and hence $f \in BV([0,1])$. On the other hand,

$$f^{-1}(\{0\}) = \{0\} \cup \left\{\tfrac{1}{n\pi} : n = 1, 2, 3, \dots\right\},$$

and so $f \notin J([0,1])$. ♥

Now, we come to the announced refinement of Proposition 1.12 which gives a precise description of all admissible changes of variables.

Proposition 1.17. *Let $\tau : [a,b] \to [c,d]$ be given. Then $g \circ \tau \in BV([a,b])$ for all $g \in BV([c,d])$ if and only if τ is pseudomonotone.*

Proof. Suppose first that $\tau : [a,b] \to [c,d]$ is pseudomonotone, and let $g : [c,d] \to \mathbb{R}$ be increasing. Similarly, as in the proof of Proposition 1.15, one may show then that $\tau^{-1}([\alpha,\beta]) \in J_n$ implies $(g \circ \tau)^{-1}([\alpha,\beta]) = \tau^{-1}(g^{-1}([\alpha,\beta])) \in J_n$, and so $g \circ \tau \in BV([a,b])$. Now, a general function $g \in BV([c,d])$ may be represented as the difference $g = p_g - n_g$ of increasing functions p_f and n_f, by Theorem 1.5, and so we have $g \circ \tau = p_g \circ \tau - n_g \circ \tau \in BV([a,b])$.

Suppose now that $\tau : [a,b] \to [c,d]$ is not pseudomonotone, and hence $\tau \notin J_n([a,b])$ for all n. We construct a function $g \in BV([c,d])$ such that $g \circ \tau \notin BV([a,b])$.

By assumption, we can find a sequence $(I_n)_n$ of intervals $I_n \subseteq [c,d]$ such that

$$\tau^{-1}(I_n) \notin \mathcal{J}_{3^n} \quad (n = 1, 2, 3, \dots)\,. \tag{1.33}$$

Define $g_n : [c,d] \to \mathbb{R}$ and $g : [c,d] \to \mathbb{R}$ by

$$g_n(x) := \frac{\chi_{I_n}(x)}{3^n}\,, \qquad g(x) := \sum_{n=1}^{\infty} g_n(x)\,.$$

Then $\operatorname{Var}(g_n;[c,d]) \leq 2/3^n$, and hence

$$\operatorname{Var}(g;[c,d]) \leq \sum_{n=1}^{\infty} \operatorname{Var}(g_n;[c,d]) \leq 2 \sum_{n=1}^{\infty} \frac{1}{3^n} = 1\,,$$

---

7 The idea of Example 1.16 is plane: pseudomonotonicity of f means that f, apart from constant pieces, does not repeat its values "too often."

8 By considering the derivative of f, one may even show that f is Lipschitz continuous on $[0,1]$, and so has bounded variation, see (1.46) in the next section.

and so $g \in BV([c,d])$. On the other hand,

$$\mathrm{Var}(g_n \circ \tau; [a,b]) = \mathrm{Var}(3^{-n}\chi_{\tau^{-1}(I_n)}; [a,b]) \geq 3^n \frac{2}{3^n} = 2\,.$$

We claim that this implies

$$\mathrm{Var}(g \circ \tau; [a,b]) \geq \frac{1}{6}\sum_{n=1}^{\infty} \mathrm{Var}(g_n \circ \tau; [a,b]) = \infty\,, \tag{1.34}$$

and so $g \circ \tau \notin BV([a,b])$. In fact, suppose that $g_n \circ \tau$ is discontinuous at some point $x_0 \in [a,b]$, but $g_1 \circ \tau, g_2 \circ \tau, \dots g_{n-1} \circ \tau$ are all continuous at x_0. Then x_0 is a point of discontinuity for $g \circ \tau$, contributing a jump of at least

$$\frac{1}{3^n} - \sum_{k=n+1}^{\infty} \frac{1}{3^k} = \frac{1}{2 \cdot 3^n}$$

in $g \circ \tau$, but contributing no more than

$$2\sum_{k=n}^{\infty} \frac{1}{3^k} = \frac{1}{3^{n-1}}$$

to the series in (1.34). From this, the assertion follows. □

We remark that other decomposition results with admissible changes of variables will be given in Theorem 1.28 and Theorem 1.41 below.

We close this section with a useful result which shows that if a function $f : [a,b] \to \mathbb{R}$ *fails* to have bounded variation, it always fails "locally:"

Proposition 1.18 (localization principle). *Suppose that $f \notin BV([a,b])$. Then there exists a point $x_0 \in [a,b]$ such that $f \notin BV([c,d])$ for each interval $[c,d] \subseteq [a,b]$ such that $c < x_0 < d$.*

Proof. Suppose that the assertion is false, which means that for each $x \in [a,b]$, there exists some open interval I_x containing x such that $f \in BV(\overline{I}_x)$. Since

$$[a,b] \subseteq \bigcup_{a \leq x \leq b} I_x$$

and $[a,b]$ is compact, we may find finitely many points $x_1, \dots, x_n \in [a,b]$ such that

$$[a,b] \subseteq \bigcup_{k=1}^{n} I_{x_k}\,.$$

However, then, Proposition 1.3 (g) implies

$$\mathrm{Var}(f; [a,b]) \leq \sum_{k=1}^{n} \mathrm{Var}(f; \overline{I}_{x_k}) < \infty\,,$$

contradicting our assumption. □

1.2 Bounded variation and continuity

Let us now return to continuity properties of functions of bounded variation. As in Chapter 0, we write $C([a,b])$ for the linear space of all continuous functions on $[a,b]$, equipped with the natural norm (0.45). As we have seen, neither of the spaces $C([a,b])$ or $BV([a,b])$ is contained in the other. In some applications, however, the function space $C([a,b]) \cap BV([a,b])$ is of some interest. This space is related to another subclass of $BV([a,b])$ which we will discuss now.

Given $f \in BV([a,b])$, consider the quantity $\mathrm{Var}_0(f)$ defined by

$$\mathrm{Var}_0(f) = \mathrm{Var}_0(f;[a,b]) := \limsup_{\delta\to 0+} \{\mathrm{Var}(f) - \mathrm{Var}(f,P) : \mu(P) \le \delta\}, \tag{1.35}$$

where $\mu(P)$ denotes the mesh size (1.2) of P. Roughly speaking, the quantity (1.35) measures the degree of uniform approximability of the total variation (1.4) by choosing partitions P in (1.3) of sufficiently small mesh size. In order to represent $\mathrm{Var}_0(f)$ in a more transparent form, note that the quantity $W(\delta)$ defined by

$$W(\delta) := \mathrm{Var}(f) - \inf\{\mathrm{Var}(f,P) : \mu(P) \le \delta\} \quad (\delta > 0)$$

is increasing on $(0,\infty)$, and so has a limit for $\delta \to 0+$, which is nothing else but (1.35).

In what follows, let us denote by $CBV([a,b])$ the subset of all functions $f \in BV([a,b])$ such that $\mathrm{Var}_0(f;[a,b]) = 0$. A straightforward calculation shows that $CBV([a,b])$ contains all monotone functions on $[a,b]$. Unfortunately, $CBV([a,b])$ is *not* a linear subspace of $BV([a,b])$:

Example 1.19. In the space $BV([0,2])$, for example, let $f = \chi_{[0,1]}$ be the characteristic function of the interval $[0,1]$, and $g = \chi_{[1,2]}$ be the characteristic function of the interval $[1,2]$. Clearly, $\mathrm{Var}(f) = \mathrm{Var}(g) = 1$ and, for any partition P of $[0,2]$, we have $\mathrm{Var}(f,P) = \mathrm{Var}(g,P) = 1$. This implies that $\mathrm{Var}_0(f) = \mathrm{Var}_0(g) = 0$, and so $f, g \in CBV([0,2])$. On the other hand, it is easily seen that $f + g = 1 + \chi_{\{1\}}$, and therefore $\mathrm{Var}(f+g) = 2$ and $\mathrm{Var}_0(f+g) = 2$, so $f + g \notin CBV([0,2])$. ♥

A similar choice of f and g shows that the set $CBV([a,b])$ is not convex, so it has a rather poor behavior from the algebraic viewpoint. Since $CBV([a,b])$ contains all monotone functions, its convex hull (equivalently, linear hull) coincides with the whole space $BV([a,b])$, by Theorem 1.5 or 1.6.

The following Proposition 1.20 provides a link between the class $CBV([a,b])$ and the *continuous* functions of bounded variation.

Proposition 1.20. *Let f be a continuous function of bounded variation on $[a,b]$. Then f belongs to $CBV([a,b])$.*

Proof. Given $\varepsilon > 0$, by definition (1.4) of the total variation, we may find a partition $P_0 = \{\tau_0, \tau_1, \ldots, \tau_n\} \in \mathcal{P}([a,b])$ such that

$$\mathrm{Var}(f,P_0) \ge \mathrm{Var}(f) - \varepsilon. \tag{1.36}$$

Since f is uniformly continuous on $[a, b]$, we may further choose a $\delta > 0$ such that for $s, t \in [a, b]$ with $|s - t| \leq \delta$, we have

$$|f(s) - f(t)| \leq \frac{\varepsilon}{2(n-1)}. \tag{1.37}$$

Now, let $P = \{t_0, t_1, \ldots, t_m\} \in \mathcal{P}([a, b])$ be an arbitrary partition satisfying $\mu(P) \leq \delta$. Consider the partition $P \cup \{\tau_i\}$ which differs from P by adding just one point τ_i from P_0. If this point lies in the interval (t_{j-1}, t_j) for some $j \in \{1, 2, \ldots, m\}$, then the enlarged partition has the form $P \cup \{\tau_i\} = \{t_0, t_1, \ldots, t_{j-1}, \tau_i, t_j, \ldots, t_m\}$, and so we have

$$\begin{aligned}
&\mathrm{Var}(f, P \cup \{\tau_i\}) \\
&= \sum_{k=1}^{j-1} |f(t_k) - f(t_{k-1})| + |f(\tau_i) - f(t_{j-1})| \\
&\quad + |f(t_j) - f(\tau_i)| + \sum_{k=j+1}^{m} |f(t_k) - f(t_{k-1})| \\
&= \mathrm{Var}(f, P) + |f(\tau_i) - f(t_{j-1})| + |f(t_j) - f(\tau_i)| - |f(t_j) - f(t_{j-1})| \\
&\leq \mathrm{Var}(f, P) + |f(\tau_i) - f(t_{j-1})| + |f(t_j) - f(\tau_i)| .
\end{aligned}$$

Since both $|\tau_i - t_{j-1}| \leq \delta$ and $|t_j - \tau_i| \leq \delta$, in view of (1.37), we get

$$\mathrm{Var}(f, P \cup \{\tau_i\}) \leq \mathrm{Var}(f, P) + \frac{\varepsilon}{n-1},$$

showing that adding one new point $\tau_i \in P_0$ to P leads to an increase of $\mathrm{Var}(f, P)$ by at most $\varepsilon/(n-1)$. Since there are not more than $n-1$ points in P_0 which are different from the points of P, we conclude that

$$\mathrm{Var}(f, P \cup P_0) - \mathrm{Var}(f, P) \leq (n-1)\frac{\varepsilon}{n-1} = \varepsilon .$$

Combining this inequality with (1.36), we obtain

$$\mathrm{Var}(f, P) \geq \mathrm{Var}(f, P \cup P_0) - \varepsilon \geq \mathrm{Var}(f, P_0) - \varepsilon \geq \mathrm{Var}(f) - 2\varepsilon .$$

This means that $\mathrm{Var}_0(f) \leq 2\varepsilon$, and so $f \in CBV([a, b])$ since $\varepsilon > 0$ is arbitrary. □

We may reformulate the statement of Proposition 1.20 as inclusion

$$C([a, b]) \cap BV([a, b]) \subseteq CBV([a, b]) \subseteq BV([a, b]) . \tag{1.38}$$

The function f in Example 1.19 shows that the first inclusion in (1.38) is strict, while the function $f + g$ in Example 1.19 shows that the second inclusion in (1.38) is strict.

To the best of our knowledge, the first characterization of the condition $\mathrm{Var}_0(f) = 0$ was apparently given by Tonelli [304] (cf. also [54]) who showed that $f \in CBV([a, b])$ if and only if

$$(f(t+) - f(t))(f(t) - f(t-)) \geq 0 \quad (a \leq t \leq b) ,$$

i.e. f keeps "jumping in the same direction" at any point of discontinuity. From this result, it follows not only that every continuous function f of bounded variation belongs to the class $CBV([a,b])$, but also every monotone function on $[a,b]$, as already observed above. Moreover, this condition again explains the geometrical meaning of Example 1.19.

We may formulate the assertion of Proposition 1.20 in the following in a slightly different way. Given a *continuous* function $f : [a,b] \to \mathbb{R}$ and a partition $P = \{t_0, t_1, \dots, t_m\} \in \mathcal{P}([a,b])$, consider the oscillation of f defined in (1.12) on each subinterval of P. Then the equality

$$\operatorname{Var}(f;[a,b]) = \lim_{\delta\to 0+} \left\{ \sum_{j=1}^{m} \operatorname{osc}(f;[t_{j-1},t_j]) : \mu(P) \le \delta \right\} \tag{1.39}$$

holds, where the variation of f on $[a,b]$ may be, of course, infinite; see Exercise 1.39.

We remark that other sufficient conditions for a function $f \in BV([a,b])$ to belong to $CBV([a,b])$ may be found in [13].

Let us now consider two important classes of functions on $[a,b]$ which are contained in $C([a,b]) \cap BV([a,b])$, and so also in $CBV([a,b])$, by (1.38). The first class will be studied in great detail in Chapter 3. We denote by $\Sigma([a,b])$ the family of all finite collections $S = \{[a_1,b_1],\dots,[a_n,b_n]\}$ of pairwise nonoverlapping subintervals[9] of $[a,b]$, and by $\Sigma_\infty([a,b])$ the family of all countably infinite collections $S_\infty = \{[a_n,b_n] : n \in \mathbb{N}\}$ of pairwise nonoverlapping subintervals of $[a,b]$.

Definition 1.21. A function $f : [a,b] \to \mathbb{R}$ is called *absolutely continuous* if it has the following property: for any $\varepsilon > 0$, there exists a $\delta > 0$ such that for all collections $S = \{[a_1,b_1],\dots,[a_n,b_n]\} \in \Sigma([a,b])$, the condition

$$\sum_{k=1}^{n} (b_k - a_k) \le \delta \tag{1.40}$$

implies that

$$\sum_{k=1}^{n} |f(b_k) - f(a_k)| \le \varepsilon. \tag{1.41}$$

Equivalently, we may require that for any $\varepsilon > 0$, there exists a $\delta > 0$ such that for all $S_\infty = \{[a_n,b_n] : n \in \mathbb{N}\} \in \Sigma_\infty([a,b])$ satisfying

$$\sum_{k=1}^{\infty} (b_k - a_k) \le \delta, \tag{1.42}$$

we have

$$\sum_{k=1}^{\infty} |f(b_k) - f(a_k)| \le \varepsilon. \tag{1.43}$$

We denote the set of all absolutely continuous functions on $[a,b]$ by $AC([a,b])$. ■

9 Two intervals I and J are called *nonoverlapping* if $I^o \cap J^o = \emptyset$, i.e. they are disjoint or have at most one boundary point in common.

Let us briefly explain the word "equivalently" in Definition 1.21. Of course, the condition involving $\Sigma_\infty([a,b])$ implies the condition involving $\Sigma([a,b])$. Conversely, suppose that the first requirement in Definition 1.21 is fulfilled, and suppose that $S_\infty = \{[a_n, b_n] : n \in \mathbb{N}\} \in \Sigma_\infty([a,b])$ satisfies (1.42). Then (1.40) holds true for any fixed n, and so also (1.41). However, this implies that (1.43) holds true as well.

Of course, absolute continuity implies (uniform) continuity on $[a,b]$. Moreover, it is not hard to see that every absolutely continuous function has bounded variation:

Proposition 1.22. *Every function $f \in AC([a,b])$ belongs to $BV([a,b])$.*

Proof. In fact, for $\varepsilon = 1$, say, choose $\delta > 0$ such that (1.40) implies (1.41), and consider the equidistant partition $P_n := \{t_0, t_1, \ldots, t_n\}$, where $n \in \mathbb{N}$ is so large that $n\delta > b-a$. On each subinterval $[t_{k-1}, t_k]$ of this partition, we then have, by construction, $\mathrm{Var}(f, P_n; [t_{k-1}, t_k]) \le 1$. By the additivity property of the variation (Proposition 1.3 (g)), we then get $\mathrm{Var}(f, P_n; [a,b]) \le n$, and thus also $\mathrm{Var}(f; [a,b]) \le n$. □

Another important class of continuous functions which is closely related to the space $BV([a,b])$ is the space of Lipschitz continuous functions introduced in Definition 0.39. Recall that the linear space $Lip([a,b])$ with norm

$$\|f\|_{Lip} := |f(a)| + lip(f)\,, \tag{1.44}$$

where

$$lip(f) := \sup_{x \neq y} \frac{|f(x) - f(y)|}{|x-y|} \tag{1.45}$$

is a Banach space. Clearly, every Lipschitz continuous function is absolutely continuous (choose $\delta := \varepsilon / lip(f)$). Thus, we may extend the chain of inclusions (1.38) to

$$Lip([a,b]) \subseteq AC([a,b]) \subseteq C([a,b]) \cap BV([a,b]) \subseteq BV([a,b]). \tag{1.46}$$

One may even show that the space $(Lip([a,b]), \|\cdot\|_{Lip})$ is continuously imbedded into the space $(BV([a,b]), \|\cdot\|_{BV})$ in the sense of Definition 0.29. In fact, suppose that $f : [a,b] \to \mathbb{R}$ satisfies a Lipschitz condition

$$|f(x) - f(y)| \le L|x-y| \quad (a \le x, y \le b)$$

for some $L > 0$, and let $P = \{t_0, t_1, \ldots, t_m\}$ be any partition of $[a,b]$. Then

$$\mathrm{Var}(f, P; [a,b]) = \sum_{j=1}^{m} |f(t_j) - f(t_{j-1})| \le L \sum_{j=1}^{m} |t_j - t_{j-1}| = L(b-a)\,,$$

and hence

$$\|f\|_{BV} = |f(a)| + \mathrm{Var}(f; [a,b]) \le |f(a)| + L(b-a), \tag{1.47}$$

and so

$$\|f\|_{BV} \le \max\{1, b-a\} \, \|f\|_{Lip} \tag{1.48}$$

since L may be taken arbitrarily close to $lip(f)$. On the other hand, the following nontrivial example shows that *none* of the Hölder spaces $Lip_\alpha([a,b])$ introduced in Definition 0.39 is contained in $BV([a,b])$ in case $\alpha < 1$.

Example 1.23. Fix $\alpha \in (0,1)$. We construct a function $f \in Lip_\alpha([0,1]) \setminus BV([0,1])$. To this end, we define a constant γ and a sequence $(t_n)_n$ in $[0,1]$ by[10]

$$\gamma := \zeta(1/\alpha, 0) = \sum_{k=1}^{\infty} \frac{1}{k^{1/\alpha}}, \quad t_n := \frac{1}{\gamma} \sum_{k=n}^{\infty} \frac{1}{k^{1/\alpha}} \tag{1.49}$$

for $n = 1, 2, 3, \ldots$, and define $f : [0,1] \to \mathbb{R}$ by

$$f(x) := \begin{cases} 0 & \text{for } x = 0\,, \\ \frac{(-1)^n}{n} & \text{for } x = t_n\,, \\ \text{linear} & \text{otherwise.} \end{cases}$$

Considering, similarly as in Example 1.8, partitions P_n which contain the points $t_1, t_2, \ldots, t_n$, we see that

$$\mathrm{Var}(f, P_n; [0,1]) \geq \sum_{k=1}^{n} \frac{1}{k} \to \infty \quad (n \to \infty)\,,$$

and so $f \notin BV([0,1])$.

Now, let $0 < x < y \leq 1$, and choose $m, n \in \mathbb{N}$ such that $t_{n+1} \leq x \leq t_n$ and $t_{m+1} \leq y \leq t_m$. We distinguish three cases.

Suppose first that $n = m$. Then

$$0 < y - x \leq t_n - t_{n+1} = \frac{1}{\gamma n^{1/\alpha}}\,,$$

and hence

$$\begin{aligned} |f(x) - f(y)| = (y-x)\frac{|f(t_n) - f(t_{n+1})|}{|t_n - t_{n+1}|} &\leq (y-x)\gamma\frac{2n^{1/\alpha}}{n} \\ &\leq 2\gamma|x-y|^\alpha |t_n - t_{n+1}|^{1-\alpha} n^{(1-\alpha)/\alpha} \\ &\leq 2\gamma|x-y|^\alpha \frac{1}{\gamma^{1-\alpha}} n^{(1-\alpha)/\alpha - (1-\alpha)/\alpha} = 2\gamma^\alpha |x-y|^\alpha\,. \end{aligned}$$

Assume now that $n = m+1$. Then $t_n = t_{m+1}$, and hence

$$\begin{aligned} |f(x) - f(y)| &\leq |f(x) - f(t_n)| + |f(t_{m+1}) - f(y)| \\ &\leq 2\gamma^\alpha \left(|x - t_n|^\alpha + |t_{m+1} - y|^\alpha\right) \leq 4\gamma^\alpha |x-y|^\alpha\,. \end{aligned}$$

---

10 In (1.49) and (1.51) below, we use the shortcut (0.17).

Finally, let $n \geq m+2$. Then we find points $s \in [t_n, t_{n-1}]$ and $t \in [t_{m+2}, t_{m+1}]$ such that[11] $f(s) = f(t) = 0$. Consequently,

$$\begin{aligned} |f(x) - f(y)| &\leq |f(x) - f(s)| + |f(s) - f(t)| + |f(t) - f(y)| \\ &\leq 4\gamma^\alpha \left(|x-s|^\alpha + |t-y|^\alpha\right) \leq 8\gamma^\alpha |x-y|^\alpha . \end{aligned}$$

In each case, we see that f satisfies the Hölder condition (0.67) with $L := 8\gamma^\alpha$ for $0 < x < y \leq 1$. Since f is continuous at 0 with $f(0) = 0$, the same is true on the interval $[0, 1]$. ♥

The function in Example 1.23 looks very much like a "mirror reversed" copy of the zigzag function (0.91) introduced in Definition 0.49. In fact, we could have also used (0.91) as an example of a function in $Lip_\alpha([0,1]) \setminus BV([0,1])$. As we have seen in Proposition 0.50, we have $Z_{C,D} \in Lip_\alpha([0,1])$ if and only if

$$S_\alpha(C, D) = \sup\{d_n c_n^{-\alpha} : n = 1, 2, 3, \ldots\} < \infty .$$

However, in this case, we have $d_n \leq S c_n^\alpha$ for all $n \in \mathbb{N}$. Therefore, from the obvious relation

$$\mathrm{Var}(Z_{C,D}; [0,1]) = \sum_{k=1}^\infty d_k, \tag{1.50}$$

it follows that for constructing such an example we should have

$$\sum_{k=1}^\infty c_k^\alpha = \infty, \quad \sum_{k=1}^\infty c_k = S_\alpha(C, D)^{1/\alpha} < \infty .$$

The simplest choice of the c_k's with this property is of course $c_k = k^{-\alpha}$, and this explains Definition (1.49).

The function f in Example 1.23 belongs to $f \in Lip_\alpha([0,1]) \setminus BV([0,1])$ with prescribed $\alpha < 1$. The following is a refinement of Example 1.23.

Example 1.24. In this example, we construct a function

$$f \in \left(\bigcap_{0<\alpha<1} Lip_\alpha([0,1]) \right) \setminus BV([0,1]) .$$

To this end, we replace (1.49) by

$$\gamma := \zeta(1,2) = \sum_{k=1}^\infty \frac{1}{k \log^2(k+1)}, \quad t_n := \frac{1}{\gamma} \sum_{k=n}^\infty \frac{1}{k \log^2(k+1)} \tag{1.51}$$

for $n = 1, 2, \ldots$, and define $f : [0,1] \to \mathbb{R}$ precisely as in Example 1.23. Again, by considering partitions containing $t_1, t_2, \ldots, t_n$, we see that $f \notin BV([0,1])$.

11 In case $n = m+2$, we take $s = t$.

Now, we show that f belongs to $Lip_\alpha([0,1])$ for *any* $\alpha < 1$. Fix $\alpha \in (0,1)$, and choose $x, y \in (0,1]$ and $m, n \in \mathbb{N}$ as in Example 1.23. As before, we distinguish the three cases $n = m$, $n = m+1$, and $n \geq m+2$.

First, suppose that $n = m$. Then

$$0 < y - x \leq t_n - t_{n+1} = \frac{1}{\gamma} \frac{1}{n \log^2(n+1)},$$

and hence

$$\begin{aligned}|f(x) - f(y)| &= (y - x)\frac{|f(t_n) - f(t_{n+1})|}{|t_n - t_{n+1}|} \\ &\leq 2\gamma|x - y| \log^2(n+1) \leq L_\alpha |x - y|^\alpha;\end{aligned}$$

here, we have used the fact that

$$\log^2(n+1)|x - y|^{1-\alpha} \leq \frac{\log^2(n+1)}{\gamma n^{1-\alpha} \log^{2(1-\alpha)}(n+1)} \to 0 \quad (n \to \infty),$$

which shows that the constant $L_\alpha > 0$ only depends on α, but not on x or y.

Assume now that $n = m+1$. Then, as before,

$$\begin{aligned}|f(x) - f(y)| &\leq |f(x) - f(t_n)| + |f(t_{m+1}) - f(y)| \\ &\leq L_\alpha \left(|x - t_n|^\alpha + |t_{m+1} - y|^\alpha\right) \leq 2L_\alpha |x - y|^\alpha .\end{aligned}$$

Finally, let $n \geq m+2$. Choosing $s \in [t_n, t_{n-1}]$ and $t \in [t_{m+2}, t_{m+1}]$ as in Example 1.23 yields

$$\begin{aligned}|f(x) - f(y)| &\leq |f(x) - f(s)| + |f(s) - f(t)| + |f(t) - f(y)| \\ &\leq 2L_\alpha \left(|x - s|^\alpha + |t - y|^\alpha\right) \leq 4L_\alpha |x - y|^\alpha .\end{aligned}$$

In each case, we see that f satisfies the Hölder condition (0.67) with $L := 4L_\alpha$, and so we are done. ♥

Observe that, by (1.46), the function f constructed in Example 1.24 belongs to $Lip_\alpha([0,1])$ for each $\alpha < 1$, but not to $Lip([0,1]) = Lip_1([0,1])$.

Our final example in this section is in a certain sense “dual” to the preceding two examples: it contains a function which belongs to $BV([0,1]) \cap C([0,1])$ (and hence also to $CBV([0,1])$, by (1.38), but not to $Lip_\alpha([0,1])$ for any $\alpha \in (0,1]$.

Example 1.25. To construct a function

$$f \in [BV([0,1]) \cap C([0,1])] \setminus \left(\bigcup_{0<\alpha<1} Lip_\alpha([0,1]) \right),$$

consider again the function (0.75) from Example 0.41. As we have seen there, this function is continuous, but does not belong to $Lip_\alpha([0,1])$ for any $\alpha \in (0,1]$. Being monotonically increasing, this function belongs to $BV([0,1])$ as well. ♥

Later (Example 3.5), we will see that the function (0.75) is not only of bounded variation, but even absolutely continuous.

Another class which is related both to the spaces $BV([a,b])$ and $C([a,b])$ consists of the functions with the intermediate value property. We say that a function $f : [a,b] \to \mathbb{R}$ has the *intermediate value property*, and write $f \in IVP([a,b])$ if for every interval $[c,d] \subseteq [a,b]$, we have $f([c,d]) \supseteq [f(c), f(d)]$ if $f(c) \le f(d)$, and $f([c,d]) \supseteq [f(d), f(c)]$ if $f(c) \ge f(d)$.[12] The most familiar example is of course that of continuous functions by the classical intermediate value theorem. However, there is also some interesting connection with functions of bounded variation: *every function $f \in BV([a,b])$ which has the intermediate value property is continuous on* $[a,b]$. In fact, having bounded variation, we know that all discontinuities of f, if there are any, are jumps, which means that $D(f) = D_0(f) \cup D_1(f)$ in the notation (0.49)–(0.51), see Exercise 1.12. However, such discontinuities are excluded by the intermediate value property, and so the assertion follows.

One may show that in this reasoning, it suffices to suppose that $|f| \in BV([a,b])$ (Exercise 1.42) which is weaker, by Exercises 1.3 and 1.4. So, if you want to present, in your first-year calculus course, a discontinuous function f with the intermediate value property, you have to ensure that $|f| \notin BV$, and so $|f|$ must have a discontinuity of a second kind somewhere. A certain modification of this is given in Exercise 1.43. We may combine the classical intermediate value theorem with the statement we just proved in the chain of inclusions

$$BV([a,b]) \cap IVP([a,b]) \subseteq C([a,b]) \subseteq IVP([a,b]) .$$

Let us now return to the variation function V_f of $f \in BV([a,b])$ defined in (1.13). The following is an extension of Proposition 1.7.

Theorem 1.26. *For* : $[a,b] \to \mathbb{R}$ *and* V_f *as in* (1.13), *the following statements are true.*
(a) *The function f is continuous if and only if the function V_f is continuous.*
(b) *The function f is of bounded variation if and only if the function V_f is of bounded variation; moreover, in this case, we have* $\mathrm{Var}(V_f; [a,b]) = \mathrm{Var}(f; [a,b])$.
(c) *The function f is Lipschitz continuous if and only if the function V_f is Lipschitz continuous; moreover, in this case, we have $lip(V_f) = lip(f)$.*
(d) *The function f is Hölder continuous of order $\alpha \in (0,1)$ if the function V_f is Hölder continuous of the same order; moreover, in this case, we have $lip_\alpha(V_f) \ge lip_\alpha(f)$.*
(e) *The function f is absolutely continuous if and only if the function V_f is absolutely continuous.*

12 In other words, for every real number η between $f(c)$ and $f(d)$, there is some $\xi \in [c,d]$ such that $f(\xi) = \eta$.

Proof. The statement (a) has already been proved in Proposition 1.7, while (b) follows from the fact that V_f is always increasing, and hence

$$\mathrm{Var}(V_f; [a,b]) = V_f(b) - V_f(a) = \mathrm{Var}(f; [a,b]).$$

To prove (c), suppose that $f \in Lip([a,b])$ and $a \le x < y \le b$, and let $P = \{t_0, t_1, \ldots, t_m\}$ be a partition of $[a,b]$. Then

$$|f(t_j) - f(t_{j-1})| \le lip(f)(t_j - t_{j-1}) \quad (j = 1, 2, \ldots, m)$$

and so

$$V_f(y) - V_f(x) = \mathrm{Var}(f; [x,y]) \le lip(f)(y - x),$$

which shows that $V_f \in Lip([a,b])$ with $lip(V_f) \le lip(f)$. The converse implication follows from the inequalities

$$|f(x) - f(y)| \le |V_f(x) - V_f(y)| \le lip(V_f)|x - y| \tag{1.52}$$

for any $x, y \in [a,b]$. Since we may replace $|x-y|$ by $|x-y|^\alpha$ for any $\alpha \in (0,1]$ in (1.52), the last part of the proof of (c) carries over to Hölder continuous functions, thus proving (d).

Now, we prove statement (e). Suppose first that $f \in AC([a,b])$, let $\varepsilon > 0$ be given, and choose $\delta > 0$ such that (1.40) implies (1.41). Since f has bounded variation, for any family $S = \{[a_1, b_1], \ldots, [a_n, b_n]\} \in \Sigma([a,b])$ satisfying (1.40), we may choose partitions

$$P_k := \{t_{0,k}, t_{1,k}, \ldots, t_{m_k,k}\} \in \mathcal{P}([a_k, b_k]) \quad (k = 1, 2, \ldots, n)$$

such that

$$V_f(b_k) - V_f(a_k) = \mathrm{Var}(f; [a_k, b_k]) \le \sum_{j=1}^{m_k} |f(t_{j,k}) - f(t_{j-1,k})| + \frac{\varepsilon}{2n}.$$

For each $k \in \{1, 2, \ldots, n\}$, the intervals $[t_{j-1,k}, t_{j,k}]$ are nonoverlapping and

$$\sum_{k=1}^{n} \sum_{j=1}^{m_k} \left(t_{j,k} - t_{j-1,k}\right) = \sum_{k=1}^{n} (b_k - a_k) \le \delta.$$

Therefore, by assumption,

$$\sum_{k=1}^{n} \left|V_f(b_k) - V_f(a_k)\right| \le \sum_{k=1}^{n} \left\{ \sum_{j=1}^{m_k} \left|f(t_{j,k}) - f(t_{j-1,k})\right| + \frac{\varepsilon}{2n} \right\} \le 2\varepsilon,$$

and hence $V_f \in AC([a,b])$. Conversely, suppose now that $V_f \in AC([a,b])$, let $\varepsilon > 0$ be given, and choose $\delta > 0$ such that (1.40) implies

$$\sum_{k=1}^{n} \left|V_f(b_k) - V_f(a_k)\right| \le \varepsilon$$

for $S = \{[a_1, b_1], \dots, [a_n, b_n]\} \in \Sigma([a, b])$. Then

$$\sum_{k=1}^{n} |f(b_k) - f(a_k)| \le \sum_{k=1}^{n} \big(V_f(b_k) - V_f(a_k)\big) \le \varepsilon$$

which shows that $f \in AC([a, b])$. The proof is complete. □

Theorem 1.26 shows that the variation function V_f inherits several properties from its parent function f. Observe, however, that there is a certain asymmetry in statement (d); in fact, the question as to if $f \in BV([a, b]) \cap Lip_\alpha([a, b])$ implies $V_f \in Lip_\alpha([a, b])$ for $0 < \alpha < 1$ seems to be open.[13] Of course, the mere requirement $f \in Lip_\alpha([a, b])$ does not imply $V_f \in Lip_\alpha([a, b])$ since V_f need not be finite, as Example 1.23 shows. However, we do not know of any example where V_f is finite for some $f \in Lip_\alpha([a, b])$, but $V_f \notin Lip_\alpha([a, b])$.

The reader may also have noticed that differentiability does not occur in Theorem 1.26. The reason is simple: differentiability does not carry over in any direction, see Exercises 1.50 and 1.51. As was pointed out in [149], if f has bounded variation on $[a, b]$, then both f and V_f are differentiable on $[a, b]$, except possibly on a nullset. However, the set of points at which $f'(x)$ exists is not necessarily the same as the set of points at which $V_f'(x)$ exists. Interestingly, in [281], it is shown that $V_f'(x) = |f'(x)|$ a.e. on $[a, b]$, see Exercise 1.52.

Now, we establish a connection between the variation of a continuous function and the Lebesgue integral of its Banach indicatrix, see Definition 0.38 or (0.106).

Proposition 1.27. *Let $f : [a, b] \to \mathbb{R}$ be continuous, and let $I_f : \mathbb{R} \to \mathbb{N}_0 \cup \{\infty\}$ denote its Banach indicatrix defined by* (0.106). *Then the equality*

$$\mathrm{Var}(f; [a, b]) = \int_{-\infty}^{\infty} I_f(y)\, dy \tag{1.53}$$

holds. In particular, $f \in BV([a, b])$ if and only if $I_f \in L_1(\mathbb{R})$.

Proof. For $n = 1, 2, 3, \dots$ and $k = 1, 2, \dots, 2^{n-1}$, we put $\delta_{k,n} := k(b - a)/2^n$ and

$$\Delta_{1,n} := [a, a + \delta_{1,n}) \,, \quad \Delta_{2,n} := [a + \delta_{1,n}, a + \delta_{2,n}) \,, \quad \dots$$
$$\dots, \quad \Delta_{2^{n-1},n} := [a + \delta_{2^{n-1}-1,n}, a + \delta_{2^{n-1},n}) \,, \quad \Delta_{2^n,n} := [a + \delta_{2^{n-1},n}, b] \,.$$

Moreover, for $k = 1, 2, \dots, 2^n$, let us denote

$$m_{k,n} := \inf\{f(x) : x \in \Delta_{k,n}\} \,, \quad M_{k,n} := \sup\{f(x) : x \in \Delta_{k,n}\} \,, \tag{1.54}$$

and define functions $g_{k,n} : \mathbb{R} \to \{0, 1\}$ and $g_n : \mathbb{R} \to \mathbb{N}_0$ by

$$g_{k,n}(y) := \chi_{f^{-1}(y) \cap \Delta_{k,n}}(y) = \begin{cases} 1 & \text{if } f(x) = y \text{ for some } x \in \Delta_{k,n} \,, \\ 0 & \text{otherwise} \end{cases}$$

13 For a related question, see Exercise 1.25.

and

$$g_n(y) := \sum_{k=1}^{2^n} g_{k,n}(y) .$$

Clearly, $g_{n+1}(y) \geq g_n(y)$ for all $y \in \mathbb{R}$. We claim that the sequence $(g_n)_n$ converges pointwise on $\mathbb{R}$ to the Banach indicatrix I_f of f.

Indeed, suppose first that $I_f(y) = m$, and let $f^{-1}(y) = \{x_1, x_2, \ldots, x_m\}$ denote the set of all solutions of the equation $f(x) = y$ in $[a, b]$. Choose $N \in \mathbb{N}$ so large that $2^{-N} < \min\{x_i - x_j : 1 \leq i, j \leq m,\ i \neq j\}$; then, for $n > N$, all elements in $f^{-1}(y)$ belong to different intervals $\Delta_{k,n}$, and so $g_n(y) = m$. Similarly, in case $I_f(y) = \infty$, an analogous reasoning shows that $g_n(y) \to \infty$ as $n \to \infty$.

Moreover, a straightforward calculation shows that

$$\int_{-\infty}^{\infty} g_n(y)\, dy = \sum_{k=1}^{2^n} (M_{k,n} - m_{k,n}) , \tag{1.55}$$

where $M_{k,n}$ and $m_{k,n}$ are given in (1.54). However, the left-hand side of (1.55) tends to the integral of I_f, by Levi's theorem (Theorem 0.3), while the right-hand side of (1.55) tends to $\mathrm{Var}(f; [a, b])$, by (1.39). □

We will use the second statement of Proposition 1.27 in the proof of an important characterization of absolutely continuous functions (Theorem 3.9).

We close this section with another two results on changes of variables for functions of bounded variation. The first result is a characterization of functions of bounded variation due to Federer ([113, Section 2.5.16]) as compositions of monotone and Lipschitz continuous functions with Lipschitz constant[14] $L = 1$. In a certain sense, this characterization may be considered as "dual" to Proposition 1.12.

Theorem 1.28. *A function f belongs to $BV([a, b])$ if and only if it may be represented as composition $f = g \circ \tau$, where $\tau : [a, b] \to [c, d]$ is increasing and $g \in Lip([c, d])$ with Lipschitz constant $L = 1$.*

Proof. Suppose that $f = g \circ \tau$, where g and τ have the mentioned properties. Given any partition $P = \{t_0, t_1, \ldots, t_m\} \in \mathcal{P}([a, b])$, we get

$$\mathrm{Var}(f, P) = \sum_{j=1}^{m} |g(\tau(t_j)) - g(\tau(t_{j-1}))| \leq \sum_{j=1}^{m} |\tau(t_j) - \tau(t_{j-1})| = |\tau(b) - \tau(a)| ,$$

and hence $f \in BV([a, b])$. Conversely, let $f \in BV([a, b])$, and let $\tau := V_f$ be the (increasing) variation function (1.13) of f. We know that τ maps $[a, b]$ into $[c, d]$, where

14 For obvious geometrical reasons, functions with Lipschitz constant 1 are often called *nonexpansive* in the literature.

$c = 0$ and $d = \mathrm{Var}(f; [a, b])$, but of course τ need not be onto. If we now define the function g on the range $\tau([a, b]) \subseteq [c, d]$ by putting $g(\tau(x)) := f(x)$, then the decomposition $f = g \circ \tau$ holds by construction. Since

$$|g(\tau(s)) - g(\tau(t))| = |f(s) - f(t)| \leq \mathrm{Var}(f; [s, t]) = |\tau(s) - \tau(t)|$$

for $a \leq s < t \leq b$, the function g is Lipschitz continuous with Lipschitz constant 1 on $\tau([a, b]) \subseteq [c, d]$.

For extending g to the whole interval $[c, d]$, we may now use the McShane extension described in Theorem 0.42. However, we may construct an extension of g from $\tau([a, b])$ to a Lipschitz continuous map $\overline{g}$ on $[c, d]$ (even on the whole real line $\mathbb{R}$) in a more explicit way. Indeed, the "convexification" $\overline{g}$ of g defined (with $0 \leq \lambda \leq 1$) by

$$\overline{g}(y) := \begin{cases} (1-\lambda)g(x-) + \lambda g(x) & \text{if } y = (1-\lambda)\tau(x-) + \lambda\tau(x)\,, \\ (1-\lambda)g(x) + \lambda g(x+) & \text{if } y = (1-\lambda)\tau(x) + \lambda\tau(x+)\,, \end{cases} \tag{1.56}$$

is easily seen to have the same Lipschitz constant as g, and so we are done. □

We will call the decomposition of $f \in BV([a, b])$ constructed in Theorem 1.28 the *Federer decomposition* in the sequel. The statement of Theorem 1.28 is somewhat surprising: although functions of bounded variation can have infinitely many discontinuities, these discontinuities may be "smoothed out" after a monotone change of variables by a Lipschitz continuous map. We illustrate this by a simple example involving just one removable discontinuity in the domain of definition f.

Example 1.29. On $[a, b] = [0, 2]$, consider the characteristic function $f = \chi_{\{1\}}$ of the singleton $\{1\}$. Let $\tau : [0, 2] \to [0, 2]$ be the variation function (1.13) of f which, in this case, has the form

$$\tau(x) = 1 + sgn(x-1) = \begin{cases} 0 & \text{for } 0 \leq x < 1\,, \\ 1 & \text{for } x = 1\,, \\ 2 & \text{for } 1 < x \leq 2\,. \end{cases} \tag{1.57}$$

Observing that $\tau([0, 2]) = \{0, 1, 2\}$ and applying the convexification (1.56) to the function $g : \{0, 1, 2\} \to \mathbb{R}$ defined by $g(0) = g(2) = 0$ and $g(1) = 1$, we end up with the peak function $\overline{g} : [0, 2] \to \mathbb{R}$ defined by $\overline{g}(y) := 1 - |y - 1|$ which is constructed from g in the simplest way by just joining the three points $(0, 0)$, $(1, 1)$ and $(2, 0)$ by straight lines. Clearly, this function is Lipschitz continuous with minimal Lipschitz constant 1 and indeed satisfies $\overline{g} \circ \tau = g \circ \tau = \chi_{\{1\}}$. ♥

A parallel result to Theorem 1.28 with $g \in Lip([a, b])$ replaced by $g \in Lip_\alpha([a, b])$ will be given in Theorem 1.41 in the next section.

The following result is, in a certain sense, sharper than Theorem 1.28: it shows that *continuous* functions of bounded variation may be "made" differentiable with bounded derivative (hence, Lipschitz continuous) after a suitable homeomorphic change of variables ([63], see also [64] and [130, Section 3.1]).

Proposition 1.30. *For a function $g : [a,b] \to \mathbb{R}$, the following three statements are equivalent.*
(a) *The function g is continuous and of bounded variation.*
(b) *There exists a homeomorphism $\tau : [a,b] \to [a,b]$ such that $g \circ \tau : [a,b] \to \mathbb{R}$ is Lipschitz continuous on $[a,b]$.*
(c) *There exists a homeomorphism $\tau : [a,b] \to [a,b]$ such that $g \circ \tau : [a,b] \to \mathbb{R}$ is differentiable with a bounded derivative on $[a,b]$.*

Proof. Without loss of generality, we take $[a,b] = [0,1]$. Suppose first that $g \in C([0,1]) \cap BV([0,1])$ and put $\mathrm{Var}(g;[0,1]) =: \omega$. To prove (b), we define $\sigma : [0,1] \to [0, 1+\omega]$ by

$$\sigma(x) := x + V_g(x) \quad (0 \le x \le 1),$$

where V_g denotes the variation function (1.13) of g. Clearly, σ is strictly increasing and satisfies $\sigma(0) = 0$, $\sigma(1) = 1 + \omega$, and

$$|g(x) - g(y)| \le |V_g(x) - V_g(y)| \le |\sigma(x) - \sigma(y)| \tag{1.58}$$

for all $x, y \in [0,1]$. So, the map $\tau : [0,1] \to [0,1]$ defined by

$$\tau(t) := \sigma^{-1}(t + \omega t) \quad (0 \le t \le 1) \tag{1.59}$$

is strictly increasing and surjective with $\tau(0) = 0$ and $\tau(1) = 1$, and thus a homeomorphism. Moreover, from (1.58), it follows that the map $f := g \circ \tau$ satisfies

$$|f(s) - f(t)| \le |V_g(\tau(s)) - V_g(\tau(t))| \le |\sigma(\tau(s)) - \sigma(\tau(t))| \le (1+\omega)|s - t|$$

for all $s, t \in [0,1]$. This shows that $f \in Lip([0,1])$ with $lip(g) \le 1 + \omega$, and so we have proved (b).

Now, we suppose that (b) holds and prove (c). We may assume that g itself is Lipschitz continuous on $[0,1]$. In particular, the set of all $t \in [0,1]$ at which $g'(t)$ does not exist is then a nullset. Choose a G_δ-set $G \supseteq N$ which is also a nullset. By Zahorski's theorem [325, 326], we may find a homeomorphism $\tau : [0,1] \to [0,1]$ which is differentiable with bounded derivative τ' on $[0,1]$ and satisfies $\tau'(t) = 0$ precisely for $t \in \tau^{-1}(G)$.

Now, again putting $f := g \circ \tau$ and taking $t \ne t_0$, the estimate

$$\left|\frac{f(t) - f(t_0)}{t - t_0}\right| = \left|\frac{f(t) - f(t_0)}{\tau(t) - \tau(t_0)}\right| \left|\frac{\tau(t) - \tau(t_0)}{t - t_0}\right| \le lip(g) \left|\frac{\tau(t) - \tau(t_0)}{t - t_0}\right|$$

shows that $f'(t_0) = 0$ whenever $\tau(t_0) \in G$. On the other hand, in case $\tau(t_0) \notin G$, the function g is differentiable at t_0 with

$$f'(t_0) = g'(\tau(t_0))\tau'(t_0), \quad |f'(t)| \le lip(g)\|\tau'\|_\infty .$$

In either case, f is differentiable with bounded derivative. Thus, we have proved that (b) implies (c).

The fact that (c) implies (a) follows from Proposition 1.12 since every differentiable map with bounded derivative is Lipschitz continuous, and hence of bounded variation, and every homeomorphism of an interval onto itself is strictly monotone. □

Proposition 1.30 (c) gives the best possible change of variables in the following sense: there exist functions $f \in C([a,b]) \cap BV([a,b])$ such that $f \circ \tau$ is not *continuously differentiable* for any homeomorphism $\tau : [a,b] \to [a,b]$, see [130].

1.3 Functions of bounded Wiener variation

Now, we consider a certain extension of the spaces $BV([a,b])$ which was introduced in 1924 by Wiener [321].

Definition 1.31. Given a real number $p \geq 1$, a partition $P = \{t_0, t_1, \ldots, t_m\} \in \mathcal{P}([a,b])$, and a function $f : [a,b] \to \mathbb{R}$, the nonnegative real number

$$\mathrm{Var}_p^W(f, P) = \mathrm{Var}_p^W(f, P; [a,b]) := \sum_{j=1}^{m} |f(t_j) - f(t_{j-1})|^p \tag{1.60}$$

is called the *Wiener variation* of f on $[a,b]$ with respect to P, while the (possibly infinite) number

$$\mathrm{Var}_p^W(f) = \mathrm{Var}_p^W(f; [a,b]) := \sup \left\{ \mathrm{Var}_p^W(f, P; [a,b]) : P \in \mathcal{P}([a,b]) \right\}, \tag{1.61}$$

where the supremum is taken over all partitions of $[a,b]$, is called the *total Wiener variation* of f on $[a,b]$. In case $\mathrm{Var}_p^W(f; [a,b]) < \infty$, we say that f has *finite Wiener variation* on $[a,b]$ and write[15] $f \in WBV_p([a,b])$.

It is useful to complete this definition by defining $WBV_\infty([a,b]) := R([a,b])$, that is, the space of regular functions (Section 0.3). As before, we will consider this space endowed with the norm (0.39). ■

In the following Proposition 1.32, which, to some extent, is parallel to Proposition 1.3, we collect some properties of the quantities (1.60) and (1.61).

Proposition 1.32. *The quantities* (1.60) *and* (1.61) *have the following properties.*

(a) *The p-th root of the variation* (1.61) *is subadditive with respect to functions, i.e.*

$$\mathrm{Var}_p^W(f + g; [a,b])^{1/p} \leq \mathrm{Var}_p^W(f; [a,b])^{1/p} + \mathrm{Var}_p^W(g; [a,b])^{1/p} \tag{1.62}$$

for $f, g : [a,b] \to \mathbb{R}$.

15 In the literature, this space is often denoted by $BV_p([a,b])$. In our notation, the letter W stands for "Wiener," to distinguish this space from another space which we will consider in Definition 2.50 in Chapter 2. Functions $f \in WBV_p([a,b])$ are sometimes said to have *finite p-variation in Wiener's sense*.

(b) *The p-th root of the variation* (1.61) *is homogeneous with respect to functions, i.e.*

$$\operatorname{Var}_p^W(\mu f; [a,b])^{1/p} = |\mu| \operatorname{Var}_p^W(f; [a,b])^{1/p} \tag{1.63}$$

for $\mu \in \mathbb{R}$.

(c) *The estimate*

$$|f(s) - f(t)| \leq \operatorname{Var}_p^W(f; [s,t])^{1/p} \tag{1.64}$$

holds for $a \leq s < t \leq b$.

(d) *Every function* $f \in WBV_p([a,b])$ *is bounded with*

$$\|f\|_\infty \leq |f(a)| + \operatorname{Var}_p^W(f; [a,b])^{1//p},$$

where the norm $\|\cdot\|_\infty$ *is given by* (0.39).

(e) *The linear space* $WBV_p([a,b])$ *equipped with the norm*

$$\|f\|_{WBV_p} := |f(a)| + \operatorname{Var}_p^W(f; [a,b])^{1/p} \tag{1.65}$$

is a Banach space.

Proof. The properties (a)–(d) are proved in exactly the same way as in Proposition 1.3, where, in the proof of (1.62), we use the Minkowski inequality.

From (a) and (b), it follows that $WBV_p([a,b])$ is a linear space. Moreover, it is not hard to see that (1.65) defines, in fact, a norm on $WBV_p([a,b])$. Let $(f_n)_n$ be a Cauchy sequence with respect to this norm. By property (d), $(f_n)_n$ is then also a Cauchy sequence with respect to the norm (0.39), and so there exists a bounded function $f : [a,b] \to \mathbb{R}$ such that $(f_n)_n$ converges uniformly on $[a,b]$ to f.

Let $\varepsilon > 0$, and let $P = \{t_0, t_1, \ldots, t_m\} \in \mathcal{P}([a,b])$. If $\|f_i - f_j\|_{WBV_p} \leq \varepsilon$ for sufficiently large i and j, then

$$\begin{aligned}\operatorname{Var}_p^W(f_i - f, P; [a,b]) &= \sum_{k=1}^m |(f_i - f)(t_k) - (f_i - f)(t_{k-1})|^p \\ &= \lim_{j\to\infty} \sum_{k=1}^m |(f_i - f_j)(t_k) - (f_i - f_j)(t_{k-1})|^p \leq \varepsilon^p,\end{aligned}$$

which together with the trivial fact $f_n(a) \to f(a)$, implies (e). □

One might ask why in Proposition 1.32 we did not state analogues to properties (f) (monotonicity with respect to partitions) and (g) (additivity with respect to intervals) of Proposition 1.3. The simple reason is that such analogues are not true, as the following example shows.

Example 1.33. Consider the function $f : [0,2] \to \mathbb{R}$ defined in Example 1.29, i.e.

$$f(x) = \begin{cases} 0 & \text{for } 0 \leq x < 1, \\ 1 & \text{for } x = 1, \\ 2 & \text{for } 1 < x \leq 2, \end{cases} \tag{1.66}$$

and consider the partitions $P := \{0, 2\}$ and $Q := \{0, 1, 2\} \supset P$. An easy calculation then shows that

$$\mathrm{Var}_p^W(f, P; [0, 2]) = 2^p, \quad \mathrm{Var}_p^W(f, Q; [0, 2]) = 2\,,$$

which shows that, in case $p > 1$, $P \subseteq Q$ does not imply $\mathrm{Var}_p^W(f, P; [a, b]) \le \mathrm{Var}_p^W(f, Q; [a, b])$, so property (f) of Proposition 1.3 has no analogue. To prove that property (g) of Proposition 1.3 fails, we consider the Wiener variation of the function (1.66) on the intervals $[0, 2]$, $[0, 1]$, and $[1, 2]$. Similarly, a straightforward computation shows that

$$\mathrm{Var}_p^W(f; [0, 1]) = \mathrm{Var}_p^W(f; [1, 2]) = 1\,,$$

but

$$\mathrm{Var}_p^W(f; [0, 2]) = 2^p\,,$$

which shows that the Wiener variation is not additive with respect to intervals if $p > 1$, and so Proposition 1.3 (g) has no analogue either. ♥

We remark that for any $p \in [1, \infty)$, the Wiener variation is *superadditive* with respect to intervals in the sense that

$$\mathrm{Var}_p^W(f; [a, b]) \ge \mathrm{Var}_p^W(f; [a, c]) + \mathrm{Var}_p^W(f; [c, b]) \tag{1.67}$$

for $a < c < b$. We will prove this in a more general context in Proposition 2.10 (a) in Chapter 2.

A comparison of Definition 1.31 with Definition 1.1 shows that

$$\mathrm{Var}_1^W(f, P; [a, b]) = \mathrm{Var}(f, P; [a, b]), \quad \mathrm{Var}_1^W(f; [a, b]) = \mathrm{Var}(f; [a, b])\,,$$

and so $WBV_1([a, b]) = BV([a, b])$. Moreover, the inclusion $Lip([a, b]) \subseteq BV([a, b])$ stated in (1.46) admits the following refinement.

Proposition 1.34. *For $1 \le p < \infty$, the inclusion*

$$Lip_{1/p}([a, b]) \subseteq WBV_p([a, b]) \tag{1.68}$$

holds.

Proof. The proof follows immediately from the estimate

$$\sum_{j=1}^{m} |f(t_j) - f(t_{j-1})|^p \le L^p \sum_{j=1}^{m} |t_j - t_{j-1}|^{p\alpha} = L^p \sum_{j=1}^{m} |t_j - t_{j-1}| = L^p(b - a)$$

for any function f which satisfies (0.67) with $\alpha = 1/p$. □

The proof of Proposition 1.34 shows that the Banach space $(Lip_{1/p}([a, b]), \|\cdot\|_{Lip_\alpha})$ is continuously imbedded into the Banach space $(WBV_p([a, b]), \|\cdot\|_{WBV_p})$ with imbedding constant $\max\{1, (b - a)^{1/p}\}$; compare this with (1.48).

One could ask whether or not the space $WBV_p([a,b])$ is contained in the space $WBV_q([a,b])$ for certain values of p and q. This is in fact true. Before proving this, we need a series of technical results about convex functions.

Let $\phi : [0,\infty) \to \mathbb{R}$ be a given function, and consider the function $\psi : (0,\infty) \to \mathbb{R}$ defined by

$$\psi(t) := \frac{\phi(t)}{t} \quad (t > 0). \tag{1.69}$$

Lemma 1.35. *If ϕ is convex on $[0,\infty)$ and $\phi(0) \le 0$, then the function (1.69) is increasing on $(0,\infty)$.*

Proof. Fix arbitrarily $x_1, x_2 \in [0,\infty)$ with $0 < x_1 < x_2$. Since ϕ is convex, we have

$$\phi(\lambda s + (1-\lambda)t) \le \lambda\phi(s) + (1-\lambda)\phi(t)$$

for $s, t \in [0,\infty)$ and $\lambda \in [0,1]$. Putting in this inequality $s = x_2$, $t = 0$, and $\lambda = x_1/x_2$, and taking into account that $\phi(0) \le 0$, we get

$$\phi(x_1) \le \lambda\phi(x_2) + (1-\lambda)\phi(0) \le \lambda\phi(x_2) = \frac{x_1}{x_2}\phi(x_2)\,.$$

Consequently,

$$\psi(x_1) = \frac{\phi(x_1)}{x_1} \le \frac{\phi(x_2)}{x_2} = \psi(x_2)$$

as claimed. □

In the next lemma, we use the concept of the Young function introduced in Definition 0.16.

Lemma 1.36. *Assume that $\phi : [0,\infty) \to [0,\infty)$ is a Young function. Then ϕ is increasing and superadditive on $[0,\infty)$, i.e.*

$$\phi(\alpha) + \phi(\beta) \le \phi(\alpha+\beta) \quad (\alpha, \beta \ge 0)\,. \tag{1.70}$$

Proof. Suppose that ϕ is not increasing; then, there exist points $x_1, x_2 \in [0,\infty)$ such that $x_1 < x_2$ and $\phi(x_1) > \phi(x_2)$. If $x_1 = 0$, then in view of our assumptions, we have $\phi(x_1) = \phi(0) = 0 < \phi(x_2)$. Thus, we may assume that $0 < x_1 < x_2$, and hence

$$\frac{1}{x_1} > \frac{1}{x_2}\,.$$

Combining this with our assumption $\phi(x_1) > \phi(x_2) > 0$, we obtain

$$\frac{\phi(x_1)}{x_1} \ge \frac{\phi(x_2)}{x_1} > \frac{\phi(x_2)}{x_2}\,.$$

However, this contradicts the fact that the function ψ defined in (1.69) is increasing on $(0,\infty)$, as we have proved in Lemma 1.35.

Now, we prove that ϕ is superadditive on $[0, \infty)$. Fix $\alpha, \beta \in [0, \infty)$. If $\alpha = 0$, then from our assumptions, we have $\phi(0) = 0$, and so

$$\phi(\alpha) + \phi(\beta) = \phi(\beta) = \phi(0 + \beta) = \phi(\alpha + \beta).$$

The case $\beta = 0$ may be treated analogously. Therefore, we may suppose without loss of generality that $\alpha > 0$ and $\beta > 0$, and hence $\alpha < \alpha + \beta$ and $\beta < \alpha + \beta$. By what we just proved, we then get

$$\frac{\phi(\alpha)}{\alpha} \le \frac{\phi(\alpha+\beta)}{\alpha+\beta}, \quad \frac{\phi(\beta)}{\beta} \le \frac{\phi(\alpha+\beta)}{\alpha+\beta},$$

and hence

$$\phi(\alpha) \le \frac{\alpha}{\alpha+\beta}\phi(\alpha+\beta), \quad \phi(\beta) \le \frac{\beta}{\alpha+\beta}\phi(\alpha+\beta).$$

Adding up these inequalities, we obtain (1.70). □

Example 1.37. The simplest example of a Young function is of course $\phi(t) = t^p$ for $p \ge 1$. The superadditivity condition (1.70) here simply reads

$$s^p + t^p \le (s+t)^p \quad (s, t \ge 0),$$

which is of course true if (and only if) $p \ge 1$. ♥

Now, we are in a position to establish a relation between WBV_p and WBV_q for suitable values of p and q.

Proposition 1.38. *Let $1 \le p \le q < \infty$. Then the inequality*

$$\mathrm{Var}_q^W(f; [a,b])^{1/q} \le \mathrm{Var}_p^W(f; [a,b])^{1/p} \tag{1.71}$$

holds. Consequently,

$$BV([a,b]) = WBV_1([a,b]) \subseteq WBV_p([a,b]) \subseteq WBV_q([a,b]) \subseteq B([a,b]) \tag{1.72}$$

for these values of p and q.

Proof. Fix an arbitrary partition $P = \{t_0, t_1, \ldots, t_m\} \in \mathcal{P}([a,b])$. Taking $\phi(t) := t^{q/p}$ in view of Lemma 1.36 and Example 1.37, we obtain

$$\begin{aligned}\sum_{j=1}^m |f(t_j) - f(t_{j-1})|^q &= \sum_{j=1}^m \left(|f(t_j) - f(t_{j-1})|^p\right)^{q/p} \\ &\le \left(\sum_{j=1}^m |f(t_j) - f(t_{j-1})|^p\right)^{q/p} \le \mathrm{Var}_p^W(f; [a,b])^{q/p},\end{aligned}$$

which proves (1.71). The inclusion (1.72) is of course an immediate consequence of (1.71). □

Choosing, in particular, $p = 1$ in (1.72), we see that $BV([a,b]) \subseteq WBV_q([a,b])$ for all $q \geq 1$. Moreover, (1.71) is

$$\mathrm{Var}_q^W(f;[a,b]) \leq \mathrm{Var}(f;[a,b])^q \quad (f \in BV([a,b])), \tag{1.73}$$

which may be proved also directly from Hölder's inequality and shows that the space $(BV([a,b]), \|\cdot\|_{BV})$ is imbedded into the space $(WBV_q([a,b]), \|\cdot\|_{WBV_q})$ for $q \geq 1$. If we admit $q = \infty$ in the sense of Definition 1.31, then the inclusion (1.72) still holds true since every function in $WBV_p([a,b])$ is regular.

At this point, let us collect some of the function classes, together with relations between them, in the following Table 1.1 where $1 < p < \infty$.

Table 1.1. Relations between function classes over $I = [a,b]$.

$Lip_{1/p}(I)$	$\subset$	$WBV_p(I)$	$\subset$	$R(I)$	$\supset$	$BV(I)$
$\cup$						$\cup$
$Lip(I)$	$\subset$	$AC(I)$	$\subset$	$BV(I) \cap C(I)$	$\subset$	$CBV(I)$
		$\cap$		$\cup$		
		$BV(I) \cap IVP(I)$	$\subset$	$C(I)$	$\subset$	$IVP(I)$

Later (see Table 2.6 in Chapter 2), we will give an enlarged version of Table 1.1 and show that all inclusions are strict.

Let us mention that the inclusion $WBV_p \subseteq WBV_q$ proved in Proposition 1.38 is also strict in case $p < q$. To show this, we consider the zigzag functions $Z_{C,D}$ introduced in Definition 0.49. It follows from the construction that

$$\mathrm{Var}_p^W(Z_{C,D};[0,1]) = \sum_{k=1}^{\infty} d_k^p \quad (1 \leq p < \infty). \tag{1.74}$$

Observe that the Jordan variation and Wiener variation of $Z_{C,D}$ is independent of the sequence C. For the particularly simple example of the special zigzag function (0.93), we get

$$\mathrm{Var}_p^W(Z_\theta;[0,1]) = \sum_{k=1}^{\infty} \frac{1}{k^{p\theta}} \quad (1 \leq p < \infty). \tag{1.75}$$

This simple observation allows us to show that $WBV_p([a,b]) \subset WBV_q([a,b])$ for $p < q$, and even more:

Example 1.39. For $p \geq 1$, consider the function $f(x) := Z_{1/p}(x)$. From (1.75), it then follows that $f \in WBV_q([0,1]) \setminus WBV_p([0,1])$ for any $q > p$. However, we can actually do better: the same function of course satisfies

$$f \in \bigcap_{q>p} WBV_q([0,1]) \setminus WBV_p([0,1]).$$

In particular, the special zigzag function Z_1 belongs to $WBV_p([0,1])$ for all $p > 1$, but not to $BV([0,1])$. ♥

We may use the zigzag function Z_θ as well to show that the inclusion (1.68) in Proposition 1.34 is strict.

Example 1.40. Choosing $\theta > 1/p$ arbitrary, by (1.75), we conclude that $Z_\theta \in WBV_p([0,1])$. On the other hand, we have already seen in Corollary 0.51 that no zigzag function Z_θ belongs to the Hölder space $Lip_{1/p}([0,1])$.

Alternatively, we could have used the function f from Example 0.41 which belongs to $WBV_p([0,1])$ for any $p \geq 1$, but not to $Lip_\alpha([0,1])$ for any $\alpha \leq 1$. ♥

We now consider the Federer decomposition of functions of bounded Wiener variation. The following is parallel to Theorem 1.28 and may be found in [65] without proof. Yet another generalization of this will be given in Exercise 2.3.

Theorem 1.41. *A function f belongs to $WBV_p([a,b])$ if and only if it may be represented as composition $f = g \circ \tau$, where $\tau : [a,b] \to [c,d]$ is increasing and $g \in Lip_{1/p}([c,d])$ with Hölder constant $L = 1$.*

Proof. The proof is very similar to that of Theorem 1.28, and thus we only sketch the idea. Suppose that $f = g \circ \tau$, where g and τ have the mentioned properties. Given any partition $P = \{t_0, t_1, \ldots, t_m\} \in \mathcal{P}([a,b])$, we get

$$\begin{aligned}\mathrm{Var}_p^W(f,P) &= \sum_{j=1}^m |g(\tau(t_j)) - g(\tau(t_{j-1}))|^p \\ &\leq \sum_{j=1}^m |\tau(t_j) - \tau(t_{j-1})| = |\tau(b) - \tau(a)|\,,\end{aligned}$$

and hence $f \in BV_p^W([a,b])$. Conversely, let $f \in BV([a,b])$ and let $\tau(t) := V_{f,p}(t)$, where $V_{f,p}$ denotes the variation function

$$V_{f,p}(x) := \mathrm{Var}_p^W(f;[a,x]) \quad (a \leq x \leq b)\,.$$

Then τ maps $[a,b]$ into $[c,d]$, where $c = 0$ and $d = \mathrm{Var}_p^W(f;[a,b])$. If we define, as in Theorem 1.28, the function g on the range $\tau([a,b]) \subseteq [c,d]$ by putting $g(\tau(x)) := f(x)$, then the decomposition $f = g \circ \tau$ holds trivially by construction and

$$|g(\tau(s)) - g(\tau(t))| = |f(s) - f(t)| \leq \mathrm{Var}_p^W(f;[s,t])^{1/p} \leq |\tau(s) - \tau(t)|^{1/p}$$

for $a \leq s < t \leq b$, by (1.64), which shows that g is Hölder continuous with exponent $\alpha = 1/p$ and Hölder constant 1 on $\tau([a,b])$. If we extend g now from $\tau([a,b])$ to $[c,d]$ by using the McShane extension (0.76), we get a function $\hat{g} \in Lip_{1/p}([c,d])$ with Hölder constant 1 such that $f = \hat{g} \circ \tau = g \circ \tau$ as claimed. □

Of course, Theorem 1.28 is contained in Theorem 1.41 in the special case $p = 1$. Also, observe the similarity of these theorems with the Sierpiński decomposition of regular functions stated in Theorem 0.36. We collect the decomposition theorems proved so far in the following Table 1.2; here, the function τ is supposed to map the interval $[a,b]$ into itself.

Table 1.2. Decomposition of functions in the spaces R, BV, and WBV_p.

$g\circ\tau\in R([a,b])$	iff	$g\in C([a,b])$	and	τ strictly monotone
$g\circ\tau\in BV([a,b])$	iff	$g\in Lip([a,b])$	and	τ monotone
$g\circ\tau\in BV([a,b])$	iff	$g\in BV([a,b])$	and	τ pseudomonotone
$g\circ\tau\in WBV_p([a,b])$	iff	$g\in Lip_{1/p}([a,b])$	and	τ monotone

Table 1.2 shows the subtle, though not surprising, "equilibrium" between the functions τ and g: weak properties of g have to be "compensated" by strong properties of τ, and vice versa.

1.4 Functions of several variables

All functions considered so far were defined on some interval $[a,b]$. In view of applications, however, functions which are defined on the Cartesian product of intervals in higher dimensional Euclidean space are also important. In this section, we discuss several types of variation for such functions and see to what extent our results carry over to higher dimensional domains of definition. For simplicity, we restrict ourselves to rectangles $[a,b]\times[c,d]$ in the xy-plane; the generalization to three or more dimensions is straightforward.

The most natural approach goes as follows. Given a function $f:[a,b]\times[c,d]\to\mathbb{R}$ and partitions $P=\{s_0,s_1,\dots,s_m\}\in\mathcal{P}([a,b])$ and $Q=\{t_0,t_1,\dots,t_n\}\in\mathcal{P}([c,d])$, consider the three expressions

$$\mathrm{Var}(f(\cdot,c),P;[a,b]) := \sum_{i=1}^{m}|f(s_i,c)-f(s_{i-1},c)|\,, \tag{1.76}$$

$$\mathrm{Var}(f(a,\cdot),Q;[c,d]) := \sum_{j=1}^{n}|f(a,t_j)-f(a,t_{j-1})|\,, \tag{1.77}$$

and

$$\begin{aligned}&\mathrm{V}_2(f,P\times Q;[a,b]\times[c,d])\\ &:= \sum_{i=1}^{m}\sum_{j=1}^{n}|f(s_i,t_j)-f(s_{i-1},t_j)-f(s_i,t_{j-1})+f(s_{i-1},t_{j-1})|\,.\end{aligned} \tag{1.78}$$

Moreover, in analogy to (1.4),

$$\mathrm{Var}(f(\cdot,c);[a,b]) := \sup\{\mathrm{Var}(f(\cdot,c),P;[a,b]) : P\in\mathcal{P}([a,b])\}\,, \tag{1.79}$$

$$\mathrm{Var}(f(a,\cdot);[c,d]) := \sup\{\mathrm{Var}(f(a,\cdot),Q;[c,d]) : Q\in\mathcal{P}([c,d])\}\,, \tag{1.80}$$

and

$$\begin{aligned} &\mathrm{V}_2(f;[a,b]\times[c,d]) \\ &\quad := \sup\{\mathrm{V}_2(f,P\times Q;[a,b]\times[c,d]) : P\in\mathcal{P}([a,b]), Q\in\mathcal{P}([c,d])\}, \end{aligned} \tag{1.81}$$

where all suprema are taken over the indicated partitions.

Definition 1.42. With the above notation, we call the (possibly infinite) number

$$\begin{aligned} &\mathrm{Var}(f;[a,b]\times[c,d]) \\ &\quad := \mathrm{Var}(f(\cdot,c);[a,b]) + \mathrm{Var}(f(a,\cdot);[c,d]) + \mathrm{V}_2(f;[a,b]\times[c,d]) \end{aligned} \tag{1.82}$$

the *total variation of f on $[a,b]\times[c,d]$*. In case $\mathrm{Var}(f;[a,b]\times[c,d]) < \infty$, we say that *$f$ has bounded variation on $[a,b]\times[c,d]$* and write $f\in BV([a,b]\times[c,d])$. ■

We now start with a list of properties of the two-dimensional variation (1.82); the following is parallel to (some part of) Proposition 1.3.

Proposition 1.43. *The quantity* (1.82) *has the following properties.*
(a) *The variation* (1.82) *is subadditive with respect to functions, i.e.*

$$\mathrm{Var}(f+g;[a,b]\times[c,d]) \le \mathrm{Var}(f;[a,b]\times[c,d]) + \mathrm{Var}(g;[a,b]\times[c,d]) \tag{1.83}$$

for $f,g:[a,b]\times[c,d]\to\mathbb{R}$.
(b) *The variation* (1.82) *is homogeneous with respect to functions, i.e.*

$$\mathrm{Var}(\mu f;[a,b]\times[c,d]) = |\mu|\,\mathrm{Var}(f;[a,b]\times[c,d]) \tag{1.84}$$

for $\mu\in\mathbb{R}$.
(c) *The estimate*

$$|f(x,y)-f(\xi,\eta)| \le \mathrm{Var}(f;[\xi,x]\times[\eta,y]) \tag{1.85}$$

holds for $a\le\xi<x\le b$ and $c\le\eta<y\le d$.
(d) *Every function $f\in BV([a,b]\times[c,d])$ is bounded with*

$$\|f\|_\infty \le |f(a,c)| + \mathrm{Var}(f;[a,b]\times[c,d])\,, \tag{1.86}$$

where

$$\|f\|_\infty := \sup\{|f(x,y)| : a\le x\le b,\ c\le y\le d\}\,. \tag{1.87}$$

Proof. The assertions (a) and (b) are obvious. To prove (c), fix $a\le\xi<x\le b$ and $c\le\eta<y\le d$. Choosing $P:=\{\xi,x\}\in\mathcal{P}([\xi,x])$ and $Q:=\{\eta,y\}\in\mathcal{P}([\eta,y])$, we then have

$$\begin{aligned} &\mathrm{Var}(f(\cdot,\eta),P;[\xi,x]) \ge |f(x,\eta)-f(\xi,\eta)|\,, \\ &\mathrm{Var}(f(\xi,\cdot),Q;[\eta,y]) \ge |f(\xi,y)-f(\xi,\eta)| \end{aligned}$$

as well as

$$V_2(f, P \times Q; [a,b] \times [c,d]) \geq |f(x,y) - f(\xi,y) - f(x,\eta) + f(\xi,\eta)|\,.$$

Consequently,

$$\begin{aligned}
&|f(x,y) - f(\xi,\eta)| \\
&\quad = |f(x,\eta) - f(\xi,\eta) + f(\xi,y) - f(\xi,\eta) + f(x,y) - f(\xi,y) - f(x,\eta) + f(\xi,\eta)| \\
&\quad < |f(x,\eta) - f(\xi,\eta)| + |f(\xi,y) - f(\xi,\eta)| + |f(x,y) - f(\xi,y) - f(x,\eta) + f(\xi,\eta)| \\
&\quad \leq \mathrm{Var}(f(\cdot,\eta), P; [\xi,x]) + \mathrm{Var}(f(\xi,\cdot), Q; [\eta,y]) + V_2(f, P \times Q; [a,b] \times [c,d]) \\
&\quad = \mathrm{Var}(f, [\xi,x] \times [\eta,y])\,.
\end{aligned}$$

Choosing, in particular, $(\xi,\eta) := (a,c)$ in (c) yields

$$|f(x,y)| \leq |f(a,c)| + \mathrm{Var}(f, [a,x] \times [c,y]) \leq |f(a,c)| + \mathrm{Var}(f, [a,b] \times [c,d])$$

which proves (d) after passing to the supremum over $[a,b] \times [c,d]$. □

We know from Proposition 1.3 (g) that the one-dimensional variations (1.79) and (1.80) are additive with respect to intervals. A similar additivity property holds for the two-dimensional variation (1.81): For $a < e < b$ and $c < f < d$, we have

$$\begin{aligned}
V_2(f; [a,b] \times [c,d]) &= V_2(f; [a,e] \times [c,f]) + V_2(f; [e,b] \times [c,f]) \\
&\quad + V_2(f; [a,e] \times [f,d]) + V_2(f; [e,b] \times [f,d])\,.
\end{aligned} \tag{1.88}$$

Proposition 1.43 (d) suggests that one equip the space $BV([a,b] \times [c,d])$ with the norm

$$\|f\|_{BV} := |f(a,c)| + \mathrm{Var}(f, [a,b] \times [c,d])\,. \tag{1.89}$$

In fact, (1.86) shows that with this norm on $BV([a,b] \times [c,d])$, the continuous imbedding $BV \hookrightarrow B$ holds with imbedding constant 1.

Proposition 1.44. *The space $BV([a,b]\times[c,d])$ equipped with the norm (1.89) is a Banach algebra satisfying*

$$\|fg\|_{BV} \leq 4\|f\|_{BV}\|g\|_{BV} \tag{1.90}$$

for $f, g \in BV([a,b] \times [c,d])$.

Proof. From Proposition 1.43 (a) and (b), it follows that $(BV([a,b] \times [c,d]), \|\cdot\|_{BV})$ is a linear space; its completeness is proved exactly as in Proposition 1.10.

To prove (1.90), recall that

$$\begin{aligned}
\|fg\|_{BV} &= |f(a,c)g(a,c)| + \mathrm{Var}(f(\cdot,c)g(\cdot,c); [a,b]) \\
&\quad + \mathrm{Var}(f(a,\cdot)g(a,\cdot); [c,d]) + V_2(fg; [a,b] \times [c,d])\,.
\end{aligned} \tag{1.91}$$

We estimate the second, third, and fourth term on the right-hand side of (1.91) separately. By (1.17) and (1.11), the second term may be estimated by

$$
\begin{aligned}
&\mathrm{Var}(f(\cdot,c)g(\cdot,c);[a,b]) \\
&\quad\le \mathrm{Var}(g(\cdot,c);[a,b]) \sup_{a\le x\le b} |f(x,c)| + \mathrm{Var}(f(\cdot,c);[a,b]) \sup_{a\le x\le b} |g(x,c)| \\
&\quad\le \mathrm{Var}(g(\cdot,c);[a,b]) \{|f(a,c)| + \mathrm{Var}(f(\cdot,c);[a,b])\} \\
&\qquad + \mathrm{Var}(f(\cdot,c);[a,b]) \{|g(a,c)| + \mathrm{Var}(g(\cdot,c);[a,b])\} \\
&\quad= |f(a,c)| \,\mathrm{Var}(g(\cdot,c);[a,b]) + |g(a,c)| \,\mathrm{Var}(f(\cdot,c);[a,b]) \\
&\qquad + 2\,\mathrm{Var}(f(\cdot,c);[a,b]) \,\mathrm{Var}(g(\cdot,c);[a,b]).
\end{aligned}
\tag{1.92}
$$

Similarly, for the third term, we get

$$
\begin{aligned}
&\mathrm{Var}(f(a,\cdot)g(a,\cdot);[c,d]) \\
&\quad\le |f(a,c)| \,\mathrm{Var}(g(a,\cdot);[c,d]) + |g(a,c)| \,\mathrm{Var}(f(a,\cdot);[c,d]) \\
&\qquad + 2\,\mathrm{Var}(f(a,\cdot);[c,d]) \,\mathrm{Var}(g(a,\cdot);[c,d]).
\end{aligned}
\tag{1.93}
$$

Estimating the fourth term in (1.91) is harder. Fix $P = \{s_0, s_1, \dots, s_m\} \in \mathcal{P}([a,b])$ and $Q = \{t_0, t_1, \dots, t_n\} \in \mathcal{P}([c,d])$. Then for $i = 1, 2, \dots, m$ and $j = 1, 2, \dots, n$, we obtain

$$
\begin{aligned}
&(fg)(s_{i-1},t_{j-1}) + (fg)(s_i,t_j) - (fg)(s_{i-1},t_j) - (fg)(s_i,t_{j-1}) \\
&\quad= \left[f(s_{i-1},t_{j-1}) + f(s_i,t_j) - f(s_{i-1},t_j) - f(s_i,t_{j-1})\right] g(s_{i-1},t_{j-1}) \\
&\qquad + f(s_i,t_j)\left[g(s_{i-1},t_{j-1}) + g(s_i,t_j) - g(s_{i-1},t_j) - g(s_i,t_{j-1})\right] \\
&\qquad + \left[f(a,t_j) - f(a,t_{j-1})\right]\left[g(s_i,c) - g(s_{i-1},c)\right] \\
&\qquad + \left[f(a,t_j) - f(a,t_{j-1})\right]\left[g(s_{i-1},c) + g(s_i,t_{j-1}) - g(s_{i-1},t_{j-1}) - g(s_i,c)\right] \\
&\qquad + \left[f(a,t_{j-1}) + f(s_i,t_j) - f(a,t_j) - f(s_i,t_{j-1})\right]\left[g(s_i,c) - g(s_{i-1},c)\right] \\
&\qquad + \left[f(a,t_{j-1}) + f(s_i,t_j) - f(a,t_j) - f(s_i,t_{j-1})\right] \\
&\qquad\quad \times \left[g(s_{i-1},c) + g(s_i,t_{j-1}) - g(s_{i-1},t_{j-1}) - g(s_i,c)\right] \\
&\qquad + \left[f(s_i,c)) - f(s_{i-1},c)\right]\left[g(a,t_j) - g(a,t_{j-1})\right] \\
&\qquad + \left[f(s_i,c) + f(s_{i-1},c)\right]\left[g(a,s_{j-1}) + g(s_{i-1},t_j) - g(a,t_j) - g(s_{i-1},t_{j-1})\right] \\
&\qquad + \left[f(s_{i-1},c)) + f(s_i,t_j) - f(s_{i-1},t_j) - f(s_i,c))\right]\left[g(a,t_j) - g(a,t_{j-1})\right] \\
&\qquad + \left[f(s_{i-1},c) + f(s_i,t_j) - f(s_{i-1},t_j) - f(s_i,c)\right] \\
&\qquad\quad \times \left[g(a,t_{j-1}) + g(s_{i-1},t_j) - g(a,t_j) - g(s_{i-1},t_{j-1})\right] \\
&\quad=: A_{ij} + B_{ij} + C_{ij} + D_{ij} + E_{ij} + F_{ij} + G_{ij} + H_{ij} + I_{ij} + J_{ij}\,.
\end{aligned}
$$

We estimate the 10 terms $A_{ij}, \dots, J_{ij}$ separately, summing over $i = 1, \dots, m$ and $j = 1, \dots, n$. By (1.87), we have

$$
\begin{aligned}
\sum_{i=1}^{m}\sum_{j=1}^{n} |A_{ij}| &= \sum_{i=1}^{m}\sum_{j=1}^{n} \left|f(a,t_j) - f(a,t_{j-1})\right| |g(s_i,c) - g(s_{i-1},c)| \\
&\le \mathrm{V}_2(f;[a,b]\times[c,d])\|g\|_\infty
\end{aligned}
$$

and

$$\sum_{i=1}^{m}\sum_{j=1}^{n}|B_{ij}|$$
$$= \sum_{i=1}^{m}\sum_{j=1}^{n} f(s_i,t_j)\left|g(s_{i-1},t_{j-1}) + g(s_i,t_j) - g(s_{i-1},t_j) - g(s_i,t_{j-1})\right|$$
$$\leq \mathrm{V}_2(g;[a,b]\times[c,d])\|f\|_\infty .$$

Clearly,

$$\sum_{i=1}^{m}\sum_{j=1}^{n}|C_{ij}| = \sum_{i=1}^{m}\sum_{j=1}^{n}\left|f(a,t_j) - f(a,t_{j-1})\right|\left|g(s_i,c) - g(s_{i-1},c)\right|$$
$$\leq \mathrm{Var}(f(a,\cdot);[c,d])\,\mathrm{Var}(g(\cdot,c);[a,b])$$

and, by symmetry,

$$\sum_{i=1}^{m}\sum_{j=1}^{n}|G_{ij}| = \sum_{i=1}^{m}\sum_{j=1}^{n}|f(s_i,c)) - f(s_{i-1},c)|\left|g(a,t_j) - g(a,t_{j-1})\right|$$
$$\leq \mathrm{Var}(f(\cdot,c);[a,b])\,\mathrm{Var}(g(a,\cdot);[c,d]) .$$

Now, from the additivity property (1.88) it follows that

$$\mathrm{V}_2(f;[a,b]\times[c,d]) = \sum_{i=1}^{m}\sum_{j=1}^{n}\mathrm{V}_2(f;[s_{i-1},s_i]\times[t_{j-1},t_j]),$$

and similarly for g, i.e.

$$\mathrm{V}_2(g;[a,b]\times[c,d]) = \sum_{i=1}^{m}\sum_{j=1}^{n}\mathrm{V}_2(g;[s_{i-1},s_i]\times[t_{j-1},t_j]).$$

From this, we conclude that

$$\sum_{i=1}^{m}\sum_{j=1}^{n}|D_{ij}|$$
$$= \sum_{i=1}^{m}\sum_{j=1}^{n}\left|f(a,t_j) - f(a,t_{j-1})\right|\left|g(s_{i-1},c) + g(s_i,t_{j-1}) - g(s_{i-1},t_{j-1}) - g(s_i,c)\right|$$
$$\leq \mathrm{Var}(f(a,\cdot);[c,d])\,\mathrm{V}_2(g;[a,b]\times[c,d]),$$

and

$$\sum_{i=1}^{m}\sum_{j=1}^{n}|E_{ij}|$$
$$= \sum_{i=1}^{m}\sum_{j=1}^{n}\left|f(a,t_{j-1}) + f(s_i,t_j) - f(a,t_j) - f(s_i,t_{j-1})\right|\left|g(s_i,c) - g(s_{i-1},c)\right|$$
$$\leq \mathrm{Var}(g(\cdot,c);[a,b])\,\mathrm{V}_2(f;[a,b]\times[c,d])$$

as well as

$$\begin{aligned}
&\sum_{i=1}^{m}\sum_{j=1}^{n}|H_{ij}| \\
&\quad = \sum_{i=1}^{m}\sum_{j=1}^{n}|f(s_i,c)+f(s_{i-1},c)|\,\big|g(a,s_{j-1})+g(s_{i-1},t_j)-g(a,t_j)-g(s_{i-1},t_{j-1})\big| \\
&\quad \le \mathrm{Var}(f(\cdot,c);[a,b])\,\mathrm{V}_2(g;[a,b]\times[c,d])\,,
\end{aligned}$$

and

$$\begin{aligned}
&\sum_{i=1}^{m}\sum_{j=1}^{n}|I_{ij}| \\
&\quad = \sum_{i=1}^{m}\sum_{j=1}^{n}\big|f(s_{i-1},c))+f(s_i,t_j)-f(s_{i-1},t_j)-f(s_i,c))\big|\,\big|g(a,t_j)-g(a,t_{j-1})\big| \\
&\quad \le \mathrm{Var}(g(a,\cdot);[c,d])\,\mathrm{V}_2(f;[a,b]\times[c,d])\,.
\end{aligned}$$

It remains to estimate the long terms F_{ij} and J_{ij}. Here, we get

$$\begin{aligned}
\sum_{i=1}^{m}\sum_{j=1}^{n}|F_{ij}| &= \sum_{i=1}^{m}\sum_{j=1}^{n}\big|f(a,t_{j-1})+f(s_i,t_j)-f(a,t_j)-f(s_i,t_{j-1})\big| \\
&\quad\times\big|g(s_{i-1},c)+g(s_i,t_{j-1})-g(s_{i-1},t_{j-1})-g(s_i,c)\big| \\
&\le \mathrm{V}_2(f;[a,b]\times[c,d])\,\mathrm{V}_2(g;[a,b]\times[c,d])\,,
\end{aligned}$$

and, by symmetry,

$$\begin{aligned}
\sum_{i=1}^{m}\sum_{j=1}^{n}|J_{ij}| &= \sum_{i=1}^{m}\sum_{j=1}^{n}\big|f(s_{i-1},c)+f(s_i,t_j)-f(s_{i-1},t_j)-f(s_i,c)\big| \\
&\quad\times\big|g(a,t_{j-1})+g(s_{i-1},t_j)-g(a,t_j)-g(s_{i-1},t_{j-1})\big| \\
&\le \mathrm{V}_2(f;[a,b]\times[c,d])\,\mathrm{V}_2(g;[a,b]\times[c,d])\,.
\end{aligned}$$

Adding up all these estimates, we obtain

$$\begin{aligned}
&\mathrm{V}_2(fg;[a,b]\times[c,d]) \\
&\quad \le |f(a,c)|\,\mathrm{V}_2(g;[a,b]\times[c,d]) + 2\,\mathrm{Var}(f(\cdot,c);[a,b])\,\mathrm{V}_2(g;[a,b]\times[c,d]) \\
&\qquad + 2\,\mathrm{Var}(f(a,\cdot);[c,d])\,\mathrm{V}_2(g;[a,b]\times[c,d]) + \mathrm{V}_2(f;[a,b]\times[c,d])|g(a,c)| \\
&\qquad + 2\,\mathrm{V}_2(f;[a,b]\times[c,d])\,\mathrm{Var}(g(\cdot,c);[a,b]) \\
&\qquad + 2\,\mathrm{V}_2(f;[a,b]\times[c,d])\,\mathrm{Var}(g(a,\cdot);[c,d]) + \mathrm{Var}(f(\cdot,c);[a,b])\,\mathrm{Var}(g(a,\cdot);[c,d]) \\
&\qquad + \mathrm{Var}(f(a,\cdot);[c,d])\,\mathrm{Var}(g(\cdot,c);[a,b]) + 4\,\mathrm{V}_2(f;[a,b]\times[c,d])\,\mathrm{V}_2(f;[a,b]\times[c,d])\,.
\end{aligned}$$

Finally, combining this with (1.91), (1.92) and (1.93), we arrive at (1.90). □

Applying Proposition 0.31 (or Exercise 0.30) to (1.90), we see that replacing (1.89) by the norm

$$|||f|||_{BV} := \|f\|_\infty + \mathrm{Var}(f, [a,b] \times [c,d]) \tag{1.94}$$

with $\|f\|_\infty$ given by (1.87), we get

$$\|f\|_{BV} \le |||f|||_{BV} \le 4\|f\|_{BV}$$

(i.e. the norms (1.89) and (1.94) are equivalent), and (1.90) may be strengthened to

$$|||fg|||_{BV} \le |||f|||_{BV}|||g|||_{BV} \tag{1.95}$$

which means that $(BV([a,b] \times [c,d]), |||\cdot|||_{BV})$ is a *normalized* Banach algebra.

Our discussion shows that Definition 1.42 is quite natural. This definition goes back to Hardy and Krause.[16] However, there are several other definitions which go back to Vitali, Fréchet, Arzelà, and Tonelli[17] and are worth being mentioned. The corresponding definitions for functions $f : [a,b] \times [c,d] \to \mathbb{R}$ read as follows.

Definition 1.45 (Vitali). A function $f : [a,b] \times [c,d] \to \mathbb{R}$ has *bounded variation in Vitali's sense* if $\mathrm{V}_2(f; [a,b]\times[c,d]) < \infty$, see (1.81). In this case, we write $f \in VBV([a,b]\times[c,d])$. ■

Definition 1.46 (Fréchet). Given a function $f : [a,b] \times [c,d] \to \mathbb{R}$ and partitions $P = \{s_0, s_1, \dots, s_m\} \in \mathcal{P}([a,b])$ and $Q = \{t_0, t_1, \dots, t_n\} \in \mathcal{P}([c,d])$, consider the expression

$$\begin{aligned} &\mathrm{V}_2^{\pm}(f, P \times Q; [a,b] \times [c,d]) \\ &:= \sum_{i=1}^{m}\sum_{j=1}^{n} \epsilon_i\epsilon_j[f(s_i,t_j) - f(s_{i-1},t_j) - f(s_i,t_{j-1}) + f(s_{i-1},t_{j-1})], \end{aligned} \tag{1.96}$$

where $\epsilon_i, \epsilon_j \in \{-1, 1\}$. If

$$\begin{aligned} &\mathrm{V}_2^{\pm}(f; [a,b] \times [c,d]) \\ &:= \sup \left\{\mathrm{V}_2^{\pm}(f, P \times Q; [a,b] \times [c,d]) : P \in \mathcal{P}([a,b]), Q \in \mathcal{P}([c,d])\right\} < \infty, \end{aligned}$$

we write $f \in FBV([a,b] \times [c,d])$ and say that f has *bounded variation in Fréchet's sense.* ■

Definition 1.47 (Arzelà). Given a function $f : [a,b] \times [c,d] \to \mathbb{R}$ and partitions[18] $P = \{s_0, s_1, \dots, s_m\} \in \mathcal{P}([a,b])$ and $Q = \{t_0, t_1, \dots, t_m\} \in \mathcal{P}([c,d])$, consider the expression

$$\mathrm{Var}^A(f, P \times Q; [a,b] \times [c,d]) := \sum_{i=1}^{m}\sum_{j=1}^{m} |f(s_i,t_j) - f(s_{i-1},t_{j-1})|. \tag{1.97}$$

---

16 Functions in $BV([a,b] \times [c,d])$ in the sense of Hardy and Krause have been studied in detail in Hildebrandt's book [147].

17 We do not cite the original papers of these authors here, but refer the reader to the books [139] or [148].

18 Here, it is important that both partitions contain the same number of points.

In case

$$\begin{aligned}&\mathrm{Var}^A(f;[a,b]\times[c,d])\\&\quad:=\sup\left\{\mathrm{Var}^A(f,P\times Q;[a,b]\times[c,d]):P\in\mathcal{P}([a,b]),Q\in\mathcal{P}([c,d])\right\}<\infty\,,\end{aligned}$$

we write $f\in ABV([a,b]\times[c,d])$ and say that f has *bounded variation in Arzelà's sense.* ■

Definition 1.48 (Tonelli). A function $f:[a,b]\times[c,d]$ has *bounded variation in Tonelli's sense* if $\mathrm{Var}(f(\cdot,y);[a,b])<\infty$ for almost all $y\in[c,d]$, see (1.79), $\mathrm{Var}(f(x,\cdot);[c,d])<\infty$ for almost all $x\in[a,b]$, see (1.80) and, in addition,

$$\int_a^b \mathrm{Var}(f(x,\cdot);[c,d])\,dx<\infty \tag{1.98}$$

and

$$\int_c^d \mathrm{Var}(f(\cdot,y);[a,b])\,dy<\infty. \tag{1.99}$$

In this case, we write $f\in TBV([a,b]\times[c,d])$. ■

Of course, it is interesting to establish some interconnections between these concepts of bounded variation. First of all, it follows directly from the definitions that

$$BV([a,b]\times[c,d])\subseteq VBV([a,b]\times[c,d])\subseteq FBV([a,b]\times[c,d])\,, \tag{1.100}$$

$$BV([a,b]\times[c,d])\subseteq ABV([a,b]\times[c,d])\,, \tag{1.101}$$

and

$$BV([a,b]\times[c,d])\subseteq TBV([a,b]\times[c,d])\,. \tag{1.102}$$

Therefore, the question arises if there are other inclusions, or even some equality, between these classes, and if the inclusions in (1.100)–(1.102) are strict. These questions have been discussed in the survey paper [100], where the authors consider the following examples over the unit square $S:=[0,1]\times[0,1]$.

Example 1.49. Let $D:=\{(x,y):0\le y\le x\le 1\}$ and $f:=\chi_D$. Then $f\in TBV(S)$, but $f\notin ABV(S)\cup VBV(S)$. ♥

Example 1.50. Let $f:S\to\mathbb{R}$ be defined by[19]

$$f(x,y):=\begin{cases}x\sin\frac{1}{x} & \text{for } 0<x\le 1\,,\\ 0 & \text{for } x=0\,.\end{cases}$$

Then $f\in ABV(S)\cap VBV(S)$, but $f\notin BV(S)$. ♥

19 So, f does not depend on the second variable.

Example 1.51. Let $M \subset [0,1]$ be a nonmeasurable set, $D := \{(x,-x) : x \in M\}$, and $f := \chi_D$. Then $f \in ABV(S)$, but $f \notin TBV(S)$. ♥

Example 1.52. Let S be divided into quarter squares, and let S_1 be the upper left-hand quarter square, i.e. $S_1 = [0,1/2] \times [1/2,1]$. Next, divide the lower right-hand square into quarter squares, and let S_2 be that quarter which has a common vertex with S_1, i.e. $S_2 = [1/2,3/4] \times [1/4,1/2]$. Continuing this process, we obtain an infinite sequence $(S_n)_n$ of square subdivisions of S converging toward the point $(1,0)$. On the square S_1, we construct a "point-rectangle function" f satisfying

$$\mathrm{V}_2(f;S_1) = 1, \quad \mathrm{V}_2^{\pm}(f;S_1) \le \frac{1}{2}.$$

Similarly, on the square S_k, we construct f as a "point-rectangle function" satisfying

$$\mathrm{V}_2(f;S_k) = 1, \quad \mathrm{V}_2^{\pm}(f;S_k) \le \frac{1}{2^k}.$$

Putting $f(x,y) := 0$ at all remaining points, we see that $f \in FBV(S)$, but $f \notin VBV(S)$. ♥

Example 1.53. Let $D := \{(x,-x) : 0 \le x \le 1\}$ and $f := \chi_D$. Then $f \in ABV(S)$, but $f \notin FBV(S)$. ♥

We summarize our results in the following Table 1.3. It is clear that no entries are possible in the diagonal and the first column because the inclusions (1.100)–(1.102) show that BV is the smallest of all considered spaces.[20] Moreover, the second inclusion in (1.100) implies that we cannot find a function which belongs to VBV, but not to FBV.

Table 1.3. BV-functions over $[a,b] \times [c,d]$.

Function f	$\in BV$	$\in VBV$	$\in FBV$	$\in ABV$	$\in TBV$
$\notin BV$	–	Example 1.50	Example 1.52	Example 1.50	Example 1.49
$\notin VBV$	–	–	Example 1.52	Example 1.53	Example 1.49
$\notin FBV$	–	–	–	Example 1.53	Example 1.53
$\notin ABV$	–	Example 1.50	Example 1.50	–	Example 1.49
$\notin TBV$	–	Example 1.50	Example 1.50	Example 1.51	–

In the paper [100], the authors also consider intersections of the five classes occurring in Table 1.3. For example, it is not hard to show that

$$\begin{aligned} &VBV([a,b]\times[c,d]) \cap ABV([a,b]\times[c,d]) \\ &\quad = VBV([a,b]\times[c,d]) \cap TBV([a,b]\times[c,d]) = BV([a,b]\times[c,d]). \end{aligned}$$

20 This explains why functions from BV have nicer properties than those from the other spaces.

On the other hand, Example 1.52 shows that the inclusion

$$BV([a,b]\times[c,d]) \subset VBV([a,b]\times[c,d]) \cap TBV([a,b]\times[c,d])$$

is strict. Other functions of this type are given in Exercises 1.61–1.64. We also remark that continuous functions in the classes VBV, FBV, ABV, and TBV have been studied in [100] and by other authors. For example, an example of a function $f \in (C\cap ABV)\setminus BV$ may be found in [174, 175].

1.5 Comments on Chapter 1

As mentioned at the beginning, Jordan's pioneering paper [153] may be considered as a starting point of the study of functions of bounded variation. Jordan also introduced the variation function (1.13) and used it to prove the decomposition theorem (Theorem 1.5) for BV-functions. The functions p_f and n_f used in the proof of Theorem 1.6 are sometimes called the *positive variation* and *negative variation* of f, respectively.

As Theorem 1.26 shows, the variation function V_f inherits many important properties from its parent function f, see [149, 281]. However, Theorem 1.26 (d) also shows that, in contrast to Lipschitz continuity, Hölder continuity is not perfectly symmetric. We state this as

Problem 1.1. *Does* $f \in BV([a,b]) \cap Lip_\alpha([a,b])$ $(0<\alpha<1)$ *imply that* $V_f \in Lip_\alpha([a,b])$*?*

By formula (1.17), $BV([a,b])$ is an algebra. Apparently, the idea to use the decomposition from Theorem 1.6 to show that $BV([a,b])$ is even a *normalized* algebra, see (1.18), is first given by Kuller in his book [173]. For the space $BV^o([a,b])$, a similar proof was later given by Bullen [73] and Russell [282] who seemed to have been unaware of Kuller's proof.

Functions of bounded variation are "intermediate" between step functions and regular functions in the sense that

$$S([a,b]) \subseteq BV([a,b]) \subseteq R([a,b])\,. \tag{1.103}$$

Although both inclusions are actually strict, the three sets in (1.103) are not "too distant" from each other; in fact, in Proposition 0.57, we have shown that the closure of the left set (in the norm of $B([a,b])$) coincides with the right set. A refinement of this may be found in Exercise 1.32.

Eduard Helly was an Austrian mathematician whose work had a profound influence on Riesz and Banach. His famous selection principle [144] (Theorem 1.11) is also known as *Helly's second theorem*.[21] Helly's selection principle may be viewed as a

21 There is another result known as *Helly's first theorem* which refers to Riemann–Stieltjes integrals, see Theorem 4.21 (c) in Chapter 4.

certain counterpart of the Arzelà–Ascoli criterion (Proposition 0.55) for BV-functions. More information on Helly's theorem and its history can be found in the books [76, 182, 238].

The harmless looking Proposition 1.12 has interesting consequences. Consider, for example, the strictly increasing homeomorphism $\tau(t) := t^\gamma$ on $[0,1]$, where $0 < \gamma < 1$. The oscillation function $g = f_{\alpha,\beta}$ defined in (0.86) belongs to $BV([0,1])$ if either $\beta > 0$ and $\alpha+\beta \geq 0$, or $\beta \leq 0$ and $\alpha+\beta > 0$, see Exercises 1.8 and 1.9. In case $\alpha+\beta \geq 1$, we even have $g \in Lip([0,1])$, as Proposition 0.48 (b) shows, and so we may use the inclusions (1.46) to deduce that also $g \in BV([0,1])$. Now, for the function $f = g \circ \tau = f_{\alpha\gamma,\beta\gamma}$, we get the same conditions for $f \in BV([0,1])$, as one could expect from Proposition 1.12. On the other hand, $f \notin Lip([0,1])$ if γ is so small that $(\alpha+\beta)\gamma < 1$, and so we cannot use (1.46) in this case.

Our discussion of the relation of the space BV with the class IVP in Section 1.2 has been inspired by the paper [179].

The following proposition is a straightforward extension of Proposition 1.12 to functions of bounded Wiener variation in Wiener's sense. The proof is the same and may be found in [232].

Proposition 1.54. *Given a function $g : [c,d] \to \mathbb{R}$, let $\tau : [a,b] \to [c,d]$ be continuous and strictly increasing with $\tau(a) = c$ and $\tau(b) = d$. Then $g \circ \tau \in WBV_p([a,b])$ if and only if $g \in WBV_p([c,d])$.*

Example 1.13 shows again that we cannot drop the continuity assumption on τ in Proposition 1.54. The following example shows that we cannot drop the monotonicity assumption either.

Example 1.55. For $p \geq 1$, define $\tau : [0,1] \to [0,1]$ by

$$\tau(t) := \begin{cases} t\left|\sin\frac{1}{t}\right|^p & \text{for } 0 < t \leq 1\,, \\ 0 & \text{for } t = 0\,. \end{cases}$$

Then τ is continuous, but of course far from being monotone. The function $g : [0,1] \to \mathbb{R}$ defined by $g(x) := x^{1/p}$ belongs to $Lip_{1/p}([0,1])$, and hence to $WBV_p([0,1])$, by (1.68). On the other hand, the function $f = g \circ \tau$ does not belong to $WBV_p([0,1])$, which can be seen by a similar reasoning as in Example 1.8. For $n \in \mathbb{N}$, consider the partition

$$P_n := \{0,1\} \cup \{s_1,\dots,s_n\} \cup \{t_1,\dots,t_n\}\,,$$

where

$$s_j := \frac{1}{4j\pi}, \quad t_j := \frac{1}{(4j+1)\pi} \quad (j = 1,2,\dots,n)\,.$$

Since $g(\tau(s_j)) = 0$ and $g(\tau(t_j)) = t_j$, the partition P_n gives the contribution

$$\mathrm{Var}_p^W(f, P_n; [0,1]) \geq \left(\frac{2}{\pi}\right)^{1/p} \sum_{k=1}^{n} \frac{1}{(4k+1)^{1/p}}, \tag{1.104}$$

and the sum in (1.104) is unbounded as $n \to \infty$ because $p \geq 1$. ♥

What we call pseudomonotone functions (Definition 1.14) was introduced and discussed by Josephy in [155] (under a different name). Proposition 1.15 shows that the class of pseudomonotone functions is situated between monotone and BV-functions, while Proposition 1.17 shows that pseudomonotone functions provide the "tailor-made" substitutions for preserving bounded variation.

In Section 1.2, we have discussed several interconnections between bounded variation and continuity. Several of our examples in this section are taken from, or at least inspired by, Chapters 14 and 15 of the excellent book [39]. The functions having the intermediate value property are interesting from an analytical viewpoint, but have "bad" algebraic properties. For example, it is not hard to see that $IVP([a,b])$ is *not* a linear space. In fact, the two (discontinuous) functions

$$f(x) := \begin{cases} \sin\frac{1}{x} & \text{for } x \neq 0 \\ 1 & \text{for } x = 0 \end{cases}$$

and

$$g(x) := \begin{cases} \sin\frac{1}{x} & \text{for } x \neq 0 \\ 0 & \text{for } x = 0 \end{cases}$$

both belong to $IVP([-1,1])$, say, but their difference $f - g = \chi_{\{0\}}$ does not. So, one could ask how the smallest vector space $X = \operatorname{span} IVP([a,b])$ looks like. Since $IVP([a,b])$ is one of the largest function classes in Table 1.1, one could expect that also X is quite large. In fact, the answer is as simple as surprising: X is the space of *all* functions on $[a,b]$ ([268, Theorem 9.5])!

Proposition 1.30 shows that a functions $g \in C([a,b]) \cap BV([a,b])$ can be transformed by a suitable homeomorphism $\tau : [a,b] \to [a,b]$ into a differentiable function $g \circ \tau$ with bounded derivative.

There is a (partial) analogue to Proposition 1.30 for functions $g \in WBV_p([a,b])$ which reads as follows:

Proposition 1.56. *For a function $g : [a,b] \to \mathbb{R}$, the following two statements are equivalent.*

(a) *The function g is continuous and of bounded Wiener p-variation.*
(b) *There exists a homeomorphism $\tau : [a,b] \to [a,b]$ such that $g \circ \tau : [a,b] \to \mathbb{R}$ is Hölder continuous on $[a,b]$ with Hölder exponent $1/p$.*

The proof of this proposition is very similar to that of Proposition 1.30 and may be found in [232]. The reader should observe the subtle difference between Proposition 1.54 and Proposition 1.56: while a function $g \in WBV_p$, in general, remains in WBV_p after a homeomorphic change of variable, a *continuous* function $g \in WBV_p$ may even become Hölder continuous of order $1/p$. So, adding continuity bridges the gap (which is essential, as Example 1.40 shows) between $Lip_{1/p}$ and WBV_p.

Functions of bounded Wiener variation are briefly treated in the Lecture Notes [65], a good survey article is [117]. A comparison of (1.46) and (1.68), or Theorem 1.28 and Theorem 1.41, shows that WBV_p-functions are related to Hölder continuous functions in rather the same way as BV-functions to Lipschitz continuous functions. Such relations are discussed in the recent paper [232].

Less is known for BV-functions $f : [a, b] \to X$, where X is a normed or even metric space. A good survey on results of this type, including extensions of Proposition 1.7, Theorem 1.11 and Theorem 1.28 to this more general setting is [85]. If X is a normed space and $\mathrm{Var}(f; [a, b], X)$ is defined in the obvious way, in [85], it is also shown that

$$\mathrm{Var}(f; [a,b], X) \leq \int_a^b \|f'(x)\|_X \, dx \tag{1.105}$$

for $f \in C^1([a, b], X)$. We will discuss further results of this type for BV-functions and absolutely continuous functions in Chapter 3.

In the special case $X = \mathbb{R}^n$ with the Euclidean norm

$$\|(\xi_1, \xi_2, \ldots, \xi_n)\| = \sqrt{\xi_1^2 + \xi_2^2 + \cdots + \xi_n^2},$$

the corresponding space $BV([a, b], \mathbb{R}^n)$ has been studied by many authors; in this case, we have equality in (1.105) for $f \in C^1([a, b], \mathbb{R}^n)$. A particularly important case is $n = 2$; here, we get

$$\mathrm{Var}(f; [a,b], \mathbb{R}^2) = \int_a^b \sqrt{f_1'(x)^2 + f_2'(x)^2} \, dx \tag{1.106}$$

for $f \in C^1([a, b], \mathbb{R}^2)$, where f_1 and f_2 are the components of f. We will come back to this special case in connection with rectifiable graphs in Section 3.4.

Set-valued maps of bounded variation (in the classical or more general sense) have been studied by many authors, among them [88, 222, 223, 231, 292, 327].

BV-functions of several (in particular, two) variables on Cartesian products of intervals (in particular, rectangles in the plane) have been studied extensively by Chistyakov [90–95]. Specifically, in [90], the author studies the Banach algebra $BV([a, b] \times [c, d])$ which we considered in Section 1.4, while BV-functions of n variables on a product $[a_1, b_1] \times [a_2, b_2] \times \cdots \times [a_n, b_n] \subset \mathbb{R}^n$ are discussed in [92, 93]. Functions of two variables which belong to the space $WBV_p([a, b] \times [c, d])$, i.e. have bounded Wiener p-variation on a rectangle in the plane, are considered in [96]. An analogue of Helly's selection principle (Theorem 1.11) for functions of several variables may be found in [178].

The other concepts of variation for functions of two variables given in Definitions 1.45–1.48 and Examples 1.49–1.53 are discussed in detail in [100], see also [3, 4]. These concepts are obviously motivated by various applications which we will, in part, consider in subsequent chapters. Our original Definition 1.42 singles out a class of functions arising in the theory of (double) Fourier series. Vitali's Definition 1.45 is sufficient to insure the existence of the Riemann–Stieltjes (double) integral for continuous functions of two variables. In particular, if such a function is of the form $f(x,y) = g(x)h(y)$, Fréchet's concept (which is weaker than Vitali's concept) is suitable. Finally, Arzelà's Definition 1.47 is modeled after the decomposition of a BV-function as the difference of two monotonically increasing functions. However, among all of these concepts, our Definition 1.42 is by far the most natural one.[22]

We remark that there are also some papers on BV-functions with weight, e.g. [54]. Needless to say, the concept of variation has been extended in many directions, leading sometimes to straightforward generalizations, and sometimes to unexpected new phenomena. Many such extensions will be discussed in detail in the next chapter.

1.6 Exercises to Chapter 1

We state some exercises on the topics covered in this chapter; exercises marked with an asterisk * are more difficult.

Exercise 1.1. Let $f, g \in BV([a,b])$ with $|g(x)| \geq c$ for some $c > 0$, i.e. g is bounded away from zero. Show that $f/g \in BV([a,b])$.

Exercise 1.2. Find functions $f, g \in BV([0,1])$ such that $g(x) > 0$ on $[0,1]$ and $f/g \notin BV([0,1])$.

Exercise 1.3. Show that $f \in BV([a,b])$ implies $|f| \in BV([a,b])$ and $\mathrm{Var}(|f|;[a,b]) \leq \mathrm{Var}(f;[a,b])$.

Exercise 1.4. Find a function $f \notin BV([0,1])$ such that $|f| \in BV([0,1])$.

Exercise 1.5. Suppose that $f \in C([a,b])$ and $|f| \in BV([a,b])$. Show that $f \in BV([a,b])$, and compare this with your example in Exercise 1.4.

Exercise 1.6. Given $f \in BV([a,b])$, show that

$$V_f(x_0+) - V_f(x_0) = |f(x_0+) - f(x_0)|$$

for each $x_0 \in [a,b)$, and

$$V_f(x_0) - V_f(x_0-) = |f(x_0) - f(x_0-)|$$

for each $x_0 \in (a,b]$. Use this to give an alternative proof of Proposition 1.7.

Exercise 1.7. Let $f \vee g$ and $f \wedge g$ be defined as in Exercise 0.70. Show that $f, g \in BV([a,b])$ implies $f \vee g \in BV([a,b])$ and $f \wedge g \in BV([a,b])$, and express $\mathrm{Var}(f \vee g;[a,b])$ and $\mathrm{Var}(f \wedge g;[a,b])$ through $\mathrm{Var}(f;[a,b])$ and $\mathrm{Var}(g;[a,b])$.

22 Another reason which makes this transparent will be contained in Theorem 3.51 in Chapter 3.

Exercise 1.8. Let $\alpha \in \mathbb{R}$ and $\beta > 0$. Show that the function (0.86) then belongs to $BV([0,1])$ if and only if $\alpha + \beta \geq 0$, and calculate V_f in this case.

Exercise 1.9. Let $\alpha \in \mathbb{R}$ and $\beta \leq 0$. Show that the function (0.86) then belongs to $BV([0,1])$ if and only if $\alpha + \beta > 0$, and calculate V_f in this case. Compare this with Example 1.8.

Exercise 1.10. Show that a function $f : \mathbb{R} \to \mathbb{R}$ is monotone if and only if $f^{-1}([\alpha, \beta])$ is an interval for each interval $[\alpha, \beta] \subseteq \mathbb{R}$.

Exercise 1.11. Obviously, the set of all monotone functions on $[a, b]$ contains a one-dimensional linear space (namely, all constant functions), as well as a two-dimensional linear space (namely, all affine functions). Does it also contain a three-dimensional linear space?

Exercise 1.12. Prove that a function $f \in BV([a, b])$ has, at most, countably many points of discontinuity, all of them being of first kind (jumps) or removable.

Exercise 1.13. Prove the following converse of Exercise 12: given a countable set $D \subset [a, b]$, there exists a monotone function f such that $D(f) = D$, see (0.49). In particular, construct such a function for $D := [a, b] \cap \mathbb{Q}$.

Exercise 1.14*. Construct a function $f \in B([0,1]) \setminus BV([0,1])$ and a sequence $(P_n)_n$ of partitions $P_n \in \mathcal{P}([0,1])$ such that $\mathrm{Var}(f, P_n; [0,1]) = 0$ for all $n \in \mathbb{N}$ and $\mu(P_n) \to 0$ as $n \to \infty$.

Exercise 1.15. Show that the variation function (1.13) of the function $f(x) = \sin x$ on $[0, 2\pi]$ has the form

$$V_f(x) = \begin{cases} \sin x & \text{for } 0 \leq x \leq \frac{1}{2}\pi, \\ 2 - \sin x & \text{for } \frac{1}{2}\pi < x \leq \frac{3}{2}\pi, \\ 4 + \sin x & \text{for } \frac{3}{2}\pi < x \leq 2\pi. \end{cases}$$

Exercise 1.16. Show that a function $f \in BV([a, b])$ and its variation function (1.13) satisfy the inequality

$$\int_a^b V_f(x) f(x)\, dx - \frac{1}{b-a}\left(\int_a^b V_f(x)\, dx\right)\left(\int_a^b f(x)\, dx\right)$$
$$\leq \int_a^b V_f(x)^2\, dx - \frac{1}{b-a}\left(\int_a^b V_f(x)\, dx\right)^2 .$$

Illustrate this by means of a (nonmonotone) function f of your choice.

Exercise 1.17. Let $(f_n)_n$ be a sequence of functions which converges pointwise on $[a, b]$ to some function f. If each f_n is increasing, show that f is also increasing. If each f_n is of bounded variation, does it follow that f is of bounded variation?

Exercise 1.18. Let $f, g \in BV([a,b])$ such that $|f(x)| \leq |g(x)|$. Does it follow that $\mathrm{Var}(f;[a,b]) \leq \mathrm{Var}(g;[a,b])$?

Exercise 1.19. Construct an example of a sequence in $BV([a,b])$ which is bounded with respect to the norm (1.16), but contains no subsequence, which is convergent in this norm. Compare with Proposition 1.9.

Exercise 1.20*. Let $M \subset NBV([a,b])$ (Definition 1.2) be bounded and closed. Solve Exercise 13.34 in [108] which gives a necessary and sufficient condition for the compactness of M in $(BV([a,b]), \|\cdot\|_{BV})$.

Exercise 1.21*. Show that $BV([a,b])$ may be decomposed as direct sum

$$BV([a,b]) = NBV([a,b]) \oplus VBV([a,b]),$$

where $VBV([a,b])$ consists of all functions in $BV([a,b])$ which vanish, except for a countable set of points. Prove that $VBV([a,b])$ is isometrically isomorphic to a certain L_1-space. Use this fact and Exercise 1.20 to give a compactness criterion in the space $BV([a,b])$.

Exercise 1.22. Using Exercise 1.20, but not general facts from functional analysis, show that the closed unit ball in $(BV([a,b]), \|\cdot\|_{BV})$ is not compact.

Exercise 1.23. The *Hellinger integral* of a function $f : [a,b] \to \mathbb{R}$ is defined as

$$\int_a^b (df)^2 := \sup\left\{\sum_{j=1}^m [f(t_j) - f(t_{j-1})]^2 : \{t_0, t_1, \ldots, t_m\} \in \mathcal{P}([a,b])\right\},$$

where the supremum is taken over all partitions of $[a,b]$. Show that if this integral exists, then $f \in BV([a,b])$ and

$$\mathrm{Var}(f;[a,b]) \leq \sqrt{(b-a)\int_a^b (df)^2}\,.$$

Is the converse also true, i.e. does every function $f \in BV([a,b])$ have a Hellinger integral?

Exercise 1.24. With the definition and notation of Exercise 1.23, prove that the Hellinger integral for $f \in BV([a,b])$ exists if and only if the Hellinger integral for its variation function V_f exists. Moreover, show that in this case, both integrals coincide.

Exercise 1.25. Suppose that the Hellinger integral (Exercise 1.23) for $f \in BV([a,b])$ exists. Prove that then both $f \in Lip_{1/2}([a,b])$ and $V_f \in Lip_{1/2}([a,b])$ with

$$lip_{1/2}(f) \leq lip_{1/2}(V_f) \leq \sqrt{\int_a^b (df)^2}\,,$$

where $lip_\alpha(f)$ is defined by (0.69).

Exercise 1.26. Determine all $\alpha, \beta \in \mathbb{R}$ for which the function (0.86) has a Hellinger integral. Compare with Exercises 1.8, 1.9 and 0.52.

Exercise 1.27. Given $f \in BV([a,b])$, consider the functions p_f and n_f constructed in the proof of Theorem 1.6. Prove that p_f and n_f are the "most slowly increasing functions" among all Jordan decompositions of f in the following sense: If $\phi, \psi : [a,b] \to \mathbb{R}$ are monotonically increasing such that $f = \phi - \psi$, then

$$p_f(y) - p_f(x) \le \phi(y) - \phi(x)$$

and

$$n_f(y) - n_f(x) \le \psi(y) - \psi(x)$$

for all x, y such that $a \le x < y \le b$. Also, show that

$$\mathrm{Var}(f; [x,y]) = \mathrm{Var}(p_f; [x,y]) + \mathrm{Var}(n_f; [x,y])$$

for $a \le x < y \le b$.

Exercise 1.28. Show that every monotone function is Riemann integrable on each compact interval. Deduce that the same is true for BV-functions.

Exercise 1.29*. Prove that every monotone function is a.e. differentiable. Deduce that the same is true for BV-functions.

Exercise 1.30*. Given $f \in BV([a,b])$, show that

$$\omega_1(f, [a,b]; \delta) = O(\delta) \quad (\delta \to 0+),$$

where $\omega_p(f, [a,b]; \delta)$ denotes the integral modulus of continuity (0.98). Compare with Proposition 0.53 and Exercise 0.76.

Exercise 1.31*. Prove the following converse of Exercise 1.30: if $f \in L_1([a,b])$ satisfies

$$\omega_1(f, [a,b]; \delta) = O(\delta) \quad (\delta \to 0+),$$

then f is equivalent to some BV-function on $[a,b]$.

Exercise 1.32*. Prove that the following three statements for a function $f : [a,b] \to \mathbb{R}$ are equivalent:
(a) f is the uniform limit of a sequence of step functions.
(b) f is the uniform limit of a sequence of BV-functions.
(c) f is regular, i.e. the unilateral limits (0.54) exist for all $x_0 \in [a,b]$.

Exercise 1.33. Let $f \in BV([a,b]) \cap C([a,b])$. Deduce from Proposition 1.7 that f may be represented as the difference of two *continuous* increasing functions on $[a,b]$. Is the same true for functions $f \in CBV([a,b])$?

Exercise 1.34. Let $\{x_1, x_2, x_3, \ldots\} \subset [a, b]$ be the set of discontinuities of a function $f \in BV([a, b])$ (Exercise 1.12). The *jump function* $\sigma_f : [a, b] \to \mathbb{R}$ of f is then defined by $\sigma_f(a) := 0$ and

$$\sigma_f(t) := \sum_{t>x_k} (f(x_k+) - f(x_k-)) + (f(t) - f(t-)) \quad (a < t \leq b).$$

Show that

$$\sum_{k=1}^{n} (|f(x_k+) - f(x_k)| + |f(x_k) - f(x_k-)|) \leq \mathrm{Var}(f; [a, b])$$

for all $n \in \mathbb{N}$. Conclude from this that the jump function σ_f is well-defined.

Exercise 1.35. Let σ_f be the jump function of a function $f \in BV([a, b])$ (Exercise 1.34). Prove that $f - \sigma_f \in BV([a, b]) \cap C([a, b])$. Also, calculate σ_f for f as in Example 1.4.

Exercise 1.36. Let $f : [a, b] \to \mathbb{R}$ be increasing, and let σ_f be the jump function of a function f (Exercise 1.34). Show that $f - \sigma_f$ is also increasing.

Exercise 1.37. Let σ_f be the jump function of a function $f \in BV([a, b])$ (Exercise 1.34). Show that $\mathrm{Var}(f; [a, b]) = \mathrm{Var}(f - \sigma_f; [a, b]) + \mathrm{Var}(\sigma_f; [a, b])$.

Exercise 1.38. Let $f : [a, b] \to \mathbb{R}$ be a bounded function and $a < c < b$. Suppose that f has local bounded variation near c, which means that there exists a $\delta > 0$ such that $f \in BV([c - \delta, c + \delta])$. Also, assume that f has a primitive on both (a, c) and (c, b), i.e. $f = F_1'$ for some differentiable function $F_1 : (a, c) \to \mathbb{R}$ and $f = F_2'$ for some differentiable function $F_2 : (c, b) \to \mathbb{R}$. Show that f has a primitive on (a, b) if and only if f is continuous at c.

Exercise 1.39. Given a continuous function $f : [a, b] \to \mathbb{R}$ and a partition $P = \{t_0, t_1, \ldots, t_m\} \in \mathcal{P}([a, b])$, let $\mathrm{osc}(f; [t_{j-1}, t_j])$ be defined as in (1.12). Show that

$$\mathrm{Var}(f, P; [a, b]) \leq \sum_{j=1}^{m} \mathrm{osc}(f; [t_{j-1}, t_j]) \leq \mathrm{Var}(f; [a, b])$$

and conclude that this implies (1.39).

Also, illustrate (1.39) by means of a function $f \in C([a, b]) \cap BV([a, b])$ and a function $f \in C([a, b]) \setminus BV([a, b])$ (e.g. the function from Example 1.8).

Exercise 1.40. Calculate the functions τ and g from Theorem 1.28 for $f = f_{\alpha,\beta}$ with α and β as in Exercises 1.8 and 1.9.

Exercise 1.41. Suppose that $f \in C([a,b])$ is injective, and $g : [a,b] \to \mathbb{R}$ has a primitive. Show that the product fg has a primitive. See what happens if you drop one of the hypotheses on f.

Exercise 1.42. Suppose that a function $f : [a,b] \to \mathbb{R}$ has the intermediate value property, and that $|f| \in BV([a,b])$. Prove that f is continuous, and compare with Exercises 1.4 and 1.5.

Exercise 1.43. Suppose that a function $f : [a,b] \to [a,b]$ has the intermediate value property, and that $f \circ f \in BV([a,b])$. Prove that $f \circ f$ is continuous. Is f itself then also continuous?

Exercise 1.44. Suppose that $f \in BV([a,b]) \cap C([a,b])$, and $g : [a,b] \to \mathbb{R}$ has a primitive. Show that the product fg has a primitive. Check what happens if you drop one of the hypotheses on f.

Exercise 1.45. Suppose that $f \in BV([a,b])$ has the property that there is a function $g : [a,b] \to \mathbb{R} \setminus \{0\}$ such that both g and the product fg have a primitive. Prove that the product fh then has a primitive for *any* function $h : [a,b] \to \mathbb{R}$ which has a primitive. See what happens if you drop one of the hypotheses on f.

Exercise 1.46. Calculate the Sierpiński decomposition (Theorem 0.36) for the function f in Example 1.25.

Exercise 1.47. Let I_f be the Banach indicatrix (Definition 0.38) of the function (1.14). Show by a direct calculation (i.e. without using Proposition 1.27) that $I_f \notin L_1(\mathbb{R})$.

Exercise 1.48. Calculate the McShane extension (0.76) of the function g in Example 1.24, and compare the result with the convexification $\overline{g}$ which we have constructed in Theorem 1.28.

Exercise 1.49. By considering the characteristic functions $\chi_{\{c\}}$ for $a \le c \le b$, show that the space $BV([a,b])$ with norm (1.16) is not separable, i.e. there is no countable dense subset. Is the space $BV([a,b]) \cap C([a,b])$ with norm (1.16) separable?

Exercise 1.50. Let $f : [-1,1] \to \mathbb{R}$ be defined by $f(x) := |x|$. Prove that $V_f \in C^1([-1,1])$, although f is not differentiable at 0.

Exercise 1.51. Let $f := f_{2,\beta} : [-1,1] \to \mathbb{R}$ be the oscillation function from (0.86), where $-2 < \beta < -1$. Prove that $V_f \notin C^1([-1,1])$, although f is differentiable at 0.

Exercise 1.52*. Given $f \in BV([a,b])$, prove that $V_f'(x) = |f'(x)|$ a.e. on $[a,b]$.

Exercise 1.53*. Given $\phi \in BV([a,b])$ and $\psi \in BV([a,b]) \cap C([a,b])$, denote by supp ψ the set of all points $x \in (a,b)$ such that $\psi(x) \neq 0$. Denoting further by (a_k, b_k) the open intervals between two subsequent zeros of ψ, i.e.

$$\operatorname{supp} \psi \subseteq \bigcup_{k=1}^{\infty} (a_k, b_k),$$

prove that

$$\operatorname{Var}(\phi\psi; [a,b]) \le \|\phi\|_\infty \operatorname{Var}(\psi; [a,b]) + \sum_{k=1}^{\infty} \operatorname{Var}(\phi\psi; [a_k, b_k]).$$

Exercise 1.54. Illustrate Exercise 1.53 by choosing for ϕ and ψ suitable oscillation functions of type (0.86).

Exercise 1.55*. Suppose that $f : [a,b] \to \mathbb{R}$ is monotone with $D(f) = [a,b] \cap \mathbb{Q}$. Prove that there is no interval $[c,d] \subseteq [a,b]$ such that $f \in IVP([c,d])$. Check this result for the function you constructed in Exercise 1.13.

Exercise 1.56. A function $f \in BV([a,b])$ is said to have *vanishing variation* at $t_0 \in [a,b]$ if

$$\lim_{\delta \to 0} \operatorname{Var}(F_{t_0,\delta}; [a,b]) = 0,$$

where $F_{t_0,\delta} : [a,b] \to \mathbb{R}$ is defined by

$$F_{t_0,\delta}(t) := \begin{cases} f(t) - f(t_0) & \text{if } |t - t_0| \le \delta, \\ 0 & \text{otherwise.} \end{cases}$$

Show that f has vanishing variation at t_0 if f is continuous at t_0.

Exercise 1.57. Show that the function (0.86) belongs to $WBV_p([0,1])$ for fixed $p \in [1,\infty)$ if and only if either $\beta > 0$ and $p\alpha + \beta \ge 0$, or $\beta \le 0$ and $p\alpha + \beta > 0$. Compare this with Exercises 1.8 and 1.9.

Exercise 1.58. Use the result of Exercise 1.57 to construct a function $f \in WBV_p([0,1]) \setminus Lip_{1/p}([0,1])$.

Exercise 1.59. Prove the following generalization of Proposition 1.34. For $1 \le p < \infty$, let $\omega(t) = O(t^{1/p})$ as $t \to 0+$, where ω is an arbitrary modulus of continuity; then, $Lip_{\omega,\infty}([a,b]) \subseteq WBV_p([a,b])$, where the space $Lip_{\omega,\infty}([a,b])$ is introduced in Definition 0.54.

Exercise 1.60*. The following result shows that the statement of Exercise 1.59 is sharp. For $1 \le p < \infty$, suppose that $\omega(t) \neq O(t^{1/p})$, but $t^{1/p} = o(\omega(t))$ as $t \to 0+$, i.e.

$$\lim_{t \to 0+} \frac{\omega(t)}{t^{1/p}} = \infty.$$

Prove that then there exists a function $f \in Lip_{\omega,\infty}([0,1]) \setminus WBV_p([0,1])$.

Exercise 1.61. Let the rational points of $[0,1]$ be enumerated, i.e. $[0,1] \cap \mathbb{Q} = \{r_1, r_2, r_3, \dots\}$, and define a function $f : [0,1] \times [0,1] \to \mathbb{R}$ by

$$f(x,y) := \begin{cases} 1 & \text{if } x = r_k,\ y \notin \mathbb{Q}, \text{ and } y > 1 - 1/2^k\,, \\ 0 & \text{otherwise.} \end{cases}$$

Check whether or not f belongs to the space $BV([0,1] \times [0,1])$ introduced in Definition 1.42.

Exercise 1.62. Check whether or not the function f from Exercise 0.61 belongs to one of the spaces $VBV([0,1] \times [0,1])$, $FBV([0,1] \times [0,1])$, $ABV([0,1] \times [0,1])$, or $TBV([0,1] \times [0,1])$ introduced in Definitions 1.45–1.48.

Exercise 1.63. Define a function $f : [0,1] \times [0,1] \to \mathbb{R}$ by

$$f(x,y) := \begin{cases} 1 & \text{if } x \in \mathbb{Q} \text{ and } y \in \mathbb{Q}\,, \\ 0 & \text{otherwise.} \end{cases}$$

Check whether or not f belongs to the space $BV([0,1] \times [0,1])$ introduced in Definition 1.42.

Exercise 1.64. Check whether or not the function f from Exercise 0.63 belongs to one of the spaces $VBV([0,1] \times [0,1])$, $FBV([0,1] \times [0,1])$, $ABV([0,1] \times [0,1])$, or $TBV([0,1] \times [0,1])$ introduced in Definitions 1.45–1.48.

Exercise 1.65. Show that both inclusions in (1.103) are strict.

2 Nonclassical BV-spaces

The function space $BV([a, b])$ discussed in the previous chapter has been generalized in various directions. N. Wiener and L. C. Young distorted the measurement of intervals in the range of functions by considering p-th powers or, more generally, continuous increasing gauge functions ϕ. Subsequently, D. Waterman and M. Schramm admitted countable families of such gauge functions in order to generalize the concept of variation. One of the most interesting generalization, however, has been introduced by F. Riesz in the classical setting, and by Yu. T. Medvedev in the setting of gauge functions because it allows an elegant characterization of absolutely continuous functions whose derivative have certain summability properties. We also consider functions of so-called Korenblum variation, as well as higher order variations in the sense of De la Vallée–Poussin and Popoviciu. All of these generalizations of the concept of bounded variation will be discussed in this chapter, and some relations between them will be illustrated by several inclusions, examples, and counterexamples.

2.1 The Wiener–Young variation

In Definition 1.31, we have introduced, for $1 \leq p < \infty$, the class $WBV_p([a, b])$ of all functions $f : [a, b] \to \mathbb{R}$, for which the p-variation in Wiener's sense

$$\mathrm{Var}_p^W(f; [a, b]) = \sup \left\{ \sum_{j=1}^{m} |f(t_j) - f(t_{j-1})|^p : \{t_0, t_1, \ldots, t_m\} \in \mathcal{P}([a, b]) \right\}$$

is finite. Now, we consider an important generalization of this concept introduced in 1937 by L. C. Young [322, 323]. To this end, we recall the notion of Young functions (or gauge functions) which we already introduced in Definition 0.16.

Definition 2.1. We call a function $\phi : [0, \infty) \to [0, \infty)$ *Young function* (or *gauge function*) if ϕ is continuous, convex[1], and satisfies[2] $\phi(0) = 0$, $\phi(t) > 0$ for $t > 0$, and $\phi(t) \to \infty$ as $t \to \infty$. ■

Typical examples of Young functions are $\phi(t) = t^p$ for $1 \leq p < \infty$, $\phi(t) = e^t - 1$, or $\phi(t) = (t + 1)\log(t + 1)$.

---

1 We remark that some authors do not require convexity of a Young function to include such examples like $\phi(t) = t^p$ for $0 < p < 1$. However, in every important result, convexity of ϕ is usually imposed as an additional condition. Sometimes, it is also required that ϕ is increasing; however, as we have shown in Lemma 1.36, this is a consequence of the other properties.

2 Here, some authors require that even $\phi(t)/t \to \infty$ as $t \to \infty$, but this excludes the example $\phi(t) = t$. Later (Definition 2.11), we will state this extra condition under the name "condition ∞_1."

Definition 2.2. Given a Young function $\phi : [0, \infty) \to [0, \infty)$, a partition $P = \{t_0, t_1, \ldots, t_m\} \in \mathcal{P}([a, b])$, and a function $f : [a, b] \to \mathbb{R}$, the nonnegative real number

$$\mathrm{Var}_\phi^W(f, P) = \mathrm{Var}_\phi^W(f, P; [a, b]) := \sum_{j=1}^{m} \phi(|f(t_j) - f(t_{j-1})|) \tag{2.1}$$

is called the *Wiener–Young variation* of f on $[a, b]$ with respect to P, while the (possibly infinite) number

$$\mathrm{Var}_\phi^W(f) = \mathrm{Var}_\phi^W(f; [a, b]) := \sup \left\{ \mathrm{Var}_\phi^W(f, P; [a, b]) : P \in \mathcal{P}([a, b]) \right\}, \tag{2.2}$$

where the supremum is taken over all partitions of $[a, b]$, is called the *total Wiener–Young variation* of f on $[a, b]$. In case $\mathrm{Var}_\phi^W(f; [a, b]) < \infty$, we say that f has *finite Wiener–Young variation* on $[a, b]$, and write $f \in V_\phi^W([a, b])$. ■

Of course, for $\phi(t) = t^p$ with $1 \le p < \infty$, the set $V_\phi^W([a, b])$ coincides with the space $WBV_p([a, b])$ from Definition 1.31. However, $V_\phi^W([a, b])$ is in general not a linear space:

Example 2.3. The function $\phi : [0, \infty) \to \mathbb{R}$ defined by

$$\phi(t) := \begin{cases} 0 & \text{for } t = 0, \\ e^{-1/t} & \text{for } 0 < t < \frac{1}{4}, \\ \frac{16t-3}{e^4} & \text{for } t \ge \frac{1}{4} \end{cases} \tag{2.3}$$

is a Young function. We claim that the corresponding set $V_\phi^W([0, 1])$ is *not* a linear space. In fact, consider the function $f : [0, 1] \to \mathbb{R}$ defined by

$$f(x) := \begin{cases} -\frac{1}{2 \log n} & \text{for } x = \frac{1}{n} \ (n = 2, 3, 4, \ldots), \\ 0 & \text{otherwise}. \end{cases}$$

Observe that the extremal partition for calculating the variation $\mathrm{Var}_\phi^W(f; [0, 1])$ is

$$P_n := \{0, s_n, t_n, s_{n-1}, t_{n-1}, \ldots, s_2, t_2, s_1\} \quad (n = 1, 2, 3, \ldots),$$

where $s_k := 1/k$ and $t_n \in (s_k, s_{k-1})$ is arbitrary. Then we have

$$\mathrm{Var}_\phi^W(f; [0, 1]) = 2 \sum_{k=2}^{\infty} \phi\left[f\left(\frac{1}{k}\right) \right] = 2 \sum_{k=2}^{\infty} e^{-2 \log k} = 2 \sum_{k=2}^{\infty} \frac{1}{k^2} < \infty,$$

and so $f \in V_\phi^W([0, 1])$. On the other hand,

$$\mathrm{Var}_\phi^W(2f; [0, 1]) \ge 2 \sum_{k=2}^{\infty} \phi\left[2f\left(\frac{1}{k}\right) \right] = 2 \sum_{k=2}^{\infty} e^{-\log k} = 2 \sum_{k=2}^{\infty} \frac{1}{k} = \infty,$$

and so $2f \notin V_\phi^W([0, 1])$. This shows that the set $V_\phi^W([0, 1])$ is not a linear space for this choice of ϕ. ♥

Proposition 2.9 below gives a criterion, both necessary and sufficient, for a Young function ϕ under which the corresponding set $V_\phi^W([a,b])$ is a linear space. Recall that the appropriate notion for solving this problem for Orlicz classes in Section 0.1 was the Δ_2-condition (for large values of t, see Definition 0.19 and Proposition 0.20). Here, we need a similar condition for small values of t:

Definition 2.4. A Young function $\phi : [0,\infty) \to [0,\infty)$ satisfies a δ_2-*condition* if

$$\phi(2t) \leq M\phi(t) \quad (0 \leq t \leq T) \tag{2.4}$$

for suitable constants $M > 0$ and $T > 0$. In this case, we write $\phi \in \delta_2$. ■

Note that the constant M appearing in the estimate (2.4) always satisfies $M \geq 1$. Indeed, since ϕ is increasing, we have

$$\phi(t) \leq \phi(2t) \leq M\phi(t)$$

for $0 \leq t \leq T$, which implies $M > 1$.

Now, we state three technical lemmas on Young functions which satisfy a δ_2-condition. The first lemma shows that the bounds M and T in Definition 2.4 are not independent.

Lemma 2.5. *A function ϕ satisfies a δ_2-condition if and only if for each $T' > 0$, there exists a number $M(T') \geq 1$ such that*

$$\phi(2t) \leq M(T')\phi(t) \quad (0 < t \leq T') .$$

Proof. Since the "if" part is trivial, we merely prove the "only if" part. Suppose that ϕ satisfies a δ_2-condition. Fix an arbitrary number T' such that $T' \geq T/2$, where $T > 0$ is as in (2.4). Since $\phi(T) \leq M(T)\phi(T/2)$, for $t \in [T/2, T']$, we obtain

$$\phi(t) \geq \frac{\phi(T)}{M(T)\phi(T/2)}\phi(t) = \frac{1}{M(T)}\frac{\phi(T)}{\phi(2t)}\frac{\phi(t)}{\phi(T/2)}\phi(2t) . \tag{2.5}$$

On the other hand, keeping the fact in mind that ϕ is increasing, from the inequalities $T/2 \leq t \leq T'$ and $2t \leq 2T'$ we get

$$\frac{\phi(2t)}{\phi(2T')} \leq 1 \leq \frac{\phi(t)}{\phi(T/2)} .$$

Consequently, by (2.5), we further obtain

$$\phi(t) \geq \frac{1}{M(T)}\frac{\phi(T)}{\phi(2t)}\frac{\phi(2t)}{\phi(2T')}\phi(2t) = \frac{1}{M(T)}\frac{\phi(T)}{\phi(2T')}\phi(2t) .$$

Thus, putting

$$M(T') := M(T)\frac{\phi(2T')}{\phi(T)}$$

and taking into account the estimate $\phi(2T') \geq \phi(T)$, we have proved the desired assertion. □

In view of Lemma 2.5, we can assign to every Young function $\phi \in \delta_2$ the function $M : (0, \infty) \to [1, \infty)$ defined by

$$M(T) := \sup_{0<t\le T} \frac{\phi(2t)}{\phi(t)} \quad (T > 0) \tag{2.6}$$

which is well-defined and increasing on $(0, \infty)$.

Lemma 2.6. *Let $\phi \in \delta_2$. Then*

$$\phi(s + t) \le M(T)[\phi(s) + \phi(t)]$$

for $0 < s, t \le T$.

Proof. Assuming without loss of generality that $s < t$, we obtain

$$\phi(s + t) = \phi\left(2\frac{s+t}{2}\right) \le M(T)\phi\left(\frac{s+t}{2}\right) \le M(T)\phi(t) \le M(T)[\phi(s) + \phi(t)]\,,$$

where we have used (2.6) and the monotonicity of ϕ. □

We remark that the more general estimate

$$\phi(t_1 + t_2 + \cdots + t_n) \le M^{n-1}((n-1)T)[\phi(t_1) + \phi(t_2) + \cdots + \phi(t_{n+1})]$$

may easily be proved by induction on n for $0 < t_i \le T$ $(i = 1, 2, \ldots, n)$. As a consequence of Lemma 2.6, we derive the following

Corollary 2.7. *Let $\phi \in \delta_2$ and let $f, g : [a, b] \to \mathbb{R}$ be bounded functions on $[a, b]$. Then*

$$\mathrm{Var}_\phi^W(f + g) \le M(2K)\left[\mathrm{Var}_\phi^W(f) + \mathrm{Var}_\phi^W(g)\right]\,,$$

where $K := \max\{\|f\|_\infty, \|g\|_\infty\}$, and $\|\cdot\|_\infty$ denotes the norm (0.39). Moreover, for $\mu \in \mathbb{R}$, we have

$$\mathrm{Var}_\phi^W(\mu f) \le (m + 1)M^m(2m\|f\|_\infty)\,\mathrm{Var}_\phi^W(f)\,,$$

where $m = \mathrm{ent}\,|\mu| = \max\{k \in \mathbb{N} : k \le |\mu|\}$ denotes the integer part of $|\mu|$. Consequently, $\phi \in \delta_2$ implies that $\mathrm{V}_\phi^W([a, b])$ is a linear space.

Now, we want to compare the function classes V_ϕ^W and V_ψ^W for two different Young functions ϕ and ψ. Let us say that a positive real sequence $(\beta_n)_n$ is *subordinate* to another positive real sequence $(\alpha_n)_n$ if the convergence of the series $\sum_{n=1}^\infty \alpha_n$ implies the convergence of the series $\sum_{n=1}^\infty \beta_n$. The following lemma is crucial for the comparison of the classes $\mathrm{V}_\phi^W([a, b])$ and $\mathrm{V}_\psi^W([a, b])$.

Lemma 2.8. *Let ϕ and ψ be two Young functions. Then, for every sequence $(t_n)_n$ of nonnegative real numbers, the sequence $(\psi(t_n))_n$ is subordinate to the sequence $(\phi(t_n))_n$ if and only if there exist numbers $A > 0$ and $B > 0$ such that*

$$\psi(t) \le B\phi(t) \quad (0 < t \le A)\,. \tag{2.7}$$

Proof. Clearly, (2.7) implies that $(\psi(t_n))_n$ is subordinate to $(\phi(t_n))_n$. Conversely, suppose that (2.7) is false. Then for each $A > 0$ and $B > 0$, we can find a number $t \in (0, A]$ such that $\psi(t) > B\phi(t)$. In particular, we may choose $B = n \in \mathbb{N}$ and $A > 0$ in such a way that $\phi(A) = 1/n^2$. Denoting by $t_n \in (0, A]$ the corresponding number satisfying $\psi(t_n) > n\phi(t_n)$, we conclude that

$$\sum_{n=1}^{\infty} \phi(t_n) \le \sum_{n=1}^{\infty} \frac{1}{n^2} < \infty .$$

Further, denote by k_n the least natural number such that

$$\frac{1}{n^2} \le k_n \phi(t_n) \le \frac{2}{n^2} .$$

For $m \in \mathbb{N}$ fixed, we choose $n = n(m) \in \mathbb{N}$ such that

$$k_1 + k_2 + \ldots + k_{m-1} < n \le k_1 + k_2 + \cdots + k_m .$$

The sequence $(t_m)_m$ then satisfies

$$\sum_{m=1}^{\infty} \phi(t_m) < \infty, \quad \sum_{m=1}^{\infty} \psi(t_m) = \infty , \tag{2.8}$$

contradicting our assumption that $(\psi(t_n))_n$ is subordinate to $(\phi(t_n))_n$. □

We point out that the hypotheses of Lemma 2.8 concerning the numbers A and B may be stated equivalently in the following way: for each $A' > 0$, we can find a constant $B(A') > 0$ such that

$$\psi(t) \le B(A')\phi(t) \quad (0 < t \le A') ; \tag{2.9}$$

the proof of this fact is similar to that of Lemma 2.5.

We are now in a position to prove two important results on the function class $V_\phi^W([a, b])$.

Proposition 2.9. *Let ϕ and ψ be two Young functions. Then the following hold.*
(a) *The inclusion $V_\phi^W([a, b]) \subseteq V_\psi^W([a, b])$ holds if and only if (2.7) is true.*
(b) *The set $V_\phi^W([a, b])$ is a linear space if and only if $\phi \in \delta_2$.*

Proof. (a) Suppose first that (2.7) is true, and so also (2.9), and assume that $f \in V_\phi^W([a, b])$. Since f is bounded, we may choose $A' := 2\|f\|_\infty$ in (2.9) and get

$$\psi(|f(s) - f(t)|) \le B(2\|f\|_\infty)\phi(|f(s) - f(t)|) .$$

Consequently, $\mathrm{Var}_\psi^W(f) \le B(2\|f\|_\infty)\,\mathrm{Var}_\phi^W(f)$, and so $V_\phi([a, b]) \subseteq V_\psi^W([a, b])$.

Conversely, suppose that (2.7) is false, i.e. for every $A > 0$ and $B > 0$, there exists $t \in (0, A]$ such that $\psi(t) > B\phi(t)$. As in Lemma 2.8, we can find a sequence $(t_m)_m$ satisfying (2.8). Denote by $(y_n)_n$ a fixed increasing sequence of points $y_n \in (a, b)$, and define $f : [a, b] \to \mathbb{R}$ by

$$f(x) := \begin{cases} t_m & \text{for } x = y_m , \\ 0 & \text{otherwise} . \end{cases}$$

We claim that $f \in V_\phi^W([a,b])$, but $f \notin V_\psi^W([a,b])$. In fact, taking an arbitrary partition $P = \{x_0, x_1, \dots, x_m\} \in \mathcal{P}([a,b])$, we obtain

$$\sum_{j=1}^{m} \phi(|f(x_j) - f(x_{j-1})|) \le 2\sum_{j=1}^{m} \phi(t_{l_j}) + \sum_{j=1}^{m} \phi(|t_j - t_{j-1}|) \le 4\sum_{j=1}^{\infty} \phi(t_j),$$

and hence $\mathrm{Var}_\phi^W(f) < \infty$ and so $f \in V_\phi^W([a,b])$. On the other hand, consider the partition $P = \{x_0, x_1, \dots, x_{2m}\}$, where

$$x_{2i+1} = y_i \quad (i = 0, 1, 2, \dots, m-2), \qquad x_{2i} = \frac{1}{2}(y_{i-1} + y_i) \quad (i = 1, 2, \dots, m-1).$$

For this partition, we have

$$\mathrm{Var}_\psi^W(f, P) = \sum_{j=1}^{2m} \psi(|f(x_j) - f(x_{j-1})|) \ge 2\sum_{j=0}^{m-2} \psi(t_j),$$

and hence $\mathrm{Var}_\psi^W(f) = \infty$ and so $f \notin V_\psi^W([a,b])$.

(b) The fact that $V_\phi^W([a,b])$ is a linear space in case $\phi \in \delta_2$ has already been proved in Corollary 2.7. Conversely, if $V_\phi^W([a,b])$ is a linear space, then $f \in V_\phi^W([a,b])$ implies $2f \in V_\phi^W([a,b])$. Thus, putting $\psi(t) := \phi(2t)$, we conclude that $V_\phi^W([a,b]) \subseteq V_\psi^W([a,b])$. However, in view of (a), this implies that

$$\phi(2t) = \psi(t) \le B\phi(t) \quad (0 < t \le A),$$

and hence $\phi \in \delta_2$, and so we are done. □

Clearly, the function $\phi(t) = t^p$ satisfies $\phi \in \delta_2$ for $0 < p < \infty$, as may be seen by choosing $M = 2^p$ and $T = 1$ in (2.4). The same is true for $\phi(t) = \log(1+t)$ since

$$\lim_{t\to 0+} \frac{\phi(2t)}{\phi(t)} = \lim_{t\to 0+} \frac{2(1+t)}{1+2t} = 2$$

by L'Hospital's rule. As we have seen in (0.22), the Young function $\phi(t) = e^t - 1$ does not satisfy a Δ_2-condition. However, it satisfies a δ_2-condition since

$$\lim_{t\to 0+} \frac{\phi(2t)}{\phi(t)} = 2\lim_{t\to 0+} e^t = 2,$$

again by L'Hospital's rule. Conversely, the Young function (2.3) satisfies $\phi \in \Delta_2$, but $\phi \notin \delta_2$ because

$$\lim_{t\to 0+} \frac{\phi(2t)}{\phi(t)} = \lim_{t\to 0+} e^{1/2t} = \infty, \qquad \lim_{t\to\infty} \frac{\phi(2t)}{\phi(t)} = 2.$$

This explains again why the set $V_\phi^W([a,b])$ in Example 2.3 is not a linear space. We remark, however, that the set $V_\phi^W([a,b])$ is always symmetric, balanced, absorbing, and convex. So, we may use the same method as for Orlicz classes and consider the set

$$B^W(\phi) := \{f \in B([a,b]) : \mathrm{Var}_\phi^W(f; [a,b]) \le 1\} \tag{2.10}$$

together with the corresponding Minkowski functional[3]

$$\|f\|_{WBV_\phi} := |f(a)| + \inf \left\{\lambda > 0 : \; f/\lambda \in B^W(\phi)\right\} \tag{2.11}$$

which is a *norm* on[4] $WBV_\phi([a,b]) = \operatorname{span} V_\phi^W([a,b])$. Moreover, the closed unit ball in the space $(WBV_\phi([a,b]), \|\cdot\|_{WBV_\phi})$ coincides with the set $B^W(\phi)$ given in (2.10).

Proposition 2.9 (b) shows that, loosely speaking, the δ_2-condition plays, for the space WBV_ϕ, the same role as the Δ_2-condition for the Orlicz space L_ϕ, see Proposition 0.20.

Let us check (2.11) for the function f from Example 2.3 in the space $WBV_\phi([0,1])$ with ϕ as in (2.3). Our calculations in Example 2.3 show that, for $\lambda > 0$,

$$\operatorname{Var}_\phi^W(f/\lambda; [0,1]) = 2\sum_{k=2}^{\infty} e^{-2\lambda \log k} = 2\sum_{k=2}^{\infty} \frac{1}{k^{2\lambda}}.$$

Thus, the infimum[5] of all $\lambda > 0$ such that $\operatorname{Var}_\phi^W(f/\lambda; [0,1])$ is finite is $1/2$. To calculate the norm of f explicitly, however, we have to find the infimum of all $\lambda > 0$ such that $\operatorname{Var}_\phi^W(f/\lambda; [0,1]) \le 1$, and this is more difficult.

In the following proposition which is parallel to Proposition 1.3, we collect some properties of the quantities (2.1) and (2.2) and the space $(WBV_\phi([a,b]), \|\cdot\|_{WBV_\phi})$.

Proposition 2.10. *The quantities* (2.1) *and* (2.2) *have the following properties.*

(a) *The variation* (2.2) *is superadditive with respect to intervals, i.e.*

$$\operatorname{Var}_\phi^W(f;[a,b]) \ge \operatorname{Var}_\phi^W(f;[a,c]) + \operatorname{Var}_\phi^W(f;[c,b])$$

for $a < c < b$.

(b) *Conversely, if* ϕ *satisfies the* δ_2*-condition* (2.4)*, then*

$$\operatorname{Var}_\phi^W(f;[a,b]) \le M(2\|f\|_\infty)\left[\operatorname{Var}_\phi^W(f;[a,c]) + \operatorname{Var}_\phi^W(f;[c,b])\right],$$

for $a < c < b$*, where* $\|\cdot\|_\infty$ *denotes the norm* (0.39)*, and* M *is the function defined in* (2.6).

(c) *If* ϕ *satisfies the* δ_2*-condition* (2.4)*, then*

$$\operatorname{Var}_\phi^W(f+g;[a,b]) \le M(2K)\left[\operatorname{Var}_\phi^W(f;[a,b]) + \operatorname{Var}_\phi^W(g,[a,b])\right],$$

where $K := \max\{\|f\|_\infty, \|g\|_\infty\}$*, and so the variation* (2.2) *is partly subadditive with respect to functions.*

3 Obviously, for $\phi(t) = t^p$ with $p \ge 1$ the infimum in (2.11) coincides with $\operatorname{Var}_p^W(f;[a,b])^{1/p}$, see (1.61). So $\|f\|_{WBV_\phi} = \|f\|_{WBV_p}$ in this case, as one should expect.

4 As usual, span M denotes the *linear hull* of M, i.e. the smallest linear space containing M.

5 Of course, this infimum is not a minimum! This is precisely the reason for the fact that $2f \notin V_\phi^W([0,1])$, as we have seen in Example 2.3.

(d) *If ϕ satisfies the δ_2-condition* (2.4), *then*

$$\mathrm{Var}_\phi^W(\mu f; [a,b]) \le (m+1)M^m(2m\|f\|_\infty)\,\mathrm{Var}_\phi^W(f; [a,b]) \quad (\mu \in \mathbb{R}),$$

where $m = \mathrm{ent}\,|\mu|$ *denotes the integer part of the number* $|\mu|$.

(e) *The estimate*

$$\mathrm{Var}_\phi^W(f; [a,b]) \le \phi(\mathrm{Var}(f; [a,b]))$$

holds, where $\mathrm{Var}(f; [a,b])$ *denotes the classical variation* (1.4). *In particular, if f is monotone on the interval* $[a,b]$, *then*

$$\mathrm{Var}_\phi^W(f; [a,b]) = \phi(|f(b) - f(a)|).$$

(f) *The space* $WBV_\phi([a,b])$ *is complete with respect to the norm* (2.11).

Proof. To prove property (a), we may assume that both variations $\mathrm{Var}_\phi^W(f; [a,c])$ and $\mathrm{Var}_\phi^W(f; [c,b])$ are finite. Given $\varepsilon > 0$, choose partitions

$$\{t_0, t_1, \dots, t_k\} \in \mathcal{P}([a,c]), \quad \{t_k, t_{k+1}, \dots, t_m\} \in \mathcal{P}([c,b])$$

such that $t_k = c$,

$$\mathrm{Var}_\phi^W(f; [a,c]) - \varepsilon \le \sum_{i=1}^{k} \phi(|f(t_i) - f(t_{i-1})|)$$

and

$$\mathrm{Var}_\phi^W(f; [c,b]) - \varepsilon \le \sum_{j=k+1}^{m} \phi(|f(t_j) - f(t_{j-1})|).$$

Then we obtain

$$\begin{aligned}&\mathrm{Var}_\phi^W(f; [a,c]) + \mathrm{Var}_\phi^W(f; [c,b]) - 2\varepsilon\\ &\le \sum_{k=1}^{m} \phi(|f(t_k) - f(t_{k-1})|) \le \mathrm{Var}_\phi^W(f; [a,b])\end{aligned}$$

which proves (a) since $\varepsilon > 0$ was arbitrary.

In order to prove property (b), let $\{t_0, t_1, \dots, t_m\} \in \mathcal{P}([a,b])$. If some point of this partition, say t_k, coincides with c, we have

$$\begin{aligned}\sum_{j=1}^{m} \phi(|f(t_j) - f(t_{j-1})|) &= \sum_{j=1}^{k} \phi(|f(t_j) - f(t_{j-1})|) + \sum_{j=k+1}^{m} \phi(|f(t_j) - f(t_{j-1})|)\\ &\le \mathrm{Var}_\phi^W(f; [a,c]) + \mathrm{Var}_\phi^W(f; [c,b]).\end{aligned}$$

Now, taking into account the fact that the function M in (2.6) satisfies $M(T) \ge 1$ for all $T > 0$ (Lemma 2.5), we derive the estimate

$$\sum_{j=1}^{m} \phi(|f(t_j) - f(t_{j-1})|) \le M(2\|f\|_\infty)\left[\mathrm{Var}_\phi^W(f; [a,c]) + \mathrm{Var}_\phi^W(f; [c,b])\right]. \tag{2.12}$$

Now, consider the opposite case, i.e. $t_{k-1} < c < t_k$ for some $k \in \{1, 2, \dots, m\}$. Then we obtain

$$\begin{aligned}
&\sum_{j=1}^{m} \phi(|f(t_j) - f(t_{j-1})|) \\
&= \sum_{j=1}^{k-1} \phi\left(|f(t_j) - f(t_{j-1})|\right) + \phi\left(|f(t_k) - f(t_{k-1})|\right) + \sum_{j=k+1}^{m} \phi\left(|f(t_j) - f(t_{j-1})|\right) \\
&= \sum_{j=1}^{k-1} \phi\left(|f(t_j) - f(t_{j-1})|\right) + \phi\left(|f(t_k) - f(c)| + |f(c) - f(t_{k-1})|\right) \\
&\quad + \sum_{j=k+1}^{m} \phi\left(|f(t_j) - f(t_{j-1})|\right).
\end{aligned}$$

From Lemma 2.5 and this estimate, we get

$$\begin{aligned}
&\sum_{j=1}^{m} \phi(|f(t_j) - f(t_{j-1})|) \\
&\le \sum_{j=1}^{k-1} \phi(f(t_j) - f(t_{j-1})) + M(2\|f\|_\infty)\phi(|f(c) - f(t_{k-1})|) \\
&\quad + M(2\|f\|_\infty)\phi(|f(t_k) - f(c)|) + \sum_{j=k+1}^{m} \phi(|f(t_j) - f(t_{j-1})|),
\end{aligned}$$

and hence

$$\sum_{j=1}^{m} \phi(|f(t_j) - f(t_{j-1})|) \le M(2\|f\|_\infty)\left[\mathrm{Var}_\phi^W(f; [a, c]) + \mathrm{Var}_\phi^W(f; [c, b])\right]. \tag{2.13}$$

Combining the estimates (2.12) and (2.13), we deduce the desired inequality.

The assertions (c) and (d) have already been proved in Corollary 2.7.

To prove (e), fix a partition $\{t_0, t_1, \dots, t_m\} \in \mathcal{P}([a, b])$. Then we obtain

$$\sum_{j=1}^{m} \phi(|f(t_j) - f(t_{j-1})|) \le \phi\left(\sum_{j=1}^{m} |f(t_j) - f(t_{j-1})|\right) \le \phi(\mathrm{Var}(f; [a, b]))$$

since ϕ is superadditive and increasing, see Lemma 1.36. This yields the estimate

$$\mathrm{Var}_\phi^W(f; [a, b]) \le \phi(\mathrm{Var}(f; [a, b])).$$

Obviously, in case of a monotone function f, we get, in particular, $\mathrm{Var}(f; [a, b]) = |f(b) - f(a)|$, so

$$\mathrm{Var}_\phi^W(f; [a, b]) \le \phi(|f(b) - f(a)|).$$

On the other hand, taking the special partition $P := \{a, b\}$, we get

$$\phi(|f(b) - f(a)|) = \mathrm{Var}_\phi^W(f, P; [a, b]) \le \mathrm{Var}_\phi^W(f; [a, b])$$

which proves (e).

It remains to prove (f). Assume that $(f_n)_n$ is a Cauchy sequence in $WBV_\phi([a, b])$ with respect to the norm (2.11). This means that for $\varepsilon > 0$, there exists a natural number n_0 such that

$$\|f_n - f_m\|_{WBV_\phi} \leq \varepsilon \quad (m, n \geq n_0).$$

Keeping in mind the definition (2.11) of the norm $\|\cdot\|_{WBV_\phi}$, this means that both

$$|f_n(a) - f_m(a)| \leq \varepsilon, \quad \mathrm{Var}_\phi^W\left(\frac{f_n - f_m}{\varepsilon}; [a, b]\right) \leq 1. \tag{2.14}$$

From the first estimate in (2.14), we deduce that the sequence $(f_n(a))_n$ converges to some real number which we denote by $f(a)$. On the other hand, the second inequality in (2.14) implies that, for any partition $\{t_0, t_1, \ldots, t_m\} \in \mathcal{P}([a, b])$, we have

$$\sum_{j=1}^{m} \phi\left(\frac{|[f_n(t_j) - f_m(t_j)] - [f_n(t_{j-1}) - f_m(t_{j-1})]|}{\varepsilon}\right) \leq 1. \tag{2.15}$$

For the particular partition $\{a, x, b\}$ (with $x \in (a, b)$ fixed), we obtain from (2.15)

$$\phi\left(\frac{|[f_n(x) - f_m(x)] - [f_n(a) - f_m(a)]|}{\varepsilon}\right) \leq 1,$$

hence

$$\frac{|[f_n(x) - f_m(x)] - [f_n(a) - f_m(a)]|}{\varepsilon} \leq \phi^{-1}(1),$$

and so

$$|f_n(x) - f_m(x)| \leq |f_n(a) - f_m(a)| + \varepsilon\phi^{-1}(1).$$

Finally, for arbitrary $x \in [a, b]$, we get

$$|f_n(x) - f_m(x)| \leq \varepsilon(1 + \phi^{-1}(1))$$

by the first estimate in (2.14). The last inequality implies that the sequence $(f_n)_n$ is uniformly convergent to some function f on the interval $[a, b]$, i.e. $f_n \to f$ in $B([a, b])$.

Now, we again take into account the estimate (2.15). Keeping n fixed and letting $m \to \infty$, we obtain

$$\sum_{j=1}^{m} \phi\left(\frac{|[f_n(t_j) - f(t_j)] - [f_n(t_{j-1}) - f(t_{j-1})]|}{\varepsilon}\right) \leq 1,$$

which means nothing else but

$$\mathrm{Var}_\phi^W\left(\frac{f_n - f}{\varepsilon}; [a, b]\right) \leq 1 \quad (n \geq n_0),$$

and so $\|f_n - f\|_{WBV_\phi} \leq 2\varepsilon$, i.e. $f_n \to f$ in $WBV_\phi([a, b])$. □

Note that, in contrast to Proposition 1.3, properties which are analogous to Proposition 1.3 (f) and (g) are not valid here. Indeed, we have shown this for the special Young function $\phi(t) = t^p$ and the space $WBV_p([a,b])$ $(p > 1)$ already in Example 1.33. Concerning the behavior of $\mathrm{Var}_p^W(f)$ with respect to subintervals, a combination of Proposition 2.10 (a) and (b) shows that

$$\mathrm{Var}_p^W(f;[a,c]) + \mathrm{Var}_p^W(f;[c,b]) \le \mathrm{Var}_p^W(f;[a,b]) \\ \le 2^p\left[\mathrm{Var}_p^W(f;[a,c]) + \mathrm{Var}_p^W(f;[c,b])\right],$$

where the second estimate follows from the fact that $M(T) \equiv 2^p$ in (2.6) for $\phi(t) = t^p$.

For further use, we now define an additional property of Young functions in the following

Definition 2.11. Let $1 \le p < \infty$. We say that a Young function $\phi : [0,\infty) \to [0,\infty)$ satisfies *condition* ∞_p if

$$\lim_{t\to\infty} \frac{\phi(t)}{t^p} = \infty. \tag{2.16}$$

In this case, we write $\phi \in \infty_p$. Similarly, we say that ϕ satisfies *condition* 0_p if

$$\lim_{t\to 0} \frac{\phi(t)}{t^p} = \infty. \tag{2.17}$$

In this case, we write $\phi \in 0_p$. ∎

For example, the Young function $\phi(t) = t^q$ satisfies condition ∞_p for $q > p$ and condition 0_p for $q < p$, while the Young function $\phi(t) = e^t - 1$ satisfies condition ∞_p for all $p \ge 1$ and condition 0_p for all $p > 1$. Finally, the Young function $\phi(t) = (t+1)\log(t+1)$ satisfies condition ∞_p for $p = 1$, but not for $p > 1$, and condition 0_p for $p > 1$, but not for $p = 1$. In the following Table 2.1, we compare the growth conditions (0.21), (2.4), (2.16) and (2.17) for some Young functions.

Table 2.1. Growth properties of Young functions.

$\phi(t) =$	$\phi \in \Delta_2$	$\phi \in \delta_2$	$\phi \in \infty_1$	$\phi \in 0_1$
t^p	yes	yes	yes	no
e^{t-1}	no	yes	no	no
$(t+1)\log(t+1)$	yes	yes	yes	no
(2.3)	yes	no	no	yes

Clearly, $\phi \in \infty_p$ implies $\phi \in \infty_q$ for all $q < p$, but not vice versa, while $\phi \in 0_p$ implies $\phi \in 0_q$ for all $q > p$, but not vice versa. So condition ∞_1 is the mildest requirement in (2.16), while condition 0_1 is the strongest requirement in (2.17).

The following proposition gives useful estimates for Young functions $\phi \in \infty_1$.

Proposition 2.12. *Suppose that a Young function $\phi : [0,\infty) \to [0,\infty)$ satisfies (2.16) for $p = 1$. Then the following is true.*

(a) *The estimate*

$$\phi\left(\frac{\sum_{k=1}^n \alpha_k u_k}{\sum_{k=1}^n \alpha_k}\right) \leq \frac{\sum_{k=1}^n \alpha_k \phi(u_k)}{\sum_{k=1}^n \alpha_k} \tag{2.18}$$

holds for all nonnegative numbers $\alpha_1, \dots, \alpha_n$ and $u_1, \dots, u_n$ satisfying

$$\sum_{k=1}^n \alpha_k > 0 .$$

(b) *The estimate*

$$\phi\left(\frac{\int_c^d \alpha(t)u(t)\,dt}{\int_c^d \alpha(t)\,dt}\right) \leq \frac{\int_c^d \alpha(t)\phi(u(t))\,dt}{\int_c^d \alpha(t)\,dt} \tag{2.19}$$

holds for all nonnegative integrable functions $u : [c,d] \to \mathbb{R}$ and $\alpha : [c,d] \to \mathbb{R}$ satisfying

$$\int_c^d \alpha(t)\,dt > 0 .$$

Proof. For $n = 2$, the estimate (2.18) is just the definition of the convexity of ϕ; for $n \geq 3$, we may prove assertion (a) easily by induction.

To prove (b), let $v > 0$ be fixed. From the convexity of ϕ, it follows that there exists a $k \in \mathbb{R}$ such that

$$\phi(u) - \phi(v) \geq k(u - v) \quad (u \geq 0) .$$

Substituting $u = u(t)$ but keeping v constant, multiplying by $\alpha(t)$ and integrating over $[c,d]$ yields

$$\int_c^d \alpha(t)\phi(u(t))\,dt - \phi(v)\int_c^d \alpha(t)\,dt \geq k\left\{\int_c^d \alpha(t)u(t)\,dt - v\int_c^d \alpha(t)\,dt\right\} . \tag{2.20}$$

Now, for the choice

$$v := \frac{\int_c^d \alpha(t)u(t)\,dt}{\int_c^d \alpha(t)\,dt},$$

(2.20) becomes

$$\int_c^d \alpha(t)\phi(u(t))\,dt - \phi\left(\frac{\int_c^d \alpha(t)u(t)\,dt}{\int_c^d \alpha(t)\,dt}\right)\int_c^d \alpha(t)\,dt \geq 0 ,$$

and the estimate (2.19) follows. □

The estimate (2.18) is usually called the *discrete Jensen inequality* and the estimate (2.19) is the *continuous Jensen inequality*.

As the definition (1.65) of $\|\cdot\|_{WBV_p}$ shows, the expression $\mathrm{Var}_p^W(f;[a,b])^{1/p}$ plays an important role since it is homogeneous in f, see Proposition 1.32 (b). Now, we provide a result concerning the relationship between two spaces WBV_ϕ and WBV_ψ which involves a similar expression for general Young functions ϕ and ψ. This result generalizes Proposition 1.38.

Proposition 2.13. *Let ϕ and ψ be Young functions such that the function $\psi \circ \phi^{-1}$ is convex. Then the inequality*

$$\psi^{-1}\left(\mathrm{Var}_\psi^W(f;[a,b])\right) \le \phi^{-1}\left(\mathrm{Var}_\phi^W(f;[a,b])\right) \tag{2.21}$$

holds. Consequently,

$$WBV_\phi([a,b]) \subseteq WBV_\psi([a,b]) \tag{2.22}$$

for such Young functions ϕ and ψ.

Proof. To prove (2.21), we may assume that $\mathrm{Var}_\phi^W(f;[a,b])$ is finite. Fix an arbitrary partition $\{t_0, t_1, \ldots, t_m\} \in \mathcal{P}([a,b])$. Then, keeping in mind the fact that both functions ψ and ϕ^{-1} are increasing, see Lemma 1.36, and taking into account (1.70) (with ϕ replaced by $\psi \circ \phi^{-1}$), we obtain

$$\begin{aligned}
\mathrm{Var}_\psi^W(f,P;[a,b]) &= \sum_{j=1}^m \psi(|f(t_j) - f(t_{j-1})|)\\
&= \sum_{j=1}^m (\psi \circ \phi^{-1} \circ \phi)(|f(t_j) - f(t_{j-1})|) = \sum_{j=1}^m (\psi \circ \phi^{-1})(\phi(|f(t_j) - f(t_{j-1})|))\\
&\le (\psi \circ \phi^{-1})\left(\sum_{j=1}^m \phi(|f(t_j) - f(t_{j-1})|)\right)\\
&= (\psi \circ \phi^{-1})\left(\mathrm{Var}_\phi^W(f,P;[a,b])\right) \le (\psi \circ \phi^{-1})\left(\mathrm{Var}_\phi^W(f;[a,b])\right).
\end{aligned}$$

Passing in the first expression to the supremum over all partitions, we get

$$\mathrm{Var}_\psi^W(f;[a,b]) \le \psi\left(\phi^{-1}\left(\mathrm{Var}_\phi^W(f;[a,b])\right)\right),$$

and applying the monotone function ψ^{-1} to both sides gives (2.21). The inclusion (2.22) is of course an immediate consequence of (2.21). □

Putting $\phi(t) = t$ in (2.21), the convexity of $\psi = \psi \circ \phi^{-1}$ implies the following

Corollary 2.14. *Let ψ be a Young function. Then the inequality*

$$\psi^{-1}\left(\mathrm{Var}_\psi^W(f;[a,b])\right) \le \mathrm{Var}(f;[a,b])$$

holds for arbitrary bounded functions $f : [a,b] \to \mathbb{R}$. Consequently, the inclusion

$$BV([a,b]) \subseteq WBV_\psi([a,b]) \tag{2.23}$$

holds true.

Observe that (2.23) means that every function of classical bounded variation on $[a, b]$ has a bounded Wiener–Young variation with respect to an arbitrary Young function ϕ. Of course, the simplest examples for Young functions satisfying the hypotheses of Proposition 2.13 are $\phi(t) = t^p$ and $\psi(t) = t^q$ with $1 \le p \le q$. In this way, we get Proposition 1.38 as a special case of Proposition 2.13.

2.2 The Waterman variation

In Section 1.2, we have introduced the family $\Sigma([a, b])$ of all finite collections $S = \{[a_1, b_1], \dots, [a_n, b_n]\}$ of nonoverlapping intervals $[a_1, b_1], \dots, [a_n, b_n] \subset [a, b]$, and the family $\Sigma_\infty([a, b])$ of all infinite collections $S_\infty = \{[a_n, b_n] : n \in \mathbb{N}\}$ of nonoverlapping intervals $[a_n, b_n] \subset [a, b]$. These families still may be used to introduce another concept of variation.

Definition 2.15. A *Waterman sequence* is a decreasing sequence $\Lambda = (\lambda_n)_n$ of positive real numbers such that $\lambda_n \to 0$ as $n \to \infty$ and[6]

$$\sum_{n=1}^{\infty} \lambda_n = \infty . \tag{2.24}$$

Given a function $f : [a, b] \to \mathbb{R}$, a set $S_\infty = \{[a_n, b_n] : n \in \mathbb{N}\} \in \Sigma_\infty([a, b])$, and a Waterman sequence $\Lambda = (\lambda_n)_n$, the positive real number

$$\mathrm{Var}_\Lambda(f, S_\infty) = \mathrm{Var}_\Lambda(f, S_\infty; [a, b]) := \sum_{k=1}^{\infty} \lambda_k |f(b_k) - f(a_k)| \tag{2.25}$$

is called the *Waterman variation* of f on $[a, b]$ with respect to S_∞, while the (possibly infinite) number

$$\mathrm{Var}_\Lambda(f) = \mathrm{Var}_\Lambda(f; [a, b]) := \sup \left\{ \mathrm{Var}_\Lambda(f, S_\infty; [a, b]) : S_\infty \in \Sigma_\infty([a, b]) \right\} , \tag{2.26}$$

where the supremum is taken over all collections $S_\infty \in \Sigma_\infty([a, b])$, is called the *total Waterman variation* of f on $[a, b]$. In case $\mathrm{Var}_\Lambda(f; [a, b]) < \infty$, we say that f has *bounded Waterman variation* (or *bounded Λ-variation in Waterman's sense*) on $[a, b]$ and write $f \in \Lambda BV([a, b])$. ■

If we drop the condition $\lambda_n \to 0$ and take $\lambda_n \equiv 1$, the space ΛBV coincides with the classical space BV. Proposition 2.17 (e) below shows that this is, in a certain sense, an "extremal" choice for Λ. We also remark that ΛBV and BV coincide if and only if the sequence $(\lambda_n)_n$ is bounded away from zero (Exercise 2.4). This is the reason why we require $\lambda_n \to 0$ as $n \to \infty$ in Definition 2.15.

6 The most natural example of such a sequence is of course $\lambda_n = n^{-q}$ for $0 < q \le 1$; we will consider this case separately in Definition 2.29 below.

Before describing some properties of the Waterman variation (2.26), we prove an auxiliary fact which is parallel to Proposition 1.18 and will be used later.

Proposition 2.16 (localization principle). *Suppose that $f \notin \Lambda BV([a,b])$. Then there exists a point $x_0 \in [a,b]$ such that $f \notin \Lambda BV([c,d])$ for each interval $[c,d] \subseteq [a,b]$ such that $x_0 \in (c,d)$.*

Proof. We successively divide the interval $I := [a,b]$ into two halves. Suppose that the series in (2.25) is uniformly bounded by some constant $M > 0$ whenever the intervals $I_n := [a_n, b_n]$ are all contained either in the lower half of I or all in the upper half of I. Then $\mathrm{Var}_\Lambda(f;[a,b]) \leq 2M$, contradicting our assumption. Therefore, the set of such series must be unbounded in one half of the interval I. If we continue this procedure we obtain a nested sequence of intervals $(J_n)_n$ converging to some point x_0. Now, if x_0 is an interior point of some subinterval $J \subseteq I$, then $J \supseteq J_n$ for n sufficiently large. From this, the conclusion follows. □

Now, we collect some simple properties of the Waterman variations (2.25) and (2.26) for further reference.

Proposition 2.17. *The quantities* (2.25) *and* (2.26) *have the following properties.*

(a) *The Waterman variation* (2.25) *is monotone with respect to collections of subintervals, i.e. $S_\infty \subseteq T_\infty$ implies*

$$\mathrm{Var}_\Lambda(f, S_\infty) \leq \mathrm{Var}_\Lambda(f, T_\infty).$$

(b) *The total Waterman variation* (2.26) *is subadditive with respect to intervals, i.e.*

$$\mathrm{Var}_\Lambda(f;[a,b]) \leq \mathrm{Var}_\Lambda(f;[a,c]) + \mathrm{Var}_\Lambda(f;[c,b]).$$

(c) *The total Waterman variation* (2.26) *is positively homogeneous, i.e. the equality*

$$\mathrm{Var}_\Lambda(\mu f;[a,b]) = |\mu|\,\mathrm{Var}_\Lambda(f;[a,b])$$

holds for all $\mu \in \mathbb{R}$.

(d) *The variation* (2.26) *is subadditive with respect to functions, i.e.*

$$\mathrm{Var}_\Lambda(f+g;[a,b]) \leq \mathrm{Var}_\Lambda(f;[a,b]) + \mathrm{Var}_\Lambda(g;[a,b]);$$

moreover,

$$|\mathrm{Var}_\Lambda(f;[a,b]) - \mathrm{Var}_\Lambda(g;[a,b])| \leq \mathrm{Var}_\Lambda(f-g;[a,b]).$$

(e) *If f is of bounded variation on the interval $[a,b]$, then $f \in \Lambda BV([a,b])$ for every Waterman sequence Λ.*

(f) *If $f \in \Lambda BV([a,b])$, then f is bounded on $[a,b]$.*

Proof. Property (a) is obvious. To prove (b), fix an arbitrary collection $S_\infty = \{[a_n,b_n] : n \in \mathbb{N}\} \in \Sigma_\infty([a,b])$. We define two index sets $N_1, N_2 \subset \mathbb{N}$ by

$$N_1 := \{k \in \mathbb{N} : [a_k,b_k] \subseteq [a,c]\}, \quad N_2 := \{k \in \mathbb{N} : [a_k,b_k] \subseteq [c,b]\}.$$

Obviously, the sets N_1 and N_2 are disjoint. Moreover, there exists at most one natural number k_0 such that $c \in (a_{k_0}, b_{k_0})$. So, we have

$$\begin{aligned}
\sum_{k=1}^{\infty} \lambda_k |f(b_k) - f(a_k)| &= \sum_{k \in N_1 \cup N_2 \cup \{k_0\}} \lambda_k |f(b_k) - f(a_k)| \\
&= \sum_{k \in N_1} \lambda_k |f(b_k) - f(a_k)| \\
&\quad + \sum_{k \in N_2} \lambda_k |f(b_k) - f(a_k)| + \lambda_{k_0} |f(b_{k_0}) - f(a_{k_0})| \\
&\le \left\{ \sum_{k \in N_1} \lambda_k |f(b_k) - f(a_k)| + \lambda_{k_0} |f(b_{k_0}) - f(c)| \right\} \\
&\quad + \left\{ \sum_{k \in N_2} \lambda_k |f(b_k) - f(a_k)| + \lambda_{k_0} |f(c) - f(a_{k_0})| \right\} \\
&\le \mathrm{Var}_\Lambda(f; [a, c]) + \mathrm{Var}_\Lambda(f; [c, b]) .
\end{aligned}$$

This yields the desired inequality from property (b). The proof of (c) is trivial and therefore omitted.

To prove (d), fix an infinite collection $S_\infty = \{[a_n, b_n] : n \in \mathbb{N}\} \in \Sigma_\infty([a, b])$. Then

$$\begin{aligned}
&\sum_{n=1}^{\infty} \lambda_n |f(b_n) + g(b_n) - f(a_n) - g(a_n)| \\
&\qquad \le \sum_{n=1}^{\infty} \lambda_n |f(b_n) - f(a_n)| + \sum_{n=1}^{\infty} \lambda_n |g(b_n) - g(a_n)| ,
\end{aligned}$$

and hence

$$\mathrm{Var}_\Lambda(f + g) \le \mathrm{Var}_\Lambda(f) + \mathrm{Var}_\Lambda(g) .$$

In the case when $f, g \in \Lambda BV([a, b])$, from this inequality, we deduce that

$$\mathrm{Var}_\Lambda(f) = \mathrm{Var}_\Lambda[(f - g) + g] \le \mathrm{Var}_\Lambda(f - g) + \mathrm{Var}_\Lambda(g) ,$$

and hence

$$\mathrm{Var}_\Lambda(f) - \mathrm{Var}_\Lambda(g) \le \mathrm{Var}_\Lambda(f - g) .$$

In the same way, keeping in mind property (c) for $\mu = -1$, we get

$$\mathrm{Var}_\Lambda(g) - \mathrm{Var}_\Lambda(f) \le \mathrm{Var}_\Lambda(f - g) ,$$

and therefore (d) is proved.

For the proof of (e), we again fix an infinite collection $S_\infty = \{[a_n, b_n] : n \in \mathbb{N}\} \in \Sigma_\infty([a, b])$. Since the sequence $\Lambda = (\lambda_n)_n$ is decreasing, we obtain

$$\sum_{n=1}^{\infty} \lambda_n |f(b_n) - f(a_n)| \le \lambda_1 \sum_{n=1}^{\infty} |f(b_n) - f(a_n)| \le \lambda_1 \,\mathrm{Var}(f; [a, b]) .$$

From this property (e) follows. Finally, to prove (f), suppose that $f \in \Lambda BV([a,b])$, but f is not bounded on the interval $[a,b]$. Then there exists a sequence $(u_n)_n$ of points $u_n \in [a,b]$ such that $|f(u_n)| \to \infty$. Since $(u_n)_n$ is bounded, it is possible to select a subsequence $(v_n)_n$ of the sequence $(u_n)_n$ such that $v_n \to x$ for some $x \in [a,b]$, but still $|f(v_n)| \to \infty$ as $n \to \infty$. Next, we can choose a monotone subsequence $(w_n)_n$ of $(v_n)_n$ with $|f(w_n)| \to \infty$ and $w_n \to x$, and then another subsequence $(z_n)_n$ of $(w_n)_n$ such that

$$|f(z_{n+1})| \geq 1 + |f(z_n)| \quad (n = 1, 2, 3, \ldots). \tag{2.27}$$

Now, the intervals $[z_1, z_2], [z_2, z_3], [z_3, z_4], \ldots$ form a collection $S_\infty \in \Sigma_\infty([a,b])$. Calculating the Waterman variation (2.25) with respect to this collection, we get

$$\begin{aligned}\mathrm{Var}_\Lambda(f, S_\infty) &= \sum_{n=1}^{\infty} \lambda_n |f(z_{n+1}) - f(z_n)| \\ &\geq \sum_{n=1}^{\infty} \lambda_n \left[|f(z_{n+1})| - |f(z_n)|\right] \geq \sum_{n=1}^{\infty} \lambda_n = \infty\end{aligned}$$

by (2.27). However, this contradicts the assumption that f belongs to $\Lambda BV([a,b])$, and the proof is complete. □

Proposition 2.17 (f) may be formulated as imbedding $\Lambda BV \hookrightarrow B$, where the imbedding constant $c(\Lambda BV, B)$ only involves the first element λ_1 of Λ, see Exercise 2.17. Note that properties (e) and (f) in Proposition 2.17 may be summarized as inclusions

$$BV([a,b]) \subseteq \bigcap_{\Lambda} \Lambda BV([a,b]), \quad \bigcup_{\Lambda} \Lambda BV([a,b]) \subseteq B([a,b]), \tag{2.28}$$

where the intersection and union in (2.28) are taken over all Waterman sequences Λ. Later (Proposition 2.24), we will prove a more precise result.

In the following proposition, we state yet another useful characterization of functions from $\Lambda BV([a,b])$.

Proposition 2.18. *The following three statements are equivalent.*
(a) *The function f belongs to $\Lambda BV([a,b])$.*
(b) *There exists a constant $M > 0$ such that*

$$\mathrm{Var}_\Lambda(f, S_\infty) = \sum_{k=1}^{\infty} \lambda_k |f(b_k) - f(a_k)| \leq M$$

for every infinite collection $S_\infty = \{[a_k, b_k] : k \in \mathbb{N}\} \in \Sigma_\infty([a,b])$.
(c) *There exists a constant $N > 0$ such that*

$$\mathrm{Var}_\Lambda(f, S) = \sum_{i=1}^{n} \lambda_i |f(b_i) - f(a_i)| \leq N$$

for any finite collection $S = \{[a_1, b_1], \ldots, [a_n, b_n]\} \in \Sigma([a,b])$.

Proof. Clearly, (a) is equivalent to (b) in view of Definition 2.15. The fact that (c) implies (b) is a simple consequence of a necessary, sufficient and well-known convergence condition for series with nonnegative terms. So, we only have to prove that (b) implies (c).

Assume that condition (b) is satisfied, but (c) is not. This means that for any number $N > 0$, there exists a finite collection $S = \{[a_1, b_1], \ldots, [a_n, b_n]\} \in \Sigma([a, b])$ such that

$$\sum_{i=1}^{n} \lambda_i |f(b_i) - f(a_i)| > N .$$

Taking $N := 2M$, where M is the constant appearing in (b), we find a finite collection $S_M := \{[a_1, b_1], \ldots, [a_n, b_n]\} \in \Sigma([a, b])$ satisfying

$$\mathrm{Var}(f, S_M) = \sum_{i=1}^{n} \lambda_i |f(b_i) - f(a_i)| > 2M .$$

Now, we distinguish two possible cases. Assume first that

$$[a_1, b_1] \cup [a_2, b_2] \cup \ldots \cup [a_n, b_n] = [a, b] .$$

Consider an infinite collection $S_\infty^* = \{[\alpha_i, \beta_i] : i = n, n + 1, n + 2, \ldots\} \in \Sigma_\infty([a_n, b_n])$ whose first element satisfies

$$\alpha_n = a_n, \quad \beta_n = \frac{a_n + b_n}{2} .$$

Adding to this collection the reduced collection $S_M' = \{[a_1, b_1], \ldots, [a_{n-1}, b_{n-1}]\}$, we get the infinite collection

$$S_M' \cup S_\infty^* = \{[a_1, b_1], \ldots, [a_{n-1}, b_{n-1}], [\alpha_n, \beta_n], [\alpha_{n+1}, \beta_{n+1}], \ldots\} \in \Sigma_\infty([a, b]) .$$

By assumption (b), we know that

$$\mathrm{Var}_\Lambda(f, S_M' \cup S_\infty^*) = \sum_{k=1}^{n-1} \lambda_k |f(b_k) - f(a_k)| + \sum_{k=n}^{\infty} \lambda_k |f(\beta_k) - f(\alpha_k)| \le M .$$

By only taking into account the first term in the last sum, we derive the estimate

$$\sum_{k=1}^{n-1} \lambda_k |f(b_k) - f(a_k)| + \lambda_n |f((a_n + b_n)/2) - f(a_n)| \le M .$$

Similarly, we now consider an infinite collection $S_\infty^{**} = \{[\gamma_i, \delta_i] : i = n, n + 1, n + 2, \ldots\} \in \Sigma_\infty([a_n, b_n])$ whose first element satisfies

$$\gamma_n = \frac{a_n + b_n}{2}, \quad \delta_n = b_n .$$

Adding to this collection again the reduced collection S_M', we get the infinite collection

$$S_M' \cup S_\infty^{**} = \{[a_1, b_1], \ldots, [a_{n-1}, b_{n-1}], [\gamma_n, \delta_n], [\gamma_{n+1}, \delta_{n+1}], \ldots\} \in \Sigma_\infty([a, b]) .$$

Using the same argument as before, we then obtain the estimate

$$\sum_{k=1}^{n-1} \lambda_k |f(b_k) - f(a_k)| + \lambda_n |f(b_n) - f(a_n + b_n)/2)| \leq M .$$

Combining both estimates obtained in this way and putting $c_n := (a_n + b_n)/2$, we end up with

$$\begin{aligned}
&2\sum_{k=1}^{n-1} \lambda_k |f(b_k) - f(a_k)| + \lambda_n |f(b_n) - f(a_n)| \\
&\quad \leq 2\sum_{k=1}^{n-1} \lambda_k |f(b_k) - f(a_k)| + \lambda_n |f(b_n) - f(c_n)| + \lambda_n |f(c_n) - f(a_n)| \\
&\quad = \sum_{k=1}^{n-1} \lambda_k |f(b_k) - f(a_k)| + \lambda_n |f(b_n) - f(c_n)| \\
&\qquad + \sum_{k=1}^{n-1} \lambda_k |f(b_k) - f(a_k)| + \lambda_n |f(c_n) - f(a_n)| \leq 2M ,
\end{aligned}$$

contradicting our choice of S_M.

Now, assume that we have the strict inclusion

$$[a_1, b_1] \cup [a_2, b_2] \cup \ldots \cup [a_n, b_n] \subset [a, b] .$$

Choose an interval

$$[\alpha, \beta] \subset [a, b] \setminus \big([a_1, b_1] \cup [a_2, b_2] \cup \ldots \cup [a_n, b_n]\big)$$

and a sequence $[a_{n+1}, b_{n+1}], [a_{n+2}, b_{n+2}], \ldots$ of nonoverlapping intervals such that $S_\infty := \{[a_j, b_j] : j = n + 1, n + 2, \ldots\} \in \Sigma_\infty([a, b])$. Then for the union

$$S_M \cup S_\infty = \{[a_k, b_k] : k \in \mathbb{N}\} \in \Sigma_\infty([a, b])$$

of the collections S_M and S_∞, we obtain

$$\begin{aligned}
\mathrm{Var}_\Lambda(f, S_M \cup S_\infty) &= \sum_{k=1}^{\infty} \lambda_k |f(b_k) - f(a_k)| \\
&= \sum_{k=1}^{n} \lambda_k |f(b_k) - f(a_k)| + \sum_{k=n+1}^{\infty} \lambda_k |f(b_k) - f(a_k)| \leq M .
\end{aligned}$$

However, this implies

$$\mathrm{Var}_\Lambda(f, S_M) = \sum_{i=1}^{n} \lambda_i |f(b_i) - f(a_i)| \leq M ,$$

again contradicting our choice S_M. Thus, in both cases, the proof is complete. □

Proposition 2.18 shows that (2.26) could be equivalently replaced by

$$\mathrm{Var}_\Lambda(f) = \mathrm{Var}_\Lambda(f;[a,b]) := \sup\left\{\mathrm{Var}_\Lambda(f,S;[a,b]) : S \in \Sigma([a,b])\right\} \tag{2.29}$$

in the definition of the Waterman space $\Lambda BV([a,b])$.

Proposition 2.19. *The set $\Lambda BV([a,b])$ is a linear space; equipped with the norm*

$$\|f\|_{\Lambda BV} := |f(a)| + \mathrm{Var}_\Lambda(f;[a,b]), \tag{2.30}$$

it is a Banach space.

Proof. The fact that $\Lambda BV([a,b])$ is a linear space follows from Proposition 2.17 (c) and (d). We show that the space $(\Lambda BV([a,b]), \|\cdot\|_{\Lambda BV})$ is complete. To this end, assume that $(f_n)_n$ is a Cauchy sequence in the norm (2.30). Given $\varepsilon > 0$, choose $n_0 \in \mathbb{N}$ such that

$$|f_n(a) - f_m(a)| \le \varepsilon \tag{2.31}$$

and

$$\mathrm{Var}_\Lambda(f_n - f_m) \le \varepsilon \tag{2.32}$$

for $m, n \ge n_0$. From (2.31), we deduce that the sequence $(f_n(a))_n$ converges to some real number, say $f(a)$. On the other hand, (2.32) implies that for any infinite collection $S_\infty = \{[a_n, b_n] : n \in \mathbb{N}\} \in \Sigma_\infty([a,b])$, we have

$$\mathrm{Var}_\Lambda(f_n - f_m, S_\infty) = \sum_{k=1}^{\infty} \lambda_k |f_n(b_k) - f_m(b_k) - f_n(a_k) + f_m(a_k)| \le \varepsilon .$$

In particular, for any $x \in [a,b]$, the inequality

$$\begin{aligned} &\lambda_1 |f_n(x) - f_m(x)| - \lambda_1 |f_n(a) - f_m(a)| \\ &\quad \le \lambda_1 |[f_n(x) - f_m(x)] - [f_n(a) - f_m(a)]| \le \varepsilon \end{aligned}$$

is satisfied. Combining this estimate with (2.31), we obtain

$$|f_n(x) - f_m(x)| \le \varepsilon \frac{1 + \lambda_1}{\lambda_1} ,$$

which shows that the sequence $(f_n)_n$ converges uniformly to some bounded function f on the interval $[a,b]$. Thus, from the inequality

$$|\,\mathrm{Var}_\Lambda(f_n) - \mathrm{Var}_\Lambda(f_m)| \le \mathrm{Var}_\Lambda(f_n - f_m) ,$$

(Proposition 2.17 (d), and (2.32)) it follows that the sequence $(\mathrm{Var}_\Lambda(f_n))_n$ has a limit in $\mathbb{R}$.

Now, we fix an arbitrary finite collection $S = \{[a_1, b_1], \dots, [a_p, b_p]\} \in \Sigma([a,b])$. From the definition of the function f, we see that, for n sufficiently large, we have

$$\sum_{i=1}^{p} \lambda_i |f(b_i) - f(a_i)| \le \sum_{i=1}^{p} \lambda_i |f_n(b_i) - f_n(a_i)| + \varepsilon \le \mathrm{Var}_\Lambda(f_n) + \varepsilon .$$

Therefore, letting $n \to \infty$, we obtain

$$\mathrm{Var}_\Lambda(f) \le \lim_{n\to\infty} \mathrm{Var}_\Lambda(f_n),$$

and so $f \in \Lambda BV([a,b])$. Using again the Cauchy property (2.32) of the sequence $(f_n)_n$, for $\varepsilon > 0$, we choose $n_0 \in \mathbb{N}$ such that given any infinite collection $S_\infty = \{[a_k, b_k] : k \in \mathbb{N}\} \in \Sigma_\infty([a,b])$, we have

$$\mathrm{Var}_\Lambda(f, S_\infty) = \sum_{k=1}^{\infty} \lambda_k |f_n(b_k) - f_m(b_k) - f_n(a_k) + f_m(a_k)| \le \varepsilon.$$

Letting $m \to \infty$ and taking the supremum over all such collections S_∞ yields

$$\mathrm{Var}_\Lambda(f_n - f) \le \varepsilon.$$

Obviously, letting in (2.31) $m \to \infty$, we also get $|f_n(a) - f(a)| \le \varepsilon$, and so $\|f - f_n\|_{\Lambda BV} \to 0$ as $n \to \infty$, and the proof is complete. □

As earlier announced, the relation (2.28) between the spaces BV and ΛBV can be made more precise. In fact, we are now going to prove the equalities

$$\bigcap_\Lambda \Lambda BV([a,b]) = BV([a,b]), \quad \bigcup_\Lambda \Lambda BV([a,b]) = R([a,b]), \tag{2.33}$$

where $R([a,b])$ denotes the space of all regular functions, see Section 0.3, and the intersection and union in (2.33) are taken over all Waterman sequences Λ. First, we need some auxiliary facts which we collect in a series of technical lemmas.

Lemma 2.20. *Let $(\alpha_n)_n$ be a sequence of positive real numbers tending to zero. Then there exists a Waterman sequence $\Lambda = (\lambda_n)_n$ such that*

$$\sum_{n=1}^{\infty} \alpha_n \lambda_n < \infty. \tag{2.34}$$

Proof. Denote $n_0 := 0$, choose a natural number n_1 such that $\alpha_n \le 1/2$ for $n \ge n_1$, and put $\lambda_n := 1/n_1$ for $n_0 < n \le n_1$. Assume that we have chosen natural numbers $n_1, n_2, \ldots, n_k$ in such a way that $n_1 < n_2 < \ldots < n_k$, $\alpha_n \le 1/2^k$ for $n \ge n_k$, and $\lambda_n := 1/(n_k - n_{k-1})$ for $n_{k-1} < n \le n_k$. Then we choose the next natural number n_{k+1} such that $n_{k+1} \ge n_k + k$, $n_{k+1} - n_k > 1/\lambda_{n_k}$ and $\alpha_n < 1/2^{k+1}$ for $n \ge n_{k+1}$, and put

$$\lambda_n := \frac{1}{n_{k+1} - n_k} \quad (n_k < n \le n_{k+1}).$$

By construction, for $n_k < n \le n_{k+1}$, we then have

$$\lambda_n = \frac{1}{n_{k+1} - n_k} < \lambda_{n_k}, \quad \lambda_n \le \frac{1}{k}.$$

Thus, $(\lambda_n)_n$ is decreasing and tends to zero. Moreover, we have

$$\sum_{n=n_k+1}^{n_{k+1}} \lambda_n = 1$$

which shows that (2.24) is true. On the other hand,

$$\sum_{n=n_k+1}^{n_{k+1}} \alpha_n \lambda_n \le \sum_{n=n_k+1}^{n_{k+1}} \frac{1}{2^k} \lambda_n = \frac{1}{2^k} \sum_{n=n_k+1}^{n_{k+1}} \lambda_n = \frac{1}{2^k},$$

which proves (2.34). □

For example, if we take $\alpha_n = 1/n$ in Lemma 2.20, then the Waterman sequence $\Lambda = (\lambda_n)_n$ constructed in the proof has the form

$$\Lambda = \left(\tfrac{1}{2}, \tfrac{1}{2}, \tfrac{1}{4}, \tfrac{1}{4}, \tfrac{1}{4}, \tfrac{1}{4}, \tfrac{1}{8}, \tfrac{1}{8}, \tfrac{1}{8}, \tfrac{1}{8}, \tfrac{1}{8}, \tfrac{1}{8}, \tfrac{1}{8}, \tfrac{1}{8}, \tfrac{1}{16}, \ldots\right),$$

and (2.34) reads

$$\sum_{n=1}^{\infty} \alpha_n \lambda_n \le \sum_{k=1}^{\infty} 2^k \frac{1}{2^{2k}} = \sum_{k=1}^{\infty} \left(\frac{1}{2}\right)^k < \infty .$$

The last auxiliary result we need is based on a combinatorial lemma.

Lemma 2.21. *Let $(\lambda_n)_n$ be a decreasing sequence of positive real numbers. If $(\delta_n)_n$ is a sequence of positive real numbers tending to zero, and $(\hat{\delta}_n)_n$ denotes the sequence $(\delta_n)_n$ arranged in decreasing order, then*

$$\sum_{k=1}^{n} \lambda_k \delta_k \le \sum_{k=1}^{n} \lambda_k \hat{\delta}_k \quad (n = 1, 2, 3, \ldots). \tag{2.35}$$

Proof. Fix $n \in \mathbb{N}$ and consider the set $\{\delta_1, \delta_2, \ldots, \delta_n\}$. Let $\delta_1' \ge \delta_2' \ge \ldots \ge \delta_n'$ denote the elements of this set arranged in decreasing order. Then from a well-known combinatorial lemma, we obtain

$$\sum_{k=1}^{n} \lambda_k \delta_k \le \sum_{k=1}^{n} \lambda_k \delta_k' .$$

Clearly, $\delta_k' \le \hat{\delta}_k$ for $k = 1, 2, \ldots, n$, and so (2.35) follows. □

The next proposition shows that the Waterman space ΛBV is "stable" under a monotone substitution of variables; compare this with Proposition 1.12.

Proposition 2.22. *Given a function $g : [c, d] \to \mathbb{R}$, let $\tau : [a, b] \to [c, d]$ be continuous and strictly increasing with $\tau(a) = c$ and $\tau(b) = d$. Then $g \circ \tau \in \Lambda BV([a, b])$ if and only if $g \in \Lambda BV([c, d])$.*

Proof. Suppose that $g \in \Lambda BV([c, d])$, and let $S_\infty = \{[c_n, d_n] : n \in \mathbb{N}\} \in \Sigma_\infty([c, d])$. By our assumption on τ, we then have $\tau(S_\infty) = \{[\tau(c_n), \tau(d_n)] : n \in \mathbb{N}\} \in \Sigma_\infty([a, b])$. Since

$g \in \Lambda BV([c,d])$, we have

$$\begin{aligned}\mathrm{Var}_\Lambda(g \circ \tau, S_\infty; [a,b]) &= \sum_{n=1}^{\infty} \lambda_n |(g \circ \tau)(d_n) - (g \circ \tau)(c_n)| \\ &= \sum_{n=1}^{\infty} \lambda_n |g[\tau(d_n)] - g[\tau(c_n)]| = \mathrm{Var}_\Lambda(g, \tau(S_\infty); [c,d]) < \infty,\end{aligned}$$

which shows that $g \circ \tau \in \Lambda BV([a,b])$ after passing to the supremum over $S_\infty \in \Sigma_\infty([a,b])$. Applying the same reasoning to the function $\tau^{-1} : [a,b] \to [c,d]$ proves the reverse implication. □

The next result states that, roughly speaking, every continuous function $f : [a,b] \to \mathbb{R}$ is contained in some appropriate Waterman space $\Lambda BV([a,b])$; as Example 1.8 shows, this is *not* true[7] for the classical space $BV([a,b])$.

Proposition 2.23. *Every function $f \in C([a,b])$ is contained in $\Lambda BV([a,b])$ for some Waterman sequence Λ.*

Proof. Note that the assertion follows from the second equality in (2.33); however, we give a direct proof which gives some insight into the method. Let $f : [a,b] \to \mathbb{R}$ be continuous; we construct a Waterman sequence $\Lambda = (\lambda_n)_n$ such that $f \in \Lambda BV([a,b])$.

For any $\delta > 0$, we denote by $\omega_\infty(f;\delta)$ the modulus of continuity of f defined in (0.97). We know that $\omega_\infty(f;\cdot)$ is increasing and $\omega_\infty(f;\delta) \to 0$ as $\delta \to 0$ since f is uniformly continuous on $[a,b]$.

Let $I_n = [a_n, b_n]$ be a sequence of nonoverlapping subintervals of $[a,b]$; as in the proof of Proposition 2.18, we use the shortcut $|f(I_n)| := |f(b_n) - f(a_n)|$. For $m \in \mathbb{N}$, put

$$E_m := \left\{ I_k : \omega_\infty\left(f; \frac{b-a}{m+1}\right) < |f(I_k)| \le \omega_\infty\left(f; \frac{b-a}{m}\right) \right\}.$$

Observe that in case $|b_k - a_k| \le (b-a)/m$, we have

$$|f(I_k)| = |f(b_k) - f(a_k)| \le \omega_\infty(f; |b_k - a_k|) \le \omega_\infty\left(f; \frac{b-a}{m+1}\right),$$

and so $I_k \in E_m$ only if $|b_k - a_k| > (b-a)/(m+1)$. Since the intervals I_k are nonoverlapping and contained in $[a,b]$, we deduce that E_m contains m intervals at most. Moreover, if $I_p \in E_r$ and $I_q \in E_{r+s}$, then

$$|f(I_q)| \le \omega_\infty\left(f; \frac{b-a}{r+s}\right) \le \omega_\infty\left(f; \frac{b-a}{r+1}\right) < |f(I_p)|.$$

7 Note that it is *not* true either that the whole space $C([a,b])$ is contained in every Waterman space $\Lambda BV([a,b])$, as the first equality in (2.33) shows. That is, for every individual continuous function f, we have to find an appropriate Waterman sequence Λ such that $f \in \Lambda BV([a,b])$.

Thus, we can choose, step by step, a sequence $(J_n)_n$ of intervals $J_k \in E_k$ satisfying

$$|f(J_1)| \geq |f(J_2)| \geq \ldots \geq |f(J_n)| \geq \ldots \to 0 \quad (n \to \infty)\,.$$

Now, we are going to show that

$$|f(J_n)| \leq \omega_\infty\left(f; \frac{b-a}{n}\right) \quad (n = 1, 2, 3, \ldots)\,. \tag{2.36}$$

To this end, assume that $m \in \mathbb{N}$ satisfies

$$|f(J_m)| > \omega_\infty\left(f; \frac{b-a}{m}\right);$$

then,

$$|f(J_1)| \geq |f(J_2)| \geq \ldots \geq |f(J_m)| > \omega_\infty\left(f; \frac{b-a}{m}\right).$$

This implies that $|J_k| > (b-a)/m$ for $k = 1, 2, \ldots, m$, which is impossible since all the intervals $J_1, J_2, \ldots, J_m$ are nonoverlapping and contained in $[a, b]$. Therefore, we have proved (2.36). Furthermore, since $\omega_\infty(f; (b-a)/k) \to 0$ as $k \to \infty$, we may apply Lemma 2.20 to the particular choice

$$\alpha_n := \omega_\infty\left(f; \frac{b-a}{n}\right) \quad (n = 1, 2, 3, \ldots)$$

and find a decreasing sequence $\Lambda = (\lambda_n)_n$ of positive real numbers tending to zero and satisfying

$$\sum_{n=1}^{\infty} \lambda_n = \infty, \quad \sum_{n=1}^{\infty} \lambda_n \omega_\infty\left(f; \frac{b-a}{n}\right) < \infty\,.$$

Subsequently, we apply Lemma 2.21 to the choice $\delta_n := |f(I_n)|$ and obtain

$$\sum_{n=1}^{\infty} \lambda_n |f(I_n)| \leq \sum_{n=1}^{\infty} \lambda_n |f(J_n)| \leq \sum_{n=1}^{\infty} \lambda_n \omega_\infty\left(f; \frac{b-a}{n}\right) < \infty\,. \tag{2.37}$$

However, this means nothing more than $f \in \Lambda BV([a, b])$ for $\Lambda = (\lambda_n)_n$, and the conclusion follows. □

We may summarize the contents of Proposition 2.23 as inclusion

$$C([a, b]) \subseteq \bigcup_{\Lambda} \Lambda BV([a, b])\,, \tag{2.38}$$

where the union is taken over all Waterman sequences Λ. This complements in some sense the second inclusion in (2.28): the union of all Waterman spaces is "intermediate" between continuous and bounded functions. We now prove the equalities (2.33) which are more precise than the inclusions (2.28).

Proposition 2.24. *The equalities* (2.33) *hold, where the intersection and union are taken over all Waterman sequences* Λ.

Proof. Observe first that, by (2.28), we only have to prove the inclusion

$$BV([a,b]) \supseteq \bigcap_{\Lambda} \Lambda BV([a,b])$$

to get the first equality in (2.33). Fix an arbitrary Waterman sequence $\Lambda = (\lambda_n)_n$ and let $f \in \Lambda BV([a,b])$. We then know that f is bounded, and so both numbers $m(f)$ and $M(f)$ in (0.61) and (0.62) are finite. Putting[8]

$$F(x) := \frac{f(x) - m(f)}{M(f) - m(f)} \quad (a \le x \le b),$$

we have $0 \le F(x) \le 1$ for $a \le x \le b$, and F clearly belongs, as f itself, to the space $\Lambda BV([a,b])$. This argument shows that we may assume, without loss of generality, that

$$0 \le f(x) \le 1 \quad (a \le x \le b). \tag{2.39}$$

Suppose that $f \notin BV([a,b])$. By Proposition 1.18, we know that then there exists a point $x \in [a,b]$ such that f is not of bounded (Jordan) variation on any neighborhood of x.

Choose a finite partition P_1 of the interval $[a,b]$ such that

$$\sum_{I \in P_1} |f(I)| \ge 3.$$

The point x is either an interior point of a single interval in P_1, or is an endpoint of two intervals in P_1. If we remove that single interval, or possibly two intervals from P_1 and denote by Q_1 the remaining collection of intervals. Then in view of (2.39), we have

$$\sum_{I \in Q_1} |f(I)| \ge 1.$$

If Q_1 contains q_1 intervals, we write $Q_1 = \{I_k^1 : k = 1, 2, \ldots, q_1\}$. Next, we define

$$\lambda_1 = \lambda_2 = \ldots = \lambda_{q_1} := 1.$$

Then

$$\sum_{k=1}^{q_1} \lambda_k |f(I_k^1)| \ge 1.$$

This completes the first step in our definition.

Suppose that we have constructed P_n and Q_n in this way; then, we proceed inductively as follows. First of all, note that one or possibly two intervals which we have removed from P_n to form Q_n create a neighborhood U_n of x. Since F is not of bounded variation on U_n, we conclude that there exists a finite partition P_{n+1} of U_n such that

$$\sum_{I \in P_{n+1}} |f(I)| \ge 3.$$

8 Here we assume, of course, that f is not constant; for constant functions there is nothing to prove.

Notice that the point x is again either an interior point of a single interval contained in P_{n+1} or an endpoint of at most two intervals in P_{n+1}. If we delete this single, or possibly two, intervals from P_{n+1} and denote by Q_{n+1} the remaining collection of intervals, then, again from (2.39), it follows that

$$\sum_{I\in Q_{n+1}} |f(I)| \geq 1\,.$$

If Q_{n+1} contains q_{n+1} intervals, we write $Q_{n+1} = \{I_k^{n+1} : k = 1, 2, \ldots, q_{n+1}\}$ and define

$$\lambda_{r_n+1} = \lambda_{r_n+2} = \ldots = \lambda_{r_n+q_n} := \frac{1}{n+1}\,,$$

where $r_n := q_1 + q_2 + \ldots + q_n$ and $q_0 = 0$. Then we obtain

$$\sum_{k=1}^{q_{n+1}} \lambda_{r_n+k}|f(I_k^{n+1})| \geq \frac{1}{n+1}\,.$$

Next, observe that all intervals of Q_{n+1} contained in U_n are pairwise nonoverlapping. Thus, we conclude that

$$\sum_{i=1}^{n+1}\sum_{k=1}^{q_i} \lambda_{r_{i-1}+k}|f(I_k^i)| \geq \sum_{i=1}^{n+1}\frac{1}{i}\,.$$

In this way, we have constructed real numbers $\{\lambda_k : k = 1, 2, 3, \ldots\}$ and nonoverlapping subintervals $\{I_k^n : k = 1, 2, \ldots, q_n;\ n = 1, 2, \ldots\}$ of $[a, b]$ such that $(\lambda_k)_k$ is decreasing and tends to zero, and

$$\sum_{k=1}^{\infty} \lambda_k = \infty, \qquad \sum_{i=1}^{\infty}\sum_{k=1}^{q_i} \lambda_{r_{i-1}+k}|f(I_k^1)| = \infty\,.$$

However, this means precisely that $\Lambda = (\lambda_k)_k$ is a Waterman sequence and f does not belong to the corresponding space $\Lambda BV([a, b])$, contradicting our assumption. Thus, we have proved the first equality in (2.33).

Now, we prove the second equality. More precisely, we show that every function $f \in \Lambda BV([a, b])$ has left limits at every point in (a, b); the proof for right limits is analogous.

Given $f \in \Lambda BV([a, b])$, assume that there exists a point $x \in (a, b)$ at which f does not have a left limit. This means that

$$l := \liminf_{t\to x-} f(t) < \limsup_{x\to x-} f(t) =: L\,.$$

Let $\delta := (L-l)/3$, and choose sequences $(p_n)_n$ and $(P_n)_n$ such that $P_1 < P_2 < \ldots \to x$, $p_1 < p_2 < \ldots \to x$, and

$$f(p_n) \to l, \quad f(P_n) \to L, \quad f(p_n) \leq l + \delta, \quad f(P_n) \geq L - \delta\,.$$

Afterwards, we choose a subsequence $(Q_n)_n$ of $(P_n)_n$ and a subsequence $(q_n)_n$ of $(p_n)_n$ such that $q_1 < Q_1 < q_2 < Q_2 < \cdots$. The intervals $[q_m, Q_m]$ and $[q_n, Q_n]$ are then disjoint for $m \neq n$ and satisfy

$$|f(Q_n) - f(q_n)| \geq f(Q_n) - f(q_n) \geq (L - \delta) - (l + \delta) \geq 3\delta - 2\delta = \delta,$$

which implies that

$$\sum_{n=1}^{\infty} \lambda_n |f(Q_n) - f(q_n)| \geq \delta \sum_{n=1}^{\infty} \lambda_n = \infty.$$

This shows that $f \notin \Lambda BV([a, b])$, contradicting our assumption.

It remains to show that every regular function belongs to $\Lambda BV([a, b])$ for some suitable Waterman sequence $\Lambda = (\lambda_n)_n$. To this end, we use the Sierpiński decomposition. By Theorem 0.36, there exist a strictly increasing function $\tau : [a, b] \to [c, d]$ and a continuous function $g : [c, d] \to \mathbb{R}$ such that $f = g \circ \tau$. The affine function ℓ given by (0.80) is a strictly increasing homeomorphism between the intervals $[c, d]$ and $[a, b]$. Therefore, $\ell \circ \tau : [a, b] \to [a, b]$ is also strictly increasing, and $g \circ \ell^{-1} : [a, b] \to \mathbb{R}$ is continuous. From Proposition 2.23, we deduce that $g \circ \ell^{-1} \in \Lambda BV([a, b])$ for some Waterman sequence Λ. On the other hand, Proposition 2.22 implies that $f = (g \circ \ell^{-1}) \circ (\ell \circ \tau)$ belongs to the space $\Lambda BV([a, b])$. This completes the proof. □

We show now that in the definition (2.26) of the total Waterman variation, we may restrict ourselves to rather special collections $S_\infty \in \Sigma_\infty([a, b])$. To this end, let us denote by $\Sigma_\infty^d([a, b])$ the family of all infinite collections $S_\infty^d = \{[a_n, b_n] : n \in \mathbb{N}\}$ of nonoverlapping subintervals of $[a, b]$ with the additional property that the sequence $(\delta_n)_n$ defined by

$$\delta_n := |f(b_n) - f(a_n)| \quad (n = 1, 2, 3, \ldots) \tag{2.40}$$

is *decreasing* and converges to 0. Before proving the announced result, we need another auxiliary lemma on regular functions, see Section 0.3.

Lemma 2.25. *Let $S_\infty = \{[a_n, b_n] : n \in \mathbb{N}\} \in \Sigma_\infty([a, b])$ be a (finite or infinite) collection of intervals, and let $f \in R([a, b])$. Then one may rearrange the intervals $[a_n, b_n] \in S_\infty$ in such a way that $|f(b_n) - f(a_n)| \geq |f(b_{n+1}) - f(a_{n+1})|$ for all $n \in \mathbb{N}$.*

Proof. Since f is regular on $[a, b]$, by Theorem 0.36, we may find a strictly increasing function $\tau : [a, b] \to [c, d]$ (where $c := \tau(a)$ and $d := \tau(b)$) and a continuous function $g : [c, d] \to \mathbb{R}$ such that $f = g \circ \tau$. Consider the collection of intervals $\tau(S_\infty) = \{[\tau(a_n), \tau(b_n)] : n \in \mathbb{N}\}$ which belongs to $\Sigma_\infty([c, d])$. The sequence $(\tau(b_n) - \tau(a_n))_n$ converges to 0. Since every nonnegative sequence converging to 0 can be arranged in decreasing order, we find a sequence $(n_k)_k$ of indices such that $(\tau(b_{n_k}) - \tau(a_{n_k}))_k$ is decreasing (and obviously also converges to 0).

Now, consider the sequence $(\eta_k)_k$ defined by

$$\eta_k := |f(b_{n_k}) - f(a_{n_k})| = |g(\tau(b_{n_k})) - g(\tau(a_{n_k}))|.$$

From the continuity of g, it follows that $\eta_k \to 0$ as $k \to \infty$. Applying now a further rearrangement, we find a sequence $(k_m)_m$ of indices such that $(\eta_{k_m})_m$ is decreasing. The corresponding sequence of intervals then meets the requirements of our assertion. □

We now state the announced proposition in terms of the family $\Sigma_\infty^d([a,b])$ introduced before Lemma 2.25.

Proposition 2.26. *Let $f \in \Lambda BV([a,b])$. Then*

$$\mathrm{Var}_\Lambda(f;[a,b]) = \sup\left\{\mathrm{Var}_\Lambda(f, S_\infty^d;[a,b]) : S_\infty^d \in \Sigma_\infty^d([a,b])\right\}. \tag{2.41}$$

Proof. Denote by $\mathrm{Var}_\Lambda^d(f;[a,b])$ the expression on the right-hand side of (2.41). The inequality $\mathrm{Var}_\Lambda^d(f;[a,b]) \le \mathrm{Var}_\Lambda(f;[a,b])$ is clear since $\Sigma_\infty^d([a,b]) \subseteq \Sigma_\infty([a,b])$, and so we only have to prove the reverse inequality. Fix an arbitrary collection $S_\infty = \{[a_n,b_n] : n \in \mathbb{N}\} \in \Sigma_\infty([a,b])$, and consider the sequence $(\delta_n)_n$ defined by (2.40). Applying Lemma 2.25, we can rearrange the sequence $(\delta_n)_n$ in decreasing order. In this way, we obtain a sequence $(\hat{\delta}_n)_n$ which, in view of Lemma 2.21, satisfies

$$\sum_{k=1}^n \lambda_k |f(b_k) - f(a_k)| \le \sum_{k=1}^n \lambda_k \hat{\delta}_k .$$

However, this implies that $\mathrm{Var}_\Lambda(f;[a,b]) \le \mathrm{Var}_\Lambda^d(f;[a,b])$ and finishes the proof. □

Now, we are going to compare the spaces ΛBV and MBV for two different Waterman sequences $\Lambda = (\lambda_n)_n$ and $M = (\mu_n)_n$. This requires a technical definition.

Definition 2.27. Given two increasing positive sequences $(L_n)_n$ and $(M_n)_n$, we write $(L_n)_n \preceq (M_n)_n$ if there exists a constant $c > 0$ such that $L_n \le cM_n$ for all $n \in \mathbb{N}$. If both $(L_n)_n \preceq (M_n)_n$ and $(M_n)_n \preceq (L_n)_n$, we write $(L_n)_n \sim (M_n)_n$. ■

Suppose that $\Lambda = (\lambda_n)_n$ and $M = (\mu_n)_n$ are two Waterman sequences. To simplify the notation, for $m, n \in \mathbb{N}$ with $m \le n$, in the sequel we will use the shortcut

$$\sum_{k=m}^n \lambda_k =: \lambda[m,n], \quad \sum_{k=m}^n \mu_k =: \mu[m,n]. \tag{2.42}$$

Note that the crucial property (2.24) of a Waterman sequence $(\lambda_n)_n$ means that $\lambda[1,n] \to \infty$ as $n \to \infty$.

Proposition 2.28. *Let $\Lambda = (\lambda_n)_n$ and $M = (\mu_n)_n$ be two Waterman sequences; then, the following is true.*
(a) *$MBV([a,b]) \subseteq \Lambda BV([a,b])$ if and only if $(\lambda[1,n])_n \preceq (\mu[1,n])_n$.*
(b) *$MBV([a,b]) = \Lambda BV([a,b])$ if and only if $(\lambda[1,n])_n \sim (\mu[1,n])_n$.*

Proof. Suppose there exists a constant $c > 0$ such that

$$\lambda[1,n] = \sum_{k=1}^n \lambda_k \le c \sum_{k=1}^n \mu_k = c\mu[1,n] \quad (n = 1, 2, \dots). \tag{2.43}$$

Fix an arbitrary function $f \in MBV([a,b])$ and an infinite collection $S_\infty = \{[a_n, b_n] : n \in \mathbb{N}\} \in \Sigma_\infty([a,b])$. Putting $\delta_n := |f(b_n) - f(a_n)|$, in view of Proposition 2.26, we can assume without loss of generality that the sequence $(\delta_n)_n$ is decreasing and tends to 0. Taking into account the identity

$$\sum_{k=1}^{n} \lambda_k \delta_k = \sum_{k=1}^{n-1} \lambda[1,k](\delta_k - \delta_{k+1}) + \delta_n \lambda[1,n],$$

by our assumptions and (2.43), we obtain

$$\sum_{k=1}^{n} \lambda_k \delta_k \le c \left(\sum_{j=1}^{n-1} \mu[1,k](\delta_k - \delta_{k+1}) + \delta_n \mu[1,k] \right) = c \sum_{k=1}^{n} \delta_k \mu_k .$$

This shows that $f \in \Lambda BV([a,b])$ and proves the "if" part in (a).

To prove the "only if" part, assume now that $(\lambda[1,n])_n \not\preceq (\mu[1,n])_n$. Then there exists a strictly increasing unbounded sequence $(n_k)_k$ of integers which satisfies

$$\lambda[n_k + 1, n_{k+1}] \ge 2^k \mu[n_k + 1, n_{k+1}],$$

where $n_0 := 0$. Given $S_\infty = \{[a_n, b_n] : n \in \mathbb{N}\} \in \Sigma_\infty([a,b])$, let $f : [a,b] \to \mathbb{R}$ be a function such that[9]

$$\delta_i := |f(b_i) - f(a_i)| = \frac{1}{2^k \mu[n_k + 1, n_{k+1}]} \qquad (i = n_k + 1, n_k + 2, \dots, n_{k+1}).$$

Observe that the sequence $(\delta_n)_n$ is then decreasing and tends to 0 as $n \to \infty$. Moreover,

$$\sum_{i=1}^{n_{k+1}} \delta_i \lambda_i = \sum_{j=0}^{k} \frac{\lambda[n_j + 1, n_{j+1}]}{2^j \mu[n_j + 1, n_{j+1}]} \ge \sum_{j=0}^{k} 2^{k-j} = 2^{k+1} - 1 \to \infty \quad (k \to \infty)$$

and

$$\sum_{i=1}^{n_{k+1}} \delta_i \mu_i = \sum_{j=0}^{k} \frac{1}{2^j} \le 2 .$$

This shows that $f \in MBV([a,b])$, but $f \notin \Lambda BV([a,b])$, and so we have proved (a). The statement (b) is of course an immediate consequence of (a). □

A particularly simple Waterman sequence is $\Lambda_q := (n^{-q})_n$ for $0 < q \le 1$. This choice of Λ is so important that we introduce a special notation.

Definition 2.29. In case $\lambda_n = n^{-q}$ for $0 < q \le 1$, we denote the corresponding Waterman space by $\Lambda_q BV([a,b])$. So, $f \in \Lambda_q BV([a,b])$ if and only if

$$\mathrm{Var}_{\Lambda_q}(f) = \mathrm{Var}_{\Lambda_q}(f;[a,b]) = \sup \left\{ \sum_{n=1}^{\infty} \frac{|f(b_n) - f(a_n)|}{n^q} \right\} < \infty ,$$

9 For example, we may take a suitable zigzag function for f, see Definition 0.49.

where the supremum is taken over all collections $S_\infty = \{[a_n, b_n] : n \in \mathbb{N}\} \in \Sigma_\infty([a,b])$. In the limit case $q = 1$ (i.e. when (2.24) becomes the harmonic series), this space will be denoted by $HBV([a,b])$ and called the space of functions of *harmonic bounded variation.* ■

Historically, the space HBV was the first Waterman-type space which occurred quite naturally in the study of the Fourier series. Formally, we could also identify[10] the Waterman space $\Lambda_0 BV$ with BV. Thus, the spaces $\Lambda_q BV$ are, for $0 < q < 1$, "intermediate" between BV and HBV. Also, from the definition, it readily follows by a simple calculation that

$$\Lambda_p BV([a,b]) \subseteq \Lambda_q BV([a,b]) \subseteq HBV([a,b]) \quad (p \le q < 1). \tag{2.44}$$

This may also be considered as a special case of Proposition 2.28 (a). In fact, for $\Lambda = (n^{-q})_n$ and $M = (n^{-p})_n$, we have

$$\lambda[1,n] = 1 + \frac{1}{2^q} + \ldots + \frac{1}{n^q}, \quad \mu[1,n] = 1 + \frac{1}{2^p} + \ldots + \frac{1}{n^p},$$

so $([\lambda[1,n])_n \preceq (\mu[1,n])_n$ if and only if $k^q \ge k^p$ for all k, and hence $p \le q$, in accordance with (2.44) and Proposition 2.28 (a).

To get an idea of these spaces, let us consider the zigzag function $Z_{C,D}$ constructed in Definition 0.49. Given a Waterman sequence $\Lambda = (\lambda_n)_n$, it is easy to see that

$$\mathrm{Var}_\Lambda(Z_{C,D}; [0,1]) = \sum_{n=1}^{\infty} d_n \lambda_n, \tag{2.45}$$

and so $Z_{C,D}$ belongs to $\Lambda BV([0,1])$ if and only if the series in (2.45) converges. This simple observation makes it possible to separate the classes $\Lambda_q BV([a,b])$, i.e. to show that inclusion (2.44) is strict for $p < q$:

Example 2.30. Let Z_θ be the special zigzag function (0.93) determined by $d_n = n^{-\theta}$ ($\theta > 0$), and let $\lambda_n = n^{-q}$ ($0 < q \le 1$) as above. Then it follows from (2.45) that $Z_\theta \in \Lambda_q BV([0,1])$ if and only if $\theta + q > 1$. Thus, if we choose $\theta := 1 - p$, the corresponding zigzag function $f = Z_{1-p}$ belongs to $\Lambda_q BV([0,1])$ for $q > p$, but not to $\Lambda_p BV([0,1])$. However, we can do better: for fixed $p < 1$, the same reasoning shows of course that $f = Z_{1-p}$ satisfies

$$f \in \left(\bigcap_{q>p} \Lambda_q BV([0,1]) \right) \setminus \Lambda_p BV([0,1]).$$

In particular, the special zigzag function Z_1 belongs to $HBV([0,1]) \setminus \Lambda_p BV([0,1])$ for $p < 1$, and also to $HBV([0,1]) \setminus BV([0,1])$. ♥

10 In the definition of BV (Definition 1.1), we only used finite partitions, while in the definition of ΛBV (Definition 2.15), we used infinite collections of nonoverlapping intervals. However, Proposition 2.18 shows that this is the same, and so we may consider BV as a special case of ΛBV with $\lambda_n \equiv 1$ for finitely many n.

In the following proposition, we need the notion of Banach indicatrix introduced in Definition 0.38 in the general setting of regular functions, i.e. functions which only have removable discontinuities or discontinuities of first kind (jumps).

Proposition 2.31. *Let $f : [a,b] \to \mathbb{R}$ be a regular function, and let $I_f : \mathbb{R} \to \mathbb{N}_0 \cup \{\infty\}$ be the Banach indicatrix of f. Suppose that $\Lambda = (\lambda_n)_n$ is a Waterman sequence, and $(\mu_n)_n$ is some positive increasing sequence such that $(\mu_n)_n \sim (\lambda[1,n])_n$ in the sense of Definition 2.27. Finally, assume that the function $\mu_f : \mathbb{R} \to \mathbb{R}$ defined by $\mu_f(y) := \mu_n$ if $n = I_f(y)$ belongs to $L_1(\mathbb{R})$, i.e.*

$$\int_{-\infty}^{\infty} \mu_f(y)\, dy < \infty .$$

Then $f \in \Lambda BV([a,b])$.

Proof. Define $m(f)$ and $M(f)$ as in (0.61) and (0.62), respectively, and consider an arbitrary sequence $(I_n)_n$ of nonoverlapping subintervals of the interval $[a,b]$. Further, define a sequence $(P_n)_n$ of functions $P_n : [m(f), M(f)] \to \mathbb{R}$ in such a way that $P_n(y) = \lambda_n$ if the equation $f(x) = y$ has at least one solution in the interval I_n, where we extended the graph of f as described in Definition 0.38. On the other hand, we put $P_n(y) := 0$ if the equation $f(x) = y$ has no solution in I_n. Denoting

$$m_n(f) := \inf\{f(x) : x \in I_n\}, \quad M_n(f) := \sup\{f(x) : x \in I_n\},$$

we see that $P_n = \lambda_n \chi_{(m_n(f), M_n(f))}$, that is,

$$P_n(y) = \begin{cases} \lambda_n & \text{for } y \in (m_n(f), M_n(f)), \\ 0 & \text{for } y \notin (m_n(f), M_n(f)). \end{cases}$$

Clearly, the function P_n is measurable. Moreover,

$$\int_{m(f)}^{M(f)} P_n(y)\, dy = \lambda_n (M_n(f) - m_n(f)).$$

Now, for arbitrary $n \in \mathbb{N}$, we have

$$\begin{aligned} \sum_{k=1}^{n} \lambda_k |f(I_k)| &\le \sum_{k=1}^{n} \lambda_k (M_k(f) - m_k(f)) \\ &\le \int_{f(I_1)\cup\ldots\cup f(I_n)} I_f(y)\, dy \le \int_{-\infty}^{\infty} \mu_f(y)\, dy < \infty . \end{aligned}$$

Since $n \in \mathbb{N}$ was arbitrary, we see that $\mathrm{Var}_\Lambda(f;[a,b]) \le \|\mu_f\|_{L_1}$, and the conclusion follows. □

Now, we return to the special Waterman space $\Lambda_q BV$ generated by the sequence $\Lambda_q = (n^{-q})$ for $q \in (0,1]$. A comparison of the inclusions (1.72) and (2.44) shows that both scales of spaces $\{WBV_p : 1 \le p < \infty\}$ and $\{\Lambda_p BV : 0 < p \le 1\}$ are increasing with respect to the index p. Thus, it is natural to ask whether or not these two scales are related. The following Propositions 2.32 and 2.33 give a complete answer to this question and are taken from the survey paper [250]. As usual, for $1 < p < \infty$, we denote by $p' := p/(p-1)$ the conjugate exponent to p, see (0.13).

Proposition 2.32. *For $p > 1$ and $\frac{1}{p'} < q \le 1$, the continuous imbedding*

$$WBV_p([a,b]) \hookrightarrow \Lambda_q BV([a,b]) \tag{2.46}$$

holds true. Moreover, the inclusion $WBV_p([a,b]) \subset \Lambda_q BV([a,b])$ is strict.

Proof. For the proof, we use our information about the convergence behavior of the series $\zeta(\alpha,\beta)$ defined in (0.17). Let $f \in WBV_p([a,b])$ and $\mathrm{Var}_p^W(f;[a,b]) =: M$. Fix $S = \{[a_1,b_1],\ldots,[a_n,b_n]\} \in \Sigma([a,b])$, and put $\eta_k := |f(b_k) - f(a_k)|$ for $k = 1,2,\ldots,n$. We know that

$$\eta_1^p + \eta_2^p + \ldots + \eta_n^p \le M,$$

and we must find a bound on $\eta_1 + \eta_2 2^{-q} + \ldots + \eta_n n^{-q}$ under the assumption $qp' > 1$, i.e. $pq > p - 1$.

Applying Hölder's inequality (0.108) to $\alpha_k := \eta_k$ and $\beta_k := k^{-q}$ $(k = 1,2,\ldots,n)$ yields

$$\sum_{k=1}^{n} \frac{\eta_k}{k^q} \le \left(\sum_{k=1}^{n} \eta_k^p\right)^{1/p} \left(\sum_{k=1}^{n} \frac{1}{k^{qp'}}\right)^{1/p'} \le M^{1/p}\zeta(qp',0)^{1/p'},$$

and so $f \in \Lambda_q BV([a,b])$ since $\zeta(qp',0)$ is finite. Moreover, our calculations show that

$$\|f\|_{\Lambda_q BV} \le \zeta(qp',0)\|f\|_{WBV_p},$$

and so (2.46) defines, in fact, a continuous imbedding.

To show that the inclusion (2.46) is strict, we assume, without loss of generality, that $[a,b] = [0,1]$. Consider the zigzag function $Z_{C,D}$ from Definition 0.49, determined by the sequences

$$c_n := \frac{1}{2^n}, \quad d_n := \frac{1}{n^{1-q}\log^2(n+1)} \quad (n = 1,2,3,\ldots).$$

Then

$$\sum_{n=1}^{\infty} \frac{d_n}{n^q} = \zeta(1,2) < \infty, \quad \sum_{n=1}^{\infty} d_n^p = \zeta(p(1-q),2p) = \infty$$

since $p(1-q) = p - pq < p - (p-1) = 1$. This shows that $Z_{C,D} \in \Lambda_q BV([0,1])$, but $Z_{C,D} \notin WBV_p([0,1])$. □

The last example shows that functions $f \in \Lambda_q BV([a,b]) \setminus WBV_p([a,b])$ for a *fixed* value of $q > 1/p'$ exist. However, we can do much better: functions

$$f \in \left(\bigcap_{q>1/p'} \Lambda_q BV([0,1]) \right) \setminus WBV_p([0,1]) \tag{2.47}$$

exist (Exercise 2.6).

Proposition 2.32 gives a complete picture for $p > 1$ and $1 < qp' \le p'$, but does not cover the case $qp' \le 1$. The next proposition shows that in that case, the inclusion goes in the other direction.

Proposition 2.33. *For $p > 1$ and $q \le \frac{1}{p'}$, the continuous imbedding*

$$\Lambda_q BV([a,b]) \hookrightarrow WBV_p([a,b]) \tag{2.48}$$

holds true. Moreover, the inclusion $\Lambda_q BV([a,b]) \subset WBV_p([a,b])$ is strict.

Proof. Since the scale of spaces $\Lambda_q BV$ is increasing in q, see (2.44), we may restrict ourselves to the case $q = 1/p'$. Thus, let $f \in \Lambda_q BV([a,b])$ and $\mathrm{Var}_{\Lambda_q}(f;[a,b]) =: M$. Fix $S = \{[a_1,b_1],\dots,[a_n,b_n]\} \in \Sigma([a,b])$, and put $\eta_k := |f(b_k) - f(a_k)|$ for $k = 1,2,\dots,n$. We know that

$$\eta_1 + \eta_2 2^{-q} + \ldots + \eta_n n^{-q} \le M,$$

and we must find a bound on $\eta_1^p + \eta_2^p + \ldots + \eta_n^p$ under the assumption $qp' = 1$, i.e. $pq = p - 1$.

By renumbering the intervals in S, if necessary, we may assume that $\eta_1 \ge \eta_2 \ge \ldots \ge \eta_n$, see Proposition 2.26. Under this additional assumption, we get

$$M \ge \sum_{j=1}^{k} \frac{\eta_j}{j^q} \ge \sum_{j=1}^{k} \frac{\eta_k}{k^q} = k\frac{\eta_k}{k^p} = k^{1-q}\eta_k = k^{1/p}\eta_k \quad (k = 1,2,\dots,n).$$

Consequently,

$$\eta_k^p = \eta_k^{p-1}\eta_k \le \eta_k^{pq}\eta_k \le \frac{M^{pq}}{k^q}\eta_k = \frac{M^{p-1}}{k^q}\eta_k,$$

and hence

$$\sum_{k=1}^{n} \eta_k^p \le M^{p-1} \sum_{k=1}^{n} \frac{\eta_k}{k^q} \le M^{p-1}M = M^p.$$

This shows that $f \in WBV_p([a,b])$ with $\mathrm{Var}_p^W(f;[a,b]) \le M^p$, and so

$$\|f\|_{WBV_p} \le \|f\|_{\Lambda_q BV}.$$

The proof of the properness of the inclusion (2.48) is left to the reader (Exercise 2.5). □

We point out that Proposition 2.32 carries over rather easily to the more general setting of the Wiener–Young space WBV_ϕ defined by the norm (2.11).

Proposition 2.34. *Let $\phi : [0, \infty) \to [0, \infty)$ be some Young function, and let ϕ^* denote its conjugate Young function* (0.23). *Let $\Lambda = (\lambda_n)_n$ be a Waterman sequence satisfying*

$$\sum_{n=1}^{\infty} \phi^*(\lambda_n) < \infty . \tag{2.49}$$

Then the inclusion $WBV_\phi([a,b]) \subseteq \Lambda BV([a,b])$ holds.

Proof. Using Young's inequality (0.24) between ϕ and ϕ^*, for any $f \in WBV_\phi([a,b])$ and any collection $S_\infty = \{[a_n, b_n] : n \in \mathbb{N}\} \in \Sigma_\infty([a,b])$, we get

$$\begin{aligned}\sum_{n=1}^{N} \lambda_n |f(b_n) - f(a_n)| &\leq \sum_{n=1}^{N} \phi(|f(b_n) - f(a_n)|) + \sum_{n=1}^{N} \phi^*(\lambda_n)\\ &\leq \mathrm{Var}_\phi^W(f; [a,b]) + \sum_{n=1}^{\infty} \phi^*(\lambda_n) < \infty ,\end{aligned}$$

and hence $\mathrm{Var}_\Lambda(f; [a,b]) < \infty$. □

Taking $\phi(t) = t^p$ and $\lambda_n = n^{-q}$ in Proposition 2.34, we get $\phi^*(t) = t^{p'}$ (up to constants), and so condition (2.49) reads

$$\sum_{n=1}^{\infty} \frac{1}{n^{qp'}} < \infty . \tag{2.50}$$

This holds precisely for $qp' > 1$, and so we recover Proposition 2.32 as a special case of Proposition 2.34.

There is another interesting function class which is related to the Waterman space ΛBV and is based on the notion of the modulus of variation of a function. First, we need another family of intervals similar to $\Sigma([a,b])$ and $\Sigma_\infty([a,b])$.

Definition 2.35. For $n \in \mathbb{N}$ *fixed*, denote by $\Sigma_n([a,b])$ the family of all collections

$$S_n := \{[a_1, b_1], \dots, [a_n, b_n]\} \tag{2.51}$$

of n nonoverlapping subintervals $[a_1, b_1], \dots, [a_n, b_n]$ of $[a,b]$. Given a bounded function $f : [a,b] \to \mathbb{R}$, the *modulus of variation* of f is the sequence $\nu(f) = (\nu(f)_n)_n$ defined by

$$\nu(f)_n = \nu(f; [a,b])_n := \sup \left\{ \sum_{k=1}^{n} |f(b_k) - f(a_k)| : S_n \in \Sigma_n([a,b]) \right\} , \tag{2.52}$$

where the supremum in (2.52) is taken over all collections S_n of the form (2.51). If $\nu := (\nu_n)_n$ is an arbitrary increasing and concave[11] real sequence, the *Chanturiya class* $\mathrm{V}_\nu = \mathrm{V}_\nu([a,b])$ is defined as a set of all bounded functions $f : [a,b] \to \mathbb{R}$ satisfying

$$\nu(f)_n = O(\nu_n) \quad (n \to \infty) , \tag{2.53}$$

11 Recall that a sequence $(\nu_n)_n$ is called *concave* if $\nu_{n+1} + \nu_{n-1} \leq 2\nu_n$ for $n \geq 2$.

with $\nu(f)_n$ given by (2.52). In the particular case of the concave sequence $\nu^r := (n^r)_n$ with $0 < r \leq 1$, we denote the corresponding Chanturiya class by $\mathrm{V}_{\nu^r} = \mathrm{V}_{\nu^r}([a,b])$. ■

It follows immediately from Definition 2.35 that

$$\nu(f)_n \leq \nu(f)_{n+1} \quad (n = 1, 2, \ldots).$$

Another interesting property of (2.52) is contained in Exercise 2.20.

The characteristic $\nu(f)$ and the function class V_ν have been introduced by Chanturiya in [79] and applied to the Fourier series in [80–84]. Roughly speaking, the Chanturiya class V_ν is modeled on the modulus of variation in a similar way as the generalized Hölder space $Lip_{\omega,p}$ (Definition 0.54) on the modulus of continuity of a function. We point out that interesting interconnections between these moduli exist. For instance, one may show that, for $f \in C([a,b])$,

$$\omega_1(f;\delta) = O\left(\delta\nu(f)_{\mathrm{ent}(1/\delta)}\right) \quad (\delta \to 0+), \tag{2.54}$$

where $\omega_1(f;\delta)$ denotes the integral modulus of continuity (0.98) of f, and $\mathrm{ent}(\xi)$ is the integer part of $\xi \in \mathbb{R}^+$, see Exercise 2.10. Vice versa, one has

$$\nu(f)_n = o\left(n\omega_\infty(f;1/n)\right) \quad (n \to \infty), \tag{2.55}$$

where $\omega_\infty(f;\delta)$ is the classical modulus of continuity (0.97), see Exercise 2.11.

The following connection of the Chanturiya class V_ν with the Wiener–Young space WBV_ϕ and the Waterman space ΛBV is due to Avdispahić [24, 25].

Proposition 2.36. (a) *For any Young function ϕ, the inclusion*

$$WBV_\phi([a,b]) \subseteq \mathrm{V}_\nu([a,b]) \tag{2.56}$$

holds true for

$$\nu_n := n\phi^{-1}\left(\frac{1}{n}\right). \tag{2.57}$$

In particular,

$$WBV_p([a,b]) \subseteq \mathrm{V}_\nu([a,b]) \quad (1 \leq p < \infty),$$

where $\nu_n = n^{1-1/p}$.

(b) *For any Waterman sequence Λ, the inclusion*

$$\Lambda BV([a,b]) \subseteq \mathrm{V}_\nu([a,b]) \tag{2.58}$$

holds true for

$$\nu_n := \frac{n}{\lambda[1,n]},$$

where $\lambda[1,n]$ is given by (2.42). In particular,

$$\Lambda_q BV([a,b]) \subseteq \mathrm{V}_{\nu^q}([a,b]) \quad (0 < q < 1),$$

where $\nu^q = (n^q)_n$.

Proof. (a) Fix $f \in WBV_\phi([a,b])$ and $S_n = \{[a_1,b_1],\dots,[a_n,b_n]\} \in \Sigma_n([a,b])$. Applying the Jensen inequality (2.18) to $\alpha_k \equiv 1$ and $u_k = |f(b_k) - f(a_k)|$ for $k = 1,\dots,n$, we get

$$\phi\left(\frac{1}{n}\sum_{k=1}^{n}|f(b_k) - f(a_k)|\right) \le \sum_{k=1}^{n}\frac{\phi(|f(b_k) - f(a_k)|)}{n} \le \frac{1}{n}\operatorname{Var}_\phi^W(f).$$

Applying ϕ^{-1} and multiplying by n yields

$$\sum_{k=1}^{n}|f(b_k) - f(a_k)| \le n\phi^{-1}\left(\frac{1}{n}\operatorname{Var}_\phi^W(f)\right) \le Cn\phi^{-1}\left(\frac{1}{n}\right) = C\nu_n$$

for some constant $C > 0$ depending on f. Choosing, in particular, $\phi(u) := |u|^p$ for $1 \le p < \infty$ in (a), the sequence $(\nu_n)_n$ in (2.57) becomes

$$\nu_n = n\frac{1}{n^{1/p}} = n^{1-1/p};$$

in particular, $\nu_n \equiv 1$ for $p = 1$.

(b) Fix $f \in \Lambda BV([a,b])$ and $S_\infty = \{[a_k,b_k] : k \in \mathbb{N}\} \in \Sigma_\infty([a,b])$. By definition of the Waterman variation (2.25), for $n \in \mathbb{N}$ fixed, we have

$$\lambda_1|f(b_1) - f(a_1)| + \lambda_2|f(b_2) - f(a_2)| + \ldots + \lambda_n|f(b_n) - f(a_n)| \le \operatorname{Var}_\Lambda(f),$$

but also

$$\begin{aligned}
\lambda_2|f(b_1) - f(a_1)| + \lambda_3|f(b_2) - f(a_2)| + \ldots + \lambda_1|f(b_n) - f(a_n)| &\le \operatorname{Var}_\Lambda(f),\\
\lambda_3|f(b_1) - f(a_1)| + \lambda_4|f(b_2) - f(a_2)| + \ldots + \lambda_2|f(b_n) - f(a_n)| &\le \operatorname{Var}_\Lambda(f),\\
&\vdots\\
\lambda_n|f(b_1) - f(a_1)| + \lambda_1|f(b_2) - f(a_2)| + \ldots + \lambda_{n-1}|f(b_n) - f(a_n)| &\le \operatorname{Var}_\Lambda(f).
\end{aligned}$$

Summing up all these estimates, we obtain

$$\lambda[1,n]\sum_{j=1}^{n}|f(b_j) - f(a_j)| = \left(\sum_{i=1}^{n}\lambda_i\right)\left(\sum_{j=1}^{n}|f(b_j) - f(a_j)|\right) \le n\operatorname{Var}_\Lambda(f),$$

and hence

$$\nu(f)_n \le \frac{n\operatorname{Var}_\Lambda(f)}{\lambda[1,n]} = O\left(\frac{n}{\lambda[1,n]}\right),$$

and the conclusion follows.

To prove the last assertion, we use the fact that in the special case $\lambda_n = n^{-q}$, we have

$$\lambda[1,n] = \sum_{k=1}^{n}\frac{1}{k^q} \ge n\frac{1}{n^q} = n^{1-q} \quad (0 < q < 1).$$

This implies that

$$\frac{n}{\lambda[1,n]} = O(n^q),$$

and the assertion follows. □

Table 2.2. Imbeddings into Chanturiya classes.

The space	*is contained in*	*for* $\nu = (\nu_n)_n$ *with*
$BV([a,b])$	$V_\nu([a,b])$	$\nu_n \equiv 1$
$WBV_p([a,b])$	$V_\nu([a,b])$	$\nu_n = n^{1-1/p}$
$WBV_\phi([a,b])$	$V_\nu([a,b])$	$\nu_n = n\phi^{-1}\left(\frac{1}{n}\right)$
$\Lambda BV([a,b])$	$V_\nu([a,b])$	$\nu_n = \dfrac{n}{\lambda_1 + \ldots + \lambda_n}$
$\Lambda_q BV([a,b])$	$V_\nu([a,b])$	$\nu_n = n^q$

From Proposition 2.36, we obtain a series of imbedding theorems for the Wiener space WBV_p and the classical space BV. We summarize these imbeddings in the following Table 2.2.

Observe that in the terminology of the last part of Definition 2.35, the first and second rows in Table 2.2 mean that

$$WBV_p([a,b]) \subseteq V_{\nu^q}([a,b]) \quad (1 \le p < \infty,\ q = 1 - 1/p).$$

Moreover, for this choice of p and q, we deduce from Proposition 2.33 the strict inclusion

$$\Lambda_q BV([a,b]) \subset WBV_p([a,b]) \subseteq V_{\nu^q}([a,b])$$

since $p'q = 1$. In this sense, the second row in Table 2.2 is sharper than the last row.

Yet another generalization of such an imbedding which contains all imbeddings in Proposition 2.36 may be found in Proposition 2.85 in Section 2.8.

The following subclass of the Waterman space ΛBV is important in the theory of summability of the Fourier series, see Section 7.2.

Definition 2.37. For $m \in \mathbb{N}$, the *m-shift* of a Waterman sequence $\Lambda = (\lambda_n)_n$ is defined by $\Lambda^m := (\lambda_{m+n})_n$. A function $f \in \Lambda BV([a,b])$ is called *continuous in Waterman variation* if

$$\lim_{m\to\infty} \operatorname{Var}_{\Lambda^m}(f;[a,b]) = 0. \tag{2.59}$$

We write $\Lambda^c BV([a,b])$ for the set of all functions $f \in \Lambda BV([a,b])$ which are continuous in Waterman variation. ■

The concept of continuity in Waterman variation was introduced by Waterman [315, 316] under the name *continuous Λ-variation*, see also [25]. In the following proposition, we give a sufficient condition for continuity in Waterman variation due to Avdispahić [24, 25] in terms of the modulus of variation (2.52) of a bounded function.

Proposition 2.38. *Let* $\Lambda = (\lambda_n)_n$ *be a Waterman sequence and* $f \in B([a,b])$. *Suppose that*

$$\sum_{k=1}^{\infty} (\lambda_k - \lambda_{k+1})\nu(f)_k < \infty, \tag{2.60}$$

where $\nu(f) = (\nu(f)_n)_n$ *denotes the modulus of variation* (2.52) *of* f. *Then* $f \in \Lambda^c BV([a,b])$.

Proof. Let $S_\infty = \{[a_n, b_n] : n \in \mathbb{N}\} \in \Sigma_\infty([a,b])$. By Abel's partial summation and the fact that $(\lambda_n)_n$ is decreasing and converges to zero, for fixed $m \in \mathbb{N}$, we obtain

$$
\begin{aligned}
&\sum_{k=1}^{n} \lambda_{n+m}|f(b_k) - f(a_k)| \\
&\quad = \sum_{k=1}^{n-1}(\lambda_{k+m} - \lambda_{k+m+1})\sum_{i=1}^{k}|f(b_i) - f(a_i)| + \lambda_{n+m}\sum_{k=1}^{n}|f(b_k) - f(a_k)| \\
&\quad \le \sum_{k=1}^{n-1}(\lambda_{k+m} - \lambda_{k+m+1})\nu(f)_k + \nu(f)_n \sum_{k=n+m}^{\infty}(\lambda_k - \lambda_{k+1}) \\
&\quad \le \sum_{k=1}^{n-1}(\lambda_{k+m} - \lambda_{k+m+1})\nu(f)_{k+m} + \sum_{k=n+m}^{\infty}(\lambda_k - \lambda_{k+1})\nu(f)_k \\
&\quad = \sum_{j=m+1}^{\infty}(\lambda_j - \lambda_{j+1})\nu(f)_j\,.
\end{aligned}
$$

However, the last expression tends to 0 as $m \to \infty$, by (2.60), and thus also

$$\sum_{k=1}^{n} \lambda_{n+m}|f(b_k) - f(a_k)| \to 0 \quad (m \to \infty)\,.$$

Taking the supremum over all collections $S_\infty \in \Sigma_\infty([a,b])$, we conclude that (2.59) holds. □

So far, we have considered imbeddings of Waterman spaces into Chanturiya classes. Proposition 2.38 allows us to derive several imbedding theorems in the opposite direction. We collect some of them in the following

Proposition 2.39.

(a) *Suppose that*

$$\sum_{k=1}^{\infty} \frac{\nu_k}{k^{1+q}} < \infty$$

for some increasing concave sequence $\nu = (\nu_k)_k$ *and some* $q \in (0,1]$. *Then the inclusion*

$$\mathrm{V}_\nu([a,b]) \subseteq \Lambda_q^c BV([a,b]) \tag{2.61}$$

holds true, where $\Lambda_q BV$ *is given in Definition 2.29. In particular,*

$$\sum_{k=1}^{\infty} \frac{\nu_k}{k^2} < \infty$$

implies that

$$\mathrm{V}_\nu([a,b]) \subseteq H^c BV([a,b])\,, \tag{2.62}$$

where HBV denotes the space of all functions of harmonic bounded variation.

(b) *Suppose that*

$$\int_0^1 \frac{1}{\phi(t)^{1-q}}\,dt < \infty$$

for some Young function ϕ and some $q \in (0,1)$. Then the inclusion

$$WBV_\phi([a,b]) \subseteq \Lambda_q^c BV([a,b]) \tag{2.63}$$

holds true. Similarly,

$$\int_0^1 \log\frac{1}{\phi(t)}\,dt < \infty$$

implies that

$$WBV_\phi([a,b]) \subseteq H^c BV([a,b])\,. \tag{2.64}$$

Proof. The statement (a) follows from the simple fact that for $\lambda_k = k^{-q}$ with $0 < q \le 1$,

$$\lambda_k - \lambda_{k+1} = \frac{(k+1)^q - k^q}{k^q(k+1)^q} \le \frac{qk^{q-1}}{k^{2q}} = \frac{q}{k^{1+q}}$$

and Proposition 2.38, while (2.62) is of course a special case of (2.61). To prove (b), observe that the integral condition given there implies

$$\sum_{k=1}^\infty \frac{1}{k^q}\phi^{-1}\left(\frac{1}{k}\right) = \sum_{k=1}^\infty \frac{\nu_k}{k^{q+1}} < \infty\,,$$

and so $\mathrm{V}_\nu([a,b]) \subseteq \Lambda_q^c([a,b])$ for the modulus of variation $(\nu_n)_n$ in Proposition 2.36 (a). However, in Proposition 2.36 (a), we have shown that $WBV_\phi \subseteq \mathrm{V}_\nu$ precisely for this choice of $\nu = (\nu_n)_n$, and so we have proved (2.63). The inclusion (2.64) follows in a similar way from the second statement in (a). □

Let us briefly show what Proposition 2.39 (b) means for the special choice $\phi(t) = t^p$ $(p > 1)$. In this case, the integral condition reads

$$\int_0^1 \frac{1}{t^{p(1-q)}}\,dt < \infty\,,$$

and so we have $WBV_p \subseteq \Lambda_q^c BV$ for $p(1-q) < 1$, i.e. $p'q > 1$. In Proposition 2.32, we have seen that the same condition on p and q ensures the inclusion $WBV_p \subseteq \Lambda_q BV$. However, since $\Lambda_q^c BV$ is a strict subspace of $\Lambda_q BV$, Proposition 2.39 (b) is better than Proposition 2.32.

Our previous discussion shows that for $0 < q < r < 1$, the especially interesting chain of inclusions

$$\Lambda_q BV([a,b]) \subseteq WBV_{1/(1-q)}([a,b]) \subseteq \mathrm{V}_{\nu^q}([a,b]) \subseteq \Lambda_r^c BV([a,b]) \tag{2.65}$$

holds, where WBV_p denotes the Wiener space from Definition 1.31, and $\nu^q = (n^q)_n$ as before. These inclusions show that the Wiener space WBV_p and the Chanturiya class V_ν are, for appropriate choices of p and ν, in a certain sense, “intermediate” spaces between the Waterman space $\Lambda_q BV$ and $\Lambda_r^c BV$ which is “slightly larger” than $\Lambda_q BV$ for $r > q$.

In fact, the first inclusion in (2.65) has been proved in Proposition 2.33, the second one follows from Proposition 2.36 (a), and the third one is a consequence of Proposition 2.39 (a) since

$$\sum_{k=1}^{\infty} \frac{\nu_k}{k^{1+r}} = \sum_{k=1}^{\infty} \frac{1}{k^{1+r-q}} < \infty$$

precisely in case $r > q$.

In Exercise 2.5, we suggest how to show that the first inclusion in (2.65) (or (2.48) is strict. The following two examples also show that the other inclusions in (2.65) are strict.

Example 2.40. Let $f : [0,1] \to \mathbb{R}$ be defined, for $n = 1, 2, 3, \ldots$, by

$$f(x) := \begin{cases} 0 & \text{if } x = 0\,, \\ \frac{1}{n^{1-q}} & \text{if } x = \frac{1}{2^{n-1}}\,, \\ 0 & \text{if } x = \frac{3}{2^{n+1}}\,, \\ \text{linear} & \text{otherwise.} \end{cases}$$

We claim that $f \in V_{\nu^q}([0,1]) \setminus WBV_{1/(1-q)}([0,1])$. In fact, choosing the intervals

$$[a_1, b_1] := \left[\tfrac{3}{4}, 1\right], [a_2, b_2] := \left[\tfrac{3}{8}, \tfrac{1}{2}\right], \ldots, [a_n, b_n] := \left[\tfrac{3}{2^{n+1}}, \tfrac{1}{2^{n-1}}\right],$$

on which the function f is increasing linearly, we have $\{[a_1, b_1], \ldots, [a_n, b_n]\} \in \Sigma_n([a,b])$ and[12]

$$\nu(f)_n = \sum_{k=1}^{n} \frac{1}{(k+1)^{1-q}} = O(n^q) \quad (0 < q < 1)\,, \tag{2.66}$$

by definition of f, and so $f \in V_{\nu^q}([0,1])$. On the other hand, taking the $(1-q)$-th root of the terms after the sum in (2.66), we essentially get the n-th partial sum of the harmonic series, and so $f \notin WBV_{1/(1-q)}([0,1])$. ♥

Example 2.41. To construct a function $f \in \Lambda_r^c BV([0,1]) \setminus V_{\nu^q}([0,1])$ for $r > q$, recall that $f \in \Lambda_r^c BV([0,1])$ means that

$$\sup \left\{ \sum_{k=1}^{\infty} \frac{|f(b_k) - f(a_k)|}{(k+m)^r} \right\} \to 0 \quad (m \to \infty)\,, \tag{2.67}$$

12 The relation (2.66) may be proved by means of the integral comparison test for series.

where the supremum in (2.67) is taken over all infinite collections $\{[a_k, b_k] : k \in \mathbb{N}\} \in \Sigma_\infty([a, b])$, while $f \in V_{\nu^q}([0, 1])$ means that

$$\sup \left\{ \sum_{k=1}^{n} |f(b_k) - f(a_k)| \right\} = O(n^q), \tag{2.68}$$

where the supremum in (2.68) is taken over all collections $\{[a_1, b_1], \ldots, [a_n, b_n]\} \in \Sigma_n([a, b])$. Thus, if we choose f in such a way that the supremum in (2.68) is $O(n^p)$ for $q < p < r$, then $f \in \Lambda_r^c BV([0, 1])$, but $f \notin V_{\nu^q}([0, 1])$. ♥

We will summarize several other imbedding theorems in Table 2.7 in Section 2.8.

2.3 The Schramm variation

In this section, we discuss the most general concept of variation which contains many of the previously considered concepts as special cases. Unfortunately, the price we have to pay for this generality is that the constructions and proofs become extremely technical and cumbersome. The following is a generalization of Definition 2.15.

Definition 2.42. A *Schramm sequence* is a decreasing sequence $\Phi = (\phi_n)_n$ of Young functions $\phi_n : [0, \infty) \to [0, \infty)$ (Definition 2.1) such that

$$\sum_{n=1}^{\infty} \phi_n(t) = \infty \quad (t > 0). \tag{2.69}$$

Given a function $f : [a, b] \to \mathbb{R}$, a collection $S_\infty = \{[a_n, b_n] : n \in \mathbb{N}\} \in \Sigma_\infty([a, b])$, and a Schramm sequence $\Phi = (\phi_n)_n$, the positive real number

$$\mathrm{Var}_\Phi(f, S_\infty) = \mathrm{Var}_\Phi(f, S_\infty; [a, b]) := \sum_{k=1}^{\infty} \phi_k(|f(b_k) - f(a_k)|) \tag{2.70}$$

is called the *Schramm variation* of f on $[a, b]$ with respect to S_∞, while the (possibly infinite) number

$$\mathrm{Var}_\Phi(f) = \mathrm{Var}_\Phi(f; [a, b]) := \sup \left\{ \mathrm{Var}_\Phi(f, S_\infty; [a, b]) : S_\infty \in \Sigma_\infty([a, b]) \right\}, \tag{2.71}$$

where the supremum is taken over all collections $S_\infty \in \Sigma_\infty([a, b])$, is called the *total Schramm variation* of f on $[a, b]$. If

$$\mathrm{Var}_\Phi(cf; [a, b]) < \infty \tag{2.72}$$

for some $c > 0$, we say that f has *bounded Schramm variation* on $[a, b]$ (or *bounded Φ-variation in Schramm's sense*) and write $f \in \Phi BV([a, b])$. ■

As we have done in Definition 1.21 for the space AC of absolutely continuous functions, we may also show here that in Definition 2.42, we get the same space ΦBV if in (2.71) we restrict ourselves to *finite* collections $S \in \Sigma([a,b])$. This means that

$$\begin{aligned}\mathrm{Var}_\Phi(f;[a,b]) &= \sup\left\{\mathrm{Var}_\Phi(f,S;[a,b]) : S \in \Sigma([a,b])\right\} \\ &= \sup\left\{\sum_{k=1}^{n} \phi_k(|f(b_k) - f(a_k)|) : \{[a_1,b_1],\dots,[a_n,b_n]\} \in \Sigma([a,b])\right\}.\end{aligned} \tag{2.73}$$

Definition 2.42 is so general that it contains many of the previously discussed spaces which may be obtained by special choices of $\Phi = (\phi_n)_n$. For further reference, we collect these spaces in the following

Proposition 2.43. *The spaces BV, WBV_p, WBV_ϕ, and ΛBV are all of type ΦBV for a suitable choice of Φ.*

Proof. In the same order as in the assertion, for $n = 1, 2, 3, \dots$, take $\phi_n(t) = t$, $\phi_n(t) = t^p$ $(p \geq 1)$, $\phi_n(t) \equiv \phi(t)$, and $\phi_n(t) = \lambda_n t$, respectively. □

As for the space $WBV_\phi([a,b])$, we may consider the set

$$B(\Phi) := \{f \in B([a,b]) : \mathrm{Var}_\Phi(f;[a,b]) \leq 1\} \tag{2.74}$$

together with the corresponding Minkowski functional

$$\|f\|_{\Phi BV} := |f(a)| + \inf\left\{\lambda > 0 : \; f/\lambda \in B(\Phi)\right\}, \tag{2.75}$$

which is a *norm* on the space $\Phi BV([a,b]) = \mathrm{span}\, B(\Phi)$. Moreover, the set (2.74) coincides then with the closed unit ball with respect to this norm. We summarize some properties of the space $\Phi BV([a,b])$ with the following

Proposition 2.44. *The number* (2.75) *and the set $\Phi BV([a,b])$ have the following properties.*

(a) *For $f \neq 0$, we have the estimate*

$$\mathrm{Var}_\Phi\left(\frac{f}{\|f\|_{\Phi BV}}\right) \leq 1. \tag{2.76}$$

(b) *From $\|f\|_{\Phi BV} \leq 1$, it follows that $\mathrm{Var}_\Phi(f) \leq \|f\|_{\Phi BV}$.*

(c) *The set $\Phi BV([a,b])$ is a linear space, and* (2.75) *defines a norm on $\Phi BV([a,b])$.*

(d) *Convergence in the norm* (2.75) *implies convergence in the norm* (0.39), *i.e. uniform convergence on $[a,b]$.*

(e) *$(\Phi BV([a,b]), \|\cdot\|_{\Phi BV})$ is a Banach space.*

Proof. By our previous discussion, in all calculations, we may restrict ourselves to finite collections $S \in \Sigma([a,b])$. The assertions (a) and (b) follow from our general results on Minkowski functionals. To prove (c), we may restrict ourselves without loss of generality to functions from the subspace

$$\Phi BV^o([a,b]) := \{f \in \Phi BV([a,b]) : \; f(a) = 0\} \tag{2.77}$$

because otherwise we may pass from the function f to the function $f - f(a)$. Clearly, the zero function has norm 0. Thus, let $x \in [a,b]$ be such that $f(x) \neq 0$. Then

$$\mathrm{Var}_{\Phi}\left(\frac{f}{\lambda}\right) \geq \phi_1\left(\frac{|f(x)|}{\lambda}\right) \to \infty \quad (\lambda \to 0)\,.$$

Thus, there is a $\lambda > 0$ such that $\mathrm{Var}_{\Phi}(f/\lambda) > 1$, implying that $\|f\|_{\Phi BV} \neq 0$. The homogeneity of (2.75) follows from

$$\begin{aligned}\|\mu f\|_{\Phi BV} &= \inf\left\{\lambda > 0 : \mathrm{Var}_{\Phi}\left(\frac{\mu f}{\lambda}\right) \leq 1\right\} = \inf\left\{\lambda > 0 : \mathrm{Var}_{\Phi}\left(\frac{|\mu f|}{\lambda}\right) \leq 1\right\} \\ &= |\mu| \inf\left\{\nu > 0 : \mathrm{Var}_{\Phi}\left(\frac{|f|}{\nu}\right) \leq 1\right\} = |\mu|\|f\|_{\Phi BV}\end{aligned}$$

for $f \in \Phi BV^o([a,b])$, $\nu := \lambda/|\lambda|$, and $\mu \in \mathbb{R}\setminus\{0\}$, while the subadditivity of (2.75) follows from the estimate[13]

$$\begin{aligned}&\sum_{k=1}^{n} \phi_k\left(\frac{|(f+g)(b_k) - (f+g)(a_k)|}{\|f+g\|_{\Phi BV}}\right) \\ &\quad\leq \sum_{k=1}^{n}\left[\frac{\|f\|_{\Phi BV}}{\|f+g\|_{\Phi BV}}\phi_k\left(\frac{|f(b_k) - f(a_k)|}{\|f\|_{\Phi BV}}\right)\right] \\ &\quad\quad + \sum_{k=1}^{n}\left[\frac{\|g\|_{\Phi BV}}{\|f+g\|_{\Phi BV}}\phi_k\left(\frac{|g(b_k) - g(a_k)|}{\|g\|_{\Phi BV}}\right)\right] \leq 1\end{aligned}$$

which holds for $f, g \in \Phi BV^o([a,b])$ and any collection $S = \{[a_1,b_1],\dots,[a_n,b_n]\} \in \Sigma([a,b])$.

Now, we prove (d). Suppose that $f \in \Phi BV^o([a,b])$ satisfies $\|f\|_{\Phi BV} \leq \varepsilon$ for some $\varepsilon > 0$. Then $\mathrm{Var}_{\Phi}(f/\varepsilon) \leq \|f(\varepsilon\|_{\Phi BV} \leq 1$ by (b). Consequently,

$$\phi_1\left(\frac{|f(x)|}{\varepsilon}\right) \leq \mathrm{Var}_{\Phi}\left(\frac{f}{\varepsilon}\right) \leq 1$$

for all $x \in [a,b]$, and so $\|f\|_{\infty} \leq \varepsilon\phi_1^{-1}(1)$, where $\|\cdot\|_{\infty}$ denotes the norm (0.39).

It remains to show that the space $\Phi BV^o([a,b])$ with the norm (2.75) is complete. Let $(f_m)_m$ be a Cauchy sequence in $\Phi BV^o([a,b])$ with respect to the norm (2.75). By (d), $(f_m)_m$ is then a Cauchy sequence in $B([a,b])$ with respect to the norm (0.39). Since $(B([a,b]), \|\cdot\|_{\infty})$ is complete, there is a function $f \in B([a,b])$ such that $f_m \to f$, as $m \to \infty$, uniformly on $[a,b]$. Now, given any collection $S = \{[a_1,b_1],\dots,[a_n,b_n]\} \in \Sigma([a,b])$, choose $n_0 \in \mathbb{N}$ such that $\|f_p - f_q\|_{\Phi BV} \leq \varepsilon$ for $p, q \geq n_0$. Then

$$\begin{aligned}&\sum_{k=1}^{n} \phi_k\left(\frac{|(f_p - f)(b_k) - (f_q - f)(a_k)|}{\varepsilon}\right) \\ &\quad= \lim_{q\to\infty}\sum_{k=1}^{n} \phi_k\left(\frac{|(f_p - f_q)(b_k) - (f_p - f_q)(a_k)|}{\varepsilon}\right) \leq 1\,.\end{aligned}$$

---

13 Here, we use the convexity of ϕ_k.

However, this implies that $\mathrm{Var}_\Phi((f_m - f)/\varepsilon) \leq 1$ for $m \geq n_0$, $f \in \Phi BV^o([a,b])$, and $\|f_m - f\|_{\Phi BV} \to 0$ as $m \to \infty$. □

Proposition 2.44 has an interesting consequence. From (d), it follows that the spaces $(\Phi BV([a,b]) \cap C([a,b]), \|\cdot\|_{\Phi BV})$ and $(\Lambda BV([a,b]) \cap C([a,b]), \|\cdot\|_{\Lambda BV})$ are also Banach spaces.

The next proposition gives a necessary and sufficient condition for two Young sequences Φ and Ψ under which one of the corresponding Schramm spaces is contained in the other.

Proposition 2.45. *Given two Schramm sequences $\Phi = (\phi_n)_n$ and $\Psi = (\psi_n)_n$, we have the inclusion*

$$\Phi BV([a,b]) \subseteq \Psi BV([a,b]) \tag{2.78}$$

if

$$\sum_{k=1}^{n} \phi_k(t) \geq c \sum_{k=1}^{n} \psi_k(t) \quad (0 \leq t \leq T; n = 1, 2, \ldots) \tag{2.79}$$

for some $T > 0$ and $c > 0$.

The proof of Proposition 2.45 is elementary and left to the reader (Exercise 2.24). Simple examples show that condition (2.79) is not necessary for the inclusion (2.78) to hold (Exercise 2.25).

Now, we are going to prove a counterpart of (2.33) for Schramm spaces. To this end, let us recall that $R([a,b])$ denotes the space of all regular functions on the interval $[a,b]$, see Section 0.3.

Proposition 2.46. *The equalities*

$$\bigcap_{\Phi} \Phi BV([a,b]) = BV([a,b]), \quad \bigcup_{\Phi} \Phi BV([a,b]) = R([a,b]) \tag{2.80}$$

hold, where the intersection and union in (2.80) *are taken over all Schramm sequences Φ.*

Proof. We start with the union in (2.80). First, we show that

$$\Phi BV([a,b]) \subseteq R([a,b]) \tag{2.81}$$

for an arbitrary Schramm sequence $\Phi = (\phi_n)_n$. So, given $f \in \Phi BV([a,b])$, we show that f has a left limit at each point $x \in (a,b]$. Suppose that this is false, which means that there exists $x_0 \in (a,b]$ such that $f(x_0-)$ does not exist, and hence

$$l := \liminf_{x \to x_0-} f(x) < \limsup_{x \to x_0-} f(x) =: L\,.$$

Take $\delta = (L - l)/3$. Then, in view of the definition of l and L, we can find strictly increasing sequences $(P_n)_n$ and $(Q_n)_n$ satisfying $f(P_n) \leq l + \delta$, $f(Q_n) \geq L - \delta$, and

$$\lim_{n\to\infty} f(P_n) = l, \quad \lim_{n\to\infty} f(Q_n) = L\,.$$

Choose a subsequence $(p_n)_n$ of $(P_n)_n$ and a subsequence $(q_n)_n$ of $(Q_n)_n$ such that

$$p_1 < q_1 < p_2 < q_2 < \ldots < p_n < q_n < \ldots,$$

and consider the collection $\{[p_n, q_n] : n \in \mathbb{N}\} \in \Sigma_\infty([a, b])$. Then we have

$$|f(q_n) - f(p_n)| \geq f(q_n) - f(p_n) \geq L - \delta - (l + \delta) = \delta,$$

and hence

$$\sum_{n=1}^{\infty} \phi_n(|f(q_n) - f(p_n)|) \geq \sum_{n=1}^{\infty} \phi_n(\delta) = \infty$$

by (2.69). However, this contradicts our assumption $f \in \Phi BV([a, b])$. Thus, we have shown that the left limit $f(x_0-)$ exists at each point $x_0 \in (a, b]$. The proof for the right limit is similar and is therefore omitted. This completes the proof of the inclusion (2.81).

Observe that (2.81) yields the inclusion

$$\bigcup_{\Phi} \Phi BV([a, b]) \subseteq R([a, b]), \tag{2.82}$$

and so in order to prove the second equality in (2.80), we have to show that every regular functions belongs to some Schramm space. However, in (2.33), we have already shown that every regular function belongs to some Waterman space, and combining this with Proposition 2.43, we obtain

$$R([a, b]) \subseteq \bigcup_{\Lambda} \Lambda BV([a, b]) \subseteq \bigcup_{\Phi} \Phi BV([a, b]),$$

which proves the second equality in (2.80).

Now, we proceed with the proof of the first equality in (2.80). First, observe that for an arbitrary fixed Waterman sequence Λ, the inclusion

$$\bigcap_{\Phi} \Phi BV([a, b]) \subseteq \Lambda BV([a, b]) \tag{2.83}$$

holds. Thus, taking the intersection over all Waterman sequences on the right-hand side of (2.83), we obtain

$$\bigcap_{\Phi} \Phi BV([a, b]) \subseteq \bigcap_{\Lambda} \Lambda BV([a, b]).$$

Combining this with (2.33), we get

$$\bigcap_{\Phi} \Phi BV([a, b]) \subseteq BV([a, b]), \tag{2.84}$$

and it remains to prove the reverse inclusion. First, we make some remarks on Schramm sequences. Given a Schramm sequence $\Phi = (\phi_n)_n$, we know that for each

$n \in \mathbb{N}$, the function $t \mapsto \phi_n(t)/t$ is increasing on the half-axis $(0, \infty)$. This implies, in particular, that

$$\frac{\phi_n(t)}{t} \le \frac{\phi_n(1)}{1} = \phi_n(1)$$

for $t \in (0, 1]$. Consequently,

$$\phi_n(t) \le \phi_n(1)t \quad (0 \le t \le 1,\ n = 1, 2, \dots).$$

Moreover,

$$\phi_n(t) \le \phi_1(t) \le \phi_1(1)t \quad (0 \le t \le 1,\ n = 1, 2, \dots)$$

since the sequence $(\phi_n)_n$ is decreasing.

Now, fix $f \in BV([a, b])$ and $S_\infty = \{[a_n, b_n] : n \in \mathbb{N}\} \in \Sigma_\infty([a, b])$. Denote by A the set of all indices $n \in \mathbb{N}$ for which $|f(b_n) - f(a_n)| \le 1$, and by $B := \mathbb{N} \setminus A$, the other indices. Then

$$\begin{aligned}
\sum_{n=1}^{\infty} \phi_n(|f(b_n) - f(a_n)|) &= \sum_{n\in A} \phi_n(|f(b_n) - f(a_n)|) + \sum_{n\in B} \phi_n(|f(b_n) - f(a_n)|) \\
&\le \sum_{n\in A} \phi_1(1)|f(b_n) - f(a_n)| + \sum_{n\in B} \phi_n(|f(b_n) - f(a_n)|) \\
&= \phi_1(1) \sum_{n\in A} |f(b_n) - f(a_n)| + \sum_{n\in B} \phi_n(|f(b_n) - f(a_n)|) \\
&\le \phi_1(1) \operatorname{Var}(f; [a, b]) + \sum_{n\in B} \phi_n(|f(b_n) - f(a_n)|).
\end{aligned}$$

However, the last sum in these estimates contains at most $N := \operatorname{ent}(\operatorname{Var}(f; [a, b]))$ terms, where $\operatorname{ent}(\xi)$ denotes the integer part of ξ. Otherwise, this sum would have at least $N + 1$ terms and we would get

$$\operatorname{Var}(f; [a, b]) > \sum_{n\in B} |f(b_n) - f(a_n)| \ge N + 1 = \operatorname{ent}(\operatorname{Var}(f; [a, b])) + 1,$$

that is, a contradiction. Thus, we obtain

$$\begin{aligned}
\sum_{n\in B} \phi_n(|f(b_n) - f(a_n)|) &\le \sum_{n\in B} \phi_n(2\|f\|_\infty) \\
&\le \sum_{n\in B} \phi_1(2\|f\|_\infty) \le \phi_1(2\|f\|_\infty) \operatorname{Var}(f; [a, b]),
\end{aligned}$$

where $\|\cdot\|_\infty$ denotes the norm (0.39). Combining these estimates yields

$$\begin{aligned}
\sum_{n=1}^{\infty} \phi_n(|f(b_n) - f(a_n)|) &\le \phi_1(1) \operatorname{Var}(f; [a, b]) + \phi_1(2\|f\|_\infty) \operatorname{Var}(f; [a, b]) \\
&\le [\phi_1(1) + \phi_1(2\|f\|_\infty)] \operatorname{Var}(f; [a, b]),
\end{aligned}$$

and hence

$$\operatorname{Var}_\Phi(f; [a, b]) \le [\phi_1(1) + \phi_1(2\|f\|_\infty)] \operatorname{Var}(f; [a, b]).$$

However, this means precisely that $f \in \Phi BV([a,b])$, and so we have shown that $BV([a,b]) \subseteq \Phi BV([a,b])$ since $f \in BV([a,b])$ was arbitrary. However, the Schramm sequence Φ was also chosen arbitrarily. Thus, we conclude that

$$BV([a,b]) \subseteq \bigcap_{\Phi} \Phi BV([a,b])$$

which together with (2.84), completes the proof of the first equality in (2.80). □

In the following proposition which is parallel to Proposition 1.7, we compare the continuity of a function f and that of its variation function.

Proposition 2.47. *Suppose that $f \in \Phi BV([a,b])$ is continuous at some point $x_0 \in [a,b]$. Then the Schramm variation function $\mathrm{V}_{f,\Phi} : [a,b] \to \mathbb{R}$ defined by*

$$\mathrm{V}_{f,\Phi}(x) := \mathrm{Var}_{\Phi}(f;[a,x]) \quad (a \le x \le b)$$

is also continuous at x_0.

Proof. We restrict ourselves to showing that the variation function is right continuous at x_0. Fix $x_0 \in [a,b)$, and take a finite collection $\{[a_1,b_1],\dots,[a_{n_1},b_{n_1}]\} \in \Sigma([x_0,b])$ of nonoverlapping subintervals of $[x_0,b]$ which are arranged in decreasing order. Moreover, we assume that

$$\sum_{n=1}^{n_1} \phi_n(|f(b_n)-f(a_n)|) > \frac{1}{2}\mathrm{Var}_{\Phi}(f;[x_0,b]).$$

Since f is right continuous at x_0, by assumption, and each ϕ_n is continuous, we may assume that $I_n \subset (x_0,b]$ for $n=1,2,\dots,n_1$. Choose a number $y_1 \in \left(x_0, \frac{1}{2}(x_0+b)\right)$ in such a way that

$$[x_0,y_1] \cap \bigcup_{n=1}^{n_1}[a_n,b_n] = \emptyset$$

and $|f(d)-f(c)| \le |f(b_{n_1})-f(a_{n_1})|$ for any interval $[c,d] \subset (x_0,y_1)$. In this way, we obtain a collection of intervals $[a_n,b_n] \subset (x_0,y_1]$ for $n=n_1+1,n_1+2,\dots,n_2$ which satisfy

$$|f(b_n)-f(a_n)| \ge |f(b_{n+1})-f(a_{n+1})| \quad (n=n_1+1,n_1+2,\dots,n_2)$$

and

$$\sum_{n=n_1+1}^{n_2} \phi_{n-n_1}(|f(b_n)-f(a_n)|) > \frac{1}{2}\mathrm{Var}_{\Phi}(f;[x_0,y_1]).$$

Proceeding with this process, for $k=1,2,3,\dots$ points y_k and intervals, we obtain

$$[a_{n_k+1},b_{n_k+1}],[a_{n_k+2},b_{n_k+2}],\dots,[a_{n_{k+1}},b_{n_{k+1}}] \subseteq [y_{k+1},y_k] \quad (k=1,2,\dots)$$

such that the sequence $(y_k)_k$ is decreasing, $y_k \to x_0$ as $k \to \infty$, the sequence $(|f(b_n)-f(a_n)|)_n$ is also decreasing, and

$$\sum_{n=n_k+1}^{n_{k+1}} \phi_{n-n_k}(|f(b_n)-f(a_n)|) > \frac{1}{2}\mathrm{Var}_{\Phi}(f;[x_0,y_k]).$$

However,

$$\sum_{n=1}^{\infty} \phi_n(|f(b_n) - f(a_n)|) < \infty$$

since f has bounded Schramm variation, and so for a given $\varepsilon > 0$, we find an index $N \in \mathbb{N}$ such that

$$\sum_{n=N+1}^{\infty} \phi_n(|f(b_n) - f(a_n)|) < \frac{\varepsilon}{2}.$$

Keeping in mind that the real sequence $(|f(b_n) - f(a_n)|)_n$ is decreasing and converges to zero, we deduce that

$$\sum_{n=N+1}^{\infty} \phi_n(|f(b_{n+j}) - f(a_{n+j})|) < \frac{\varepsilon}{2}$$

for any $j \in \mathbb{N}$. Consequently, there exists a natural number j_ε such that

$$\sum_{n=1}^{N} \phi_n(|f(b_{n+j}) - f(a_{n+j})|) < \frac{\varepsilon}{2}$$

for $j > j_\varepsilon$. Taking, in particular, $j = n_k$, we have

$$\frac{1}{2}\operatorname{Var}_\Phi(f; [x_0, y_k]) < \sum_{n=1}^{\infty} \phi_n(|f(b_{n+n_k}) - f(a_{n+n_k})|) < \varepsilon$$

for k sufficiently large. This implies that $\operatorname{Var}_\Phi(f; [x_0, y_k]) \to 0$ as $k \to \infty$, and so

$$\lim_{y \to x_0+} \operatorname{Var}_\Phi(f; [x_0, y]) = 0$$

since the map $y \mapsto \operatorname{Var}_\Phi(f; [x_0, y])$ is monotone. □

A converse of Proposition 2.47 may be found in Exercise 2.22. Moreover, a "one-sided" version of Proposition 2.47 is given in Exercise 2.23.

Now, we provide an analogue of Helly's selection principle (Theorem 1.11) for the space $\Phi BV([a, b])$. First, we introduce some notation. Given a Schramm sequence $\Phi = (\phi_n)_n$, an index $p \in \mathbb{N}$, and real numbers $M > 0$ and $\delta > 0$, we introduce the numbers

$$k(M, \delta) := \min\left\{n \in \mathbb{N} : \sum_{j=1}^{n} \phi_j(\delta/2) > M\right\} \in \mathbb{N} \tag{2.85}$$

and

$$\beta(p, \delta) := \min\left\{1 - \frac{\phi_j(\delta/2)}{\phi_j(\delta)} : j = 1, 2, \dots, p\right\} \in \mathbb{R}. \tag{2.86}$$

Note that $M_1 \geq M_2$ implies $k(M_1, \delta) \geq k(M_2, \delta)$, and $p_1 \geq p_2$ implies $\beta(p_1, \delta) \leq \beta(p_2, \delta)$. Moreover, $\beta(p, \delta) > 0$ for any $p \in \mathbb{N}$ and $\delta > 0$.

In what follows, we will utilize the following technical lemma for the Schramm variation function introduced in Proposition 2.47.

Lemma 2.48. *Let $f \in \Phi BV([a,b])$ and $\delta > 0$. Suppose that $a \le x < y \le b$ such that $|f(y) - f(x)| \ge \delta$. Then*

$$\mathrm{V}_{f,\Phi}(y) - \mathrm{V}_{f,\Phi}(x) \ge \beta_0 \phi_{k_0}(\delta),$$

where $k_0 := k(\mathrm{V}_{f,\Phi}(x), \delta)$ and $\beta_0 := \beta(k_0, \delta)$ are defined as in (2.85) and (2.86).

We do not prove this lemma since the proof is extremely technical (see Lemma 2.5 in [286]). Instead, we use this result to state a Helly-type theorem for sequences in $\Phi BV([a,b])$.

Theorem 2.49. *Let $(f_n)_n$ be a sequence in $\Phi BV([a,b])$ such that*

$$\|cf_n\|_\infty \le M, \quad \mathrm{Var}_\Phi(cf_n; [a,b]) \le M \quad (n = 1, 2, \ldots)$$

for some constants $c > 0$ and $M > 0$. Then $(f_n)_n$ contains a subsequence which converges pointwise on $[a,b]$ to some $f \in \Phi BV([a,b])$; moreover, $\mathrm{Var}_\Phi(cf; [a,b]) \le M$.

Proof. For $n = 1, 2, 3, \ldots$, put

$$v_n(x) := \mathrm{V}_{cf_n,\Phi}(x) = \mathrm{Var}_\Phi(cf_n; [a,x]) \quad (a \le x \le b).$$

Being monotonically increasing and bounded on $[a,b]$, each function v_n belongs to $BV([a,b])$. Thus, we may apply the classical Helly selection principle (Theorem 1.11) and obtain a subsequence $(v_{n_k})_k$ of the sequence $(v_n)_n$ and a function v such that $v_{n_k}(x) \to v(x)$ for all $x \in [a,b]$ as $k \to \infty$. Clearly, v is increasing on $[a,b]$, see Exercise 1.17, and satisfies $0 \le v(x) \le M$ for $a \le x \le b$. Using the usual diagonalization procedure, we may find a subsequence $(f_j)_j$ of the sequence $(f_{n_k})_k$ which converges at the endpoints a and b as well as at all rational points in $[a,b]$. Denote by $(v_j)_j$ the corresponding subsequence of $(v_{n_k})_k$. Since v is increasing, it is continuous on $[a,b] \setminus N$, where $N \subset [a,b]$ is at most countable. Let $x_0 \in [a,b) \setminus N$ be irrational. Then we find a rational number $y \in (x_0, b)$ such that

$$0 \le v(y) - v(x_0) < \eta := \frac{\beta_1 \phi_{k_1}(\varepsilon)}{3},$$

where $k_1 := k(M, \varepsilon)$ and $\beta_1 := \beta(k_1, \varepsilon)$ are defined according to (2.85) and (2.86). We can also find an index J such that $|v_j(y) - v(y)| < \eta$ and $|v_j(x_0) - v(x_0)| < \eta$ for $j < J$. For these j, we obtain[14]

$$0 \le v_j(y) - v_j(x_0) < 3\eta = \beta_1 \phi_{k_1}(\varepsilon) \le \beta_0 \phi_{k_0}(\varepsilon),$$

where $k_0 := k(v_j(x_0), \varepsilon)$ and $\beta_0 = \beta(k_0, \varepsilon)$. In view of Lemma 2.48, we have $|f_j(y) - f_j(x_0)| < \varepsilon$ for $j > J$. Since the sequence $(f_j(y))_j$ is convergent, there exists an index J_1 such that $|f_j(y) - f_k(y)| < \varepsilon$ for $j, k > J_1$. Thus, we get

$$|f_j(x_0) - f_k(x_0)| < \varepsilon \quad (j, k > \max\{J, J_1\}),$$

14 The last inequality is a consequence of the monotonicity of the quantities (2.85) and (2.86) in M and p.

which implies that the sequence $(f_j(x_0))_j$ is convergent. Our discussion shows that the sequence $(f_j(x))_j$ *fails* to converge only for $x \in N \cup [(a,b) \cap \mathbb{Q}]$, which is a countable set.

Applying again a diagonalization procedure, we may find a subsequence $(f_m)_m$ of $(f_j)_j$ which converges for all $x \in [a,b]$; we denote the corresponding limit by $f(x)$.

Let $(v_m)_m$ be the corresponding subsequence of $(v_j)_j$, and choose an arbitrary collection $S = \{[a_1,b_1],\dots,[a_{n_0},b_{n_0}]\} \in \Sigma([a,b])$. Then, for any $\varepsilon > 0$, we may find $m_0 \in \mathbb{N}$ such that $v(b) > v_m(b) - \varepsilon$ and

$$\left|\sum_{n=1}^{n_0} \phi_n(c|f_m(b_n) - f_m(a_n)|) - \sum_{n=1}^{n_0} \phi_n(c|f(b) - f(a)|)\right| < \varepsilon$$

for $m > m_0$. Then, for these m, we get

$$\begin{aligned} M \geq v(b) > v_m(b) - \varepsilon &\geq \sum_{n=1}^{n_0} \phi_n(c|f_m(b_n) - f_m(a_n)|) - \varepsilon \\ &\geq \sum_{n=1}^{n_0} \phi_n(c|f(b) - f(a)|) - 2\varepsilon\,. \end{aligned}$$

However, this implies that

$$\mathrm{Var}_\Phi(cf;[a,b]) < M + 2\varepsilon\,,$$

and hence $\mathrm{Var}_\Phi(cf;[a,b]) \leq M$ since $\varepsilon > 0$ was arbitrary, and completes the proof. □

Choosing, in particular, $\phi_n(t) = \lambda_n t$ for $t \geq 0$, where $\Lambda = (\lambda_n)_n$ is some Waterman sequence, we obtain a Helly-type selection principle in ΛBV from Theorem 2.49.

2.4 The Riesz–Medvedev variation

In this section, we study yet another concept of variation which goes back to Riesz [264, 265], contains the Jordan variation as special case and has particularly interesting applications.

Definition 2.50. Given a real number $p \geq 1$, a partition $P = \{t_0, t_1, \dots, t_m\} \in \mathcal{P}([a,b])$, and a function $f : [a,b] \to \mathbb{R}$, the nonnegative real number

$$\mathrm{Var}_p^R(f,P) = \mathrm{Var}_p^R(f,P;[a,b]) := \sum_{j=1}^{m} \frac{|f(t_j) - f(t_{j-1})|^p}{(t_j - t_{j-1})^{p-1}} \tag{2.87}$$

is called the *Riesz variation* of f on $[a,b]$ with respect to P, while the (possibly infinite) number

$$\mathrm{Var}_p^R(f) = \mathrm{Var}_p^R(f;[a,b]) := \sup\left\{\mathrm{Var}_p^R(f,P;[a,b]) : P \in \mathcal{P}([a,b])\right\}, \tag{2.88}$$

where the supremum is again taken over all partitions of $[a, b]$, is called the *total Riesz variation* of f on $[a, b]$. In case $\mathrm{Var}_p^R(f; [a, b]) < \infty$, we say that f has *bounded Riesz variation* (or *bounded p-variation in Riesz's sense*) on $[a, b]$ and write[15] $f \in RBV_p([a, b])$. ■

From Hölder's inequality (0.108), it follows that

$$(b-a)^{1-1/p}\,\mathrm{Var}_p^R(f; [a, b])^{1/p} \geq \mathrm{Var}(f; [a, b])\,, \tag{2.89}$$

which shows that the inclusion $RBV_p([a, b]) \subseteq BV([a, b])$ holds. Moreover, similarly as we have done this in Proposition 1.3 (d) and Proposition 1.10 for BV, we may prove the following result for RBV_p.

Proposition 2.51. *The set $RBV_p([a, b])$ equipped with the norm*

$$\|f\|_{RBV_p} := |f(a)| + \mathrm{Var}_p^R(f; [a, b])^{1/p} \quad (f \in RBV_p([a, b])) \tag{2.90}$$

is a Banach space which is for $p > 1$ continuously imbedded into the space $C([a, b])$ with norm (0.45) *as well as into the space $BV([a, b])$ with norm* (1.16). *Moreover, $RBV_p([a, b])$ is an algebra with*

$$\mathrm{Var}_p^R(fg) \leq \|f\|_\infty \mathrm{Var}_p^R(g) + \|g\|_\infty \mathrm{Var}_p^R(f) \tag{2.91}$$

for all $f, g \in RBV_p([a, b])$. Finally, the set $RBV_p^o([a, b])$ of all $f \in RBV_p([a, b])$ satisfying $f(a) = 0$ is a subalgebra.

Proof. It is not hard to prove that $RBV_p([a, b])$ is a linear space, and that every function $f \in RBV_p([a, b])$ for $p > 1$ is (uniformly) continuous on $[a, b]$. The fact that $(RBV_p([a, b]), \|\cdot\|_{RBV_p})$ is continuously imbedded into both $(C([a, b]), \|\cdot\|_\infty)$ and $(BV([a, b]), \|\cdot\|_{BV})$ follows from Proposition 1.3 (d) and (2.89).

To prove the completeness of $(RBV_p([a, b]), \|\cdot\|_{RBV_p})$, let $(f_n)_n$ be a Cauchy sequence in the norm (2.90). Given $\varepsilon > 0$, choose $n_0 \in \mathbb{N}$ such that $m, n \geq n_0$ implies

$$\sum_{j=1}^{k} \frac{|(f_m - f_n)(t_j) - (f_m - f_n)(t_{j-1})|^p}{(t_j - t_{j-1})^{p-1}} \leq \mathrm{Var}_p^R(f_m - f_n, P) \leq \varepsilon^p \tag{2.92}$$

for every partition $P = \{t_0, t_1, \dots, t_k\} \in \mathcal{P}([a, b])$. From what we have observed above, it follows that $(f_n)_n$ is a Cauchy sequence in the norm (0.45) of $C([a, b])$. Thus, $f_n \to f$ uniformly on $[a, b]$, where f is continuous on $[a, b]$. Letting $m \to \infty$ in (2.92), we obtain

$$\sum_{j=1}^{k} \frac{|(f - f_n)(t_j) - (f - f_n)(t_{j-1})|^p}{(t_j - t_{j-1})^{p-1}} \leq \varepsilon^p\,,$$

15 Here, the letter R stands of course for "Riesz," to distinguish this space from the space introduced in Definition 1.31.

and so $\mathrm{Var}_p^R(f - f_n)^{1/p} \le \varepsilon$ for $n \ge n_0$. Consequently, $f \in RBV_p([a,b])$ and

$$\|f_n - f\|_{RBV_p} \to 0 \quad (n \to \infty).$$

Formula (2.91) is proved in the same way as formula (1.17) in Proposition 1.10. The last assertion is obvious. □

A comparison with Definition 1.1 shows that

$$\mathrm{Var}_1^R(f, P; [a,b]) = \mathrm{Var}(f, P; [a,b]), \quad \mathrm{Var}_1^R(f; [a,b]) = \mathrm{Var}(f; [a,b]),$$

and so $RBV_1 = BV$. Consequently, both spaces WBV_p (Definition 1.31) and RBV_p generalize the classical space BV in different directions. However, Definition 2.50 seems more interesting than Definition 1.31 since the space RBV_p has two remarkable properties. Firstly, it has a natural characterization by means of absolutely continuous functions (see Theorem 3.34 in the next chapter); secondly, it can be identified[16] in a natural way with the dual space of the Lebesgue space $L_p([a,b])$, as we will show in Section 4.2.

In contrast to the space $BV([a,b])$ (or the spaces $WBV_p([a,b])$), all functions $f \in RBV_p([a,b])$ are *continuous* in case $p > 1$. This implies, in particular, that $RBV_p([a,b])$ is contained, for $p > 1$, in the set $CBV([a,b])$ introduced in Section 1.2. The question arises if the spaces $RBV_p([a,b])$ are connected to the Lipschitz space $Lip([a,b])$ or the Hölder spaces $Lip_\alpha([a,b])$, in a similar way as the spaces $WBV_p([a,b])$ (Proposition 1.34). The following result is parallel to the chain of inclusions (1.46); in particular, it shows that RBV_p is intermediate between Lip and AC.

Proposition 2.52. *The inclusions*

$$Lip([a,b]) \subseteq RBV_p([a,b]) \subseteq AC([a,b]) \subseteq BV([a,b]) \tag{2.93}$$

hold for $p > 1$.

Proof. The proof of the first inclusion follows immediately from the estimate

$$\begin{aligned}\sum_{j=1}^{m} \frac{|f(t_j) - f(t_{j-1})|^p}{(t_j - t_{j-1})^{p-1}} &\le L^p \sum_{j=1}^{m} (t_j - t_{j-1})^{p-(p-1)} \\ &= L^p \sum_{j=1}^{m} (t_j - t_{j-1}) = L^p(b-a)\end{aligned} \tag{2.94}$$

for any function f which satisfies (0.66). To prove the second inclusion, fix $f \in RBV_p([a,b])$, $f(x) \not\equiv 0$, and let $\{[a_1,b_1], \ldots, [a_n,b_n]\} \in \Sigma([a,b])$ be a finite collection of pairwise nonoverlapping subintervals of $[a,b]$. Then

$$\sum_{k=1}^{n} |f(b_k) - f(a_k)| = \sum_{k=1}^{n} \frac{|f(b_k) - f(a_k)|}{(b_k - a_k)^{1/p'}} (b_k - a_k)^{1/p'},$$

16 To be precise, to get the mentioned duality, we have to consider the subspace of all $f \in RBV_p([a,b])$ satisfying $f(a) = 0$, see Theorem 4.35.

where $p' := p/(p-1)$ as usual. Now, applying Hölder's inequality (0.108) to

$$\alpha_k := \frac{|f(b_k) - f(a_k)|}{(b_k - a_k)^{1-1/p}}, \quad \beta_k := (b_k - a_k)^{1/p'}$$

yields

$$\begin{aligned}
&\sum_{k=1}^{n} \frac{|f(b_k) - f(a_k)|}{(b_k - a_k)^{1/p'}} (b_k - a_k)^{1/p'} \\
&\leq \left[\sum_{k=1}^{n} \frac{|f(b_k) - f(a_k)|^p}{(b_k - a_k)^{p-1}}\right]^{1/p} \left[\sum_{k=1}^{n} (b_k - a_k)\right]^{1/p'} \\
&\leq \operatorname{Var}_p^R(f; [a,b])^{1/p} \left[\sum_{k=1}^{n} (b_k - a_k)\right]^{1/p'} \leq \|f\|_{RBV_p} \left[\sum_{k=1}^{n} (b_k - a_k)\right]^{1/p'} .
\end{aligned}$$

Therefore, if, for given $\varepsilon > 0$, we choose $\delta \leq \varepsilon^{p'}/\|f\|_{RBV_p}^{p'}$, then (1.40) implies (1.41), which means that $f \in AC([a,b])$ as claimed.

The fact that every absolutely continuous function has bounded variation has already been proved in Proposition 1.22. □

In Theorem 3.34 in the next chapter, we will give a precise characterization of those functions $f \in AC([a,b])$, in terms of their derivatives a.e. which belong to $RBV_p([a,b])$.

Observe that (2.90) and (2.94) show that

$$\|f\|_{RBV_p} = |f(a)| + \operatorname{Var}_p^R(f; [a,b])^{1/p} \leq |f(a)| + L$$

if f satisfies (0.66). Taking L as close as we want to the minimal Lipschitz constant (0.68) of f, we conclude that $(Lip([a,b]), \|\cdot\|_{Lip})$ is continuously imbedded into $(RBV_p([a,b]), \|\cdot\|_{RBV_p})$. Similarly, the norm estimate $\|f\|_{BV} \leq \|f\|_{RBV_p}$ for all $f \in RBV_p([a,b])$ means that $RBV_p([a,b]) \hookrightarrow BV([a,b])$ with $c(RBV_p, BV) = 1$. Moreover, letting $p \to \infty$ in (2.87), we see that we may identify the space $RBV_\infty([a,b])$ formally with the space $Lip([a,b])$. In this sense, we may consider (2.93) as "interpolation inclusion"

$$RBV_\infty([a,b]) \subset RBV_p([a,b]) \subset RBV_1([a,b]) \quad (1 < p < \infty). \tag{2.95}$$

Since $RBV_p \subseteq AC$ for $p > 1$, see (2.93), one cannot expect that the space RBV_p contains some Hölder space[17] Lip_α for a suitable choice of $\alpha < 1$. The next example shows this in explicit form.

Example 2.53. Let $f : [0,1] \to \mathbb{R}$ be defined as in Example 1.24. Then f belongs to *each* Hölder space $Lip_\alpha([0,1])$ for $\alpha < 1$. However, f cannot belong to $RBV_p([0,1])$ for any choice of $p \geq 1$, by (2.93), since $f \notin BV([0,1])$. ♥

17 So, an analogue to Proposition 1.34 is not true for WBV_p replaced by RBV_p.

One could also ask if a parallel result as Proposition 1.38 holds for the spaces RBV_p, i.e. whether or not the space RBV_p is contained in some space RBV_q for suitable values of p and q. In fact, the following is true.

Proposition 2.54. *Let $1 \le p \le q < \infty$. Then the inclusion $RBV_p([a,b]) \supseteq RBV_q([a,b])$ holds.*

Proof. Without loss of generality, let $1 < p < q$; then,

$$r := \frac{q}{p} > 1, \quad s := \frac{q}{q-p} > 1, \quad \frac{1}{r} + \frac{1}{s} = 1 .$$

Consequently, given $f \in RBV_q([a,b])$ and any partition $P := \{t_0, t_1, \dots, t_m\} \in \mathcal{P}([a,b])$, from Hölder's inequality (0.108) for

$$\alpha_j := \frac{|f(t_j) - f(t_{j-1})|^p}{(t_j - t_{j-1})^{p-1+1/s}}, \quad \beta_j := (t_j - t_{j-1})^{1/s}$$

(and (p, p') replaced by (r, s)), we get

$$\begin{aligned}
&\sum_{j=1}^{m} \frac{|f(t_j) - f(t_{j-1})|^p}{(t_j - t_{j-1})^{p-1}} \\
&= \sum_{j=1}^{m} \frac{|f(t_j) - f(t_{j-1})|^p}{(t_j - t_{j-1})^{p-1+1/s}} (t_j - t_{j-1})^{1/s} \le \left(\sum_{j=1}^{m} \frac{|f(t_j) - f(t_{j-1})|^{pr}}{(t_j - t_{j-1})^{(p-1+1/s)r}} \right)^{1/r} \left(\sum_{j=1}^{m} (t_j - t_{j-1}) \right)^{1/s} \\
&= \left(\sum_{j=1}^{m} \frac{|f(t_j) - f(t_{j-1})|^q}{(t_j - t_{j-1})^{q-1}} \right)^{p/q} (b-a)^{(q-p)/q} \le \operatorname{Var}_q^R(f; [a,b])^{p/q} (b-a)^{(q-p)/q} ,
\end{aligned}$$

and so $f \in RBV_p([a,b])$ as claimed. □

Another method for proving Proposition 2.54 will be given in Theorem 3.34 in the next chapter. As the inclusion in Proposition 1.38, the inclusion in Proposition 2.54 is strict in case $p < q$, see Exercise 2.26.

In spite of their similarity, the Wiener spaces WBV_p introduced in Definition 1.31 and the Riesz spaces RBV_p introduced in Definition 2.50 have quite different properties for $p > 1$. We compare some of them in Table 2.3 below.

We now come to a new function class which has been introduced by Medvedev [211] and which generalizes the space RBV_p in rather the same way as the class WBV_ϕ generalizes the space WBV_p, see Definition 2.2.

Definition 2.55. Given a Young function ϕ, a partition $P = \{t_0, t_1, \dots, t_m\} \in \mathcal{P}([a,b])$, and a function $f : [a,b] \to \mathbb{R}$, the nonnegative real number

$$\operatorname{Var}_\phi^R(f, P) = \operatorname{Var}_\phi^R(f, P; [a,b]) := \sum_{j=1}^{m} \phi\left(\frac{|f(t_j) - f(t_{j-1})|}{t_j - t_{j-1}} \right) (t_j - t_{j-1}) \tag{2.96}$$

Table 2.3. WBV_p vs. RBV_p $(1 < p < \infty)$.

Space	$WBV_p([a,b])$	$RBV_p([a,b])$
definition	Wiener (1924)	Riesz (1910)
dependence on p	increasing	decreasing
for $p = 1$ *coincides with*	$BV([a,b])$	$BV([a,b])$
for $p = \infty$ *coincides with*	$R([a,b])$	$Lip([a,b])$
is contained in $C([a,b])$	no	yes
is contained in $BV([a,b])$	no	yes
contains $Lip([a,b])$	yes	yes
contains $Lip_\alpha([a,b])$	yes (for $\alpha \le 1/p$)	no

is called the *Riesz–Medvedev variation* of f on $[a,b]$ with respect to P, while the (possibly infinite) number

$$\mathrm{Var}^R_\phi(f) = \mathrm{Var}^R_\phi(f;[a,b]) := \sup\left\{\mathrm{Var}^R_\phi(f,P;[a,b]) : P \in \mathcal{P}([a,b])\right\}, \tag{2.97}$$

where the supremum is taken over all partitions of $[a,b]$, is called the *total Riesz–Medvedev variation* of f on $[a,b]$. In case $\mathrm{Var}^R_\phi(f;[a,b]) < \infty$, we say that f has *bounded Riesz–Medvedev variation* (or *bounded ϕ-variation in Riesz's sense*) on $[a,b]$ and write $f \in V^R_\phi([a,b])$. ∎

Similarly as in (2.10) and (2.11), we put

$$B^R(\phi) := \{f \in B([a,b]) : \mathrm{Var}^R_\phi(f;[a,b]) \le 1\}, \tag{2.98}$$

and[18]

$$\|f\|_{RBV_\phi} := |f(a)| + \inf\left\{\lambda > 0 : f/\lambda \in B^R(\phi)\right\}, \tag{2.99}$$

and denote its linear hull span $B^R(\phi)$ by $RBV_\phi([a,b])$).

Proposition 2.56. *The inclusions*

$$Lip([a,b]) \subseteq RBV_\phi([a,b]) \subseteq BV([a,b]) \tag{2.100}$$

are true.

Proof. Let $f \in Lip([a,b])$. From $|f(s) - f(t)| \le L|s-t|$, it then follows that

$$\sum_{j=1}^m \phi\left(\frac{|f(t_j) - f(t_{j-1})|}{t_j - t_{j-1}}\right)(t_j - t_{j-1}) \le \sum_{j=1}^m \phi(L)(t_j - t_{j-1}) = \phi(L)(b-a),$$

which shows that $f \in RBV_\phi([a,b])$ with

$$\|f\|_{RBV_\phi} \le |f(a)| + \phi(lip(f))(b-a),$$

18 Clearly, for $\phi(t) = t^p$ with $p \ge 1$, the infimum in (2.99) coincides with $\mathrm{Var}^R_p(f;[a,b])^{1/p}$, see (2.88).

where $lip(f)$ denotes the smallest Lipschitz constant (0.68) of f. To prove the second inclusion in (2.100), assume that $f \in RBV_\phi([a,b])$. Then

$$\begin{aligned}\sum_{j=1}^{m} |f(t_j) - f(t_{j-1})| &= \sum_{j=1}^{m} \frac{|f(t_j) - f(t_j - 1)|}{t_j - t_{j-1}}(t_j - t_{j-1}) \\ &= \phi^{-1}\left(\phi\left[\sum_{j=1}^{m} \frac{|f(t_j) - f(t_{j-1})|}{t_j - t_{j-1}}(t_j - t_{j-1})\right]\right) \\ &= \phi^{-1}\left(\phi\left[\sum_{j=1}^{m} (b-a)\frac{|f(t_j) - f(t_{j-1})|}{t_j - t_{j-1}}\frac{t_j - t_{j-1}}{b-a}\right]\right).\end{aligned}$$

Hence, using the fact that ϕ^{-1} is increasing, we obtain

$$\begin{aligned}\sum_{j=1}^{m} |f(t_j) - f(t_{j-1})| &\le \phi^{-1}\left(\sum_{j=1}^{m} \phi\left[(b-a)\frac{|f(t_j) - f(t_{j-1})|}{t_j - t_{j-1}}\right]\frac{t_j - t_{j-1}}{b-a}\right) \\ &\le \phi^{-1}\left(\frac{1}{b-a}\sum_{j=1}^{m} \phi\left[(b-a)\frac{|f(t_j) - f(t_{j-1})|}{t_j - t_{j-1}}\right](t_j - t_{j-1})\right) \\ &\le \phi^{-1}\left(\frac{1}{b-a}\operatorname{Var}_\phi^R((b-a)f;[a,b])\right) < \infty .\end{aligned}$$

This shows that $f \in BV([a,b])$ and proves the second inclusion in (2.100). □

Let us return for a moment to the condition ∞_p introduced in Definition 2.11. This condition was important in the discussion of the Wiener–Young space WBV_ϕ. It turns out that it is also useful in the discussion of the Riesz–Medvedev space RBV_ϕ. In fact, one can show that if the second inclusion in (2.100) is strict, then the Young function ϕ necessarily satisfies condition ∞_1.

To see this, suppose that $\phi \notin \infty_1$ which means that

$$\lim_{t\to\infty} \frac{\phi(t)}{t} < \infty , \tag{2.101}$$

see (2.16). Then there exist constants $T > 0$ and $M > 0$ such that $\phi(t) \le Mt$ for $t \ge T$. Given an arbitrary partition $P = \{t_0, t_1, \dots, t_m\} \in \mathcal{P}([a,b])$, we define an index set $A(T) \subseteq \{1, 2, \dots, m\}$ by

$$A(T) := \{j \in \{1, 2, \dots, m\},\ |f(t_j) - f(t_{j-1})| \ge T(t_j - t_{j-1})\}$$

and put $B(T) := \{1, 2, \dots, m\} \setminus A(T)$. Then we have

$$\begin{aligned}
\mathrm{Var}_\phi^R(f, P : [a,b]) &= \sum_{j=1}^{m} \phi\left(\frac{|f(t_j) - f(t_{j-1})|}{t_j - t_{j-1}}\right)(t_j - t_{j-1}) \\
&= \sum_{j \in A(T)} \phi\left(\frac{|f(t_j) - f(t_{j-1})|}{t_j - t_{j-1}}\right)(t_j - t_{j-1}) \\
&\quad + \sum_{j \in B(T)} \phi\left(\frac{|f(t_j) - f(t_{j-1})|}{t_j - t_{j-1}}\right)(t_j - t_{j-1}) \\
&\le \sum_{j \in A(T)} M \frac{|f(t_j) - f(t_{j-1})|}{t_j - t_{j-1}}(t_j - t_{j-1}) + \sum_{j \in B(T)} \phi(T)(t_j - t_{j-1}) \\
&\le M \,\mathrm{Var}(f, P : [a,b]) + \phi(T)(b-a) \le M \,\mathrm{Var}(f; [a,b]) + \phi(T)(b-a)\,.
\end{aligned}$$

Passing to the supremum with respect to $P \in \mathcal{P}([a,b])$, we conclude that

$$\mathrm{Var}_\phi^R(f; [a,b]) \le M \,\mathrm{Var}(f; [a,b]) + \phi(T)(b-a)\,,$$

and so $BV([a,b]) \subseteq RBV_\phi([a,b])$. The converse inclusion has already been proved in Proposition 2.56 for arbitrary Young functions ϕ. We summarize our discussion with the following

Proposition 2.57. *Let ϕ be a Young function which does not satisfy the condition ∞_1. Then the space $RBV_\phi([a,b])$ coincides with $BV([a,b])$.*

We do not know whether or not the following generalization of Proposition 2.57 holds true: if ϕ is a Young function which does not satisfy the condition ∞_p, see (2.16), then the space $RBV_\phi([a,b])$ coincides with the space $RBV_p([a,b])$ introduced in Definition 2.50.

To conclude this section, let us again consider our favorite families of functions which we introduced in the last part of Section 0.3, namely, the oscillatory functions (0.86) and the special zigzag functions (0.93). It is illuminating to determine all values of $\alpha, \beta \in \mathbb{R}$ for which the function (0.86) belongs to one of the function classes considered so far, and also all values of $\theta \in \mathbb{R}$ for which the function (0.93) belongs to such classes. In the synoptic Table 2.4 below, which essentially extends Table 0.1, we do this for the classes C, C^1, L_1, Lip, Lip_γ $(0 < \gamma < 1)$, BV, WBV_p $(1 < p < \infty)$, RBV_p $(1 < p < \infty)$, and AC.

Let us explain the new entries in Table 2.4. The admissible values for α and β for the function (0.86) have already been collected in Proposition 0.48 and Exercises 0.52, 1.8, 1.9, 1.57, and 2.27. From (0.17) and (1.75), it follows that $Z_\theta \in WBV_p([0,1])$ if and only if the series

$$\zeta(k\theta, 0) = \sum_{k=1}^{\infty} \frac{1}{k^{p\theta}}$$

Table 2.4. Oscillation functions and zigzag functions.

	The function $f_{\alpha,\beta}$	*The function Z_θ*
belongs to $C([0,1])$ *iff*	$\alpha > 0$ or $\alpha \le 0$ and $\alpha + \beta > 0$	always
belongs to $C^1([0,1])$ *iff*	$\alpha + \beta > 1$	never
belongs to $L_1([0,1])$ *iff*	$\alpha + \beta \ge 1$	always
belongs to $Lip([0,1])$ *iff*	$\alpha + \beta \ge 1$	never
belongs to $Lip_\gamma([0,1])$ *iff*	see Exercise 0.52	never
belongs to $BV([0,1])$ *iff*	$\beta > 0$ and $\alpha + \beta \ge 0$ or $\beta \le 0$ and $\alpha + \beta > 0$	$\theta > 1$
belongs to $WBV_p([0,1])$ *iff*	$\beta > 0$ and $p\alpha + \beta \ge 0$ or $\beta \le 0$ and $p\alpha + \beta > 0$	$p\theta > 1$
belongs to $RBV_p([0,1])$ *iff*	see Exercise 2.27	$p = 1$ and $\theta > 1$
belongs to $AC([0,1])$ *iff*	$\alpha + \beta > 0$	$\theta > 1$

converges, i.e. precisely for $p\theta > 1$; in particular, $Z_\theta \in BV([0,1])$ if and only if $\theta > 1$. A similar computation shows that

$$\mathrm{Var}_p^R(Z_\theta; [0,1]) = \sum_{k=1}^{\infty} \frac{2^{k(p-1)}}{k^{p\theta}} . \tag{2.102}$$

Since $p \ge 1$, elementary convergence criteria show that $Z_\theta \in RBV_p([0,1])$ if $p = 1$ and $\theta > 1$. Finally, we have already shown in Corollary 0.51 that $Z_\theta \notin Lip_\gamma([0,1])$ for any $\theta > 0$ and $\gamma \in (0,1]$.

Table 2.4 shows that the zigzag function (0.93) exhibits a more interesting behavior in spaces of (generalized) bounded variation than in the other spaces occurring in Table 0.1.

2.5 The Korenblum variation

As we have seen in this and the preceding chapter, the concept of variation has been generalized in many directions.

Wiener [321] distorted the measurement of intervals in the range using powers $|f(t_j) - f(t_{j-1})|^p$, see Definition 1.31, while Young [322] used more general distortions of the form $\phi(|f(t_j) - f(t_{j-1})|)$, see Definition 2.2. The most general concept by Schramm [286], see Definition 2.42, replaced such distortion functions ϕ by countable families Φ. All of these extensions have the advantage of allowing us to define quite general Riemann–Stieltjes integrals, as we shall see in Chapter 4. On the other hand, a flaw is the loss of an effective decomposition of a function from the corresponding classes WBV_p, WBV_ϕ, and ΦBV into, hopefully, simpler functions, such as in Theorem 1.5 for the case of functions from the classical space BV.

In 1975, Korenblum [163] considered a new kind of variation which he called *κ-variation*, introducing a function κ for distorting the expression $|t_j - t_{j-1}|$ in the partition itself, rather than the expression $|f(t_j) - f(t_{j-1})|$ in the range. Subsequently, this class of functions has been studied in detail by Cyphert and Kalingos [102]. One advantage of this alternate approach is that a *function of bounded κ-variation* may be decomposed into the difference of two simpler functions called *κ-decreasing functions* (for the precise definitions, see below). We will follow the article [102] in the sequel and will mostly consider, without loss of generality, functions over $[a, b] = [0, 1]$.

Definition 2.58. A function $\kappa : [0,1] \to [0,1]$ is called a *distortion function* if κ is increasing, concave, and satisfies $\kappa(0) = 0$, $\kappa(1) = 1$, and

$$\lim_{t\to 0+} \frac{\kappa(t)}{t} = \infty\,, \tag{2.103}$$

i.e. has infinite slope at the origin. ■

Note that from the estimate

$$\frac{\kappa(s+t) - \kappa(t)}{(s+t) - t} \le \frac{\kappa(s) - \kappa(0)}{s - 0},$$

it follows that a distortion function is always subadditive in the sense that

$$\kappa(s+t) \le \kappa(s) + \kappa(t) \quad (0 \le s, t \le 1)\,. \tag{2.104}$$

Moreover, the definition implies that a distortion function is continuous[19] on $(0, 1]$. Here are some typical examples of distortion functions.

Example 2.59. The simplest example is of course

$$\kappa(t) = t^\alpha \quad (0 < \alpha < 1)\,.$$

Since $\kappa'(t) = \alpha t^{\alpha-1} > 0$ and $\kappa''(t) = \alpha(\alpha - 1)t^{\alpha-2} < 0$ on $(0, 1)$, κ is increasing and concave. Another example is

$$\kappa(t) = \begin{cases} t(1 - \log t) & \text{for } 0 < t \le 1\,, \\ 0 & \text{for } t = 0\,; \end{cases}$$

here, we have $\kappa'(t) = -\log t > 0$ and $\kappa''(t) = -\frac{1}{t} < 0$ on $(0, 1)$. Finally, the function

$$\kappa(t) = \begin{cases} \frac{2}{2-\log t} & \text{for } 0 < t \le 1\,, \\ 0 & \text{for } t = 0 \end{cases}$$

is a distortion function which arises in entropy theory, see Section 2.8. ♥

19 A comparison with Definition 0.52 shows that distortion functions are similar to moduli of continuity; in particular, they are both increasing and subadditive. However, the main difference is that a modulus is continuous at zero, while a distortion function is continuous everywhere except, perhaps, at zero.

Building on the concept of distortion functions, we may now introduce a new class of functions of bounded variation.

Definition 2.60. Given a distortion function $\kappa : [0,1] \to [0,1]$, a partition $P = \{t_0, t_1, \dots, t_m\} \in \mathcal{P}([0,1])$, and a function $f : [0,1] \to \mathbb{R}$, the nonnegative real number

$$\operatorname{Var}_\kappa(f,P) = \operatorname{Var}_\kappa(f,P;[0,1]) := \frac{\sum_{j=1}^m |f(t_j) - f(t_{j-1})|}{\sum_{j=1}^m \kappa(t_j - t_{j-1})} \tag{2.105}$$

is called the *Korenblum variation* of f on $[0,1]$ with respect to P, while the (possibly infinite) number

$$\operatorname{Var}_\kappa(f) = \operatorname{Var}_\kappa(f;[0,1]) := \sup\left\{\operatorname{Var}_\kappa(f,P;[0,1]) : P \in \mathcal{P}([0,1])\right\}, \tag{2.106}$$

where the supremum is taken over all partitions of $[0,1]$, is called the *total Korenblum variation* of f on $[0,1]$. In case $\operatorname{Var}_\kappa(f;[0,1]) < \infty$, we say that f has *bounded Korenblum variation* (or *bounded κ-variation*) on $[0,1]$ and write $f \in \kappa BV([0,1])$. ■

Of course, Definition 2.60 may also be formulated for functions on an arbitrary interval $[a,b]$ by defining that f belongs to $\kappa BV([a,b])$ if the function $x \mapsto f((b-a)x+a)$ belongs to $\kappa BV([0,1])$. This amounts to replacing (2.105) by

$$\operatorname{Var}_\kappa(f,P) = \operatorname{Var}_\kappa(f,P;[0,1]) := \frac{\sum_{j=1}^m |f(t_j) - f(t_{j-1})|}{\sum_{j=1}^m \kappa\left(\frac{t_j - t_{j-1}}{b-a}\right)}. \tag{2.107}$$

From (2.104), it then follows that

$$\sum_{j=1}^m \kappa\left(\frac{t_j - t_{j-1}}{b-a}\right) \ge \kappa\left(\sum_{j=1}^m \frac{t_j - t_{j-1}}{b-a}\right) = \kappa(1) = 1.$$

In the following proposition, we compare $\kappa BV([0,1])$ with other function classes.

Proposition 2.61. *For any distortion function κ, the inclusion*

$$BV([0,1]) \subseteq \kappa BV([0,1]) \subseteq R([0,1]) \tag{2.108}$$

holds, where $R([a,b])$ denotes the set of regular functions introduced in Section 0.3.

Proof. By (2.104), we have

$$\sum_{j=1}^m \kappa(t_j - t_{j-1}) \ge \kappa\left(\sum_{j=1}^m (t_j - t_{j-1})\right) = \kappa(1) = 1,$$

and hence

$$\operatorname{Var}_\kappa(f,P;[0,1]) \le \sum_{j=1}^m |f(t_j) - f(t_{j-1})| = \operatorname{Var}(f,P;[0,1])$$

for every partition $P \in \mathcal{P}([0,1])$. This proves the first inclusion in (2.108). To prove the second inclusion, we have to show that every function $f \in \kappa BV([0,1])$ is bounded and has unilateral limits at each point $x_0 \in [0,1]$.

By considering the partition $P_x := \{0, x, 1\}$, we see that

$$|f(x) - f(0)| \leq \frac{3}{2}\operatorname{Var}_\kappa(f, P_x; [0,1]) \leq \frac{3}{2}\operatorname{Var}_\kappa(f; [0,1]) \quad (0 \leq x \leq 1),$$

and so

$$\|f\|_\infty \leq |f(0)| + \frac{3}{2}\operatorname{Var}_\kappa(f), \tag{2.109}$$

where $\|\cdot\|_\infty$ denotes the norm (0.39). To show the existence of the right limit at $x_0 \in [0,1)$, say, suppose that

$$l := \liminf_{x \to x_0+} f(x) < \limsup_{x \to x_0+} f(x) =: L.$$

Then for each sufficiently large $n \in \mathbb{N}$, one can choose points $t_1, t_2, \ldots, t_n, t_{n+1}$ such that

$$x_0 < t_1 < t_2 < \ldots < t_n < t_{n+1} \leq x_0 + \frac{1}{n}$$

and

$$|f(t_{j+1}) - f(t_j)| \geq \frac{L-l}{2} \quad (j = 1, 2, \ldots, n).$$

Using the partition $P = \{0, t_1, t_2, \ldots, t_n, t_{n+1}, 1\} \in \mathcal{P}([0,1])$ and the definition (2.105), we obtain

$$\begin{aligned}\frac{n(L-l)}{2} &\leq |f(t_1) - f(0)| + \sum_{j=1}^{n} |f(t_{j+1}) - f(t_j)| + |f(1) - f(t_{n+1})| \\ &\leq \operatorname{Var}_\kappa(f, P)[\kappa(t_1) + \sum_{j=1}^{n} \kappa(t_{j+1} - t_j) + \kappa(1 - t_{n+1})] \leq \operatorname{Var}_\kappa(f, P)\,[n\kappa(1/n) + 2]\,.\end{aligned}$$

Dividing by n and letting $n \to \infty$ yields $L - l = 0$, that is, a contradiction. The existence of the left-hand limit at any $x_0 \in (0,1]$ is proved in the same way. □

We point out that the slope condition (2.104) ensures that the first inclusion in (2.108) is strict. In fact, if the limit in (2.103) is finite, one may show that $\kappa BV([0,1]) = BV([0,1])$.

To get an idea of how to calculate the Korenblum variation, we consider a very simple class of functions.

Example 2.62. Suppose that $f : [0,1] \to \mathbb{R}$ only takes two values, say

$$f(x) := \begin{cases} A & \text{for } 0 \leq x \leq \xi, \\ B & \text{for } \xi < x \leq 1 \end{cases}$$

for some fixed $\xi \in (0,1)$. For each partition $P = \{t_0, t_1, \ldots, \xi, \ldots, t_m\} \in \mathcal{P}([0,1])$, we obtain, by (2.104),

$$\operatorname{Var}_\kappa(f, P; [0,1]) = \frac{\sum_{j=1}^m |f(t_j) - f(t_{j-1})|}{\sum_{j=1}^m \kappa(t_j - t_{j-1})} \leq |A - B|,$$

and hence $\mathrm{Var}_\kappa(f;[0,1]) \le |A-B|$. More generally, this implies that κBV contains all step functions. ♥

At this point, one could ask whether or not the class $\kappa BV([0,1])$ is related to monotone functions as the class $BV([0,1])$. It is easy to see that every monotone function $f : [0,1] \to \mathbb{R}$ belongs to $\kappa BV([0,1])$ with

$$\mathrm{Var}_\kappa(f;[0,1]) \le |f(1)-f(0)|. \tag{2.110}$$

This is completely analogous to Proposition 1.3 (e) and may be proved by using the trivial partition $P = \{0,1\} \in \mathcal{P}([0,1])$ which is optimal for (2.105) for monotone f. However, to formulate a decomposition theorem for functions of bounded Korenblum variation, one has to replace monotonicity by the following notion which has been studied in [CyK].

Definition 2.63. Given a distortion function $\kappa : [0,1] \to [0,1]$, a function $f : [0,1] \to \mathbb{R}$ is said to be *κ-decreasing* if there exists a constant $C \ge 0$ such that

$$f(y) - f(x) \le C\kappa(y-x) \quad (0 \le x < y \le 1). \tag{2.111}$$

We write

$$C_\kappa(f) = C_\kappa(f;[0,1]) := \sup\left\{\frac{f(y)-f(x)}{\kappa(y-x)} : 0 \le x < y \le 1\right\} \tag{2.112}$$

for the smallest $C \ge 0$ satisfying (2.111). ■

Let us make some remarks about this definition. Clearly, every decreasing function f is κ-decreasing, for any distortion function κ, with $C_\kappa(f) = 0$. However, the class of κ-decreasing functions is much larger. To see this, observe that *any* function $f \in Lip_\alpha([0,1])$ is κ-decreasing for $\kappa(t) = t^\alpha$ $(0 < \alpha < 1)$ and $C_\kappa(f) = lip_\alpha(f)$, see (0.69). This shows, in particular, that continuous *nowhere differentiable* functions which are κ-decreasing exist. Intuitively, a function f is κ-decreasing if f is either decreasing or, at least locally, not increasing faster than some fixed multiple of κ itself.

The following proposition provides an important link between κ-decreasing functions and functions of bounded κ-variation. Here and in what follows, we use the notation

$$a^+ := \max\{a,0\}, \quad a^- := \max\{-a,0\} \quad (a \in \mathbb{R}); \tag{2.113}$$

so, we have $a^+ + a^- = |a|$ and $a^+ - a^- = a$.

Proposition 2.64. *Every κ-decreasing function $f : [0,1] \to \mathbb{R}$ has bounded κ-variation with*

$$\mathrm{Var}_\kappa(f;[0,1]) \le 2C_\kappa(f) + |f(1)-f(0)|. \tag{2.114}$$

Proof. Suppose first that $f(0) = f(1)$, and let $P = \{t_0, t_1, \dots, t_m\} \in \mathcal{P}([0,1])$ be fixed. From (2.112), it follows that

$$f(t_j) - f(t_{j-1}) \le C_\kappa(f)\kappa(t_j - t_{j-1}) \quad (j = 1,2,\dots,m).$$

So, the equality $|a| = a^+ + a^-$ implies that

$$\sum_{j=1}^{m} |f(t_j) - f(t_{j-1})| = \sum_{j=1}^{m} [f(t_j) - f(t_{j-1})]^+ + \sum_{j=1}^{m} [f(t_j) - f(t_{j-1})]^-$$
$$\leq 2C_\kappa(f) \sum_{j=1}^{m} \kappa(t_j - t_{j-1}),$$

showing that $f \in \kappa BV([0,1])$ with $\mathrm{Var}_\kappa(f;[0,1]) \leq 2C_\kappa(f)$. The proof in the general case is similar and left to the reader (Exercise 2.33). □

Observe that (2.114) generalizes (2.110) since $C_\kappa(f) = 0$ for monotone functions f.

Now, we present some examples which give some indication as to the flavor of the subject, and, in particular, of the abstract Definitions 2.60 and 2.63. The first example shows that not every function $f \in \kappa BV([0,1])$ is κ-decreasing.

Example 2.65. Let $\kappa : [0,1] \to [0,1]$ be an arbitrary distortion function, and define $f : [0,1] \to \mathbb{R}$ by $f(x) := \sqrt{\kappa(x)}$. Then f is monotonically increasing, and so has bounded κ-variation with $\mathrm{Var}_\kappa(f;[0,1]) = 1$ by (2.110). On the other hand,

$$\lim_{x\to 0+} \frac{f(x) - f(0)}{\kappa(x-0)} = \lim_{x\to 0+} \frac{1}{\sqrt{\kappa(x)}} = \infty,$$

and so f cannot satisfy (2.111) for any $C > 0$. ♥

Let us return to the inclusions (2.108). To find an example of a regular function which does not belong to $\kappa BV([0,1])$, for a given distortion function, κ is easy (Exercise 2.34). In the next example, we also show that the first inclusion in (2.108) is strict by constructing a function $f \in \kappa BV([0,1]) \setminus BV([0,1])$.

Example 2.66. Let $\kappa : [0,1] \to [0,1]$ be defined by $\kappa(t) = t^\alpha$ for some $\alpha \in (0,1)$. Using the notation (0.17), put

$$t_n := \frac{1}{\zeta(1/\alpha, 0)} \sum_{k=1}^{n} \frac{1}{k^{1/\alpha}} \quad (n = 1, 2, 3, \ldots).$$

We define a function $f : [0,1] \to \mathbb{R}$ by

$$f(x) := \begin{cases} 0 & \text{for } x = 0 \text{ or } x = 1, \\ (x - t_n)^\alpha & \text{for } t_n \leq x < t_{n+1}. \end{cases}$$

Considering partitions containing the points $t_0, t_1, \ldots, t_n$, one easily sees that

$$\mathrm{Var}(f;[0,1]) \geq 2\sum_{k=1}^{\infty} \kappa\left(\frac{1}{k^{1/\alpha}}\right) = 2\sum_{k=1}^{\infty} \frac{1}{k} = \infty,$$

and so $f \notin BV([0,1])$. On the other hand, f is κ-decreasing. To see this, fix x and y with $0 \leq x < y \leq 1$. Then there exists unique integers $m, n \in \mathbb{N}$, $m \leq n$, such that

$x \in [t_n, t_{n+1})$ and $y \in [t_m, t_{m+1})$. Then $t_n \le t_m$, by construction, and the definition of f implies that

$$f(y) - f(x) = \kappa(y - t_m) - \kappa(x - t_n) \le \kappa(y - t_n) - \kappa(x - t_n) \le \kappa(y - x),$$

where we have used the monotonicity and subadditivity of κ. This shows that f is κ-decreasing with $C_\kappa(f) = 1$. By Proposition 2.64, $f \in \kappa BV([0,1])$ with $\mathrm{Var}_\kappa(f;[0,1]) \le 2$. ♥

Our final example, which is left to the reader (Exercise 2.32), shows that the *Korenblum variation function* $V_{f,\kappa}$ defined in analogy to (1.13) by

$$V_{f,\kappa}(x) := \mathrm{Var}_\kappa(f;[0,x]) \quad (0 \le x \le 1) \tag{2.115}$$

need not be monotonically increasing. This explains some of the difficulties one encounters when replacing Jordan variation by Korenblum variation.

There is also a Helly-type selection theorem for κ-decreasing functions; compare this with Theorem 1.11.

Theorem 2.67. *Let $(f_n)_n$ be a uniformly bounded and uniformly κ-decreasing sequence of functions $f_n : [0,1] \to \mathbb{R}$. Then there exists a subsequence of $(f_n)_n$ which converges pointwise on $[0,1]$ to a κ-decreasing function.*

Proof. The hypotheses mean that

$$C := \sup_n \max\{\|f_n\|_\infty, C_\kappa(f_n)\} < \infty; \tag{2.116}$$

in particular,

$$f_n(y) - f_n(x) \le C\kappa(y - x) \quad (0 \le x < y \le 1) \tag{2.117}$$

for all $n \in \mathbb{N}$. Since the sequence $(f_n)_n$ is bounded, we can assume, without loss of generality, that the sequence $(f_n(x))_n$ converges[20] at each rational point $x \in [0,1]$ to some number which we call $f(x)$. Then the function f also satisfies

$$f(y) - f(x) \le C\kappa(y - x) \quad (0 \le x < y \le 1), \tag{2.118}$$

at least for rational x and y. We extend f to the whole interval $[0,1]$ by putting

$$f(x) := \lim_{n\to\infty} f(y_n), \tag{2.119}$$

where $(y_n)_n$ is a rational sequence converging to x. The fact that this limit exists can be seen as follows. Suppose that

$$l := \liminf_{y\to x-} f(y) < \limsup_{y\to x-} f(y) =: L,$$

20 To see this, we may use the standard Cantor diagonalization technique.

where $y \in [0,1] \cap \mathbb{Q}$, and let $(y_n)_n$ and $(y_n')_n$ be two rational sequences which converge to x and are interlacing in the sense that

$$y_1 < y_1' < y_2 < y_2' < \ldots < y_n < y_n' < \ldots < x$$

and satisfy

$$\lim_{n\to\infty} f(y_n) = l, \quad \lim_{n\to\infty} f(y_n') = L\,.$$

Then (2.118) implies that, for $y = y_n'$ and $x = y_n$,

$$f(y_n') - f(y_n) \le C\kappa(y_n' - y_n) \quad (n = 1, 2, 3, \ldots)\,.$$

Passing to the limit $n \to \infty$ yields $L - l \le 0$, that is, a contradiction.

The same argument shows that (2.118) also holds for irrational $x, y \in [0,1]$, which means that f is κ-decreasing with $C_\kappa(f) \le C$. By Proposition 2.64, f has bounded κ-variation with $\mathrm{Var}_\kappa(f;[0,1]) \le 2C$, and therefore its discontinuity set $D(f)$, see (0.49), is at most countable, by (2.108). Using, if necessary, another Cantor diagonalization process, we may assume that the sequence $(f_n)_n$ converges to f at each point of $D(f)$.

So, it remains to show that $f_n(x) \to f(x)$ also at each point of continuity of f. Let $x_0 \in [0,1] \setminus D(f)$ be such a point, and let $\varepsilon > 0$ be given. Fix $x_1, x_2 \in [0,1] \cap \mathbb{Q}$ such that $x_1 < x_0 < x_2$ and

$$|f(x_i) - f(x_0)| < \frac{\varepsilon}{3}, \quad \kappa(|x_i - x_0|) < \frac{\varepsilon}{3C} \quad (i = 1, 2)\,, \tag{2.120}$$

which is possible by the continuity of κ. Since $f_n(x) \to f(x)$ at each rational point x, we may further choose $n_0 \in \mathbb{N}$ such that

$$|f_n(x_i) - f(x_i)| < \frac{\varepsilon}{3} \quad (i = 1, 2) \tag{2.121}$$

for $n \ge n_0$. Now, combining (2.118), (2.120) and (2.121) with (2.117) (for $y := x_2$ and $x := x_0$), we get

$$\begin{aligned} f(x_0) - f_n(x_0) &\le |f(x_0) - f_n(x_0)| \\ &\le |f(x_0) - f(x_2)| + |f(x_2) - f_n(x_2)| + f_n(x_2) - f_n(x_0) \\ &< \frac{\varepsilon}{3} + \frac{\varepsilon}{3} + C\kappa(x_2 - x_0) < \frac{\varepsilon}{3} + \frac{\varepsilon}{3} + C\frac{\varepsilon}{3C} = \varepsilon \quad (n \ge n_0)\,. \end{aligned}$$

A similar calculation shows that $f_n(x_0) - f(x_0) < \varepsilon$ for $n \ge n_0$. Therefore, we have shown that $(f_n)_n$ converges pointwise on the whole interval $[0,1]$ to f, and the proof is complete. □

We are now in a position to state a decomposition theorem for functions of bounded κ-variation which is parallel to the classical Jordan decomposition (Theorem 1.5) for functions of bounded variation.

Theorem 2.68. *Every function $f : [0,1] \to \mathbb{R}$ of bounded κ-variation may be represented in the form $f = g - h$, where both g and h are κ-decreasing functions, and*

$$\max\{C_\kappa(g), C_\kappa(h)\} \le \frac{5}{4}\operatorname{Var}_\kappa(f;[0,1]). \tag{2.122}$$

In case $f(0) = f(1)$, one can choose g and h in such a way that also $g(0) = h(0) = g(1) = h(1)$, and in this case, (2.122) *may be improved to*

$$\max\{C_\kappa(g), C_\kappa(h)\} \le \frac{1}{2}\operatorname{Var}_\kappa(f;[0,1]). \tag{2.123}$$

Proof. We split the proof into three steps, where the first step is the most technical one. Without loss of generality, we may assume throughout that $f(0) = 0$.

1st step. Here, we assume in addition that also $f(1) = 0$, and that the function f is piecewise linear, i.e. we find a partition $\hat{P} = \{\tau_0, \tau_1, \dots, \tau_{n+1}\} \in \mathcal{P}([0,1])$ such that f is linear on each subinterval $[\tau_{i-1}, \tau_i]$ for $i = 1, 2, \dots, n+1$. Suppose that $\operatorname{Var}_\kappa(f;[0,1]) \le C$, and thus

$$\sum_{j=1}^{m} |f(t_j) - f(t_{j-1})| \le C \sum_{j=1}^{m} \kappa(t_j - t_{j-1}) \tag{2.124}$$

for any partition $P = \{t_0, t_1, \dots, t_m\} \in \mathcal{P}([0,1])$. In particular, (2.124) is valid for every subpartition of $\hat{P}$, and there are precisely 2^n such subpartitions.[21] Thus, we have 2^n inequalities of type (2.124).

Our task is to construct two suitable functions g and h such that $f = g - h$. More precisely, we want g and h to be continuous on $[0,1]$ and vanishing at 0 and 1, to be linear on the subintervals $[\tau_{i-1}, \tau_i]$ for $i = 1, 2, \dots, n+1$, and to satisfy (2.111), with C replaced by $C/2$, for each choice of points $x, y \in [0,1]$ with $x < y$. However, using the piecewise linearity of g and h and the concavity of κ, we see that it suffices to verify this only for the subintervals of the partition $\hat{P}$, i.e. to show that

$$g(\tau_j) - g(\tau_i) \le \frac{C}{2}\kappa(\tau_j - \tau_i), \quad h(\tau_j) - h(\tau_i) \le \frac{C}{2}\kappa(\tau_j - \tau_i) \tag{2.125}$$

for $0 \le i < j \le n+1$. This means that we have to actually solve a finite-dimensional problem: find $2n$ numbers $g(\tau_1), g(\tau_2), \dots, g(\tau_n), h(\tau_1), h(\tau_2), \dots, h(\tau_n)$ satisfying

$$g(0) = h(0) = g(1) = h(1) = 0, \quad f(\tau_i) = g(\tau_i) - h(\tau_i) \quad (i = 1, 2, \dots, n),$$

and (2.125). Replacing $g(\tau_k)$ by $f(\tau_k) + h(\tau_k)$ in (2.125), we may rephrase the problem as follows: find n numbers $h(\tau_1), h(\tau_2), \dots, h(\tau_n)$ satisfying the set of inequalities

$$\begin{cases} h(\tau_j) - h(\tau_i) \le \dfrac{C}{2}\kappa(\tau_j - \tau_i) \\[2ex] h(\tau_j) - h(\tau_i) \le \dfrac{C}{2}\kappa(\tau_j - \tau_i) - f(\tau_j) + f(\tau_i) \end{cases} \tag{2.126}$$

21 Here, we use the fact that there are $\binom{n}{i}$ subpartitions of $\hat{P}$ containing i interior points in $(0,1)$, and a well-known identity for binomial coefficients.

for $0 \le i < j \le n+1$. Using the notation (2.113), we may combine the two sets of inequalities (2.126) to the one set

$$h(\tau_j) - h(\tau_i) \le \frac{C}{2}\kappa(\tau_j - \tau_i) - \left[f(\tau_j) - f(\tau_i)\right]^+ \quad (0 \le i < j \le n+1). \tag{2.127}$$

An admissible range for each $h(\tau_i)$ can be found by first setting $0 = \tau_i < \tau_j < 1$ and then setting $0 < \tau_i < \tau_j = 1$ into (2.127). The result is, after some rearrangement,

$$f(\tau_i)^- - \frac{C}{2}\kappa(1-\tau_i) \le h(\tau_i) \le \frac{C}{2}\kappa(\tau_i) - f(\tau_i)^+. \tag{2.128}$$

Writing down (2.124) for the special partition $\{0, \tau_i, 1\} \in \mathcal{P}([0,1])$ and using the hypothesis $f(0) = f(1) = 0$, we get

$$\begin{aligned} 2|f(\tau_i)| &= |f(\tau_i) - f(0)| + |f(1) - f(\tau_i)| \\ &\le C\kappa(\tau_i) + C\kappa(1-\tau_i) \le 2f(\tau_i)^+ + 2f(\tau_i)^- \end{aligned}$$

which implies that the closed intervals

$$I_i := \left[f(\tau_i)^- - \tfrac{C}{2}\kappa(1-\tau_i), \tfrac{C}{2}\kappa(\tau_i) - f(\tau_i)^+\right] \quad (i = 1, 2, \ldots, n)$$

all have nonempty interior, and (2.128) precisely means that $h(\tau_i) \in I_i$.

We now define the numbers $h(\tau_1), h(\tau_2), \ldots, h(\tau_n)$ using the following recursive algorithm. We start with

$$h(\tau_1) := \frac{C}{2}\kappa(\tau_1) - f(\tau_1)^+,$$

and afterwards for each $j = 2, 3, \ldots, n$, we define $h(\tau_j)$ as the smallest of all expressions

$$\eta_i := \min\left\{\frac{C}{2}\kappa(\tau_j) - f(\tau_j)^+, \frac{C}{2}\kappa(\tau_j - \tau_i) - \left[f(\tau_j) - f(\tau_i)\right]^+\right\} \tag{2.129}$$

for $i = 1, 2, \ldots, j-1$. It follows then from the definition that each $h(\tau_j)$ satisfies (2.127). The construction will be completed if we can show that each $h(\tau_j)$ satisfies (2.128) as well, i.e. they belongs to the interval I_j. Since the first term in (2.129) is the right endpoint of I_j, the point $h(\tau_j)$ automatically satisfies the right inequality of (2.128). Thus, we only have to show that

$$h(\tau_i) + \tfrac{C}{2}\kappa(\tau_j - \tau_i) - \left[f(\tau_j) - f(\tau_i)\right]^+ \ge f(\tau_i)^- - \tfrac{C}{2}\kappa(1-\tau_i).$$

However, this estimate reduces exactly to inequality (2.124) for a certain subpartition of $\hat{P}$, depending upon which choice of the minima was made in (2.129) for each $h(\tau_i)$ $(i = 1, 2, \ldots, j-1)$. The details are left to the reader (Exercise 2.37).

2nd step. Now, we drop the assumption on the piecewise linearity of f, but keep the assumption that $f(0) = f(1) = 0$. So, let $f \in \kappa BV([0,1])$ be arbitrary with $\mathrm{Var}_\kappa(f; [0,1]) \le C$. By (2.108), we know that the set of discontinuity points is countable, say $D(f) = \{x_1, x_2, x_3, \ldots\}$. For each $n \in \mathbb{N}$, choose a partition $P_n \in \mathcal{P}([0,1])$ which contains the points $x_1, x_2, \dots, x_n$ and satisfies $\mu(P_n) \to 0$ as $n \to \infty$, see (1.2). The functions $f_n : [0,1] \to \mathbb{R}$ defined by

$$f_n(x) := \begin{cases} f(x_n) & \text{if } x = x_n\,, \\ \text{linear} & \text{otherwise} \end{cases} \tag{2.130}$$

is then piecewise linear and satisfies

$$\lim_{n\to\infty} \|f_n - f\|_\infty = 0, \quad \sup_n \operatorname{Var}_\kappa(f_n; [0,1]) \le C\,. \tag{2.131}$$

By what we just proved in the first step, there exist κ-decreasing functions $g_n, h_n : [0,1] \to \mathbb{R}$ satisfying $f_n = g_n - h_n$ and

$$\max\{C_\kappa(g_n), C_\kappa(h_n)\} \le \frac{1}{2}C, \quad \max\{\|g_n\|_\infty, \|h_n\|_\infty\} \le \frac{3}{2}C\,, \tag{2.132}$$

where the second estimate follows from (2.109). Helly's selection principle for κ-decreasing functions (Theorem 2.67) implies that we can find subsequences $(g_{n_k})_k$ and $(h_{n_k})_k$ converging to some κ-decreasing functions g and h, respectively, such that

$$\max\{C_\kappa(g), C_\kappa(h)\} \le \frac{1}{2}C, \quad \max\{\|g\|_\infty, \|h\|_\infty\} \le \frac{3}{2}C\,,$$

and $f = g - h$. So, we have proved the assertion and (2.123) in case $f(0) = f(1) = 0$.

3rd step. Now, we also drop the assumption $f(1) = 0$ (but keep our general hypothesis $f(0) = 0$) and prove (2.122). Define $\overline{f} : [0,1] \to \mathbb{R}$ by

$$\overline{f}(x) := \begin{cases} f(x) & \text{for } 0 \le x < 1\,, \\ 0 & \text{for } x = 1\,. \end{cases} \tag{2.133}$$

So, for any partition $\{t_0, t_1, \dots, t_m\} \in \mathcal{P}([0,1])$, we have

$$\begin{aligned}\sum_{j=1}^{m} |\overline{f}(t_j) - \overline{f}(t_{j-1})| &= \sum_{j=1}^{m-1} |f(t_j) - f(t_{j-1})| + |f(t_{m-1})| \\ &\le C\sum_{j=1}^{m-1} \kappa(t_j - t_{j-1}) + \frac{3}{2}C \le \frac{5}{2}C\sum_{j=1}^{m} \kappa(t_j - t_{j-1})\,,\end{aligned}$$

where we have used (2.109). We conclude that $\overline{f}$ is of bounded κ-variation with

$$\operatorname{Var}_\kappa(\overline{f}; [0,1]) \le \frac{5}{2}C\,.$$

From what we have shown in the second step, it follows that $\overline{f} = \overline{g} - \overline{h}$, where $\overline{g}(0) = \overline{h}(0) = 0$ and both $C_\kappa(\overline{g}) \le 5C/4$ and $C_\kappa(\overline{h}) \le 5C/4$. Now, we distinguish three cases. If $f(1) = 0$, the assertion has already been proved in the second step. If $f(1) < 0$, we define $g : [0,1] \to \mathbb{R}$ by

$$g(x) := \begin{cases} \overline{g}(x) & \text{for } 0 \le x < 1\,, \\ f(1) & \text{for } x = 1 \end{cases} \tag{2.134}$$

and set $h := \overline{h}$. If $f(1) > 0$, we define $h : [0,1] \to \mathbb{R}$ by

$$h(x) := \begin{cases} \overline{h}(x) & \text{for } 0 \le x < 1\,, \\ -f(1) & \text{for } x = 1 \end{cases} \tag{2.135}$$

and set $g := \overline{g}$. In either case $f = g - h$, so it only remains to prove that g and h are κ-decreasing.

Consider, for definiteness, the function g from (2.134). We have to show that (2.124) holds for g, where $P = \{t_0, t_1, \dots, t_m\} \in \mathcal{P}([x,y])$ is arbitrary. However, the definition (2.134) implies that

$$g(1) - g(t_j) = g(1) + \overline{g}(1) - \overline{g}(t_j) \le f(1) + \frac{5}{4}C\kappa(1 - t_j) \le \frac{5}{4}C\kappa(1 - t_j)$$

since $f(1) < 0$, and so g is κ-decreasing with $C_\kappa(g) \le 5C/4$. The proof for the function h is analogous, and so we are done. □

Since $\mathrm{Var}_\kappa(f; [0,1]) = 0$ if and only if f is constant on $[0,1]$, it is easy to see that

$$\|f\|_{\kappa BV} := |f(0)| + \mathrm{Var}_\kappa(f; [0,1]) \tag{2.136}$$

defines a norm on $\kappa BV([0,1])$. By (2.109), the space $(\kappa BV([0,1]), \|\cdot\|_{\kappa BV})$ is continuously imbedded into the space of bounded functions with norm (0.39). In [161], it was shown that $\kappa BV([0,1])$ with norm (2.136) is a Banach space. The following proposition shows that $\kappa BV([0,1])$ is also a Banach algebra.

Proposition 2.69. *From $f, g \in \kappa BV([0,1])$, it follows that also $fg \in \kappa BV([0,1])$ with*

$$\|fg\|_{\kappa BV} \le \|f\|_{\kappa BV}\|g\|_\infty + \|f\|_\infty\|g\|_{\kappa BV}\,. \tag{2.137}$$

Consequently, the norm

$$|||f|||_{\kappa BV} := \|f\|_\infty + \mathrm{Var}_\kappa(f; [0,1]) \tag{2.138}$$

is equivalent to the norm (2.136), and $(\kappa BV([0,1]), |||\cdot|||_{\kappa BV})$ is a normalized Banach algebra.

Proof. The proof of (2.137) follows from the estimate

$$\begin{aligned} &\frac{\sum_{j=1}^m |f(t_j)g(t_j) - f(t_{j-1})g(t_{j-1})|}{\sum_{j=1}^m \kappa(t_j - t_{j-1})} \\ &= \frac{\sum_{j=1}^m |[f(t_j) - f(t_{j-1})]g(t_j) + f(t_{j-1})[g(t_j) - g(t_{j-1})]|}{\sum_{j=1}^m \kappa(t_j - t_{j-1})} \\ &\le \|g\|_\infty \frac{\sum_{j=1}^m |f(t_j) - f(t_{j-1})|}{\sum_{j=1}^m \kappa(t_j - t_{j-1})} + \|f\|_\infty \frac{\sum_{j=1}^m |g(t_j) - g(t_{j-1})|}{\sum_{j=1}^m \kappa(t_j - t_{j-1})} \end{aligned}$$

in the same way as the proof of (1.17) in Proposition 1.10. The last statement follows from Proposition 0.31. □

At this point, we collect the various definitions of variation with respect to some partition $\{t_0, t_1, \ldots, t_m\} \in \mathcal{P}([a,b])$ or collections $\{[a_n, b_n] : n \in \mathbb{N}\} \in \Sigma_\infty([a,b])$ in the following Table 2.5.

Table 2.5. Concepts of variation ($1 < p < \infty$).

Definition	*Variation*	*Space*
Jordan (1881)	$\sum_{j=1}^{m} \lvert f(t_j) - f(t_{j-1})\rvert$	BV
Wiener (1924)	$\sum_{j=1}^{m} \lvert f(t_j) - f(t_{j-1})\rvert^p$	WBV_p
Young (1937)	$\sum_{j=1}^{m} \phi(\lvert f(t_j) - f(t_{j-1})\rvert)$	WBV_ϕ
Schramm (1985)	$\sum_{k=1}^{\infty} \phi_k(\lvert f(b_k) - f(a_k)\rvert)$	ΦBV
Riesz (1910)	$\sum_{j=1}^{m} \frac{f(t_j) - f(t_{j-1})\rvert^p}{(t_j - t_{j-1})^{p-1}}$	RBV_p
Medvedev (1953)	$\sum_{j=1}^{m} \phi\left(\frac{f(t_j) - f(t_{j-1})\rvert}{t_j - t_{j-1}}\right)(t_j - t_{j-1})$	RBV_ϕ
?????	$\sum_{k=1}^{\infty} \phi_k\left(\frac{\lvert f(b_k) - f(a_k)\rvert}{b_k - a_k}\right)(b_k - a_k)$	?????
Korenblum (1975)	$\frac{\sum_{j=1}^{m} \lvert f(t_j) - f(t_{j-1})\rvert}{\sum_{j=1}^{m} \kappa(t_j - t_{j-1})}$	κBV

We remark that a variation of type

$$\mathrm{Var}_\Phi^R(f, S_\infty; [a,b]) := \sum_{k=1}^{\infty} \phi_k\left(\frac{\lvert f(b_k) - f(a_k)\rvert}{b_k - a_k}\right)(b_k - a_k)$$

with respect to a Schramm sequence $\Phi = (\phi_n)_n$ and a collection $S_\infty = \{[a_n, b_n] : n \in \mathbb{N}\} \in \Sigma_\infty([a,b])$ would generalize Schramm's definition (2.70) in the same way as the Riesz p-variation (2.87) or the Medvedev variation (2.96) generalizes Jordan's definition (1.3). In fact, in case $\phi_n(t) \equiv t$, for all $n \in \mathbb{N}$, the above expression becomes

$$\mathrm{Var}_\Phi^R(f, S_\infty; [a,b]) = \sum_{k=1}^{\infty} \lvert f(b_k) - f(a_k)\rvert \,.$$

As far as we know, such a variation has not been defined and studied so far[22] in the literature; that is why we have put the question marks ????? in the corresponding row in Table 2.5.

22 The only contribution in this direction seems to be the forthcoming paper [59].

In the following Table 2.6, which complements Table 1.1, we collect some of the function spaces, together with relations between them, that have been introduced in this chapter; here, $1 < p < \infty$.

Table 2.6. Relations between function classes over $I = [a, b]$.

$Lip(I)$	$\subset$	$AC(I)$	$\subset$	$BV(I)$	$\subset$	$WBV_p(I)$
$\cap$		$\cap$		$\cap$		$\cap$
$RBV_p(I)$		$C(I)$	$\subset$	$\Lambda BV(I)$	$\subset$	$\Phi BV(I)$
$\cap$				$\cap$		
$BV(I)$	$\subset$	$\kappa BV(I)$	$\subset$	$R(I)$		

Let us recall some examples over $I = [0, 1]$ which show that the inclusions in Table 2.6 are in fact strict.

Example 2.70. In Example 1.39, we have shown that $Z_1 \in WBV_p([0,1]) \setminus BV([0,1])$, see (0.93). For the function $f_{\alpha,\beta}$ defined by (0.86), we have $f_{\alpha,\beta} \in BV([0,1]) \setminus RBV_p([0,1])$ if

$$-\alpha < \beta < p(1-\alpha) - 1\,,$$

and $f_{\alpha,\beta} \in C([0,1]) \setminus RBV_p([0,1])$ if, in addition, $\alpha \geq 0$. Moreover, we have $f_{\alpha,\beta} \in RBV_p([0,1]) \setminus Lip([0,1])$ if

$$p(1-\alpha) < \beta < 2 - \alpha\,.$$

All of these statements concerning the function $f_{\alpha,\beta}$ follow from Table 2.4. ♥

A function $f \in \kappa BV([0,1]) \setminus BV([0,1])$ has been constructed in Example 2.66. In Example 3.6 in the next chapter, we will construct a function $f \in BV([0,1]) \setminus AC([0,1])$. Examples of functions $f \in AC([0,1]) \setminus Lip([0,1])$, or $f \in R([0,1]) \setminus \kappa BV([0,1])$, are easily found.

2.6 Higher order Wiener-type variations

Already beginning with De la Vallée Poussin's work about 100 years ago, so-called *higher order variations* have been introduced and studied which build on divided differences of functions. In this section, we follow the presentation in Sections 1.5 and 1.6 of the monograph [226].

Definition 2.71. Given a function $f : [a,b] \to \mathbb{R}$ and points $t_0, t_1, \ldots, t_k \in [a,b]$, the higher order *divided differences* for are defined recursively by

$$
\begin{aligned}
f[t_0] &:= f(t_0), \quad f[t_0,t_1] := \frac{f(t_1)-f(t_0)}{t_1-t_0}, \\
f[t_0,t_1,t_2] &:= \frac{f[t_1,t_2]-f[t_0,t_1]}{t_2-t_0}, \quad \ldots \\
\ldots, \quad f[t_0,t_1,\ldots,t_k] &:= \frac{f[t_1,t_2,\ldots,t_k]-f[t_0,t_1,\ldots,t_{k-1}]}{t_k-t_0}.
\end{aligned}
\tag{2.139}
$$

Given a partition $P = \{t_0, t_1, \ldots, t_m\} \in \mathcal{P}([a,b])$, $k \in \mathbb{N}$, and $p \in [1,\infty)$, we set

$$
\mathrm{Var}^W_{k,p}(f,P;[a,b]) := \sum_{j=1}^{m-k+1} |f[t_j,\ldots,t_{j+k-1}] - f[t_{j-1},\ldots,t_{j+k-2}]|^p \tag{2.140}
$$

and call (2.140) the *(k, p)-variation* of f with respect to P (in Wiener's sense) on $[a,b]$. ■

Definition 2.71 contains many notions of variation previously defined in the literature. In the special case $k = 1$ and $p = 1$, (2.140) reduces to Jordan's classical definition [153]

$$
\mathrm{Var}^W_{1,1}(f,P;[a,b]) := \sum_{j=1}^{m} |f(t_j) - f(t_{j-1})| = \mathrm{Var}(f,P;[a,b]),
$$

while in the special case $k = 1$ and $p > 1$, (2.140) reduces to Wiener's p-variation [321]

$$
\mathrm{Var}^W_{1,p}(f,P;[a,b]) := \sum_{j=1}^{m} |f(t_j) - f(t_{j-1})|^p = \mathrm{Var}^W_p(f,P;[a,b])
$$

which we have studied in Section 1.3. Moreover, for $k = 2$ and $p = 1$, we get the second variation

$$
\begin{aligned}
\mathrm{Var}^W_{2,1}(f,P;[a,b]) &:= \sum_{j=1}^{m-1} |f[t_j,t_{j+1}] - f[t_{j-1},t_j]| \\
&= \sum_{j=1}^{m-1} \left| \frac{f(t_{j+1})-f(t_j)}{t_{j+1}-t_j} - \frac{f(t_j)-f(t_{j-1})}{t_j-t_{j-1}} \right|
\end{aligned}
\tag{2.141}
$$

which has been introduced by De la Vallée Poussin [103] in 1908, and for general $k \in \mathbb{N}$ and $p = 1$, we get the variation

$$
\mathrm{Var}^W_{k,1}(f,P;[a,b]) := \sum_{j=1}^{m-k+1} |f[t_j,\ldots,t_{j+k-1}] - f[t_{j-1},\ldots,t_{j+k-2}]| \tag{2.142}
$$

introduced in 1934 by Popoviciu [252]. For the most general setting of $k \in \mathbb{N}$ and $p \geq 1$, we have the following definition.

Definition 2.72. Given $k \in \mathbb{N}$ and $p \in [1, \infty)$, we call the (possibly infinite) number

$$\mathrm{Var}^W_{k,p}(f; [a,b]) := \sup \{\mathrm{Var}^W_{k,p}(f, P; [a,b]) : P \in \mathcal{P}([a,b])\},$$

where the supremum is taken over all partitions of $[a, b]$, the *total* (k, p)*-variation* of f (in Wiener's sense) on $[a, b]$. The set

$$WBV_{k,p}([a,b]) := \{f \in B([a,b]) : \mathrm{Var}^W_{k,p}(f; [a,b]) < \infty\}$$

is called the *space of all functions of bounded* (k, p)*-variation* (in Wiener's sense) on $[a, b]$. ■

By what we observed above, we have the special cases

$$WBV_{1,1}([a,b]) = BV([a,b]), \quad WBV_{1,p}([a,b]) = WBV_p([a,b]).$$

In rather the same way as we have done this for the spaces BV and WBV_p in Propositions 1.10 and 1.32, one may show that the linear space $WBV_{k,p}([a,b])$ with norm

$$\|f\|_{WBV_{k,p}} := \sum_{i=0}^{k-1} |f^{(i)}(a)| + \mathrm{Var}^W_{k,p}(f; [a,b])^{1/p} \tag{2.143}$$

is a Banach space [278]. Moreover, in [280], it is shown that the subclass

$$WBV^o_{k,1}([a,b]) := \{f \in WBV_{k,1}([a,b]) : f(a) = f'(a) = \ldots = f^{(k-1)}(a) = 0\},$$

equipped with the norm[23]

$$|||f|||_{WBV_{k,1}} := 2^{k-1}(k-1)!(b-a)^{k-1}\,\mathrm{Var}^W_{k,1}(f; [a,b])$$

is a normalized algebra in the sense of Definition 0.30.

It is interesting to compare the spaces $WBV_{k,p}([a,b])$ for different values of k and p. The following proposition shows that $WBV_{k,p}([a,b])$ is "decreasing" with respect to $k \in \mathbb{N}$ and "increasing" with respect to $p \in [1, \infty)$.

Proposition 2.73. *The inclusions*

$$WBV_{k,p}([a,b]) \supseteq WBV_{k+1,p}([a,b]) \quad (k = 1, 2, 3, \ldots; 1 \le p < \infty) \tag{2.144}$$

and

$$WBV_{k,p}([a,b]) \subseteq WBV_{k,q}([a,b]) \quad (k = 1, 2, 3, \ldots; 1 \le p < q < \infty) \tag{2.145}$$

hold, and both are strict.

---

23 Of course, for $k = 1$, this norm reduces to the usual variation $\mathrm{Var}(f; [a,b])$, and we recover Proposition 1.10 for the normalized algebra $BV([a,b])$

Proof. For $p = 1$, the idea of the proof of (2.144) is to use the estimate

$$\mathrm{Var}^W_{k,1}(f;[a,b]) \le k(b-a)\left[\mathrm{Var}^W_{k+1,1}(f;[a,b]) + \inf |f[t_0,t_1,\dots,t_k]|\right], \tag{2.146}$$

where the infimum is taken over all partitions $\{t_0,t_1,\dots,t_k\} \in \mathcal{P}([a,b])$. For $p > 1$, the proof is similar with changes in the constants. Details may be found in [274]. The inclusion (2.145) is proved in the same way as in the special case $k = 1$, see Proposition 1.38. □

The spaces $WBV_{k,p}([a,b])$ have been particularly well-studied for $p = 1$. For example, De la Vallée Poussin [103] has proved the following representation theorem for functions in $WBV_{2,1}([a,b])$, which is a natural analogue to Jordan's representation theorem (Theorem 1.5):

Theorem 2.74 (De la Vallée Poussin). *A function $f : [a,b] \to \mathbb{R}$ has bounded second variation $\mathrm{Var}^W_{2,1}(f;[a,b])$ if and only if it may be represented in the form $f = P_f - N_f$, where both P_f and N_f are convex functions.*

Proof. To prove the "if" part, it obviously suffices to show that every convex function $f : [a,b] \to \mathbb{R}$ belongs to $WBV_{2,1}([a,b])$. Now, if f is convex, then the map $f[\cdot,y] : [a,y) \to \mathbb{R}$ defined by

$$x \mapsto f[x,y] = \frac{f(y)-f(x)}{y-x} \quad (a \le x < y)$$

is increasing on $[a,y)$ for any $y \in (a,b]$. Moreover, the unilateral derivatives f'_+ and f'_- of f exist on $[a,b]$ and satisfy

$$f'_+(x) \le f[x,y] \le f'_-(y) \quad (a \le x < y) \tag{2.147}$$

by definition of the second divided difference in (2.139). Given any partition $P = \{t_0,t_1,\dots,t_m\} \in \mathcal{P}([a,b])$, we have

$$\begin{aligned}\mathrm{Var}^W_{2,1}(f,P;[a,b]) &= \sum_{j=1}^{m-1} |f[t_j,t_{j+1}] - f[t_{j-1},t_j]| \\ &= \sum_{j=1}^{m-1} f[t_j,t_{j+1}] - f[t_{j-1},t_j] = f[t_{m-1},b] - f[a,t_1] \le f'_-(b) - f'_+(a),\end{aligned}$$

where we may drop the absolute value after the sum, since $f[\cdot,y]$ is increasing. Passing to the supremum with respect to all partitions $P \in \mathcal{P}([a,b])$ we conclude that

$$\mathrm{Var}^W_{2,1}(f;[a,b]) \le f'_-(b) - f'_+(a),$$

and so $f \in WBV_{2,1}([a,b])$ which proves the "if" part of the assertion.

Now suppose that $f \in WBV_{2,1}([a,b])$, which means that the second variation

$$\begin{aligned}&\mathrm{Var}^W_{2,1}(f;[a,b]) \\ &\quad = \sup\left\{\sum_{j=1}^{m-1} |f[t_j,t_{j+1}] - f[t_{j-1},t_j]| : \{t_0,t_1,\dots,t_m\} \in \mathcal{P}([a,b])\right\}\end{aligned} \tag{2.148}$$

of f on $[a, b]$ is finite. Without loss of generality we may assume that $f(a) = 0$. We use the fact that every function $f \in WBV_{2,1}([a, b])$ is Lipschitz continuous (Exercise 2.44), and so we can find a nullset $N \subset [a, b]$ such that f' exists on $[a, b] \setminus N$. We define $p_f, n_f : [a, b] \setminus N \to \mathbb{R}$ by

$$p_f(x) := \frac{1}{2}\left[\mathrm{Var}^W_{2,1}(f; [a, x]) + f'(x)\right], \quad n_f(x) := \frac{1}{2}\left[\mathrm{Var}^W_{2,1}(f; [a, x]) - f'(x)\right],$$

and so we have $p_f - n_f = f'$ on $[a, b] \setminus N$. We claim that both functions p_f and n_f are increasing. In fact, for $a \le x < y \le b$, we have

$$\mathrm{Var}^W_{2,1}(f; [x, y]) \le \mathrm{Var}^W_{2,1}(f; [a, y]) - \mathrm{Var}^W_{2,1}(f; [a, x]),$$

and hence

$$\begin{aligned} 2p_f(y) - 2p_f(x) &= \mathrm{Var}^W_{2,1}(f; [a, y]) - \mathrm{Var}^W_{2,1}(f; [a, x]) + f'(y) - f'(x) \\ &\ge \mathrm{Var}^W_{2,1}(f; [x, y]) + f'(y) - f'(x) \ge 0, \end{aligned}$$

and similarly for the function n_f. Since both functions p_f and n_f are increasing and bounded on $[a, b] \setminus N$, we may extend them to increasing bounded functions on the whole interval $[a, b]$ which we still denote by p_f and n_f. It is now a natural idea to define $P_f, N_f : [a, b] \to \mathbb{R}$ by

$$P_f(x) := \int_a^x p_f(t)\, dt, \quad N_f(x) := \int_a^x n_f(t)\, dt.$$

Then we obtain

$$P_f(x) - N_f(x) = \int_a^x \left[p_f(t) - n_f(t)\right] dt = \int_a^x f'(t)\, dt = f(x)$$

since $f(a) = 0$. Moreover, both functions P_f and N_f are convex because their derivatives are monotonically increasing, and so we have proved the "only if" part of the assertion. □

Note that we have constructed the convex functions P_f and N_f satisfying $P_f - N_f = f$ in Theorem 2.74 from the increasing functions p_f and n_f satisfying $p_f - n_f = f'$ for the derivative (up to nullsets). Thus, one might ask if there is a connection between functions belonging to $WBV_{2,1}([a, b])$ and functions having derivatives in $WBV_{1,1}([a, b]) = BV([a, b])$. In fact, we have seen that every function $f \in WBV_{2,1}([a, b])$ admits unilateral derivatives f'_+ and f'_- on $[a, b]$. Moreover, in [274], it is shown that there exists a nullset $N \subset [a, b]$ such that $f \in C^1([a, b] \setminus N)$. This means that the derivative f' of a function $f \in WBV_{2,1}([a, b])$ exists and is *continuous* a.e. on $[a, b]$. One cannot claim, however, that the derivative f' is even *absolutely continuous* a.e. on $[a, b]$, as the following example shows.

Example 2.75. Let $\varphi : [0,1] \to \mathbb{R}$ be the classical Cantor function, see Example 3.6 in the next chapter, and let $f : [0,1] \to \mathbb{R}$ be defined by

$$f(x) := \int_0^x \varphi(t)\,dt \quad (0 \le x \le 1)\,. \tag{2.149}$$

Since φ is monotonically increasing, f is convex, and so $f \in WBV_{2,1}([a,b])$, by Theorem 2.74. On the other hand, we have $f' = \varphi$ a.e. on $[0,1]$, and φ is not absolutely continuous because it fails to have the so-called Luzin property, as we will also show in the next chapter. ♥

Remarkably, if we *require* a function $f \in WBV_{2,1}([a,b])$ to have an absolutely continuous derivative, we get an interesting integral representation for the second variation of f which was proved by Russell in [276].

Theorem 2.76 (Russell). *Suppose that a function $f \in WBV_{2,1}([a,b])$ has an absolutely continuous derivative f' a.e. on $[a,b]$. Then the second variation of f has the integral representation*

$$\mathrm{Var}^W_{2,1}(f;[a,b]) = \int_a^b |f''(t)|\,dt\,. \tag{2.150}$$

We postpone the proof of Theorem 2.76 to Chapter 3, where we will give a more general result (Theorem 3.39).

Theorem 2.76 shows again that the function (2.149) from Example 2.75 cannot have an absolutely continuous derivative. Indeed, for f as in (2.149), the left-hand side of (2.150) is 1, while the right-hand side of (2.150) is 0 because $f'(t) = \varphi(t) = 0$ a.e. on $[0,1]$.

2.7 Comments on Chapter 2

Among the various generalized variations discussed in this chapter, the Waterman variation (2.26) is probably the most important one, both from its theoretical interest and from an application-oriented point of view.

As pointed out several times before, passing from the Wiener space WBV_p to the Wiener–Young space WBV_ϕ has the same advantages and drawbacks as passing from the Lebesgue space L_p to the Orlicz space L_ϕ: the range of applications increases, but the technical expenditure increases as well.

It is not surprising that all functions $f \in WBV_\phi([a,b])$, although not being continuous, are always regular in the sense of our definition in Section 0.3; so they share this important property with classical BV-functions (in particular, monotone functions). The following Theorem [255] gives a refinement (and certain converse) of this: every function $f \in R([a,b])$ is contained in some space $WBV_\phi([a,b])$ for a suitable Young function ϕ.

Theorem 2.77. *The equalities*

$$\bigcup_{\phi} WBV_{\phi}([a,b]) = R([a,b]), \quad \bigcap_{\phi} WBV_{\phi}([a,b]) = FR([a,b]) \tag{2.151}$$

hold where $R([a,b])$ *denotes the space of all regular functions,* $FR([a,b])$ *denotes the space of all regular functions which assume only a finite number of values, and the intersection and union in* (2.151) *are taken over all Young functions* ϕ.

Many interesting structural theorems about WBV_{ϕ}-functions have been given by Prus–Wiśniowski in a series of papers [254–261]. For example, in [256], the author analyzes the class of all functions $f \in WBV_{\phi}([a,b])$ satisfying

$$\limsup_{\delta \to 0+} \{\mathrm{Var}_{\phi}^{W}(f, P; [a,b]) : \mu(P) \le \delta\} = 0\,, \tag{2.152}$$

where $\mu(P)$ denotes the mesh size (1.2) of the partition P. This is analogous to what we have done in the first part of Section 1.2 for classical BV-functions. In [99], a necessary and sufficient condition is given under which a sequence of functions $(h_n \circ f)_n$ is bounded, where $f \in WBV_{\phi}([a,b])$ with $f([a,b]) \subseteq [c,d]$, and $h_n \in WBV_{\psi}([c,d])$ for each $n \in \mathbb{N}$.

The spaces $RBV_p([a,b])$ have been introduced and studied by Riesz [265], and the more general space $RBV_{\phi}([a,b])$ by Medvedev [211]. Both spaces will play a prominent role in Section 3.5 in the next chapter.

One of the important properties of the Riesz space $RBV_p([a,b])$ is its close relation to (scalar) Sobolev spaces; we shall discuss this in the next chapter. In Proposition 2.51, we have shown that $RBV_p([a,b])$ is a Banach algebra with respect to multiplication. The following elementary counterexample shows that $RBV_p([a,b])$ is not stable with respect to the composition of functions.

Example 2.78. For $0 < \tau < 1$, define $f_{\tau} : [0,1] \to \mathbb{R}$ by $f_{\tau}(x) := x^{\tau}$. A somewhat cumbersome calculation[24] shows that

$$f_{\tau} \in RBV_p([0,1]) \iff \frac{1}{p} > 1 - \tau\,; \tag{2.153}$$

moreover,

$$\|f_{\tau}\|_{RBV_p} = \frac{\tau}{[1-(1-\tau)p]^{1/p}}$$

in this case. So, if we take $1 - \tau < \frac{1}{p} \le 1 - \tau^2$, then $f_{\tau} \in RBV_p([0,1])$, but $f_{\tau} \circ f_{\tau} = f_{\tau^2} \notin RBV_p([0,1])$. ♥

Another example of this kind will be given in Chapter 5 (Example 5.8). The following proposition due to Szigeti [299, 300] shows how to choose the “right” indices p, q,

24 In the next chapter, we will provide a much more elegant method to prove (2.153) and to calculate $\|f_{\tau}\|_{RBV_p}$, see Example 3.35.

and r to guarantee that compositions of Riesz space functions remain in some other Riesz space. An analogous result for several variables has been proved in [216].

Proposition 2.79. *Suppose that $p, q, r \in (1, \infty)$ satisfy*

$$\left(1 - \frac{1}{p}\right)\left(1 - \frac{1}{q}\right) = 1 - \frac{1}{r}. \tag{2.154}$$

Let $f \in RBV_p([a, b])$ be monotone with $f([a, b]) \subseteq [c, d]$, and $g \in RBV_q([c, d])$. Then $g \circ f \in RBV_r([a, b])$ and

$$\mathrm{Var}_r^R(g \circ f; [a, b]) \le C\, \mathrm{Var}_q^R(g; [a, b])^{r/q}\, \mathrm{Var}_p^R(f; [a, b])^{1-r/q}$$

for some $C > 0$.

Example 2.78 shows that condition (2.154) is sharp. In fact, by choosing $f := f_\alpha \in RBV_p([0, 1])$ and $g := f_\beta \in RBV_q([0, 1])$ $(0 < \alpha, \beta < 1)$, we obtain, by (2.153),

$$\left(1 - \frac{1}{p}\right)\left(1 - \frac{1}{q}\right) < \alpha\beta$$

as a necessary and sufficient condition, and $g \circ f = f_{\alpha\beta}$ implies that

$$1 - \frac{1}{r} < \alpha\beta$$

to ensure that $g \circ f \in RBV_r([0, 1])$.

Passing from the Riesz space RBV_p to the Riesz–Medvedev space RBV_ϕ, we of course encounter similar phenomena. We will come back to both classes of spaces in Section 3.5. Many interesting facts about the space RBV_ϕ are discussed in [209]. Proposition 2.57 shows that when studying the space RBV_ϕ, we have to assume that $\phi \in \infty_1$ if we want to obtain something new. In this connection, the following question is of interest:

Problem 2.1. *Suppose that ϕ is a Young function satisfying $\phi \notin \infty_p$ for some $p > 1$. Does this imply that $RBV_\phi = RBV_p$?*

In view of their importance, let us now spend some time to comment on our results about the Waterman spaces ΛBV and $\Lambda_q BV$. The proof of Proposition 2.17 (f) shows that convergence in the norm (2.30) implies uniform convergence on $[a, b]$. Thus, the set $C([a, b]) \cap \Lambda BV([a, b])$ of continuous functions in $\Lambda BV([a, b])$ forms a closed subspace of $\Lambda BV([a, b])$, and thus is also a Banach space with respect to the norm (2.30).

The space ΛBV has been introduced by Waterman in [315, 316] and afterwards has been studied and further developed by several authors. Without aiming for a complete list, we mention [24, 25, 213, 214, 246, 247, 250, 251, 260, 262, 287, 288, 312, 313].

We remark that the definition of ΛBV is somewhat different in the literature: the sequence $\Lambda = (\lambda_n)_n$ is supposed to be *increasing* and unbounded, (2.24) is replaced by

$$\sum_{n=1}^{\infty} \frac{1}{\lambda_n} = \infty\,,$$

and all expressions of the form $|f(b_k) - f(a_k)|$ are not multiplied, but divided by λ_k.

The equalities (2.33) have been proved by Perlman in [246], where it is also shown that (2.33) is not true if the intersection and union are taken only over a countable family of Waterman sequences Λ. The second equality in (2.33) shows that a function $f \in \Lambda BV$ only has removable discontinuities or jumps. Perlman and Waterman [247] prove that whenever

$$\liminf_{x\to x_0} f(x) \le f(x_0) \le \limsup_{x\to x_0} f(x)$$

at each point of discontinuity x_0 of f, then the total Waterman variation (2.26) of f is actually independent of the value $f(x_0)$. Our Proposition 2.28 which gives a necessary and sufficient condition for inclusion between different Waterman spaces is also proved in the paper [247]. In this connection, the following result [262] is interesting: for every Waterman sequence $\Lambda = (\lambda_n)_n$, one may find a Waterman sequence $M = (\mu_n)_n$ (called "regularization" of Λ) such that

$$\limsup_{n\to\infty} \frac{\mu_{n+1}}{\mu_n} = 1$$

and $(\lambda[1,n])_n \sim (\mu[1,n])_n$ (in the notation (2.42)), and hence $\Lambda BV = MBV$. The Propositions 2.32 and 2.33, as well as the idea of using zigzag functions to separate the function classes WBV_p and $\Lambda_q BV$, are due to Pierce and Velleman [250].

The subspace $\Lambda^c BV$ of ΛBV has been introduced in 1976 by Waterman [315, 316]. For $\lambda_n \equiv 1$ (i.e. $\Lambda BV = BV$), the space $\Lambda^c BV$ is not interesting because it only contains constant functions. However, if $\Lambda = (\lambda_n)_n$ is not bounded away from zero, the following question seems to be important:

Problem 2.2. *How can we characterize the elements of $\Lambda^c BV$ in case $\lambda_n \to 0$ (i.e. $\Lambda BV \neq BV$)?*

We mentioned that an analogous subspace of the Schramm space ΦBV has been defined in the literature, namely, the class $\Phi^c BV([a,b])$ of all $f \in \Phi BV([a,b])$ satisfying

$$\lim_{\lambda\to\infty} \mathrm{Var}_\Phi(f/\lambda;[a,b])) = 0\,,$$

where $\mathrm{Var}_\Phi(f;[a,b])$ denotes the Schramm variation (2.71). Here, a similar question arises:

Problem 2.3. *How can we characterize the elements of $\Phi^c BV$ for arbitrary Schramm sequences Φ?*

Observe that, as Exercise 2.1 shows, this question does not make sense in the Wiener–Young space WBV_ϕ (which may be considered as a Schramm space with $\phi_n(t) \equiv \phi(t)$) because *every* function $f \in WBV_\phi$ is continuous in the Wiener–Young variation (2.2).

Interestingly, the Waterman space ΛBV is also intimately related to the generalized Hölder space $Lip_{\omega,p}$ introduced in Definition 0.54. Imbedding theorems between these spaces (in both directions) have been established by various authors; for example, the following result ([133], see also [312]) holds.

Proposition 2.80. *Let $\Lambda = (\lambda_n)_n$ be a Waterman sequence, $\omega : [0,\infty) \to [0,\omega)$ a modulus of continuity, and $1 \le p < \infty$. Then the inclusion $\Lambda BV \subseteq Lip_{\omega,p}$ holds if and only if*

$$\max_{k=1,\dots,n} \frac{k^{1/p}}{\lambda[1,k]} = O\left(n^{1/p}\omega\left(\frac{1}{n}\right)\right) \quad (n \to \infty),$$

where we have used the notation (2.42). In particular, for $0 < q < 1$ and $0 < \alpha \le 1$, we have $\Lambda_q BV \subseteq Lip_{\alpha,p}$ if and only if

$$\alpha \le \min\left\{\frac{1}{p}, 1 - q\right\},$$

where $\Lambda_q BV$ is given in Definition 2.29, and $Lip_{\alpha,p}$ in Definition 0.54.

Imbedding theorems for Hölder spaces into spaces of functions of (generalized) bounded variation have also been given in the literature. The simplest example is Proposition 1.34 which states that $Lip_{1/p}([a,b]) \hookrightarrow WBV_p([a,b])$ for $1 \le p < \infty$. The following more general and precise result involving the space $Lip_{\omega,\infty}([a,b])$ from Definition 0.54 has been proved by Medvedev in [212], see also [213].

Proposition 2.81. *Let ϕ be a Young function which satisfies the δ_2-condition (2.4). Then $Lip_{\omega,\infty}([a,b]) \hookrightarrow WBV_\phi([a,b])$ if and only if*

$$\omega(t) = O(\phi^{-1}(t)) \quad (t \to 0+). \tag{2.155}$$

In the special case $\omega(t) = t^\alpha$ (and hence $Lip_{\omega,\infty} = Lip_\alpha$) and $\phi(t) = t^p$ (and hence $WBV_\phi = WBV_p$), condition (2.155) holds precisely for $\alpha \ge 1/p$. Therefore, we recover Proposition 1.34 for the sufficiency of (2.155) for the imbedding $Lip_\alpha([a,b]) \hookrightarrow WBV_p([a,b])$. Observe also that Exercises 1.59 and 1.60 are contained in Proposition 2.79 for the special choice $\phi(t) = t^p$.

The following result which provides a certain converse of Proposition 2.81 is also due to Medvedev [212]:

Proposition 2.82. *Let $\omega : [0,\infty) \to [0,\infty)$ be a modulus of continuity, and let $\Lambda = (\lambda_n)_n$ be a Waterman sequence. Then the inclusion $Lip_{\omega,\infty} \subseteq \Lambda BV$ holds if and only if there exists a nonnegative sequence $(t_n)_n$ satisfying*

$$\sum_{n=1}^{\infty} t_n \le 1, \quad \sum_{n=1}^{\infty} \lambda_n \omega(t_n) < \infty. \tag{2.156}$$

Again, let us check what Proposition 2.82 means for $\omega(t) = t^\alpha$ (and hence $Lip_{\omega,\infty} = Lip_\alpha$) and $\lambda_n = n^{-q}$ (and hence $\Lambda BV = \Lambda_q BV$). Choosing in this case $t_n := 2^{-n}$, the second condition in (2.156) reads

$$\sum_{n=1}^{\infty} \frac{1}{n^q 2^{n\alpha}} < \infty\,, \tag{2.157}$$

and the classical quotient criterion implies that (2.157) holds for *any* $q \geq 0$ and $\alpha > 0$. So, we have the imbedding $Lip_\alpha \hookrightarrow \Lambda_q$ in this case.[25] More general sufficient conditions for imbeddings of type $Lip_{\omega,\infty} \hookrightarrow \Lambda BV$ (which are sometimes also necessary) may be found in [213].

As we have shown in Proposition 1.34, the Wiener space $WBV_{1/(1-q)}$ contains the Hölder space Lip_{1-q}, see (1.68). However, the following example shows that the smaller space $\Lambda_q BV([0,1])$ does *not* contain this Hölder space.

Example 2.83. This example is quite similar to Example 1.24, where we constructed a function which belongs to Lip_α for any $\alpha < 1$, but not to BV. Given $q \in (0,1)$, let

$$\gamma := \zeta(1,q) = \sum_{k=1}^{\infty} \frac{1}{k \log^q(k+1)}\,, \qquad t_n := \frac{1}{\gamma}\sum_{k=1}^{n} \frac{1}{k\log^q(k+1)}\,,$$

and define $f : [0,1] \to \mathbb{R}$ by

$$f(x) := \begin{cases} 0 & \text{for } x = 0\,, \\ \frac{1}{\gamma^q}\sum_{k=1}^{n} \frac{(-1)^{k+1}}{k^q \log(k+1)} & \text{for } x = t_n\,, \\ \frac{1}{\gamma^q}\sum_{k=1}^{\infty} \frac{(-1)^{k+1}}{k^q \log(k+1)} & \text{for } x = 1\,, \\ \text{linear} & \text{otherwise}\,. \end{cases}$$

As in Example 1.24, one may show that $f \in Lip_{1-q}([0,1])$; we claim that $f \notin \Lambda_q BV([0,1])$. In fact, on the interval $[a_k, b_k] := [t_{k-1}, t_k]$, we have

$$\begin{aligned} \sum_{k=1}^{n} \frac{|f(b_k) - f(a_k)|}{k^q} &= \frac{1}{\gamma^p}\sum_{k=1}^{n}\left[\sum_{j=1}^{k} \frac{(-1)^{j+1}}{j^q \log(j+1)} - \sum_{j=1}^{k-1} \frac{(-1)^{j+1}}{j^q\log(j+1)}\right] \\ &= \frac{1}{\gamma^q}\sum_{k=1}^{n} \frac{1}{k^q \log(k+1)} = \frac{\zeta(1,q)}{\gamma^q} \to \infty \quad (n \to \infty)\,, \end{aligned}$$

and thus $\mathrm{Var}_{\Lambda_q}(f;[0,1]) = \infty$ and so $f \notin \Lambda_q BV([0,1])$. ♥

We remark that sufficient conditions for the inclusions

$$C([a,b]) \subseteq \Lambda^c BV([a,b]), \quad Lip_\alpha([a,b]) \subseteq \Lambda_q^c BV([a,b])$$

---

25 Observe that we may combine (1.68) and (2.46) to deduce that $Lip_\alpha \hookrightarrow WBV_{1/\alpha} \hookrightarrow \Lambda_q$; however, according to Proposition 2.32, the last imbedding is true only for $q > \frac{1}{p'} = 1 - \alpha$, while (2.157) does not require a link between α and q.

are given in Exercises 2.12 and 2.13. Relations between spaces of functions of bounded variation, generalized Hölder spaces, and Orlicz spaces are studied in the survey paper [101].

In Proposition 1.27, we have shown how bounded variation of f reflects in terms of the corresponding Banach indicatrix I_f: we have $f \in BV([a,b])$ if and only if $I_f \in L_1(\mathbb{R})$. This admits an interesting generalization which provides a link between moduli of continuity, Waterman sequences, and the Banach indicatrix.

Suppose that $\omega : [0,\infty) \to [0,\infty)$ is a modulus of continuity which satisfies $\omega(t) = o(t)$ as $t \to \infty$. Then the sequence $\Lambda = (\lambda_n)_n$ defined by $\lambda_n := \omega(n) - \omega(n-1)$ is a Waterman sequence.[26] In [25, Theorem 5.2], it is shown that the condition

$$\int_{-\infty}^{\infty} \omega(N_f(y))\,dy < \infty$$

for a regular function $f : [a,b] \to \mathbb{R}$ implies that $f \in \Lambda^c BV([a,b])$, where Λ is as before. A particularly interesting case is $\omega(t) = t^\alpha$ with $0 < \alpha < 1$; here, the condition[27]

$$\int_{-\infty}^{\infty} N_f(y)^\alpha\,dy < \infty$$

implies that $f \in \Lambda^c_{1-\alpha} BV([a,b])$. By Proposition 2.33, this, in turn, implies $f \in WBV_{1/\alpha}([a,b])$ which is a result by Zerekidze [328].

In this chapter, we have discussed various imbedding theorems between the Wiener–Young space WBV_ϕ, the generalized Hölder space $Lip_{\omega,\infty}$, the Waterman space ΛBV, and the Chanturiya class V_ν. It is always illuminating to illustrate such imbeddings by means of the special choices

$$\phi(t) = t^p, \quad \omega(t) = t^\alpha, \quad \lambda_n = \frac{1}{n^q}, \quad \nu_n = n^r$$

which lead to the function classes WBV_p, Lip_α, $\Lambda_q BV$, and V_{ν^r}, respectively. In Table 2.7 below, we summarize our previous results for these special cases. In this table, we tacitly assume that $1 \le p < \infty$, $0 < \alpha \le 1$, $0 < q \le 1$, and $0 < r \le 1$. Note that the class WBV_p is increasing in p, the class Lip_α is decreasing in α, the class $\Lambda_q BV$ is increasing in q, and the class V_{ν^r} is increasing in r.

While the Waterman spaces ΛBV are extremely useful in the theory of Fourier series (see Section 7.2 below), the Schramm space ΦBV seems to be only of very limited interest. However, Proposition 2.43 shows that it provides a unified approach to many spaces we considered so far. A particularly simple example of a Schramm space is

26 Indeed, from the concavity of ω, it follows that $(\lambda_n)_n$ is decreasing, from $\omega(n) = o(n)$ as $n \to \infty$ it follows that $(\lambda_n)_n$ tends to zero, and from $\lambda[1,n] = \omega(n) \to \infty$ as $n \to \infty$ it follows that (2.24) is true.
27 Observe that our hypothesis $\omega(t) = o(t)$ as $t \to \infty$ excludes the case $\alpha = 1$.

Table 2.7. Imbeddings between function classes.

The function class	*is imbedded into*	*for*
$Lip_\alpha([a,b])$	$WBV_p([a,b])$	$p\alpha \le 1$
$\Lambda_q([a,b])$	$WBV_p([a,b])$	$q \le 1-1/p$
$WBV_p([a,b])$	$\Lambda_q BV([a,b])$	$q > 1-1/p$
$WBV_p([a,b])$	$\mathrm{V}_{\nu^r}([a,b])$	$r \ge 1-1/p$
$\mathrm{V}_{\nu^r}([a,b])$	$\Lambda_q^c BV([a,b])$	$q > r$
$\Lambda_q BV([a,b])$	$\mathrm{V}_{\nu^r}([a,b])$	$q \le r$

when Φ is a constant sequence, i.e. $\phi_n(t) \equiv \phi(t)$ for some Young function ϕ. In this case the variation (2.70) has the form

$$\mathrm{Var}_\Phi(f, S_\infty) = \mathrm{Var}_\Phi(f, S_\infty; [a,b]) = \sum_{k=1}^{\infty} \phi(|f(b_k) - f(a_k)|),$$

and the corresponding space $\Phi BV([a,b])$ is sometimes denoted by $\phi BV([a,b])$. Observe, however, that in the definition of this space, we may restrict ourselves to *finite* collections $S \in \Sigma([a,b])$. This means that we may define, in analogy to (2.73), the variation by

$$\begin{aligned}\mathrm{Var}_\Phi(f; [a,b]) &= \sup\left\{\mathrm{Var}_\Phi(f, S; [a,b]) : S \in \Sigma([a,b])\right\} \\ &= \sup\left\{\sum_{k=1}^{n} \phi(|f(b_k) - f(a_k)|) : \{[a_1,b_1],\dots,[a_n,b_n]\} \in \Sigma([a,b])\right\}.\end{aligned}$$

The proof of formula (2.54) can be found in [80, Theorem 1], the proof of formula (2.55) in [79, Theorem 4]. The related Proposition 2.36 has been proved in [24]. Our Propositions 2.38 and 2.39, as well as Examples 2.40 and 2.41, are taken from [25]. Interestingly, the discontinuity behavior of a function $f : [a,b] \to \mathbb{R}$ can be described by means of the characteristic (2.52). In fact, in [79] (see also [80, 81]), it is shown that in order for f to only have removable discontinuities or discontinuities of the first kind (jumps), it is necessary and sufficient that

$$\nu(f)_n = o(n) \quad (n \to \infty);$$

compare this with (2.53). Since

$$\lim_{n\to\infty} \phi^{-1}\left(\frac{1}{n}\right) = 0$$

and

$$\lim_{n\to\infty} \sum_{k=1}^{n} \lambda_k = \infty,$$

Table 2.2 shows that this is true for functions from WBV_ϕ and ΛBV, and thus also for all special cases like BV, WBV_p, or $\Lambda_q BV$.

The following definition [287] gives a unified approach to the Wiener space WBV_ϕ introduced in Definition 2.2 and the Waterman space ΛBV introduced in Definition 2.15. Later, this class of functions was studied again by Kim [158] who seems to have been unaware of the paper [287].

Definition 2.84. Let $\phi : [0,\infty) \to [0,\infty)$ be a Young function and $\Lambda = (\lambda_n)_n$ be a Waterman sequence. Given a function $f : [a,b] \to \mathbb{R}$ and a collection $S_\infty = \{[a_n, b_n] : n \in \mathbb{N}\} \in \Sigma_\infty([a,b])$, the positive real number

$$\mathrm{Var}_{\phi,\Lambda}(f, S_\infty) = \mathrm{Var}_{\phi,\Lambda}(f, S_\infty; [a,b]) := \sum_{k=1}^{\infty} \lambda_k \phi(|f(b_k) - f(a_k)|) \tag{2.158}$$

is called the (ϕ, Λ)*-variation* of f on $[a,b]$ with respect to S_∞, while the (possibly infinite) number

$$\begin{aligned}\mathrm{Var}_{\phi,\Lambda}(f) &= \mathrm{Var}_{\phi,\Lambda}(f; [a,b]) \\ &:= \sup\left\{\mathrm{Var}_{\phi,\Lambda}(f, S_\infty; [a,b]) : S_\infty \in \Sigma_\infty([a,b])\right\},\end{aligned}$$

where the supremum is taken over all collections $S_\infty \in \Sigma_\infty([a,b])$, is called the *total* (ϕ, Λ)*-variation* of f on $[a,b]$. In case $\mathrm{Var}_{\phi,\Lambda}(cf; [a,b]) < \infty$, for some $c > 0$, we say that f has *bounded* (ϕ, Λ)*-variation* (or *Schramm–Waterman variation*) on $[a,b]$ and write $f \in \Lambda BV_\phi([a,b])$. ∎

Clearly, Definition 2.84 contains many of the spaces considered before. Thus, choosing $\phi(t) = t$ in (2.158), we get

$$\mathrm{Var}_{\phi,\Lambda}(f; [a,b]) = \mathrm{Var}_\Lambda(f; [a,b]),$$

and hence $\Lambda BV_\phi = \Lambda BV$. By choosing $\lambda_k \equiv 1$ in (2.158), we get

$$\mathrm{Var}_{\phi,\Lambda}(f; [a,b]) = \mathrm{Var}_\phi^W(f; [a,b]),$$

and hence $\Lambda BV_\phi = WBV_\phi$. Moreover, combining the proof of Proposition 2.36 (a) and (b), one may prove the following connection of the space $\Lambda BV_\phi([a,b])$ with the Chanturiya class $\mathrm{V}_\nu([a,b])$, see [25].

Proposition 2.85. *For any Young function ϕ and any Waterman sequence Λ, the inclusion*

$$\Lambda BV_\phi([a,b]) \subseteq \mathrm{V}_\nu([a,b]) \tag{2.159}$$

holds true for

$$\nu_n := n\phi^{-1}\left(\frac{1}{\lambda[1,n]}\right),$$

where $\lambda[1,n] = \lambda_1 + \ldots + \lambda_n$.

Observe that Proposition 2.85 contains all inclusions occurring in Table 2.2 as special cases. In fact, for arbitrary ϕ and $\lambda_n \equiv 1$, we get

$$\nu_n = n\phi^{-1}\left(\frac{1}{n}\right),$$

in accordance with the third row in Table 2.2, while for arbitrary Λ and $\phi(t) = t$, we get

$$\nu_n = \frac{n}{\lambda[1,n]},$$

in accordance with the fourth row in Table 2.2.

Spaces of functions of bounded variation have also been considered with weight. A quite general definition is given in [27], where the authors define, for a given strictly increasing weight function $w : [a,b] \to \mathbb{R}^+$ and some Young function $\phi : [0,\infty) \to [0,\infty)$, the weighted ϕ-variation of $f : [a,b] \to \mathbb{R}$ in Riesz's sense by

$$\mathrm{Var}^R_{\phi,w}(f;[a,b]) := \sup\left\{\sum_{j=1}^{m} \phi\left(\frac{|f(t_j) - f(t_{j-1})|}{w(t_j) - w(t_{j-1})}\right)\left(w(t_j) - w(t_{j-1})\right)\right\}, \tag{2.160}$$

where the supremum is taken over all partitions $\{t_0, t_1, \dots, t_m\} \in \mathcal{P}([a,b])$. The corresponding space of functions of bounded weighted ϕ-variation is denoted by $RBV_{\phi,w}([a,b])$. Still, more generally, Chistyakov [87] replaces (2.160) by

$$\mathrm{Var}_{\Phi,w}(f;[a,b]) := \sup\left\{\sum_{j=1}^{m} \phi_j\left(\frac{|f(t_j) - f(t_{j-1})|}{w(t_j) - w(t_{j-1})}\right)\left(w(t_j) - w(t_{j-1})\right)\right\},$$

with $\Phi = (\phi_n)_n$ being a Schramm sequence, see Definition 2.42. Equipped with the norm

$$\|f\|_{\Phi BV_w} := |f(a)| + \inf\{\lambda > 0 : \mathrm{Var}_{\Phi,w}(f/\lambda;[a,b])) \le 1\},$$

the corresponding set $\Phi BV_w([a,b])$ of functions of bounded weighted Schramm variation then becomes a Banach space.

Given a Schramm sequence $\Phi = (\phi_n)_n$, some kind of second Schramm variation for $f \in B([a,b])$ is defined by

$$\mathrm{Var}_{2,\Phi}(f) = \mathrm{Var}_{2,\Phi}(f;[a,b]) := \sup\left\{\sum_{k=1}^{\infty} \phi_k(|f[t_{k+1}, t_{k+2}] - f[t_k - t_{k+1}]|)\right\},$$

with the supremum being taken over all collections $S_\infty = \{[a_n, b_n] : n \in \mathbb{N}\} \in \Sigma_\infty([a,b])$, and studied in the recent paper [121]. Of course, this is inspired by the Wiener variation $\mathrm{Var}^W_{2,1}(f)$ introduced in Definition 2.71. The paper [121] also contains an integral representation for functions with finite second Schramm variation; this generalizes formula (2.150) from Theorem 2.76.

The Korenblum variation $\mathrm{Var}_\kappa(f;[a,b])$ and the Banach space $\kappa BV([a,b])$ have been introduced in 1975 by Korenblum [163] in connection with the Poisson integral representation of certain classes of harmonic functions on the complex unit disc. In our presentation in Section 2.5, we mainly followed the paper [102] which contains the main decomposition theorem and a Helly-type selection principle for κBV-functions. Further results in this direction may be found in [158] and [161]. Given a distortion function $\kappa : [0,1] \to [0,1]$ and a Schramm sequence $\Phi = (\phi_n)_n$, in the paper [160], the authors define the (κ, Φ)-variation of $f : [0,1] \to \mathbb{R}$ in the natural way by

$$\mathrm{Var}_{\kappa,\Phi}(f; [0,1]) := \sup \left\{ \frac{\sum_{k=1}^{\infty} \phi_k(|f(b_k) - f(a_k)|)}{\sum_{k=1}^{\infty} \kappa(b_k - a_k)} \right\},$$

where the supremum is taken over all collections $S_\infty = \{[a_n, b_n] : n \in \mathbb{N}\} \in \Sigma_\infty([0,1])$, thus providing a unified approach to the Schramm variation (2.71) and the Korenblum variation (2.106). The corresponding space $\kappa\Phi BV([a,b])$ of functions of bounded (κ, Φ)-variation has been systematically studied afterwards by Park and others in a series of papers [241–245, 296].

Distortion functions and Korenblum variation occur in entropy theory. In the paper [164], the author defines the *κ-entropy* of a partition $P = \{t_0, t_1, \dots, t_m\} \in \mathcal{P}([a,b])$ by

$$\kappa(P) := \sum_{j=1}^{m} \kappa\left(\frac{t_j - t_{j-1}}{b-a}\right).$$

The κ-entropy of an arbitrary nonempty closed set $F \subseteq [a,b]$ is then defined by

$$\kappa(F) = \kappa(F; [a,b]) := \sup\{\kappa(P) : P \in \mathcal{P}(F)\},$$

where the supremum is taken over all partitions in the set F. This entropy has a number of natural properties, see Exercises 2.38 and 2.39.

For particular choices of κ, such entropies are well known. Thus, for $\kappa(t) = t^\alpha$, we get the so-called *Hölder entropy*, for $\kappa(t) = t(1 - \log t)$, the *Shannon entropy*, and for $\kappa(t) = 2/(2 - \log t)$, the *Dini entropy*.

So-called *modular functionals* and *modular spaces* of functions of bounded variation have been studied by Musielak and Orlicz [235, 236], and later by Leśniewicz and Orlicz [180] and Herda [145].

Theorem 2.74 is due to De la Vallée Poussin [103], see also [272] for a generalization to so-called u-convex functions.[28] We point out that Theorem 2.74 carries over to the higher variation $\mathrm{Var}^W_{k,1}(f; [a,b])$, see Exercise 2.45 or [273, 274]. Theorem 2.76 was proved in [276] for functions f having a bounded second derivative f'' on $[a,b]$, and extended to functions $f \in AC^1([a,b])$ in [279]. Of course, the equality (2.150) is a perfect analogue to the classical formula

$$\mathrm{Var}(f; [a,b]) = \int_a^b |f'(t)|\,dt$$

which holds for functions $f \in AC([a,b])$; we will come back to this and similar formulas in the next chapter.

28 A self-contained account of convex functions and their properties can be found in the monographs [253, 267].

Exercise 2.46 shows that functions from $WBV_{k,1}^W([a,b])$ have better smoothness properties if k increases. Therefore, the following result [274] is not too surprising.

Proposition 2.86. *The equalities*

$$\bigcap_{k=1}^{\infty} WBV_{k,1}([a,b]) = C^{\infty}([a,b]), \quad \bigcup_{k=1}^{\infty} WBV_{k,1}([a,b]) = BV([a,b]) \tag{2.161}$$

are true.

It would be interesting to have a similar explicit description of the classes

$$A_p := \bigcap_{k=1}^{\infty} WBV_{k,p}([a,b]), \quad B_p := \bigcup_{k=1}^{\infty} WBV_{k,p}([a,b]),$$

see also Problem 2.7 below.

In this chapter, we have obtained several results about unions and intersections of various spaces of functions of bounded variation, such as Waterman spaces, Schramm spaces, and related spaces. We collect these results in the following Table 2.8.

Table 2.8. Unions and intersections of function classes over $I = [a,b]$.

		$\bigcap_{k=1}^{\infty} WBV_{k,1}(I)$	$=$	$C^{\infty}(I)$		
				$\cap$		
$\bigcup_{k=1}^{\infty} WBV_{k,1}(I)$	$=$	$\bigcap_{\Lambda} \Lambda BV(I)$	$=$	$BV(I)$	$=$	$\bigcap_{\Phi} \Phi BV(I)$
				$\cap$		
		$\bigcup_{p>1} WBV_p(I)$	$\subset$	$HBV(I)$		
				$\cap$		
$\bigcup_{\phi} WBV_{\phi}(I)$	$=$	$\bigcup_{\Lambda} \Lambda BV(I)$	$=$	$R(I)$	$=$	$\bigcup_{\Phi} \Phi BV(I)$
				$\cup$		
		$\bigcap_{\phi} WBV_{\phi}(I)$	$=$	$FR(I)$		

Given a partition $P = \{t_0, t_1, \ldots, t_m\} \in \mathcal{P}([a,b])$, let us recall the Popoviciu variation

$$\begin{aligned}
\mathrm{Var}_{k,1}^W(f, P; [a,b]) &= \sum_{j=1}^{m-k+1} |f[t_j, \ldots, t_{j+k-1}] - f[t_{j-1}, \ldots, t_{j+k-2}]| \\
&= |f[t_1, \ldots, t_k] - f[t_0, \ldots, t_{k-1}]| + |f[t_2, \ldots, t_{k+1}] - f[t_1, \ldots, t_k]| \\
&\quad + \ldots + |f[t_{m-k+1}, \ldots, t_m] - f[t_{m-k}, \ldots, t_{m-1}]|
\end{aligned}$$

defined in (2.142). A similar second variation was defined in [226], where the authors start, for fixed $k \in \mathbb{N}$, with a “block partition” of the form

$$Q := \{\tau_1^1, \ldots, \tau_1^{2k}, \tau_2^1, \ldots, \tau_2^{2k}, \ldots, \tau_n^1, \ldots, \tau_n^{2k}\} \tag{2.162}$$

and associate to this $Q \in \mathcal{P}([a,b])$ the second variation

$$\begin{aligned}\mathrm{Var}^{MR}_{k,1}(f,Q;[a,b]) &= \sum_{i=1}^{n} |f[\tau_i^{k+1},\dots,\tau_i^{2k}] - f[\tau_i^1,\dots,\tau_i^k]| \\ &= |f[\tau_1^{k+1},\dots,\tau_1^{2k}] - f[\tau_1^1,\dots,\tau_1^k]| + |f[\tau_2^{k+1},\dots,\tau_2^{2k}] - f[\tau_2^1,\dots,\tau_2^k]| \\ &\quad + \dots + |f[\tau_n^{k+1},\dots,\tau_n^{2k}] - f[\tau_n^1,\dots,\tau_n^k]| . \end{aligned} \tag{2.163}$$

We call the expression

$$\mathrm{Var}^{MR}_{k,1}(f;[a,b]) = \sup\{\mathrm{Var}^{MR}_{k,1}(f,Q;[a,b]) : Q \in \mathcal{P}([a,b])\}, \tag{2.164}$$

where the supremum in (2.164) is taken over all block partitions of the form (2.162), the second variation in the sense of Merentes and Rivas, and denote the corresponding space of all functions f satisfying $\mathrm{Var}^{MR}_{k,1}(f;[a,b]) < \infty$ by $MRBV_{k,1}([a,b])$. The following proposition shows that in this way, one does not obtain a new space, though they do gain something we already know:

Proposition 2.87. *For* $\mathrm{Var}^{W}_{k,1}(f;[a,b])$ *as in* (2.142) *and* $\mathrm{Var}^{MR}_{k,1}(f;[a,b])$ *as in* (2.164), *the estimates*

$$\mathrm{Var}^{MR}_{k,1}(f;[a,b]) \le \mathrm{Var}^{W}_{k,1}(f;[a,b]) \le 2(k+1)\,\mathrm{Var}^{MR}_{k,1}(f;[a,b]) \tag{2.165}$$

are true. Consequently, the spaces $WBV_{k,1}([a,b])$ *and* $MRBV_{k,1}([a,b])$ *coincide.*

A sketch of the proof of Proposition 2.87 can be found in [226].

In the following Table 2.9, we compare the definitions of higher order variations in Wiener's sense introduced in Section 2.7, together with corresponding representation theorems.

Table 2.9. Higher order variations and Wiener spaces $WBV_{k,p}$.

$WBV_{k,p}([a,b])$	$p = 1$	$1 < p < \infty$
$k = 1$ *characterization of* $f \in WBV_{1,p}$	Jordan 1881 $f = f_1 - f_2$ f_1, f_2 increasing	Wiener 1924 –
$k = 2$ *characterization of* $f \in WBV_{2,p}$	De la Vallée Poussin 1915 $f = f_1 - f_2$ f_1, f_2 convex	Merentes/Rivas 1996 –
$k \in \mathbb{N}$ *characterization of* $f \in WBV_{k,p}$	Popoviciu 1934 $f = f_1 - f_2$ $f_1^{(k-1)}, f_2^{(k-1)}$ increasing	Merentes/Rivas 1996 –

We remark that an elementary construction for the variation $\mathrm{Var}^{W}_{k,1}(f)$ of a *continuous* function f is given in [62], together with an analogue of Jordan's classical decomposition, which reads as follows. For $k = 1, 2, 3, \dots$, denote by A_k the class of all functions $f \in C([a,b])$ whose first k divided differences (2.139) are nonnegative. Then the difference algebra $A_k - A_k$ consists precisely of all continuous functions $f : [a,b] \to \mathbb{R}$ for which $\mathrm{Var}^W_{k,1}(f;[a,b]) < \infty$.

The following generalizes Definition 2.71 in the same way as Definition 2.2 generalizes Definition 1.31.

Definition 2.88. Given a Young function $\phi : [0,\infty) \to [0,\infty)$, a function $f : [a,b] \to \mathbb{R}$, a partition $P = \{t_0, t_1, \ldots, t_m\} \in \mathcal{P}([a,b])$, and $k \in \mathbb{N}$, we set

$$\mathrm{Var}^W_{k,\phi}(f, P; [a,b]) := \sum_{j=1}^{m-k+1} \phi\left(|f[t_j, \ldots, t_{j+k-1}] - f[t_{j-1}, \ldots, t_{j+k-2}]|\right) \tag{2.166}$$

and call (2.166) the *(k,ϕ)-variation* of f with respect to P on $[a,b]$. Moreover, we call the (possibly infinite) number

$$\mathrm{Var}^W_{k,\phi}(f;[a,b]) := \sup\left\{\mathrm{Var}^W_{k,\phi}(f, P; [a,b]) : P \in \mathcal{P}([a,b])\right\},$$

where the supremum is taken over all partitions of $[a,b]$, the *total (k,ϕ)-variation* of f on $[a,b]$. The elements of the set

$$\mathrm{V}^W_{k,\phi}([a,b]) := \{f \in B([a,b]) : \mathrm{Var}^W_{k,\phi}(f;[a,b]) < \infty\}$$

are called *functions of bounded (k,ϕ)-variation* (in the Wiener–Young sense) on $[a,b]$. ■

It is clear how we may define a norm on $\mathrm{V}^W_{k,\phi}([a,b])$ similarly as for $\mathrm{V}^W_{\phi}([a,b])$ in (2.10) and (2.11). Putting

$$B^W_k(\phi) := \{f \in B([a,b]) : \mathrm{Var}^W_{k,\phi}(f;[a,b]) \leq 1\},$$

the corresponding Minkowski functional

$$\|f\|_{WBV_{k,\phi}} = |f(a)| + \inf\left\{\lambda > 0 : f/\lambda \in B^W_k(\phi)\right\} \tag{2.167}$$

is then a norm on the linear space $WBV_{k,\phi}([a,b]) = \mathrm{span}\ \mathrm{V}^W_{k,\phi}([a,b])$. Clearly, Definition 2.88 contains the Definitions 2.71 and 2.72 for the special case $\phi(u) = |u|^p$. While the space $WBV_{k,p}([a,b])$ has been studied for $k = 2$ in [220] and for general $k \in \mathbb{N}$ in [229], the space $WBV_{k,\phi}([a,b])$ has been studied for $k = 2$ in [218] and for general $k \in \mathbb{N}$ in [228]. More general results, also for generalized variations and functions of two variables, can be found in [34, 58–60, 109, 120]. In this connection, the following problems seem to be open:

Problem 2.4. *Equipped with the norm (2.167), is $(WBV_{k,\phi}([a,b]), \|\cdot\|_{WBV_{k,\phi}})$ a Banach space?*

Problem 2.5. *Find conditions on two Young functions ϕ and ψ which imply that*

$$\mathrm{V}^W_{k,\phi}([a,b]) \subseteq \mathrm{V}^W_{k,\psi}([a,b])$$

or

$$WBV_{k,\phi}([a,b]) \subseteq WBV_{k,\psi}([a,b]).$$

Applying Proposition 0.31 *and the continuous imbedding* $WBV_\phi \hookrightarrow B$, *we see that* $WBV_\phi([a,b])$ *equipped with the norm*

$$|||f|||_{WBV_\phi} := \|f\|_\infty + \|f\|_{WBV_\phi},$$

where $\|\cdot\|_\infty$ *denotes the norm* (0.39), *is a Banach algebra. In [192], it is shown that the same is true for the norm*

$$|||f|||_{WBV_\phi} := 2\phi^{-1}(1)\|f\|_{WBV_\phi}.$$

In fact, from $\phi(|f(x)|) \leq \mathrm{Var}_\phi^W(f;[a,b])$ *for* $a \leq x \leq b$, *it follows that* $\|f\|_\infty \leq \phi^{-1}(1)\|f\|_{WBV_\phi}$ *for* $f \in WBV_\phi([a,b])$. *For higher order variations, we have the following*

Problem 2.6. *Does there exist a norm* $|||\cdot|||_{WBV_{k,\phi}}$ *which is equivalent to the norm* (2.167) *and makes* $(WBV_{k,\phi}([a,b]), |||\cdot|||_{WBV_{k,\phi}})$ *a Banach algebra?*

Problem 2.7. *From* (2.161), *it follows that*

$$A := \bigcap_{k=1}^{\infty}\bigcap_{\phi} WBV_{k,\phi}([a,b]) \subseteq C^\infty([a,b]) \tag{2.168}$$

and

$$B := \bigcup_{k=1}^{\infty}\bigcup_{\phi} WBV_{k,\phi}([a,b]) \supseteq BV([a,b]), \tag{2.169}$$

where the intersection in (2.168) *and the union in* (2.169) *are taken over all Young functions* ϕ. *Can you give an explicit description of the sets A and B?*

Finally, concerning intersections and unions of Korenblum spaces, the inclusions (2.108) suggest the following

Problem 2.8. *Similarly as in* (2.33) *and* (2.80), *is it true that*

$$\bigcap_{\kappa} \kappa BV([a,b]) = BV([a,b]), \quad \bigcup_{\kappa} \kappa BV([a,b]) = R([a,b]),$$

where the intersection and the union are taken over all distortion functions κ?

Generalized variations of functions of several variables have been only considered in a few papers, e.g. [97, 111]. As a sample result, we mention that in [111], it is shown that the class $\Phi BV([a,b]\times[c,d])$ of all functions of two variables which have bounded Schramm variation on a rectangle $[a,b]\times[c,d] \subset \mathbb{R}^2$ (defined in the obvious way) has the structure of a Banach algebra.

2.8 Exercises to Chapter 2

We state some exercises on the topics covered in this chapter; exercises marked with an asterisk * are more difficult.

Exercise 2.1. Show that

$$\lim_{\lambda\to\infty} \mathrm{Var}_\phi^W(f/\lambda;[a,b]) = 0$$

for each $f \in WBV_\phi([a,b])$, where $\mathrm{Var}_\phi^W(f;[a,b])$ denotes the variation (2.2).

Exercise 2.2. Given $f \in WBV_\phi([a,b])$, show that the set

$$E_\phi^W(f) := \{\lambda > 0 : \mathrm{Var}_\phi^W(f/\lambda;[a,b]) \le 1\}$$

is always a nonempty interval.[29] More precisely, prove that

$$E_\phi^W(f) = \begin{cases} [0,\infty] & \text{if } f(x) \equiv 0\,,\\ (\|f\|_{WBV_\phi},\infty) & \text{if } f(x) \not\equiv 0\,. \end{cases}$$

Exercise 2.3. Using Exercise 0.75, prove the following analogue to Theorem 1.41: a function f belongs to $WBV_\phi([a,b])$ if and only if it may be represented as composition $f = g \circ \tau$, where $\tau : [a,b] \to [c,d]$ is increasing and $g \in Lip_{\phi^{-1},\infty}([c,d])$ with $lip_{\phi^{-1},\infty}([c,d]) \le 1$, see (0.101).

Exercise 2.4. Show that $\Lambda BV([a,b]) = BV([a,b])$ if and only if the sequence $\Lambda = (\lambda_n)_n$ is bounded away from zero, i.e. $\lambda_n \ge \delta$ for all $n \in \mathbb{N}$ and some $\delta > 0$.

Exercise 2.5. Let $Z_{C,D}$ be the zigzag function (0.91) determined by

$$c_n := \frac{1}{2^n}\,,\quad d_n := \frac{1}{n^{1-q}\log(n+1)}\quad (n = 1,2,3,\dots)\,.$$

Use this function to show that the inclusion in the imbedding (2.48) is strict.

Exercise 2.6. Let Z_θ be the special zigzag function (0.93) determined by $\theta := 1/p$. Show that this function satisfies (2.47).

Exercise 2.7. By means of Proposition 2.32, Proposition 2.33, Exercise 2.5 and Exercise 2.6, show that

$$\bigcup_{p>1} WBV_p([a,b]) \subset \Lambda_1 BV([a,b]) = HBV([a,b])$$

as well as

$$\bigcup_{1<p<p_0} WBV_p([a,b]) \subset \Lambda_{q_0} BV([a,b]) = WBV_{p_0}([a,b]) \subset \bigcap_{q_0<q\le 1} \Lambda_q BV([a,b])$$

for $p_0 > 1$ and $q_0 = 1 - 1/p_0 - 1$.

29 Observe that (2.10) shows that $1 \in E_\phi^W(f)$ if and only if $f \in B^W(\phi)$.

Exercise 2.8. Use Proposition 2.31 to show that $Z_{C,D} \in \Lambda_q BV([0,1])$ if

$$\sum_{n=1}^{\infty} \frac{d_n}{n^q} < \infty,$$

and compare this with (2.45).

Exercise 2.9. Generalizing Proposition 2.33, find a condition on a Waterman sequence $\Lambda = (\lambda_n)_n$ and a Young function ϕ such that $\Lambda BV([a,b]) \subseteq WBV_\phi([a,b])$. Compare with Proposition 2.34.

Exercise 2.10*. Prove (2.54).

Exercise 2.11*. Prove (2.55).

Exercise 2.12*. Let $f \in C([a,b])$, and let $\omega_\infty(f;\delta)$ denote its modulus of continuity (0.97). Suppose that $\Lambda = (\lambda_n)_n$ is a Waterman sequence satisfying

$$\sum_{k=1}^{\infty} (\lambda_k - \lambda_{k+1}) k\omega(f; 1/k) < \infty.$$

Use Proposition 2.38 and (2.54) to show that $f \in \Lambda^c BV([a,b])$.

Exercise 2.13*. Let $f \in Lip_\alpha([a,b])$ for $0 < \alpha \leq 1$, and let $1 - \alpha < q \leq 1$. Deduce from Exercise 2.12 that $f \in \Lambda_q^c BV([a,b])$, where $\Lambda_q BV$ is given in Definition 2.29.

Exercise 2.14*. In analogy to (1.13), define the *Waterman variation function* of $f \in \Lambda BV([a,b])$ by

$$\mathrm{V}_{f,\Lambda}(x) := \mathrm{Var}_\Lambda(f;[a,x]) \quad (a \leq x \leq b).$$

Show that

$$\mathrm{V}_{f,\Lambda}(a+) = \lim_{x \to a+} \mathrm{V}_{f,\Lambda}(x) = 0$$

if f is right-continuous at a, and

$$\mathrm{V}_{f,\Lambda}(b-) = \lim_{x \to b-} \mathrm{V}_{f,\Lambda}(x) = \mathrm{Var}_\Lambda(f;[a,b])$$

if f is left-continuous at b.

Exercise 2.15. Conclude from Exercise 2.14 that $\mathrm{Var}_\Lambda(f;[x,y]) \to 0$ as x and y together approach either a or b.

Exercise 2.16. Given $f \in \Lambda BV([a,b])$, show that the Waterman variation function $\mathrm{V}_{f,\Lambda}$ defined in Exercise 2.14 is continuous at $x_0 \in (a,b)$ if and only if f is continuous at x_0.

Exercise 2.17. Let $\Lambda = (\lambda_n)_n$ be a Waterman sequence and $f \in \Lambda BV([a,b])$. By considering the special collection $S = \{[a,x],[x,b]\} \in \Sigma([a,b])$, show that $\lambda_1|f(x) - f(a)| \leq \mathrm{Var}_\Lambda(f;[a,b])$. Deduce that $\Lambda BV([a,b]) \hookrightarrow B([a,b])$ with imbedding constant $c = \max\{1/\lambda_1, 1\}$. Is this the sharp imbedding constant?

Exercise 2.18. What does Proposition 2.31 mean for $\lambda_n \equiv 1$, i.e. $cn \le \mu_n \le Cn$ with $c, C > 0$ independent of n?

Exercise 2.19. Let V_μ and V_ν denote the Chanturiya classes defined by two sequences $\mu = (\mu_n)_n$ and $\nu = (\nu_n)_n$, see Definition 2.35. Show that $\mathrm{V}_\mu = \mathrm{V}_\nu$ if and only if

$$c\mu_n \le \nu_n \le C\mu_n \quad (n = 1, 2, \ldots)$$

for two constants $c, C > 0$ independent of n.

Exercise 2.20. Show that the modulus of variation (2.52) satisfies the property

$$\nu(f)_n \le \nu(f)_m + \nu(f)_{n-m} \quad (m = 1, 2, \ldots, n).$$

Exercise 2.21*. Given a regular function $f : [a, b] \to \mathbb{R}$, prove that the condition

$$\int_{-\infty}^{\infty} \log(1 + N_f(y))\, dy < \infty$$

implies that $f \in H^c BV([a, b])$. Compare with Proposition 2.31.

Exercise 2.22*. Prove the following converse to Proposition 2.47. Suppose that the variation function $\mathrm{V}_{f,\Phi}(x) = \mathrm{Var}_\Phi(f; [a, x])$ of $f \in \Phi BV([a, b])$ is continuous at some point $x_0 \in (a, b)$. Show that f is then also continuous at x_0.

Exercise 2.23. Prove the following "one-sided version" of Proposition 2.47: if $f \in \Phi BV([a, b])$ is right continuous at a [respectively left continuous at b], then $\mathrm{Var}_\Phi(f; [a, x]) \to 0$ as $x \to a+$ [respectively $\mathrm{Var}_\Phi(f; [x, b]) \to 0$ as $x \to b-$].

Exercise 2.24. Prove Proposition 2.45.

Exercise 2.25. Construct two Schramm sequences $\Phi = (\phi_n)_n$ and $\Psi = (\psi_n)_n$ such that $\Phi BV([0, 1]) \subseteq \Psi BV([0, 1])$, but (2.79) does not hold.

Exercise 2.26. For $1 < p < q < \infty$, find a function $f \in RBV_p([a, b]) \setminus RBV_q([a, b])$ using Table 2.4.

Exercise 2.27. Show that the oscillation function (0.86) belongs to $RBV_p([0, 1])$ for fixed $p \in [1, \infty)$ if and only if either $\beta > 0$ and $p\alpha + \beta \ge p - 1$, or $\beta \le 0$ and $p\alpha + \beta > p - 1$.

Exercise 2.28. Use Table 2.4 to construct a function $f \in RBV_p([0, 1]) \setminus Lip([0, 1])$.

Exercise 2.29. Is the space $RBV_p([a, b])$ with norm (2.90) separable? Compare your answer with Exercises 0.6 and 1.49.

Exercise 2.30. Define $RBV_p^1([a, b])$ from $RBV_p([a, b])$ as in Definition 0.32, equipped with the norm (0.44), i.e.

$$\|f\|_{RBV_p^1} = |f(a)| + \|f'\|_{RBV_p} = |f(a)| + |f'(a)| + \mathrm{Var}_p^R(f'; [a, b])^{1/p}.$$

Show that $(RBV_p^1([a, b]), \|\cdot\|_{RBV_p^1})$ is a Banach space.

Exercise 2.31. With the notation of Exercise 2.30, show that $RBV_p^1([a,b]) \hookrightarrow Lip([a,b])$ for $p > 1$, and calculate the sharp imbedding constant $c(RBV_p^1, Lip)$.

Exercise 2.32. For fixed $\alpha \in (0,1)$, let $f : [0,1] \to \mathbb{R}$ be defined by

$$f(x) := \begin{cases} 2x & \text{for } 0 \le x \le \frac{1}{2}, \\ (2b_\alpha - 2)x + 2 - b_\alpha & \text{for } \frac{1}{2} < x \le 1, \end{cases}$$

where $b_\alpha := 2/(1 + 2^{1-\alpha})$. Since $2^{1-\alpha} > 1$, the function f is increasing on $[0,1/2]$ with $f(0) = 0$ and $f(1/2) = 1$, and decreasing on $[1/2,1]$ with $f(1) = b_\alpha < 1$.

(a) Show that $f \in \kappa BV([0,1])$, where $\kappa(t) = t^\alpha$.
(b) Calculate $\mathrm{Var}_\kappa(f;[0,1])$ by only considering the special partitions $P_1 := \{0,1\}$ and $P_2 := \{0,1/2,1\}$ of $[0,1]$.
(c) Prove that the variation function $\mathrm{V}_{f,\kappa}$ defined in (2.115) coincides with f and is therefore not monotonically increasing.

Exercise 2.33. Prove Proposition 2.64 without the additional assumption that $f(0) = f(1)$.

Exercise 2.34*. Given a distortion function $\kappa : [0,1] \to [0,1]$, construct a function $f \in R([0,1]) \setminus \kappa BV([0,1])$. Can you also construct a function

$$f \in R([0,1]) \setminus \left(\bigcup_\kappa \kappa BV([0,1]) \right),$$

where the union is taken over all distortion functions κ?

Exercise 2.35. Given an arbitrary distortion function $\kappa : [0,1] \to [0,1]$, imitate the construction in Example 2.66 to find a function $f \in \kappa BV([0,1]) \setminus BV([0,1])$.

Exercise 2.36. Combine Theorem 2.67 with Theorem 2.68 to derive and prove a Helly-type selection theorem for sequences of functions of bounded Korenblum variation.

Exercise 2.37. Given an $(n+2)$-tuple $(a_0, a_1, \ldots, a_n, a_{n+1}) \in \mathbb{R}^{n+2}$ with $a_0 = a_{n+1} = 0$, show that

$$\sum_{i=1}^{n} (a_i - a_{i-1})^+ + a_n^- = \frac{1}{2} \sum_{i=1}^{n+1} |a_i - a_{i-1}|,$$

where a^+ and a^- are defined by (2.113). Use this to carry out the details of the first step in the proof of Theorem 2.68.

Exercise 2.38. Show that the κ-entropy $\kappa(P)$ defined in Section 2.8 has the following property: if $P = \{t_0, t_1, \ldots, t_m\} \in \mathcal{P}([a,b])$, then $\kappa(P) \le m/\kappa(1/m)$. Moreover, this estimate is sharp and attained for equidistant partitions P.

Exercise 2.39. Show that the κ-entropy $\kappa(F)$ defined in Section 2.8 satisfies $\kappa(F) \le \kappa(G)$ for $F \subseteq G$ as well as $\kappa(F \cup G) \le \kappa(F) + \kappa(G)$ for arbitrary closed sets F and G.

Exercise 2.40. In analogy to Exercise 1.56, we say that a function $f \in \kappa BV([a,b])$ has *vanishing κ-variation* at $t_0 \in [a,b]$ if

$$\lim_{\delta\to 0} \mathrm{Var}_\kappa(F_{t_0,\delta};[a,b]) = 0\,,$$

where $F_{t_0,\delta}$ is defined as in Exercise 1.56. Show that, in contrast to BV-functions, a function $f \in \kappa BV$ which is continuous at t_0 does not necessarily have vanishing κ-variation at t_0.

Exercise 2.41. For points $t_0, t_1, \dots, t_k \in [a,b]$ and indices $j = 0, 1, \dots, k$, let

$$P_j(t_0,\dots,t_k) := \prod_{i\neq j}(t_i - t_j) = (t_0 - t_j)\cdots(t_{j-1}-t_j)(t_{j+1}-t_j)\cdots(t_k - t_j)\,.$$

Prove by induction that

$$f[t_0,t_1,\dots,t_k] = \sum_{j-0}^{k} \frac{f(t_j)}{P_j(t_0,\dots,t_k)}\,,$$

where $f[t_0,t_1,\dots,t_k]$ is given by (2.139).

Exercise 2.42. Let $f \in C^k([a,b])$ and $a < t_0 < b$. Show that

$$\lim_{h\to 0} f[\underbrace{t_0,\dots,t_0}_{k\text{ times}},t_0+h] = \frac{f^{(k)}(t_0)}{k!}\,,$$

where $f[t_0,t_1,\dots,t_k]$ is given by (2.139).

Exercise 2.43. Prove the following weaker form of Exercise 2.42. Let $f \in C^{k-1}([a,b])$ and $a < t_0 < b$. Suppose that the unilateral derivatives $f_-^{(k)}$ and $f_+^{(k)}$ exist everywhere in $[a,b]$, and the usual derivative $f^{(k)}$ exists in $[a,b]\setminus N$, where N is countable. Show that

$$\lim_{h\to 0-} f[\underbrace{t_0,\dots,t_0}_{k\text{ times}},t_0+h] = \frac{f_-^{(k)}(t_0)}{k!}$$

and

$$\lim_{h\to 0+} f[\underbrace{t_0,\dots,t_0}_{k\text{ times}},t_0+h] = \frac{f_+^{(k)}(t_0)}{k!}\,,$$

where $f[t_0,t_1,\dots,t_k]$ is given by (2.139).

Exercise 2.44*. Prove that every function $f \in WBV_{2,1}([a,b])$ is Lipschitz continuous.

Exercise 2.45*. Prove the following higher order analogue of Theorem 2.74: a function $f : [a,b] \to \mathbb{R}$ has bounded k-th variation $\mathrm{Var}^W_{k,1}(f;[a,b])$ if and only if its derivative $f^{(k-2)}$ may be represented as a difference of two convex functions.

Exercise 2.46. Prove that every function $f \in WBV_{k,1}([a,b])$ admits unilateral derivatives $f_+^{(k-1)}$ and $f_-^{(k-1)}$ on $[a,b]$. Moreover, show that there exists a nullset $N \subset [a,b]$ such that $f \in C^{k-1}([a,b]\setminus N)$.

Exercise 2.47. Show that the inclusions (2.144) and (2.145) are strict.

Exercise 2.48. Show by means of the function $f : [0,2] \to \mathbb{R}$ defined by $f(x) := \max\{0, x-1\}$ that the estimate (2.146) is sharp.

Exercise 2.49. Suppose that $f : [a,b] \to \mathbb{R}$ is k-times differentiable, and $f^{(k)}$ is Riemann integrable over $[a,b]$. Prove that

$$f[t_0, t_1, \dots, t_k] = \int_a^b \int_0^{t_1} \cdot \int_0^{t_{k-1}} f^{(k)}(\tau_k(t_k - t_{k-1}) + \dots + \tau_1(t_1 - t_0) + t_0)\, d\tau_k \dots d\tau_1 \,,$$

where $f[t_0, t_1, \dots, t_k]$ is given by (2.139).

Exercise 2.50. Show that Theorem 2.77 also holds true if we replace, in (2.151), the space $WBV_\phi([a,b])$ by the class $\mathrm{V}_\phi^W([a,b])$ introduced in Definition 2.2.

Exercise 2.51. Prove Proposition 2.79.

Exercise 2.52*. Prove the “only if” part in Proposition 2.80.

Exercise 2.53. Using Exercise 0.75, prove Proposition 2.81.

3 Absolutely continuous functions

Absolutely continuous functions are intimately related to functions of bounded variation in several respects. First, absolute continuity is equivalent to the combination of three properties, namely, continuity, bounded variation, and invariance of nullsets; this is the assertion of the Vitali–Banach–Zaretskij theorem which we will prove in Section 3.2. Second, the absolutely continuous functions on $[a, b]$ are precisely those functions $f \in BV([a, b])$ whose derivatives f' (which exist a.e. by Lebesgue's differentiation theorem) belong to $L_1([a, b])$; moreover, the fundamental theorem of calculus (in the Lebesgue integral version) holds in this case. This shows that the definition of absolute continuity is both important and natural. If one replaces the condition $f \in AC([a, b])$ with the stronger condition $f \in RBV_p([a, b])$ $(1 < p < \infty)$ in this statement, one gets precisely the condition $f' \in L_p([a, b])$; this is the statement of the famous Riesz theorem which we will discuss, together with some generalization due to Yu. T. Medvedev, in Section 3.5. Section 3.4 is concerned with some relations between bounded variation, absolute continuity, and functions with rectifiable graphs.

3.1 Continuity and absolute continuity

Although we have introduced the concept of absolute continuity in Definition 1.21, for the reader's ease, we repeat the definition here.

Definition 3.1. Given a compact interval $[a, b]$, by $\Sigma([a, b])$, we denote the family of all finite systems $S := \{[a_1, b_1], \dots, [a_n, b_n]\}$ $(n \in \mathbb{N}$ variable) of pairwise nonoverlapping subintervals $[a_k, b_k] \subseteq [a, b]$. For $S \in \Sigma([a, b])$ and $f : [a, b] \to \mathbb{R}$, we put

$$\Theta(S) := \sum_{k=1}^{n} |b_k - a_k|, \quad \Gamma(f; S) := \sum_{k=1}^{n} |f(b_k) - f(a_k)| . \tag{3.1}$$

We call the function f *absolutely continuous* on $[a, b]$ and write $f \in AC([a, b])$ if for each $\varepsilon > 0$, there exists some $\delta > 0$ such that $\Gamma(f; S) \le \varepsilon$ for any $S \in \Sigma([a, b])$ satisfying $\Theta(S) \le \delta$. ∎

Since the number n in (3.1) is arbitrary, we could replace the finite sums in Definition 3.1 by infinite series, see Exercise 3.7.

We point out that it is very important to only choose mutually nonoverlapping subintervals $[a_k, b_k] \subseteq [a, b]$. If we drop this assumption, we get an essentially smaller class of functions, see Exercise 3.8.

For further use, we collect some natural algebraic properties of the function class $AC([a, b])$.

Proposition 3.2. *Let $f, g \in AC([a, b])$ and $\mu \in \mathbb{R}$. Then $f + g$, μf and fg also belong to $AC([a, b])$. Moreover, in case $g(x) \neq 0$ for $a \le x \le b$, we also have $f/g \in AC([a, b])$.*

Proof. Let $\varepsilon > 0$, and choose $\delta > 0$ such that both $\Gamma(f; S) \le \varepsilon$ and $\Gamma(g; S) \le \varepsilon$ for any $S \in \Sigma([a, b])$ satisfying $\Theta(S) \le \delta$. Then

$$\Gamma(f + g; S) \le \Gamma(f; S) + \Gamma(g; S) \le 2\varepsilon,$$

which shows that $f + g \in AC([a, b])$. Proving that $f \in AC([a, b])$ implies $\mu f \in AC([a, b])$ goes similarly by using the fact that $\Gamma(\mu f; S) = |\mu|\Gamma(f; S)$ for all $\mu \in \mathbb{R}$ and $S \in \Sigma([a, b])$.

Since every absolutely continuous function is continuous, we know that $\|f\|_C \le c$ and $\|g\|_C \le c$ for some $c > 0$, where $\|\cdot\|_C$ denotes the norm (0.45). Given $\varepsilon > 0$, choose again $\delta > 0$ such that both $\Gamma(f; S) \le \varepsilon$ and $\Gamma(g; S) \le \varepsilon$ for any $S \in \Sigma([a, b])$ satisfying $\Theta(S) \le \delta$. For any subinterval $[a_k, b_k] \subseteq [a, b]$, we then have

$$\begin{aligned} |f(a_k)g(a_k) - f(b_k)g(b_k)| &\le |f(a_k)|\,|g(a_k) - g(b_k)| + |f(a_k) - f(b_k)|\,|g(b_k)| \\ &\le c\,|g(a_k) - g(b_k)| + c\,|f(a_k) - f(b_k)|, \end{aligned}$$

and hence

$$\Gamma(fg; S) \le c(\Gamma(f; S) + \Gamma(g; S)) \le 2c\varepsilon$$

for all $S \in \Sigma([a, b])$ satisfying $\Theta(S) \le \delta$. Thus, we have proved that $fg \in AC([a, b])$.

It remains to prove that $1/g \in AC([a, b])$ for all $g \in AC([a, b])$ satisfying $g(x) \neq 0$ on $[a, b]$. Given $\varepsilon > 0$, choose $\delta > 0$ such that $\Gamma(g; S) \le \varepsilon$ for any $S \in \Sigma([a, b])$ satisfying $\Theta(S) \le \delta$. Again, from the continuity of g, we deduce that $|g(x)| \ge c$ on $[a, b]$ for some $c > 0$. Thus, for any subinterval $[a_k, b_k] \in S$, we have

$$\left| \frac{1}{g(a_k)} - \frac{1}{g(b_k)} \right| = \frac{|g(a_k) - g(b_k)|}{|g(a_k)g(b_k)|} \le \frac{|g(a_k) - g(b_k)|}{c^2},$$

and hence

$$\Gamma(1/g; S) \le \frac{\Gamma(g; S)}{c^2} \le \frac{\varepsilon}{c^2}$$

which proves the assertion. □

It is again illuminating to compare the conditions $f \in AC([a, b])$ and $|f| \in AC([a, b])$ (Exercises 3.1–3.3), as we have done for the function space $BV([a, b])$ in Exercises 1.3–1.5 in Chapter 1.

The following elementary result on continuous functions is taught in every first-year calculus course: if $f : [a, b] \to [f(a), f(b)]$ is continuous and strictly increasing, then $f^{-1} : [f(a), f(b)] \to [a, b]$ exists and is also continuous and strictly increasing, and so f is a homeomorphism. A corresponding result may be stated for absolutely continuous functions, but it is more delicate, see Exercise 3.11.

Since every function $f \in AC([a, b])$ has bounded variation, by (1.46), we may consider its variation function V_f defined in (1.13). Theorem 1.26 (e) shows that the absolute continuity of f carries over to V_f; this immediately implies the following

Proposition 3.3. *Every function $f \in AC([a, b])$ may be represented as a difference of two absolutely continuous increasing functions.*

Since $AC([a,b])$ is a linear subspace of $BV([a,b])$, the question arises if this subspace is closed in $BV([a,b])$ in the norm (1.16), i.e. if $(AC([a,b]), \|\cdot\|_{BV})$ is a Banach space. This is in fact true, as we will show in the next section (Proposition 3.24).

The next proposition shows that, roughly speaking,“indefinite integrals” of L_1-functions are always absolutely continuous.

Proposition 3.4. *Let $g \in L_1([a,b])$ and*

$$f(x) := \int_a^x g(t)\,dt \quad (a \le x \le b)\,. \tag{3.2}$$

Then $f \in AC([a,b])$.

Proof. Let $\varepsilon > 0$. From the absolute continuity of the Lebesgue integral (Theorem 0.6), it follows that there exists a $\delta > 0$ such that

$$\int_M |g(x)|\,dx \le \varepsilon$$

for all subsets $M \subseteq [a,b]$ with $\lambda(M) \le \delta$. Consider $S = \{[a_1,b_1],\dots,[a_n,b_n]\} \in \Sigma([a,b])$ with $\Theta(S) \le \delta$, and put $M := [a_1,b_1] \cup \ldots \cup [a_n,b_n]$. Then

$$\Gamma(f;S) = \sum_{k=1}^n |f(b_k) - f(a_k)| = \sum_{k=1}^n \left| \int_{a_k}^{b_k} g(t)\,dt \right| \le \int_M |g(x)|\,dx \le \varepsilon$$

since $\lambda(M) \le \Theta(S) \le \delta$, by construction. □

Recall that the analogue to Proposition 3.4 for the Riemann integral reads as follows: *if $g : [a,b] \to \mathbb{R}$ is Riemann integrable and bounded, the function f defined in* (3.2) *is Lipschitz continuous.*

A certain converse of Proposition 3.4 will be proved in the following section (Theorem 3.18). Here, we use Proposition 3.4 to give an example of an absolutely continuous function which is not Hölder continuous for any $\alpha \in (0,1]$.

Example 3.5. Let $f : [0,1] \to \mathbb{R}$ be the function from Example 1.25 (or 0.41). As we have seen there, $f \notin Lip_\alpha([0,1])$ for any $\alpha > 0$. Now, the derivative $f' = g$ exists everywhere and has the form

$$g(x) = \begin{cases} \frac{2}{x \log^2 \frac{2}{x}} & \text{for } 0 < x \le 1\,, \\ 0 & \text{for } x = 0\,. \end{cases}$$

Since the (improper) Riemann integral of g over $(0,1)$ exists and $g(x) > 0$ on $(0,1)$, we conclude that $g \in L_1([0,1])$. Thus, from Proposition 3.4, it follows that f satisfies (3.2) and is absolutely continuous on $[0,1]$. ♥

Later (Theorem 3.19), we will show that not only an absolutely continuous function, but also its total variation may be expressed through an integral.

3.2 The Vitali–Banach–Zaretskij theorem

As we have seen, any absolutely continuous function is both continuous and of bounded variation. We now show by means of a famous example that the converse is not true.

Recall that the *Cantor set* $C \subset [0,1]$ may be defined as intersection

$$C := \bigcap_{n=0}^{\infty} C_n, \tag{3.3}$$

where

$$C_0 := [0,1], \quad C_1 := \left[0, \frac{1}{3}\right] \cup \left[\frac{2}{3}, 1\right], \quad C_2 := \left[0, \frac{1}{9}\right] \cup \left[\frac{2}{9}, \frac{3}{9}\right] \cup \left[\frac{6}{9}, \frac{7}{9}\right] \cup \left[\frac{8}{9}, 1\right], \ldots$$

$$C_n := [0, 3^{-n}] \cup [2 \cdot 3^{-n}, 3 \cdot 3^{-n}] \cup \ldots \cup [(3^n - 1) \cdot 3^{-n}, 1]\,. \tag{3.4}$$

Alternatively, C may be defined as a set of all elements $x \in [0,1]$ whose ternary representation[1]

$$x = 0.x_1x_2x_3 \ldots = \sum_{k=1}^{\infty} x_k 3^{-k} \quad (x_k \in \{0,1,2\}) \tag{3.5}$$

only contains the numbers 0 and 2. This set has many interesting properties: for instance, it is a perfect, compact and uncountable nullset.[2] Furthermore, the function $\varphi : C \to [0,1]$ defined by

$$\varphi\left(\sum_{n=1}^{\infty} x_n 3^{-n}\right) = \sum_{n=1}^{\infty} \frac{x_n}{2} 2^{-n}, \tag{3.6}$$

where the element on the right-hand side of (3.6) is represented in the binary system, is monotonically increasing and surjective, i.e. satisfies $\varphi(C) = [0,1]$. Since the function φ takes the same values at the endpoints of two adjacent intervals in (3.4), φ admits an increasing *continuous* extension from C to the whole interval $[0,1]$ which we still denote by φ. This function will be called the *Cantor function* in what follows.[3]

The map $\psi : [0,1] \to [0,1]$ defined by

$$\psi(x) := \frac{1}{2}(x + \varphi(x)) \tag{3.7}$$

which we call *strict Cantor function* is also of interest. Since ψ is, in contrast to φ, strictly increasing, it is even a *homeomorphism* of $[0,1]$ onto itself. These functions have the required property which we announced before:

1 Observe that the representation (3.5) is not unique: for example, $\frac{1}{3} = 0.100000\ldots = 0.022222\ldots$; however, for our purposes, this is not relevant.

2 Recall that a set is *perfect* if it only consists of accumulation points.

3 Since the continuous extension of φ from C to $[0,1]$ is constant on $[0,1] \setminus C$, it is even *differentiable a.e.* on $[0,1]$ because C is a nullset; more precisely, $\varphi'(x) \equiv 0$ on $[0,1] \setminus C$.

Example 3.6. Being monotone and continuous, the Cantor function $\varphi : [0,1] \to [0,1]$ belongs to $BV([0,1]) \cap C([0,1])$. However, φ is *not* absolutely continuous on $[0,1]$. To see this, we choose, as intervals in (3.1), precisely the 2^n disjoint intervals of length 3^{-n} remaining in the n-th construction step of the Cantor set C, i.e. the intervals occurring in the union C_n in (3.4). Collecting these intervals (in their natural order) in the family

$$S := \{[a_1, b_1], \dots, [a_{2^n}, b_{2^n}]\},$$

we get, on the one hand,

$$\Theta(S) = \sum_{k=1}^{2^n} |b_k - a_k| = 1 - \frac{1}{3} - \frac{2}{9} - \frac{4}{27} - \dots - \frac{2^{n-1}}{3^n} = \left(\frac{2}{3}\right)^n,$$

and this may be made arbitrarily small by choosing n sufficiently large. On the other hand,

$$\Gamma(f; S) = \sum_{k=1}^{2^n} |\varphi(b_k) - \varphi(a_k)| - \varphi(1) - \varphi(0) = 1$$

since by construction, all terms cancel out, except for the first and the last one. This shows that $\varphi \notin AC([0,1])$ as claimed. Of course, the strict Cantor function ψ defined in (3.7) may also serve as an example. ♥

In a moment, we will give an alternative and rather elegant proof of the fact that the function (3.6) is not absolutely continuous, see Theorem 3.9.

Functions like the Cantor function φ discussed in Example 3.6 are so important in the theory of real functions that they have a special name:

Definition 3.7. A nonconstant function $f \in BV([a,b]) \cap C([a,b])$ is called *singular* if it is differentiable a.e. on $[a,b]$ with $f'(x) = 0$. ■

In view of Example 3.6, the question arises as to what property is "missing" if a function is continuous and of bounded variation, but not absolutely continuous, as the Cantor functions (3.6) and (3.7). Here, the following definition is crucial.

Definition 3.8. We say that a function $f : [a,b] \to \mathbb{R}$ satisfies the *Luzin condition*[4] if f maps every nullset into a nullset. We denote the set of all functions satisfying the Luzin condition on $[a,b]$ by $Lu([a,b])$. ■

The most prominent example of a function which fails to satisfy the Luzin condition is the Cantor function (3.6). Indeed, C is a nullset, but $\varphi(C) = [0,1]$ is not.

Using the Luzin condition, we are now in the position to formulate and prove the main result of this section which is usually referred to as the *Vitali–Banach–Zaretskij theorem*.

4 This condition was introduced by the Soviet mathematician N. N. Luzin; in the literature, it is sometimes called the *property* (N), where N stands for "nullset."

Theorem 3.9 (Vitali–Banach–Zaretskij). *A function $f : [a,b] \to \mathbb{R}$ is absolutely continuous if and only if f is continuous, has bounded variation, and satisfies the Luzin condition.*

Proof. To prove the "if" part, we suppose that f is continuous, has bounded variation, and satisfies the Luzin condition; we have to show that f is absolutely continuous.

Suppose that $f \notin AC([a,b])$. Then there exists $\varepsilon_0 > 0$ such that for every $n \in \mathbb{N}$, we find a collection $S_n := \{[a_{1,n}, b_{1,n}], [a_{2,n}, b_{2,n}], \dots, [a_{k_n,n}, b_{k_n,n}]\} \in \Sigma([a,b])$ ($k_n \in \mathbb{N}$) with $\Theta(S_n) \le 1/n^2$ and $\Gamma(f; S_n) > \varepsilon_0$. Let

$$m_{i,n} := \inf\{f(x) : a_{i,n} \le x \le b_{i,n}\}, \tag{3.8}$$

$$M_{i,n} := \sup\{f(x) : a_{i,n} \le x \le b_{i,n}\} \tag{3.9}$$

and

$$E_n := \bigcup_{i=1}^{k_n} (a_{i,n}, b_{i,n}), \quad N := \bigcap_{m=1}^{\infty} \bigcup_{n=m}^{\infty} E_n. \tag{3.10}$$

Then N is a nullset, and thus also $f(N)$ since f satisfies the Luzin condition. In addition to (3.8) and (3.9), we need the numbers $m(f)$ and $M(f)$ defined in (0.61) and (0.62), respectively. For $n = 1, 2, 3, \dots$ and $i = 1, 2, \dots, k_n$, consider the functions $g_{i,n} : \mathbb{R} \to \{0, 1\}$ and $g_n : \mathbb{R} \to \mathbb{N}_0$ defined by

$$g_{i,n}(y) := \begin{cases} 1 & \text{if } f(x) = y \text{ for some } x \in (a_{i,n}, b_{i,n}), \\ 0 & \text{otherwise} \end{cases}$$

and

$$g_n(y) := g_{1,n}(y) + g_{2,n}(y) + \dots + g_{k_n,n}(y),$$

respectively. Then we have

$$\int_{-\infty}^{\infty} g_n(y)\,dy = \sum_{i=1}^{k_n} (M_{i,n} - m_{i,n}) \ge \sum_{i=1}^{k_n} |f(b_{i,n}) - f(a_{i,n})| > \varepsilon_0. \tag{3.11}$$

Moreover, $g_n(y) \le I_f(y)$ for all n, where $I_f : [m(f), M(f)] \to \mathbb{R}$ denotes the Banach indicatrix of f (Definition 0.38). Consider the sets

$$N_\infty(f) := \{y \in \mathbb{R} : I_f(y) = \infty\}, \quad N_0(f) := \{y \in \mathbb{R} : g_n(y) \not\to 0 \ (n \to \infty)\}.$$

Since $f \in C([a,b]) \cap BV([a,b])$, we know that $I_f \in L_1(\mathbb{R})$, see Proposition 1.27, and so $N_\infty(f)$ is a nullset. Let $y \in N_0 \setminus N_\infty$; in particular, $I_f(y) < \infty$. Then we find a subsequence $(g_{n_k})_k$ of $(g_n)_n$ such that $g_{n_k}(y) \ge 1$ for all k. For each k, choose $x_k \in E_{n_k}$ with $f(x_k) = y$. Since the set $f^{-1}(y) \cap [a,b]$ is finite, we find at least one point x_0 in this set which belongs to infinitely many of the sets E_{n_k}, and so $x_0 \in N$, see (3.10) and $y = f(x_0) \in f(N)$. However, we have already seen above that $f(N)$ is a nullset, and so $N_0(f) \subseteq f(N) \cup N_\infty(f)$ is also a nullset. By definition of $N_0(f)$, this means that $g_n(y) \to 0$, as $n \to \infty$, for almost all $y \in [m(f), M(f)]$. Since $g_n(y) \leq I_f(y)$ and $I_f \in L_1(\mathbb{R})$, from Lebesgue's dominated convergence theorem (Theorem 0.4) we conclude that

$$\lim_{n\to\infty} \int_{-\infty}^{\infty} g_n(y)\, dy = 0,$$

contradicting (3.11). This contradiction shows that our assumption was false, and so $f \in AC([a,b])$.

Now, we prove the converse implication, i.e. the "only if" part. We already know from Proposition 1.22 that the inclusion

$$AC([a,b]) \subseteq C([a,b]) \cap BV([a,b])$$

is true. Consequently, to prove the "only if" part, we merely have to show that an absolutely continuous function satisfies the Luzin condition. Therefore, let $N \subset [a,b]$ be a nullset and $\varepsilon > 0$. Exercise 3.7 shows that we may find a $\delta > 0$ such that whenever $\{[a_n, b_n] : n \in \mathbb{N}\} \in \Sigma_\infty([a,b])$ is a sequence of nonoverlapping subintervals of $[a,b]$, the condition

$$\sum_{k=1}^{\infty} |b_k - a_k| \leq \delta$$

implies the condition

$$\sum_{k=1}^{\infty} |f(b_k) - f(a_k)| \leq \varepsilon .$$

Since N is a nullset, we may further choose an open set $G \supset N$ with $\lambda(G) \leq \delta$. Now, we use the well-known fact that being an open set, G may be represented as a countable disjoint union of open intervals $(a_1, b_1), (a_2, b_2), (a_3, b_3), \ldots$ and obtain

$$\sum_{k=1}^{\infty} |b_k - a_k| = \lambda\left(\bigcup_{k=1}^{\infty} (a_k, b_k)\right) = \lambda(G) \leq \delta .$$

For each k, we choose points[5] x_k and y_k with $f([a_k, b_k]) = [f(x_k), f(y_k)]$, and let $m_k := \min\{x_k, y_k\}$ and $M_k := \max\{x_k, y_k\}$. Then the intervals $[m_k, M_k]$ are pairwise nonoverlapping and satisfy

$$\sum_{k=1}^{\infty} |x_k - y_k| = \sum_{k=1}^{\infty} (M_k - m_k) \leq \sum_{k=1}^{\infty} |b_k - a_k| \leq \delta .$$

However,

$$f(N) \subseteq \bigcup_{k=1}^{\infty} [f(x_k), f(y_k)], \quad \sum_{k=1}^{\infty} |f(x_k) - f(y_k)| \leq \varepsilon$$

since f is absolutely continuous, by assumption. Since $\varepsilon > 0$ was arbitrary, we conclude that $f(N)$ is a nullset, and so the proof is complete. □

5 Here, we use the continuity of f, so f attains its supremum and infimum on each compact interval.

The statement of Theorem 3.9 may be summarized as the equality of sets

$$AC([a,b]) = C([a,b]) \cap BV([a,b]) \cap Lu([a,b]). \tag{3.12}$$

Theorem 3.9 explains again why the function f from Example 3.5 is absolutely continuous: in Example 1.25, we have seen that f is continuous and of bounded variation, and it is easy to see that it also satisfies the Luzin condition.

None of the three function classes on the right-hand side of (3.12) may be omitted. Indeed, the Cantor function φ and the strict Cantor function ψ may serve as examples of functions in $(C \cap BV) \setminus AC$. Examples of functions in $(BV \cap Lu) \setminus AC$ and $(C \cap Lu) \setminus AC$ are easily found:

Example 3.10. Let $f : [0,2] \to \mathbb{R}$ be defined by $f := \chi_{[0,1]}$. Clearly, f satisfies the Luzin condition and, being monotone, also has bounded variation, but is of course discontinuous.

On the other hand, let $f : [0,1] \to \mathbb{R}$ be the function (1.14) from Example 1.8. As we have seen there, f is continuous, but not of bounded variation. Now, if $N \subset [0,1]$ is a nullset, then

$$f(N) \subseteq \{0\} \cup \bigcup_{n=1}^{\infty} f(N \cap [1/n, 1]),$$

and the set on the right-hand side is a nullset since the function (1.14) is of class C^∞ on $[1/n, 1]$. So, f satisfies the Luzin condition. ♥

We summarize these examples in the following Table 3.1. It is clear that by (3.12), no entries are possible in the diagonal and the last column.

Table 3.1. The Vitali–Banach–Zaretskij theorem.

Function f	$\in C$	$\in BV$	$\in Lu$	$\in AC$
$\notin C$	–	Example 3.10	Example 3.10	–
$\notin BV$	Example 1.8	–	Example 1.8	–
$\notin Lu$	Example 3.6	Example 3.6	–	–
$\notin AC$	Example 3.6	Example 3.6	Example 1.8	–

In Exercise 3.9, we consider a more sophisticated function $f \in C([0,1]) \cap Lu([0,1])$ whose total variation may be made finite or infinite by a suitable choice of free parameters.

The same reasoning as in Example 3.10 shows that the function $f_{\alpha,\beta}$ defined in (0.86) satisfies the Luzin property for all choices of α and β. Combining this with Exercises 1.8 and 1.9 allows us to precisely determine those values of α and β for which the function (0.86) is absolutely continuous:

Example 3.11. For $\alpha, \beta \in \mathbb{R}$, let $f_{\alpha,\beta} : [0,1] \to \mathbb{R}$ be defined as in (0.86). From the above reasoning, we know that $f_{\alpha,\beta}$ satisfies the Luzin property for all choices of α

Table 3.2. Spaces $AC_p(I)$ and $WBV_p(I)$ over $I = [a, b]$.

$AC(I)$	$=$	$AC_1(I)$	$\subset$	$AC_p(I)$	$\subset$	$AC_q(I)$	$\subset$	$C(I)$
$\cap$		$\cap$		$\cap$		$\cap$		$\cap$
$BV(I)$	$=$	$WBV_1(I)$	$\subset$	$WBV_p(I)$	$\subset$	$WBV_q(I)$	$\subset$	$B(I)$

and β, and from Exercises 1.8 and 1.9 we know that $f_{\alpha,\beta} \in BV([0,1])$ if and only if $\beta > 0$ and $\alpha + \beta \geq 0$, or $\beta \leq 0$ and $\alpha + \beta > 0$. Moreover, Proposition 0.48 shows that $f_{\alpha,\beta}$ is continuous at zero (and so on the whole interval $[0,1]$) if and only if $\alpha > 0$ and β is arbitrary, or $\alpha \leq 0$ and $\alpha + \beta > 0$. Thus, from Theorem 3.9, we conclude that $f_{\alpha,\beta} \in AC([0,1])$ if and only if $\alpha + \beta > 0$. ♥

Now, we consider a generalization of absolute continuity due to Love [183] which is usually called p-absolute continuity.

Definition 3.12. Given a collection $S = \{[a_1, b_1], \ldots, [a_n, b_n]\} \in \Sigma([a,b])$ and a function $f : [a,b] \to \mathbb{R}$, for $p \geq 1$, we put, in analogy to (3.1),

$$\Theta_p(S) := \left\{ \sum_{k=1}^{n} |b_k - a_k|^p \right\}^{1/p}, \quad \Gamma_p(f;S) := \left\{ \sum_{k=1}^{n} |f(b_k) - f(a_k)|^p \right\}^{1/p}. \tag{3.13}$$

We call the function f *p-absolutely continuous* on $[a,b]$ and write $f \in AC_p([a,b])$ if for each $\varepsilon > 0$, there exists some $\delta > 0$ such that $\Gamma_p(f;S) \leq \varepsilon$ for any $S \in \Sigma([a,b])$ satisfying $\Theta_p(S) \leq \delta$. ■

Of course, for $p = 1$, we get the old Definition 3.1, which means that $AC_1 = AC$. In the same way as we have proved the inclusion $AC([a,b]) \subseteq BV([a,b])$ in Proposition 1.22, one may show that[6]

$$C([a,b]) \cap WBV_p([a,b]) \subseteq AC_q([a,b]) \subseteq C([a,b]) \cap WBV_q([a,b]) \tag{3.14}$$

for $1 \leq p < q < \infty$, where $WBV_p([a,b])$ denotes the Wiener space introduced in Definition 1.31. Moreover, the spaces $AC_p([a,b])$ are increasing in p; more precisely,

$$AC([a,b]) = AC_1([a,b]) \subseteq AC_p([a,b]) \subseteq AC_q([a,b]) \subseteq C([a,b]) \tag{3.15}$$

for $1 \leq p \leq q < \infty$ which is parallel to (1.72). We summarize all of these inclusions in Table 3.2 above.

Following [183] (see also [184, 185]), we now consider a class of functions which is similar to the class CBV introduced in Section 1.2.

6 Exercise 3.25 shows that the first inclusion in (3.14) is strict in case $p < q$, while Exercise 3.26 shows that the second inclusion in (3.15) is strict.

Definition 3.13. For $p > 1$, we denote by $CWBV_p([a, b])$ the set of all continuous functions $f : [a, b] \to \mathbb{R}$ satisfying

$$\inf \left\{ \sum_{j=1}^{m} \mathrm{Var}_p^W(f; [t_{j-1}, t_j]) : \{t_0, t_1, \ldots, t_m\} \in \mathcal{P}([a, b]) \right\} = 0\,, \tag{3.16}$$

where $\mathrm{Var}_p^W(f; [\alpha, \beta])$ denotes the Wiener p-variation (1.61) of f on $[\alpha, \beta]$, and the infimum in (3.16) is taken over all partitions of $[a, b]$. ∎

This definition is not made for $p = 1$ as it would admit only constant functions. However, for $p > 1$, it turns out to be something we already know:

Proposition 3.14. *The equality*

$$CWBV_p([a, b]) = AC_p([a, b]) \tag{3.17}$$

holds for $p > 1$.

Proof. Suppose first that $f \in CWBV_p([a, b])$, and let $\varepsilon > 0$. By definition, we may then find a partition $\{t_0, t_1, \ldots, t_m\} \in \mathcal{P}([a, b])$ satisfying

$$\sum_{j=1}^{m} \mathrm{Var}_p^W(f; [t_{j-1}, t_j]) \le \left(\frac{\varepsilon}{2}\right)^p .$$

Let $\delta := \min\{t_1 - t_0, t_2 - t_1, \ldots, t_m - t_{m-1}\}$. Now, if $S = \{[a_1, b_1], \ldots, [a_n, b_n]\} \in \Sigma([a, b])$ is any collection satisfying $\Theta_p(S) \le \delta$, see (3.13), not more than one t_j lies in each of the subintervals of S. Let τ_k be the t_j in $[a_k, b_k]$ if there is one. Otherwise, let $\tau_k := a_k$. Then we obtain

$$\begin{aligned}
\Gamma_p(f; S)^p &= \sum_{k=1}^{n} |f(b_k) - f(a_k)|^p \\
&\le 2^{p-1} \sum_{k=1}^{n} |f(b_k) - f(\tau_k)|^p + 2^{p-1} \sum_{k=1}^{n} |f(\tau_k) - f(a_k)|^p \\
&\le 2^p \sum_{j=1}^{m} \mathrm{Var}_p^W(f; [t_{j-1}, t_j]) \le \varepsilon^p ,
\end{aligned}$$

which shows that $f \in AC_p([a, b])$.

Conversely, suppose now that $f \in AC_p([a, b])$ (so $f \in C([a, b])$), and let $\varepsilon > 0$. Choose $\delta > 0$ such that $\Gamma_p(f; S) \le \varepsilon$ for all collections $S \in \Sigma([a, b])$ satisfying $\Theta_p(S) \le (b - a)^{1/p}\delta^{1-1/p}$.

Let $P = \{t_0, t_1, \ldots, t_m\} \in \mathcal{P}([a, b])$ be a fixed partition with $\mu(P) \le \delta$, where $\mu(P)$ denotes the mesh size (1.2) of P. Taking $a_k := t_{k-1}$ and $b_k := t_k$ for $k = 1, 2, \ldots, m$, we then have

$$\sum_{j=1}^{m} (t_j - t_{j-1})^p \le \sum_{k=1}^{m} \delta^{p-1}(b_k - a_k) \le \delta^{p-1}(b - a)$$

which means that $S = \{[t_0, t_1], [t_1, t_2] \dots, [t_{m-1}, t_m]\} \in \Sigma([a, b])$ satisfies $\Theta_p(S)^p \leq (b-a)\delta^{p-1}$. By assumption, we then have

$$\sum_{j=1}^{m} \mathrm{Var}_p^W(f; [t_{j-1}, t_j]) = \Gamma_p(f; S)^p \leq \varepsilon^p$$

and so $f \in CWBV_p([a, b])$. □

Another characterization of the class AC_p for $p > 1$ may be found in Theorem 3.41 in Section 3.7 below.

3.3 Reconstructing a function from its derivative

This section will be dedicated to the following two questions:

– Given $f : [a, b] \to \mathbb{R}$, when may we write f as an "indefinite integral," i.e. in the form

$$f(x) = c + \int_a^x g(t)\, dt$$

for some constant $c \in \mathbb{R}$ and some function $g \in L_1([a, b])$?

– Given $f : [a, b] \to \mathbb{R}$ with $f' \in L_1([a, b])$, is it true that f and the function

$$\tilde{f}(x) := \int_a^x f'(t)\, dt$$

differ only by some constant?

It turns out that absolute continuity is precisely what we need here. In particular, the answer to the first question is suggested by Proposition 3.4, while the answer to the second question is closely related to the problem of reconstructing a function from its derivative. To put things in the right framework, let us recall how this may be done very easily for C^1 functions. Given a continuously differentiable function $f : [a, b] \to \mathbb{R}$, the *fundamental theorem of calculus for the Riemann integral* states that

$$\int_a^x f'(t)\, dt = f(x) - f(a) \tag{3.18}$$

for each $x \in [a, b]$, where the integral in (3.18) is the Riemann integral; in particular,

$$\int_a^b f'(t)\, dt = f(b) - f(a)\,. \tag{3.19}$$

Conversely, given a Riemann integrable (in particular, bounded) function $g: [a,b] \to \mathbb{R}$, the function f defined by

$$f(x) := \int_a^x g(t)\,dt \quad (a \le x \le b) \tag{3.20}$$

is Lipschitz continuous; moreover, in case of a continuous integrand g we even have $f \in C^1([a,b])$ and $f'(x) = g(x)$ for all $x \in [a,b]$.

However, the two questions stated above refer to the *Lebesgue integral*, and here the corresponding problem is more delicate. First of all, if the integral in (3.18) (or (3.19) is the Lebesgue integral, it suffices that f' exists only a.e. on $[a,b]$. However, the equality (3.19) may then *fail* for two different reasons: the derivative f' need not be integrable, and even if it is, (3.19) need not hold. We illustrate both phenomena by an example.

Example 3.15. Consider the function (0.86) for $\alpha = 2$ and $\beta = -2$, i.e.

$$f_{2,-2}(x) := \begin{cases} x^2 \sin \frac{1}{x^2} & \text{for } 0 < x \le 1\,, \\ 0 & \text{for } x = 0\,. \end{cases} \tag{3.21}$$

Then $f_{2,-2}$ is *everywhere* differentiable on $[0,1]$ with

$$f'_{2,-2}(x) := \begin{cases} 2x \sin \frac{1}{x^2} - \frac{2}{x} \cos \frac{1}{x^2} & \text{for } 0 < x \le 1\,, \\ 0 & \text{for } x = 0\,. \end{cases} \tag{3.22}$$

The first term in (3.22) is continuous and therefore belongs to $L_1([0,1])$. However, the second term is of the form[7] (0.86) with $\alpha = -1$ and $\beta = -2$, and Exercise 0.7 shows that this term does not belong to $L_1([0,1])$. ♥

The behavior of the function in Example 3.15 is explained by the fact that the choice $(\alpha, \beta) = (2,-2)$ is not covered by the condition given in Example 3.11, i.e. $f_{2,-2} \notin AC([0,1])$. We will come back to this in a moment (see Theorem 3.18 below). First, we consider another example, where the integral in (3.19) exists, but is not equal to $f(b) - f(a)$.

Example 3.16. Let $\varphi : [0,1] \to \mathbb{R}$ be the Cantor function from Example 3.6. As we have seen there, $\varphi \notin AC([0,1])$. Since $\varphi'(t) = 0$ a.e. on $[0,1]$ (namely, for all $t \in [0,1] \setminus C$), we have $\varphi' \in L_1([0,1])$ and

$$\int_0^1 \varphi'(t)\,dt = 0\,.$$

On the other hand, $\varphi(1) - \varphi(0) = 1$, by definition of φ. ♥

7 In (0.86), we have the sine function instead, but the cosine function (0.88) of course has the same behavior near zero.

The following important theorem shows that a one-sided estimate holds in (3.19) if f is monotonically increasing.

Theorem 3.17. *If $f : [a,b] \to \mathbb{R}$ is increasing, then the derivative f exists a.e. on $[a,b]$, belongs to $L_1([a,b])$, and satisfies*

$$\int_a^b f'(t)\,dt \le f(b) - f(a)\,. \tag{3.23}$$

Proof. By Exercise 1.29, f is a.e. differentiable on $[a,b]$. We extend f to the larger domain $[a-1,b+1]$ by putting $f(t) = f(a)$ for $a-1 \le t < a$ and $f(t) = f(b)$ for $b < t \le b+1$, and define functions f_n, g_n, and g by

$$f_n(t) := f(t + \tfrac{1}{n}), \quad g_n(t) := n\left[f_n(t) - f(t)\right], \tag{3.24}$$

and

$$g(t) := \lim_{h\to 0} \frac{f(t+h) - f(t)}{h}\,. \tag{3.25}$$

Since f is differentiable a.e., g is a.e. defined on $[a,b]$. Moreover, g is measurable since both f and f_n are increasing, and $(g_n)_n$ converges a.e. on $[a,b]$ to g. By Fatou's theorem (Theorem 0.5), we have

$$\int_a^b g(t)\,dt \le \liminf_{n\to\infty} \int_a^b g_n(t)\,dt = \liminf_{n\to\infty} n \int_a^b \left[f(t + \tfrac{1}{n}) - f(t)\right] dt\,. \tag{3.26}$$

After a simple change of variables for the last integral, we obtain

$$\begin{aligned} \int_a^b \left[f(t + \tfrac{1}{n}) - f(t)\right] dt &= \int_{a+1/n}^{b+1/n} f(t)\,dt - \int_a^b f(t)\,dt \\ &= \int_b^{b+1/n} f(t)\,dt - \int_a^{a+1/n} f(t)\,dt = \frac{f(b)}{n} - \int_a^{a+1/n} f(t)\,dt\,. \end{aligned} \tag{3.27}$$

Since $f(t) \ge f(a)$ for $t \in [a, a+1/n]$, we further get

$$\int_a^{a+1/n} f(t)\,dt \ge \int_a^{a+1/n} f(a)\,dt = \frac{f(a)}{n}\,,$$

and combining this with (3.26) and (3.27) yields

$$\int_a^b g(t)\,dt \le \liminf_{n\to\infty} n\left(\frac{f(b)}{n} - \frac{f(a)}{n}\right) = f(b) - f(a)\,.$$

The proof is complete. □

As we have seen, an example for strict inequality in (3.23) is the monotone Cantor function φ. The next theorem shows that this is not accidental.

Theorem 3.18. *If $f : [a,b] \to \mathbb{R}$ is absolutely continuous, then the derivative f' exists a.e. on $[a,b]$, belongs to $L_1([a,b])$, and satisfies the equality* (3.19).

Proof. Since every absolutely continuous function has bounded variation, the derivative f' of $f \in AC([a,b])$ exists a.e. in $[a,b]$. Moreover, by Proposition 3.3, we may assume without loss of generality that f is increasing.

We define $g_n : [a,b] \to \mathbb{R}$ as in (3.24); then, g_n is continuous, and again $g_n(t) \to f'(t)$ a.e. on $[a,b]$. As in (3.27), we get

$$\int_a^b g_n(t)\,dt = f(b) - n \int_a^{a+1/n} f(t)\,dt\,,$$

where the integral on the right-hand side is now the Riemann integral. The continuity of f implies that

$$\int_a^b g_n(t)\,dt \to f(b) - f(a) \quad (n \to \infty)\,,$$

by the mean value theorem for Riemann integrals. So, for proving the theorem, it suffices to show that also

$$\int_a^b g_n(t)\,dt \to \int_a^b f'(t)\,dt \quad (n \to \infty)\,. \tag{3.28}$$

Let $\varepsilon > 0$. Since $f \in AC([a,b])$, we find a $\delta > 0$ such that $\Gamma(f;S) \le \varepsilon$ for any $S \in \Sigma([a,b])$ with $\Theta(S) \le \delta$. Moreover, by Theorem 0.6, we can choose $\delta' \in (0,\delta)$ with the property that

$$\int_M f'(t)\,dt \le \varepsilon \tag{3.29}$$

for any set $M \subset [a,b]$ satisfying $\lambda(M) \le \delta'$. Denote by D the set of all points in $[a,b]$ where f is differentiable, so $[a,b] \setminus D$ is a nullset. By Egorov's theorem (Theorem 0.1), we find a measurable set $M \subseteq D$ such that $\lambda(M) \le \delta'$ and $(g_n)_n$ converges uniformly to f' on $D \setminus M$. This implies, in particular, that there exists $n_0 \in \mathbb{N}$ such that

$$\int_{D\setminus M} |g_n(t) - f'(t)|\,dt \le \varepsilon \tag{3.30}$$

for $n \ge n_0$. So, for these n, we have, by (3.29) and (3.30),

$$\begin{aligned}\left|\int_a^b g_n(t)\,dt - \int_a^b f'(t)\,dt\right| &= \left|\int_D g_n(t)\,dt - \int_D f'(t)\,dt\right| \le \int_D |g_n(t) - f'(t)|\,dt \\ &= \int_{D\setminus M} |g_n(t) - f'(t)|\,dt + \int_M |g_n(t) - f'(t)|\,dt \le 2\varepsilon + \int_M |g_n(t)|\,dt\,.\end{aligned}$$

Thus, to prove (3.28), it remains to show that the last integral tends to zero as $n \to \infty$. Since $\lambda(M) \leq \delta' < \delta$, there exists an open set $O \subset (a, b)$ such that $M \subseteq O$ and $\lambda(O) < \delta$. We again use the fact that we may write O as a countable union of disjoint open intervals

$$O = \bigcup_{k=1}^{\infty} (a_k, b_k) .$$

Now, for any $\tau \in [0, 1]$ and any $m \in \mathbb{N}$, the collection of intervals

$$S := \{[a_1 + \tau, b_1 + \tau], [a_2 + \tau, b_2 + \tau], \ldots, [a_m + \tau, b_m + \tau]\} \in \Sigma([a, b])$$

satisfies

$$\Theta(S) = \sum_{k=1}^{m} |b_k - a_k| \leq \delta .$$

Consequently,

$$\Gamma(f; S) = \sum_{k=1}^{m} [f(b_k + \tau) - f(a_k + \tau)] \leq \varepsilon ,$$

by our choice of δ. However, for $k = 1, 2, \ldots, m$, we have

$$\begin{aligned} \int_{a_k}^{b_k} g_n(t)\, dt &= n \left(\int_{b_k}^{b_k+1/n} f(t)\, dt - \int_{a_k}^{a_k+1/n} f(t)\, dt \right) \\ &= n \int_0^{1/n} [f(b_k + \tau) - f(a_k + \tau)] \; d\tau . \end{aligned}$$

Consequently,

$$\begin{aligned} \int_M g_n(t)\, dt \leq \int_O g_n(t)\, dt &= \sum_{k=1}^{m} \int_{a_k}^{b_k} g_n(t)\, dt \\ &= n \int_0^{1/n} \left(\sum_{k=1}^{m} [f(b_k + \tau) - f(a_k + \tau)] \right) d\tau \leq \varepsilon , \end{aligned}$$

from which (3.28) follows. □

We may summarize Proposition 3.4 and Theorem 3.18 in the following way: *a function* $f : [a, b] \to \mathbb{R}$ *is absolutely continuous if and only if there exists* $g \in L_1([a, b])$ *such that*

$$f(x) = f(a) + \int_a^x g(t)\, dt . \tag{3.31}$$

Moreover, $f'(x) = g(x)$ *a.e. on* $[a, b]$ *in this case.* This is the *fundamental theorem of calculus for the Lebesgue integral.*

We have even more to say on the subject. An absolutely continuous function f cannot only be expressed through the integral (3.18) of its derivative f', but also its total variation (which is finite by the inclusion $AC \subseteq BV$) may be expressed by the integral of $|f'|$. This is the contents of the following

Theorem 3.19. *Let $f \in BV([a,b])$. Then*

$$\mathrm{Var}(f;[a,b]) \geq \int_a^b |f'(t)|\,dt\,, \tag{3.32}$$

where $\mathrm{Var}(f;[a,b])$ *denotes the total variation* (1.4) *of f on $[a,b]$. In case $f \in AC([a,b])$, we even have equality*

$$\mathrm{Var}(f;[a,b]) = \int_a^b |f'(t)|\,dt\,. \tag{3.33}$$

Proof. Given $\varepsilon > 0$, we choose $\delta > 0$ such that

$$\int_M |f'(t)|\,dt \leq \varepsilon$$

for any set $M \subset [a,b]$ satisfying $\lambda(M) \leq \delta$, which is possible by Theorem 0.6. Denoting

$$D_+ := \{t : a < t < b,\, f'(t) \geq 0\}$$

and

$$D_- := \{t : a < t < b,\, f'(t) < 0\},$$

we may find an open set $O \subset (a,b)$ satisfying[8]

$$O = \bigcup_{j=1}^m (a_j, b_j), \qquad \lambda(O \Delta D_+) \leq \delta\,.$$

Here, we may assume without loss of generality that $a_1 < b_1 < a_2 < \ldots < b_{m-1} < a_m < b_m$. Adding, if necessary, the points $b_0 := a$ and $a_{m+1} := b$, we thus obtain a partition $P := \{b_0, a_1, b_1, a_2, \ldots, b_{m-1}, a_m, b_m, a_{m+1}\} \in \mathcal{P}([a,b])$. Calculating the variation (1.3) of f with respect to this partition and observing that

$$D_-\Delta((a,b) \setminus O) = (D_\cap O) \cup [(a,b) \setminus O] \setminus D_- = O\Delta D_+,$$

8 As usual, $A\Delta B = (A \cup B) \setminus (A \cap B) = (A \setminus B) \cup (B \setminus A)$ denotes the *symmetric difference* of A and B.

we get

$$\begin{aligned}
\mathrm{Var}(f,P;[a,b]) &= \sum_{j=1}^{m}\left|\int_{a_j}^{b_j} f'(t)\,dt\right| + \sum_{k=1}^{m+1}\left|\int_{b_{k-1}}^{a_k} f'(t)\,dt\right| \\
&\geq \left|\int_{O} f'(t)\,dt\right| + \left|\int_{(a,b)\setminus O} f'(t)\,dt\right| \\
&\geq \left|\int_{D_+} f'(t)\,dt\right| - \int_{D_+\triangle O} |f'(t)|\,dt + \left|\int_{D_-} f'(t)\,dt\right| - \int_{D_-\triangle((a,b)\setminus O)} |f'(t)|\,dt \\
&\geq \int_a^b |f'(t)|\,dt - \int_{O\triangle D_+} |f'(t)|\,dt \geq \int_a^b |f'(t)|\,dt - \varepsilon\,.
\end{aligned}$$

Since $\varepsilon > 0$ was arbitrary, we conclude that

$$\mathrm{Var}(f;[a,b]) \geq \mathrm{Var}(f,P;[a,b]) \geq \int_a^b |f'(t)|\,dt\,. \tag{3.34}$$

The converse inequality follows for $f \in AC([a,b])$ from (3.18). In fact, for arbitrary partitions $P = \{t_0, t_1, \dots, t_m\} \in \mathcal{P}([a,b])$, we obtain

$$\mathrm{Var}(f,P;[a,b]) = \sum_{j=1}^{m}\left|\int_{t_{j-1}}^{t_j} f'(t)\,dt\right| \leq \sum_{j=1}^{m}\int_{t_{j-1}}^{t_j} |f'(t)|\,dt = \int_a^b |f'(t)|\,dt\,,$$

and hence $\mathrm{Var}(f;[a,b]) \leq \|f'\|_{L_1}$, which, together with (3.34), proves the statement. □

Theorem 3.19 admits a certain refinement which states that we have equality in (3.32) *precisely* for $f \in AC([a,b])$, see Exercise 3.33. Obviously, in the same way as we proved formula (3.32), we may show that

$$V_f(x) \geq \int_a^x |f'(t)|\,dt \quad (a \leq x \leq b) \tag{3.35}$$

if $f \in BV([a,b])$, where V_f denotes the variation function (1.13). As before, in case $f \in AC([a,b])$, we even have equality

$$V_f(x) = \int_a^x |f'(t)|\,dt \quad (a \leq x \leq b)\,. \tag{3.36}$$

As a consequence of Theorem 3.18, we may also obtain the following characterization of Lipschitz continuous functions.

Theorem 3.20. *A function f belongs to $Lip([a,b])$ if and only if f may be written in the form*

$$f(x) = c + \int_a^x g(t)\,dt\,, \tag{3.37}$$

where $c \in \mathbb{R}$ and $g \in L_\infty([a,b])$.

Proof. For $f \in Lip([a,b]) \subseteq AC([a,b])$, we conclude from Theorem 3.18 that (3.19) holds true, so we have (3.37) with $c := f(a)$ and $g := f'$. Moreover, from $|f(x) - f(y)| \leq L|x - y|$, it follows that $|f'(t)| \leq L$ at any point $t \in [a,b]$ where f is differentiable, and so $f' \in L_\infty([a,b])$ with $\|f'\|_{L_\infty} \leq lip(f;[a,b])$.

Conversely, suppose that f satisfies (3.37) for some $c \in \mathbb{R}$ and $g \in L_\infty([a,b])$. Then for $a \leq x < y \leq b$, we obtain

$$|f(x) - f(y)| \leq \left| \int_x^y g(t)\,dt \right| \leq \|g\|_{L_\infty} |x - y|\,,$$

and so $f \in Lip([a,b])$ with $lip(f;[a,b]) \leq \|g\|_{L_\infty}$ as claimed. □

Finally, we mention another consequence of Theorem 3.18 which we will need in what follows and which uses the concept of singular function in the sense of Definition 3.7.

Proposition 3.21. *Let $f \in BV([a,b]) \cap C([a,b])$. Then f may be represented as sum*

$$f(x) = f_{ac}(x) + f_{sg}(x) \quad (a \leq x \leq b)\,, \tag{3.38}$$

where f_{ac} is absolutely continuous and f_{sg} is singular or $f_{sg}(x) \equiv 0$. Moreover, these functions are uniquely determined within additive constants, and so the representation (3.38) may be made unique by requiring that $f(a) = f_{ac}(a)$.

Proof. Being of bounded variation, the function f admits a derivative a.e. on $[a,b]$, see Exercise 1.29. From Proposition 3.4, we know that the function

$$f_{\mathrm{ac}}(x) = f(a) + \int_a^x f'(t)\,dt \tag{3.39}$$

is absolutely continuous; moreover, $f_{\mathrm{ac}}(a) = f(a)$. The function $f_{\mathrm{sg}} : [a,b] \to \mathbb{R}$ defined by $f_{\mathrm{sg}}(x) := f(x) - f_{\mathrm{ac}}(x)$ belongs to $BV([a,b]) \cap C([a,b])$ and satisfies $f'_{\mathrm{sg}}(x) := f'(x) - f'_{\mathrm{ac}}(x) = 0$ a.e. on $[a,b]$. So, f_{sg} is a singular function in the sense of Definition 3.7, unless $f_{\mathrm{sg}}(x) \equiv 0$.

To prove uniqueness, suppose that

$$f(x) = f_{\mathrm{ac}}(x) + f_{\mathrm{sg}}(x) = \hat{f}_{\mathrm{ac}}(x) + \hat{f}_{\mathrm{sg}}(x) \quad (a \leq x \leq b)\,, \tag{3.40}$$

where both f_{ac} and $\hat{f}_{\mathrm{ac}}$ are absolutely continuous with $f_{\mathrm{ac}}(a) = \hat{f}_{\mathrm{ac}}(a) = f(a)$, and both f_{sg} and $\hat{f}_{\mathrm{sg}}$ are singular. Then $(f'_{\mathrm{ac}} - \hat{f}'_{\mathrm{ac}})(x) \equiv 0$, and so $f_{\mathrm{ac}} - \hat{f}_{\mathrm{ac}}$ is constant, see Proposition 3.33 below. However, $f_{\mathrm{ac}}(a) = \hat{f}_{\mathrm{ac}}(a)$ implies that $f_{\mathrm{ac}}(x) \equiv \hat{f}_{\mathrm{ac}}(x)$, and so $f_{\mathrm{sg}}(x) \equiv \hat{f}_{\mathrm{sg}}(x)$ as well. □

In Theorem 1.26, we have established some relations between the properties of the variation function V_f given in (1.13) and its parent function f. Now, we come back to this problem with respect to differentiability. As we have seen in Exercises 1.50 and 1.51, it is not true that in analogy to Proposition 1.7, the differentiability of f at some point $x_0 \in (a,b)$ implies the differentiability of V_f at x_0, or vice versa. However, the following result was proved in [149] and generalized in [281].

Proposition 3.22. *Suppose that f is differentiable on $[a,b]$ with bounded derivative. Then V_f is differentiable a.e. on $[a,b]$, and the equality*

$$V_f'(x) = |f'(x)| \tag{3.41}$$

holds at all points where V_f' exists.

Proof. Having a bounded derivative on $[a,b]$, by Theorem 3.20, we have $f \in Lip([a,b]) \subseteq AC([a,b])$. Moreover, (3.33) implies that

$$V_f(x) = \int_a^x |f'(t)|\,dt$$

for each $x \in [a,b]$, and so (3.41) holds a.e. in $[a,b]$. □

If f is of bounded variation, then both f and V_f are differentiable a.e. on $[a,b]$. However, the following example shows that the set of points $x \in [a,b]$ at which $f'(x)$ exists is not necessarily the same as the set of points $x \in [a,b]$ at which $V_f'(x)$ exists; compare this with Exercise 1.51.

Example 3.23. Consider the function (0.86) for $\alpha = 2$ and $\beta = -1$, i.e.

$$f_{2,-1}(x) := \begin{cases} x^2 \sin \frac{1}{x} & \text{for } 0 < x \le 1\,, \\ 0 & \text{for } x = 0\,. \end{cases}$$

Then $f_{2,-1}$ is everywhere differentiable on $[0,1]$ with

$$f_{2,-1}'(x) := \begin{cases} 2x \sin \frac{1}{x} - \cos \frac{1}{x} & \text{for } 0 < x \le 1\,, \\ 0 & \text{for } x = 0\,. \end{cases}$$

Since the derivative $f_{2,-1}'$ is bounded, but discontinuous on $[0,1]$, we have $f_{2,-1} \in Lip([0,1]) \subseteq AC([0,1])$, by Theorem 3.20, but $f_{2,-1} \notin C^1([0,1])$. From (3.36), it follows that

$$V_f(x) = \int_0^x |f'(t)|\,dt = \int_0^x \left|2t \sin \frac{1}{t} - \cos \frac{1}{t}\right| dt\,.$$

However, considering suitable sequences $(x_n)_n$ of "maximal oscillation," one may show that V_f is not differentiable at zero. ♥

As a further application of our results, we consider now two norms on AC which both make AC complete.

Proposition 3.24. *The linear space $AC([a,b])$, equipped with either the norm*

$$\|f\|_{AC} := \|f\|_{BV} \tag{3.42}$$

or the norm

$$|||f|||_{AC} := \|f\|_{L_1} + \|f'\|_{L_1} \tag{3.43}$$

is a Banach space.

Proof. To prove the first statement, we have to show that $AC([a,b])$ is closed in $BV([a,b])$ with respect to the norm (1.16). So, let $(f_n)_n$ be a sequence in $AC([a,b])$ satisfying $\|f_n - f\|_{BV} \to 0$, as $n \to \infty$, for some $f \in BV([a,b])$, where without loss of generality, $f_n(a) = f(a) = 0$. Writing f in the form (3.38), with f_{ac} being absolutely continuous and f_{sg} being singular (or zero), we get

$$\mathrm{Var}(f_{sg}; [a,b]) + \mathrm{Var}(f_{ac} - f_n; [a,b]) = \mathrm{Var}(f - f_n; [a,b]) \to 0 \quad (n \to \infty)\,.$$

This implies that $\mathrm{Var}(f_{sg}; [a,b]) = 0$, and hence $f = f_{ac} \in AC([a,b])$ as claimed.

To prove the second statement, let $(f_n)_n$ be a Cauchy sequence in $AC([a,b])$ with respect to the norm (3.43); then, $(f_n)_n$ converges in $L_1([a,b])$ to some function $f \in L_1([a,b])$, and $(f_n')_n$ is Cauchy with respect to the L_1-norm. We show that f is equivalent to an absolutely continuous function and $f_n' \to f'$, as $n \to \infty$, in $L_1([a,b])$.

Since $L_1([a,b])$ is complete with respect to the natural norm (0.11), we find some $g \in L_1([a,b])$ such that $f_n' \to g$ in the L_1-norm. Moreover, since each f_n is absolutely continuous, we have

$$f_n(x) = f_n(a) + \int_a^x f_n'(t)\,dt\,,$$

by Theorem 3.18. Define $h : [a,b] \to \mathbb{R}$ by

$$h(x) := \int_a^x g(t)\,dt\,.$$

Then $\|f_n' - g\|_{L_1} \to 0$ implies that

$$h(x) = \lim_{n\to\infty} (f_n(x) - f_n(a))$$

for every $x \in [a,b]$. Finally, from $\|f_n - f\|_{L_1} \to 0$, we conclude that there exists a subsequence of $(f_n)_n$ which converges pointwise a.e. to f. Consequently, $f = h + c$ for some constant c, and $f' = g$ a.e. on $[a,b]$. Therefore, $\|f_n' - h'\|_{L_1} \to 0$, as $n \to \infty$, and we are done. □

Observe that, by (3.33), the norm (3.42) may be written in the form

$$\|f\|_{AC} = |f(a)| + \|f'\|_{L_1} .$$

We close this section with a geometric characterization of bounded variation and absolute continuity of a function $g : [a, b] \to \mathbb{R}$ taken from the paper [127]. Without loss of generality, we take $[a, b] = [0, 1]$. This characterization is related to the *integral mean* g_h of a function $g \in L_1([0, 1])$ which is defined for $0 < h < 1$ by[9]

$$g_h(x) := \begin{cases} \frac{1}{h}\int_0^h g(x+t)\,dt & \text{for } 0 \le x \le 1-h, \\ \frac{1}{h}\int_0^h g(1-h+t)\,dt & \text{for } 1-h < x \le 1. \end{cases} \tag{3.44}$$

Thus, g_h is defined to be constant on the interval $[1-h, 1]$. By a simple change of variables, we may rewrite (3.44) equivalently in the form

$$g_h(x) = \begin{cases} \frac{1}{h}\int_x^{x+h} g(s)\,ds = \frac{f(x+h)-f(x)}{h} & \text{for } 0 \le x \le 1-h, \\ \frac{1}{h}\int_{1-h}^1 g(s)\,ds \equiv \frac{f(1)-f(1-h)}{h} & \text{for } 1-h < x \le 1, \end{cases} \tag{3.45}$$

where f is the "primitive a.e." of g given in (3.20).

Theorem 3.25. *For $g \in L_1([0, 1])$, the function (3.44) has the following properties.*
(a) *For each $h \in (0, 1)$, the function g_h is absolutely continuous.*
(b) *We have*

$$\limsup_{h\to 0+} |||g_h|||_{AC} < \infty , \tag{3.46}$$

where $|||\cdot|||_{AC}$ denotes the norm (3.43), if and only if g is equivalent to a function of bounded variation; moreover,

$$\mathrm{Var}(g_h; [0, 1]) \le \mathrm{Var}(g; [0, 1]) \tag{3.47}$$

in this case.
(c) *We have*

$$\lim_{h\to 0+} |||g_h - g|||_{AC} = 0 , \tag{3.48}$$

if and only if g is equivalent to an absolutely continuous function.

Proof. The assertion (a) follows from (3.45) and Proposition 3.4. To prove (b), suppose that $g \in BV([0, 1])$, and fix $h \in (0, 1)$. By the fundamental theorem of calculus and (3.45), we have

$$g_h'(x) = \frac{g(x+h) - g(x)}{h} \qquad (0 \le x \le 1-h) \tag{3.49}$$

9 In a similar form, such integral means occur in the definition of the integral modulus of continuity and the formulation of compactness criteria, see (0.98) and Proposition 0.55.

a.e. on $[0, 1-h]$, and $g_h'(x) \equiv 0$ on $[1-h, 1]$. Choose $m \in \mathbb{N}$ such that $(m-1)h \le 1-h < mh$. Then from Theorem 3.19, we deduce that

$$\begin{aligned} \operatorname{Var}(g_h; [0,1]) &= \int_0^{1-h} |g_h'(t)|\,dt = \frac{1}{h}\int_0^{1-h} |g(t+h) - g(t)|\,dt \\ &= \frac{1}{h}\int_0^h \sum_{j=1}^m |g(t+jh) - g(t+(j-1)h)|\,dt\,. \end{aligned} \tag{3.50}$$

For fixed t, consider the partition $P_t := \{t, t+h, \dots, t+mh\} \in \mathcal{P}([t, t+mh])$. We may then estimate the last integrand in (3.50) by

$$\sum_{j=1}^m |g(t+jh) - g(t+(j-1)h)| = \operatorname{Var}(g, P_t; [t, t+mh]) \le \operatorname{Var}(g; [0,1])\,, \tag{3.51}$$

and so (3.47) follows. Moreover, given $\varepsilon > 0$, we may find a $\delta > 0$ such that

$$\sum_{j=1}^m |g(t+jh) - g(t+(j-1)h)| > \operatorname{Var}(g; [0,1]) - \varepsilon$$

provided that $0 < h < \delta$. Combining this with (3.51), we conclude that

$$\operatorname{Var}(g; [0,1]) - \varepsilon < \operatorname{Var}(g_h; [0,1]) \le \operatorname{Var}(g; [0,1]),$$

showing that the upper limit in (3.46) exists and is finite.

Conversely, suppose that g is not equivalent to any function of bounded variation. Then for each $\omega > 0$, there is a $\delta > 0$ such that $\operatorname{Var}(g_h; [0,1]) > \omega$ for all $h \in (0, \delta)$, which implies that the upper limit (3.46) cannot be finite.

Now, we prove assertion (c). Suppose first that $g \in AC([0,1])$. Since $\|g - g_h\|_{L_1} \to 0$ as $h \to 0+$, for every function $g \in L_1([0,1])$ (Proposition 0.53), we only have to show that also

$$\lim_{h\to 0+} \|g' - g_h'\|_{L_1} = 0\,. \tag{3.52}$$

Since g is absolutely continuous, by (3.33) and (3.49), we have

$$\operatorname{Var}(g; [0,1]) = \int_0^1 |g'(t)|\,dt, \quad \operatorname{Var}(g_h; [0,1]) = \frac{1}{h}\int_0^{1-h} |g(t+h) - g(t)|\,dt\,.$$

Let $\varepsilon > 0$. As in the proof of Theorem 3.18, we find a $\delta \in (0,1)$ such that

$$\int_M |g'(x)|\,dx \le \varepsilon$$

for any set $M \subset [0,1]$ satisfying $\lambda(M) \le \delta$. Consider any sequence $(h_n)_n$ of positive real numbers converging to zero as $n \to \infty$. The equality (3.49) then shows that $(g_{h_n}'(x))_n$

converges to $g'(x)$ at any point x of differentiability of g, i.e. a.e. in $[0,1]$. By Egorov's theorem (Theorem 0.1), we find a set $T \subseteq [0,1]$ of measure $\lambda(T) > 1-\delta$ such that $(g'_{h_n})_n$ converges even uniformly on T to g' as $n \to \infty$. Choose $\eta > 0$ such that $0 < h_n < \eta$ implies

$$\int_T |g'_{h_n}(x) - g'(x)|\,dx \le \varepsilon\,.$$

Now, the estimate

$$\int_{[0,1]\setminus T} |g'_{h_n}(x)|\,dx + \int_T |g'_{h_n}(x)|\,dx \le \int_{[0,1]\setminus T} |g'(x)|\,dx + \int_T |g'(x)|\,dx$$

shows that

$$\begin{aligned}\int_{[0,1]\setminus T} |g'_{h_n}(x)|\,dx &\le \varepsilon + \int_T |g'(x)|\,dx - \int_T |g'_{h_n}(x)|\,dx\\ &\le \varepsilon \int_T |g'(x) - g'_{h_n}(x)|\,dx \le 2\varepsilon\,.\end{aligned}$$

It follows that

$$\begin{aligned}\int_0^1 |g'(x) - g'_{h_n}(x)|\,dx &\le \int_T |g'(x) - g'_{h_n}(x)|\,dx\\ &\quad + \int_{[0,1]\setminus T} |g'(x)|\,dx + \int_{[0,1]\setminus T} |g'_{h_n}(x)|\,dx \le 4\varepsilon\,.\end{aligned}$$

We conclude that (3.52) is true, and so $\|\!|g - g_h|\!\|_{AC} \to 0$ as $h \to 0+$. The converse argument is proved as in part (b). □

We illustrate Theorem 3.25 by means of a simple example involving a function $g \in BV([0,1]) \setminus AC([0,1])$.

Example 3.26. Let $g = \chi_{[1/2,1]} : [0,1] \to \mathbb{R}$ be the characteristic function of the interval $[1/2,1]$. Clearly, this function has bounded variation, being monotonically increasing, but is not equivalent to an absolutely continuous function. By Theorem 3.25, this means that

$$\limsup_{h\to 0+} \|\!|g_h|\!\|_{AC} < \infty, \quad \|\!|g - g_h|\!\|_{AC} \not\to 0 \quad (h \to 0+)\,. \tag{3.53}$$

Since only small values of h are important for proving (3.53), we take $h < 1/2$. A simple computation shows that

$$f(x) = \begin{cases} 0 & \text{for } 0 \le x < \frac{1}{2}\,,\\ x - \frac{1}{2} & \text{for } \frac{1}{2} \le x \le 1\end{cases}$$

and

$$f(x+h) = \begin{cases} 0 & \text{for } 0 \le x < \frac{1}{2} - h, \\ x + h - \frac{1}{2} & \text{for } \frac{1}{2} - h \le x < 1 - h, \\ \frac{1}{2} & \text{for } 1 - h \le x \le 1. \end{cases}$$

So, the function (3.45) and its derivative have the form

$$g_h(x) = \begin{cases} 0 & \text{for } 0 \le x < \frac{1}{2} - h, \\ \frac{2x-1}{2h} + 1 & \text{for } \frac{1}{2} - h \le x < \frac{1}{2}, \\ 1 & \text{for } \frac{1}{2} \le x \le 1 \end{cases}$$

and

$$g_h'(x) = \begin{cases} 0 & \text{for } 0 \le x < \frac{1}{2} - h, \\ \frac{1}{h} & \text{for } \frac{1}{2} - h \le x < \frac{1}{2}, \\ 0 & \text{for } \frac{1}{2} \le x \le 1, \end{cases}$$

respectively. Clearly, $\|g_h\|_{L_1} = \frac{1}{2}(1+h)$ and $\|g_h'\|_{L_1} \equiv 1$ for $0 < h < 1/2$, so the upper limit in (3.53) exists and has the value 3/2. On the other hand,

$$\|g - g_h\|_{L_1} = \frac{h}{2} \to 0, \quad \|g' - g_h'\|_{L_1} = \frac{2}{h} \to \infty \quad (h \to 0+),$$

which illustrates the second statement in (3.53). ♥

Other examples of this kind involving functions $f \in BV \setminus AC$ may be found in Exercises 3.34 and 3.37.

3.4 Rectifiable functions

A geometrical notion which is closely related to bounded variation and absolute continuity is that of the length of the graph

$$\Gamma(f) = \{(x, f(x)) : a \le x \le b\} \tag{3.54}$$

of a function $f : [a, b] \to \mathbb{R}$. This length is defined, in accordance with our geometric intuition, as supremum of the lengths of all polygons whose knots lie on the graph of f. Since such polygons are defined with respect to suitable partitions, it is not surprising that functions with finite graph length (over a compact interval $[a, b]$) are precisely the elements of $BV([a, b])$; we will prove this in Proposition 3.28 below.

First, we have to make this geometric idea more precise. Let $P = \{t_0, t_1, \ldots, t_m\} \in \mathcal{P}([a, b])$, and let

$$(t_0, f(t_0)), (t_1, f(t_1)), \ldots, (t_m, f(t_m)) \in \Gamma(f)$$

denote the corresponding knots on the graph of f. The total Euclidean length of this polygon on $[a, b]$ is then given by

$$L(\Gamma(f), P) = L(\Gamma(f), P; [a, b]) = \sum_{j=1}^{m} \sqrt{\left(t_j - t_{j-1}\right)^2 + \left(f(t_j) - f(t_{j-1})\right)^2}. \tag{3.55}$$

Therefore, the following definition seems reasonable.

Definition 3.27. The *graph length* of f over $[a, b]$ is defined by

$$L(\Gamma(f)) = L(\Gamma(f); [a, b]) := \sup \{L(\Gamma(f), P; [a, b]) : P \in \mathcal{P}([a, b])\}, \tag{3.56}$$

where the supremum in (3.56) is taken over all partitions P of the interval $[a, b]$. In case $L(\Gamma(f)) < \infty$, the function f (or its graph $\Gamma(f)$) is called *rectifiable*. In this case, we call the function $L_f : [a, b] \to \mathbb{R}$, defined by

$$L_f(x) := L(\Gamma(f); [a, x]) \quad (a \leq x \leq b), \tag{3.57}$$

the *length function* of f. ■

Loosely speaking,[10] the length function L_f may be used for "scanning" the graph of f in the same way as the variation function V_f may be used for "scanning" the oscillations of f. For further reference, we summarize some useful properties of the graph length (3.56) in the following

Proposition 3.28. *The graph length* (3.56) *has the following properties.*
(a) *For any $c \in (a, b)$, the additivity formula*

$$L(\Gamma(f); [a, b]) = L(\Gamma(f); [a, c]) + L(\Gamma(f); [c, b]) \tag{3.58}$$

is true.
(b) *The estimate*

$$\mathrm{Var}(f; [a, b]) \leq L(\Gamma(f); [a, b]) \leq \mathrm{Var}(f; [a, b]) + b - a \tag{3.59}$$

holds.
(c) *A function $f : [a, b] \to \mathbb{R}$ has finite graph length if and only if it is of bounded variation.*

Proof. The assertion (a) is obvious. To prove (b), we use the trivial estimate $|B| \leq \sqrt{A^2 + B^2}$ for $A, B \in \mathbb{R}$ and get

$$\sum_{j=1}^{m} |f(t_j) - f(t_{j-1})| \leq \sum_{j=1}^{m} \sqrt{\left(t_j - t_{j-1}\right)^2 + \left(f(t_j) - f(t_{j-1})\right)^2} = L(\Gamma(f), P)$$

10 A precise relation between the length function L_f and the variation function V_f is given in Exercise 3.44.

which gives $\mathrm{Var}(f;[a,b]) \leq L(\Gamma(f);[a,b])$ after passing to the supremum over $P \in \mathcal{P}([a,b])$. On the other hand, using the estimate $\sqrt{A^2+B^2} \leq |A|+|B|$, we obtain

$$\begin{aligned}\sum_{j=1}^{m} \sqrt{(t_j - t_{j-1})^2 + (f(t_j) - f(t_{j-1}))^2} &\leq \sum_{j=1}^{m} \left[|t_j - t_{j-1}| + |f(t_j) - f(t_{j-1})|\right] \\ &= \sum_{j=1}^{m} |t_j - t_{j-1}| + \sum_{j=1}^{m} |f(t_j) - f(t_{j-1})| \\ &= b - a + \mathrm{Var}(f, P; [a,b]),\end{aligned}$$

and passing again to the supremum over $P \in \mathcal{P}([a,b])$ yields

$$L(\Gamma(f);[a,b]) \leq b - a + \mathrm{Var}(f;[a,b]).$$

So, we have shown that f is rectifiable if and only if $f \in BV([a,b])$ which is (c). □

As Example 1.8 shows, a continuous function may be nonrectifiable. On the other hand, any C^1-function is rectifiable by (1.46) and the fact that $C^1([a,b]) \subseteq Lip([a,b])$. Moreover, for C^1-functions, we may easily calculate their graph length:

Proposition 3.29. *For $f \in C^1([a,b])$, we have*[11]

$$L(\Gamma(f);[a,b]) = \int_a^b \sqrt{1 + f'(x)^2}\, dx. \tag{3.60}$$

Proof. Since $f \in C^1([a,b])$, the function $x \mapsto \sqrt{1+f'(x)^2}$ is continuous, and so the (Riemann) integral on the right-hand side of (3.60) exists; we denote it by I. First, we show that, given an arbitrary partition $P = \{t_0, t_1, \ldots, t_m\} \in \mathcal{P}([a,b])$, we have

$$\sqrt{(t_j - t_{j-1})^2 + (f(t_j) - f(t_{j-1}))^2} \leq I. \tag{3.61}$$

To see this, we choose $r_j > 0$ and $\theta_j \in [0, 2\pi)$ in such a way that[12]

$$t_j - t_{j-1} =: r_j \cos\theta_j, \quad f(t_j) - f(t_{j-1}) = r_j \sin\theta_j.$$

Then

$$\sqrt{(t_j - t_{j-1})^2 + (f(t_j) - f(t_{j-1}))^2} = r_j = \int_{t_{j-1}}^{t_j} [\cos\theta_j + f'(x)\sin\theta_j]\, dx,$$

11 Observe that (3.60) may be considered as a direct analogue to formula (3.33) in Theorem 3.19 if one considers functions with values in the plane and replaces the absolute value of $f'(t)$ by the Euclidean norm of the pair $(1, f'(t))$, see also (1.106).

12 This is possible since the function $t \mapsto \tan t$ is a bijection between the interval $(-\pi/2, \pi/2)$ and the real axis.

and so

$$L(\Gamma(f), P; [a,b]) \le \sum_{j=1}^{m} \int_{t_{j-1}}^{t_j} [\cos\theta_j + f'(x)\sin\theta_j]\,dx$$

$$\le \sum_{j=1}^{m} \int_{t_{j-1}}^{t_j} \sqrt{1+f'(x)^2}\,dx = \int_a^b \sqrt{1+f'(x)^2}\,dx = I$$

by the Cauchy Schwarz inequality, and hence $L(\Gamma(f); [a,b]) \le I$.

To prove the converse inequality, we consider, for $n = 1,2,3,\ldots$, the equidistant partitions $P_n := \{t_{0,n}, t_{1,n}, \ldots, t_{n,n}\}$ given by

$$t_{0,n} := a,\ t_{1,n} := a + \frac{1}{n}(b-a),\ \ldots\ t_{n-1,n} := a + \frac{n-1}{n}(b-a),\ t_{n,n} := b. \tag{3.62}$$

Moreover, we define piecewise constant functions $g_n : [a,b] \to \mathbb{R}$ by

$$g_n(x) := \begin{cases} 0 & \text{for } x = t_{k,n}\ (k = 0,2,\ldots,n), \\ \frac{f(t_{k,n})-f(t_{k-1,n})}{t_{k,n}-t_{k-1,n}} & \text{for } t_{k-1,n} < x < t_{k,n}\ (k = 1,2,\ldots,n). \end{cases}$$

From $f \in C^1([a,b])$, it follows that $g_n(x) \to f'(x)$ for $n \to \infty$ and $a \le x \le b$, and so

$$I = \int_a^b \sqrt{1+f'(x)^2}\,dx \le \sup\left\{\int_a^b \sqrt{1+g_n(x)^2}\,dx : n = 1,2,3,\ldots\right\}.$$

However, by construction of g_n, we see that

$$\int_a^b \sqrt{1+g_n(x)^2}\,dx$$

$$= \sum_{k=1}^{n} \sqrt{\left(t_{k,n}-t_{k-1,n}\right)^2 + \left(f(t_{k,n}) - f(t_{k-1,n})\right)^2} = L(\Gamma(f), P_n; [a,b]),$$

and all real numbers $L(\Gamma(f), P_n)$ are of course bounded above by $L(\Gamma(f))$. Therefore, we have proved the estimate $I \le L(\Gamma(f); [a,b])$, and so (3.60) holds. □

The hypothesis $f \in C^1([a,b])$ in Proposition 3.29 is unnecessarily strong: one may show that (3.60) also holds if f is differentiable and f' is integrable[13], see Exercise 3.43.

One might hope that formula (3.60) also holds for monotone functions which are still differentiable almost everywhere. However, this is not true:

13 Remember that we have to *assume* the integrability of f' since it does not follow from the differentiability of f, see Example 3.15.

Example 3.30. Let $\varphi : [0,1] \to \mathbb{R}$ be the Cantor function (3.6). We claim that

$$L(\Gamma(\varphi); [0,1]) = 2\,. \tag{3.63}$$

To prove this, we cannot use formula (3.60) since φ is not everywhere differentiable on $[0,1]$; instead, we have to go back to Definition 3.27.

Obviously, the length of any polygon with knots on the graph of φ cannot be larger than the sum of all horizontal and vertical projections of the graph of φ and all its "gaps," which is 2. So, we trivially have $L(\Gamma(\varphi); [0,1]) \le 2$.

To prove the converse estimate, we consider the polygon which starts at $(0,0)$, ends at $(1,1)$, and has as intermediate knots the endpoints of the (finitely many) intervals which we have canceled at the n-th step in the construction (3.4) of the Cantor set. The lengths of the horizontal pieces of this polygon sum up to

$$\sum_{k=1}^{n} \frac{2^{k-1}}{3^k} = 1 - \frac{2^n}{3^n}\,.$$

On the other hand, the oblique pieces of this polygon all have the same length, namely,

$$\lambda_n := \sqrt{\frac{1}{2^{2n}} + \frac{1}{3^{2n}}} = \frac{1}{2^n}\sqrt{1 + \frac{2^{2n}}{3^{2n}}}\,.$$

Since the polygon precisely contains 2^n such oblique pieces, the total length of the polygon becomes

$$L_n = 1 - \frac{2^n}{3^n} + 2^n\lambda_n = 1 - \frac{2^n}{3^n} + \sqrt{1 + \frac{2^{2n}}{3^{2n}}}\,.$$

However, we can take the last expression arbitrarily close to 2 by choosing n sufficiently large, and so (3.63) follows. ♥

There is a general principle behind Example 3.30 which is discussed in Exercise 3.41. We point out again that we could not use (3.60) for calculating the graph length of φ in Example 3.30. Indeed, the derivative of φ at all points, where it exists, is 0, and so we would obtain the wrong value 1 for the integral if we blindly adopted formula (3.60) in this example.

Our discussion shows that existence everywhere and continuity of f' is too strong, while existence a.e. and integrability of f' is too weak to ensure the equality (3.60). The following Theorem 3.31 gives the "correct" condition.

Theorem 3.31. *For $f \in BV([a,b]) \cap C([a,b])$, the lower estimate*

$$L(\Gamma(f); [a,b]) \ge \int_a^b \sqrt{1 + f'(x)^2}\, dx \tag{3.64}$$

is true, while for $f \in AC([a,b])$, the equality (3.60) *holds.*

Proof. We use Proposition 3.21 which asserts that a function $f \in BV([a,b]) \cap C([a,b])$ may be represented in the form (3.38), where f_{ac} is absolutely continuous and f_{sg} is singular (or zero). Thus, we may write the graph (3.54) of f in the form

$$\Gamma(f) = \{(x, f_{ac}(x)) + (0, f_{sg}(x)) : a \leq x \leq b\}.$$

The length of the absolutely continuous part is

$$L_a = \int_a^b \sqrt{1 + f_{ac}'(x)^2}\, dx = \int_a^b \sqrt{1 + f'(x)^2}\, dx, \tag{3.65}$$

while the length of the singular part is simply

$$L_s = \mathrm{Var}(f_{sg}; [a,b]). \tag{3.66}$$

Thus, the total graph length of f adds up to

$$L(\Gamma(f); [a,b]) = L_a + L_s = \int_a^b \sqrt{1 + f'(x)^2}\, dx + \mathrm{Var}(f_{sg}; [a,b]), \tag{3.67}$$

from which both statements follow. □

Theorem 3.31 shows that in our Example 3.30, we have intentionally chosen a function $\varphi \in BV([0,1]) \setminus AC([0,1])$ to get strict inequality in (3.64). Since the Cantor function φ is even monotone, one may also calculate the area under its graph "by hand," see Exercise 3.12.

In addition to Propositions 1.7 and 3.22, we give another result which establishes a link between a function $f \in BV([a,b])$ and its variation function V_f given in (1.13).

Proposition 3.32. *A function f is rectifiable if and only if its variation function V_f given in* (1.13) *is rectifiable; moreover, in this case, we have*

$$L(\Gamma(V_f); [a,b]) = L(\Gamma(f); [a,b]). \tag{3.68}$$

Proof. The first assertion follows from Theorem 1.26 (b) and the fact that the functions $f \in BV$ are precisely those with a rectifiable graph. To prove (3.68), let $P = \{t_0, t_1, \ldots, t_m\} \in \mathcal{P}([a,b])$. Then Proposition 3.28 (b) implies

$$\sqrt{\left(V_f(t_j) - V_f(t_{j-1})\right)^2 + (t_j - t_{j-1})^2} \leq L_f(t_j) - L_f(t_{j-1}) \quad (j = 1, 2, \ldots, m),$$

where L_f denotes the length function (3.57) of f. Consequently,

$$\begin{aligned}&\sum_{j=1}^m \sqrt{\left(V_f(t_j) - V_f(t_{j-1})\right)^2 + (t_j - t_{j-1})^2}\\&\qquad \leq \sum_{j=1}^m (L_f(t_j) - L_f(t_{j-1})) = L(\Gamma(f), P; [a,b]),\end{aligned}$$

and so $L(\Gamma(V_f);[a,b]) \le L(\Gamma(f);[a,b])$. On the other hand, for each $j \in \{1,2,\dots,m\}$, we have, by Proposition 1.3 (c),

$$
\begin{aligned}
L(\Gamma(f),P;[a,b]) &= \sum_{j=1}^{m} \sqrt{\big(f(t_j)-f(t_{j-1})\big)^2 + (t_j - t_{j-1})^2} \\
&\le \sum_{j=1}^{m} \sqrt{(V_f(t_j)-V_f(t_{j-1})) + (t_j - t_{j-1})^2} \le L(\Gamma(V_f),P;[a,b])\,,
\end{aligned}
$$

which proves the converse estimate. □

A certain refinement of Proposition 3.32 is given in Exercise 3.44. We collect in the following Table 3.3 the properties of a function $f:[a,b]\to\mathbb{R}$ which carry over to its variation function, or vice versa.

Table 3.3. Relations between f and V_f.

$f \in C([a,b])$	$\Leftrightarrow$	$V_f \in C([a,b])$	(Theorem 1.26 (a))
$f \in BV([a,b])$	$\Leftrightarrow$	$V_f \in BV([a,b])$	(Theorem 1.26 (b))
$f \in Lip([a,b])$	$\Leftrightarrow$	$V_f \in Lip([a,b])$	(Theorem 1.26 (c))
$f \in Lip_\alpha([a,b])$	$\Leftarrow$	$V_f \in Lip_\alpha([a,b])$	(Theorem 1.26 (d))
$f \in AC([a,b])$	$\Leftrightarrow$	$V_f \in AC([a,b])$	(Theorem 1.26 (e))
f differentiable a.e.	$\Leftrightarrow$	V_f differentiable a.e.	(Proposition 3.22)
f rectifiable	$\Leftrightarrow$	V_f rectifiable	(Proposition 3.32)

At this point, it is time to take a breath and to recall all the estimates for functions of bounded variation which, as a rule, turn into equalities for absolutely continuous functions.

- The estimate

$$V_f(x) \ge \int_a^x |f'(t)|\,dt$$

  for the variation function (1.13) holds for every function $f \in BV([a,b])$, with equality in case $f \in AC([a,b])$.
- The estimate

$$\mathrm{Var}(f;[a,b]) \ge \int_a^b |f'(t)|\,dt$$

  for the total variation (1.4) holds for every function $f \in BV([a,b])$, with equality[14] in case $f \in AC([a,b])$.

---

14 Taking into account the definition (1.16) of the BV-norm, we even see that $\|f\|_{BV} = \|f'\|_{L_1}$ for $f \in AC^o([a,b]) = \{f \in AC([a,b]),\, f(a)=0\}$.

- The estimate
$$L_f(x) \geq \int_a^x \sqrt{1 + f'(t)^2}\, dt$$
for the length function (3.57) holds for every function $f \in BV([a, b])$, with equality in case $f \in AC([a, b])$.
- The estimate
$$L(\Gamma(f); [a, b]) \geq \int_a^b \sqrt{1 + f'(t)^2}\, dt$$
for the graph length (3.56) holds for every function $f \in BV([a, b])$, with equality in case $f \in AC([a, b])$.
- The estimate
$$f(b) - f(a) \geq \int_a^b f'(t)\, dt$$
holds for every increasing function $f : [a, b] \to \mathbb{R}$, with equality in case $f \in AC([a, b])$.
- The equality[15]
$$lip(f; [a, b]) = \operatorname{esssup} \{|f'(t)| : a \leq t \leq b\}$$
for the smallest Lipschitz constant (0.68) holds for $f \in Lip([a, b])$.

The monotonically increasing continuous Cantor function φ which is not absolutely continuous may be used throughout as an example for strict inequalities in this list. The reason for this is that $\varphi'(x) = 0$ a.e. on $[0, 1]$, although φ is not constant. For absolutely continuous functions, this is impossible:

Proposition 3.33. *If $f \in AC([a, b])$ satisfies $f'(x) = 0$ a.e. on $[a, b]$, then f is constant.*

Proof. Given a fixed $c \in (a, b]$, we show that $f(c) = f(a)$. Let $M := \{x : a \leq x \leq c, f'(x) = 0\}$. Since $f \in AC([a, b])$, for $\varepsilon > 0$, we find $\delta > 0$ such that $\Gamma(f; S) \leq \varepsilon$ for any $S \in \Sigma([a, b])$ satisfying $\Theta(S) \leq \delta$. Furthermore, for fixed $x \in M$, we may choose, by definition of M, a small interval $[a_x, c_x] \subset [a, c]$ such that
$$a_x \leq x \leq c_x, \quad |f(c_x) - f(a_x)| \leq \varepsilon(c_x - a_x).$$

Now, by the Vitali covering theorem, we can find a finite collection
$$S := \{[a_{x,1}, c_{x,1}], \dots, [a_{x,n}, c_{x,n}]\}$$
of nonoverlapping intervals of this sort such that
$$\lambda\left(M \setminus [a_{x,1}, c_{x,1}] \cup \dots \cup [a_{x,n}, c_{x,n}]\right) \leq \delta.$$

15 Taking into account the definition (0.70) of the *Lip*-norm, we even see that $\|f\|_{Lip} = \|f'\|_{L_\infty}$ for $f \in Lip^o([a, b]) = \{f \in Lip([a, b]), f(a) = 0\}$.

Since $\lambda([a,c] \setminus M) = 0$, by assumption, we have

$$\begin{aligned}\lambda\left([a,c] \setminus [a_{x,1}, c_{x,1}] \cup \ldots \cup [a_{x,n}, c_{x,n}]\right) \\ = \lambda\left([a,c] \setminus [a_{x,1}, c_{x,1}] \cup \ldots \cup [a_{x,n}, c_{x,n}]\right) \leq \delta .\end{aligned}$$

Suppose, without loss of generality, that $a \leq a_{x,1} < c_{x,1} \leq a_{x,2} < \ldots < c_{x,n-1} \leq a_{x,n} < c_{x,n} \leq c$, and put $c_{x,0} := a$ and $a_{x,n+1} := c$. Then

$$\Theta(S \cup \{a,c\}) = \sum_{k=0}^{n} (a_{x,k+1} - c_{x,k}) = \lambda\left([a,c] \setminus \bigcup_{k=1}^{n} [a_{x,k}, c_{x,k}] \right) \leq \delta ,$$

and hence

$$\sum_{k=0}^{n} |f(a_{x,k+1}) - f(c_{x,k})| \leq \varepsilon .$$

Furthermore, the definition of the intervals $[a_{x,k}, c_{x,k}]$ implies that

$$\sum_{k=1}^{n} |f(c_{x,k}) - f(a_{x,k})| \leq \varepsilon \sum_{k=1}^{n} (c_{x,k}) - a_{x,k}) \leq \varepsilon(c-a) .$$

Combining these estimates, we obtain

$$|f(c) - f(a)| \leq \sum_{k=0}^{n} |f(a_{x,k+1}) - f(c_{x,k})| + \sum_{k=1}^{n} |f(c_{x,k}) - f(a_{x,k})| \leq \varepsilon(c - a + 1) .$$

Since $\varepsilon > 0$ was arbitrary, we conclude that $f(c) = f(a)$ as claimed. □

Proposition 3.33 shows that a singular function in the sense of Definition 3.7 cannot be absolutely continuous.[16] So Theorem 3.9 implies that a singular function *must* fail to have the Luzin property. The Cantor function (3.6) is of course a standard example. In general, we may formulate a "Golden Rule" which states that whenever one wants to see whether or not a result for absolutely continuous functions carries over to functions of bounded variation, it is a useful device to take the Cantor function (3.6) as a "test animal."

Observe that we could have deduced Proposition 3.33 also very easily from Theorem 3.18: indeed, our hypothesis implies that the integral in (3.18) is zero, and so $f(x) \equiv f(a)$. We have given an independent proof which gives some insight and uses an interesting compactness argument.

Conversely, we also note that Proposition 3.33 may be used to prove the equality (3.18) for $f \in AC$ quite easily once we know that $f' \in L_1$. In fact, the function $\tilde{f} : [a,b] \to \mathbb{R}$ defined by

$$\tilde{f}(x) := f(a) + \int_a^x f'(t)\, dt \quad (a \leq x \leq b)$$

16 It is precisely this fact which shows that the decomposition (3.38) makes sense.

is absolutely continuous, by Proposition 3.4, and satisfies $\tilde{f}' = f'$ a.e. on $[a, b]$. So, from Proposition 3.33, we conclude that $\tilde{f} - f$ is constant on $[a, b]$. However, $\tilde{f}(a) = f(a)$, and (3.18) follows. In this way, we have obtained a complete answer to the two questions raised at the beginning of Section 3.3.

3.5 The Riesz–Medvedev theorem

As we have seen in Section 3.3, we may characterize Lipschitz continuous and absolutely continuous functions through properties of their derivatives in the following way:

- Lipschitz continuous functions are precisely those with L_∞-derivatives a.e. (Theorem 3.20).
- Absolutely continuous functions are precisely those with L_1-derivatives a.e. (Proposition 3.4 and Theorem 3.18).

Now, the question arises if one may also characterize the functions with L_p-derivatives for $1 < p < \infty$; by the inclusions (0.15), such functions should be intermediate between Lipschitz continuous and absolutely continuous functions. It turns out that in this way, we get nothing else but the space RBV_p which, by (2.93), is in fact intermediate between the spaces *Lip* and *AC*. This is the contents of the famous *Riesz theorem* which reads as follows.

Theorem 3.34 (Riesz). *Let $1 < p < \infty$. Then a function f belongs to $RBV_p([a, b])$ if and only if $f \in AC([a, b])$ and $f' \in L_p([a, b])$. Moreover, in this case, the equality*

$$\mathrm{Var}_p^R(f; [a, b]) = \|f'\|_{L_p}^p = \int_a^b |f'(x)|^p \, dx \tag{3.69}$$

holds, where $\mathrm{Var}_p^R(f; [a, b])$ denotes the p-variation (2.88) of f in Riesz's sense.

Proof. To prove the "if" part, we suppose that f is absolutely continuous with $f' \in L_p([a, b])$. For $x, y \in [a, b]$ with $x < y$, we then get, by Hölder's inequality (0.14),

$$\begin{aligned} |f(y) - f(x)|^p &= \left| \int_x^y f'(t) \, dt \right|^p \leq \left(\int_x^y |f'(t)| \, dt \right)^p \\ &\leq \left[\left(\int_x^y dt \right)^{1-1/p} \right]^p \int_x^y |f'(t)|^p \, dt = (y - x)^{p-1} \int_x^y |f'(t)|^p \, dt \, . \end{aligned}$$

Consequently,

$$\frac{|f(x) - f(y)|^p}{|x - y|^{p-1}} \leq \int_x^y |f'(t)|^p \, dt \leq \int_a^b |f'(t)|^p \, dt \, ,$$

and hence $\mathrm{Var}_p^R(f;[a,b]) \le \|f'\|_{L_p}^p$, which shows that $f \in RBV_p([a,b])$ and proves one estimate for (3.69).

We already know from Proposition 2.52 that the inclusion $RBV_p([a,b]) \subseteq AC([a,b])$ is true for $p > 1$, and so $f'(x)$ exists a.e. on $[a,b]$. Consequently, to prove the "only if" part, we merely have to show that $f' \in L_p([a,b])$.

As in the proof of Proposition 3.29, for every $n \in \mathbb{N}$, we consider the equidistant partition (3.62), and define piecewise constant functions $g_n : [a,b] \to \mathbb{R}$ for $n = 1,2,3,\dots$ and $k = 0,1,\dots,n$ by

$$g_n(t) := \begin{cases} \frac{f(t_{k+1,n})-f(t_{k,n})}{t_{k+1,n}-t_{k,n}} & \text{for } t_{k,n} \le t < t_{k+1,n}, \\ 0 & \text{for } t = b. \end{cases}$$

A somewhat cumbersome, but straightforward calculation, then shows that $g_n(x) \to f'(x)$ a.e. in $[a,b]$. From Fatou's theorem (Theorem 0.5), we conclude that

$$\int_a^b |f'(t)|^p\,dt \le \lim_{n\to\infty} \int_a^b |g_n(t)|^p\,dt = \lim_{n\to\infty} \sum_{k=1}^n \frac{|f(t_{k+1,n}) - f(t_{k,n})|^p}{|t_{k+1,n} - t_{k,n}|^{p-1}} \le \mathrm{Var}_p^R(f;[a,b]).$$

This shows that $f' \in L_p([a,b])$ and proves the other estimate for (3.69). □

We make some comments on Theorem 3.34. First of all, we point out that an analogous result is not true in case $p = 1$ since functions $f \in RBV_1 = BV$ are generally not continuous, let alone absolutely continuous. However, a certain analogue of Theorem 3.34 for $p = \infty$ *is* true: since $RBV_\infty = Lip$, Theorem 3.20 may be considered as such an analogue.

Moreover, Theorem 3.34 may be used to give another proof of Proposition 2.54. In fact, given $f \in RBV_q([a,b])$, we have $f' \in L_q([a,b])$, and so $f' \in L_p([a,b])$. Therefore, $f \in RBV_p([a,b])$ for any $p \le q$.

Another interesting consequence of Theorem 3.34 is the following. If we define the subspace RBV_p^o, as in Proposition 2.51, by

$$RBV_p^o([a,b]) := \{f \in RBV_p([a,b]) : f(a) = 0\}, \tag{3.70}$$

the differential operator D defined by $Df := f'$ is a linear surjective isometry between $RBV_p^o([a,b])$ and $L_p([a,b])$. This is not only of theoretical interest, but also of practical use: it makes it possible to calculate the norm of functions $f \in RBV_p$ quite easily. Here is a simple example.

Example 3.35. For $0 < \tau < 1$, let $f_\tau : [0,1] \to \mathbb{R}$ be defined as in Example 2.78. Then f_τ' exists and belongs to $L_p([0,1])$ if and only if $p(1-\tau) < 1$, which reconfirms (2.153). Moreover, in this case, we have

$$\|f_\tau'\|_{L_p}^p = \int_0^1 \tau^p x^{p(\tau-1)}\,dx = \frac{\tau^p}{1-(1-\tau)p},$$

in accordance with our result for $\|f_\tau\|_{RBV_p}$ in Example 2.78. ♥

The meaning of Theorem 3.34 becomes clearer if we write the Riesz variation (2.87) with respect to $P = \{t_0, t_1, \ldots, t_m\} \in \mathcal{P}([a,b])$ in the form

$$\operatorname{Var}_p^R(f, P; [a,b]) = \sum_{j=1}^{m} \left(\frac{|f(t_j) - f(t_{j-1})|}{t_j - t_{j-1}} \right)^p (t_j - t_{j-1}) . \tag{3.71}$$

Indeed, the expression (3.71) may then be viewed as a Riemann sum, and for $\mu(P) \to 0$, see (1.2), the summation becomes an integral, while the term after the summation approaches, roughly speaking, $|f'(t)|^p\, dt$. The notation (3.71) also motivates the definition (2.96) for the Riesz variation with respect to an arbitrary Young function ϕ. In this more general setting, the expression after the summation in (2.96) approaches $\phi(|f'(t)|)\, dt$ as $\mu(P) \to 0$. Therefore, the following important generalization of Theorem 3.34 which involves the Orlicz class $\mathcal{L}_\phi([a,b])$ (Definition 0.16) and is due to Medvedev [211] appears very natural.

Theorem 3.36 (Medvedev). *Let ϕ be a Young function which satisfies condition ∞_1, see (2.16). Then a function f belongs to $RBV_\phi([a,b])$ if and only if $f \in AC([a,b])$ and $f' \in \mathcal{L}_\phi([a,b])$. Moreover, in this case, the equality*

$$\operatorname{Var}_\phi^R(f; [a,b]) = \int_a^b \phi(|f'(x)|)\, dx \tag{3.72}$$

holds, where $\operatorname{Var}_\phi^R(f; [a,b])$ denotes the ϕ-variation (2.97) of f in Riesz's sense.

Proof. Suppose first that $f \in RBV_\phi([a,b])$, and fix $S = \{[a_1,b_1], \ldots, [a_n,b_n]\} \in \Sigma([a,b])$. Let $\varepsilon > 0$. From our hypothesis $\phi \in \infty_1$, it follows that for each $\varepsilon > 0$ and $k > 0$, we can find a $\delta > 0$ such that $0 \le c \le \delta$ implies

$$c\phi^{-1}\left(\frac{k}{c}\right) \le \varepsilon .$$

Taking, in particular, $c := \Theta(S)$ and $k := \operatorname{Var}_\phi^R(f; [a,b])$, we deduce that $\Theta(S) \le \delta$ which implies

$$\Theta(S)\phi^{-1}\left(\frac{\operatorname{Var}_\phi^R(f; [a,b])}{\Theta(S)}\right) \le \varepsilon . \tag{3.73}$$

Now, applying the discrete Jensen inequality (2.18) to

$$\alpha_k := b_k - a_k, \quad u_k := \frac{|f(b_k) - f(a_k)|}{b_k - a_k} \quad (k = 1, 2, \ldots, n)$$

and using the notation (3.1), we get the estimates

$$\phi\left(\frac{\Gamma(f;S)}{\Theta(S)}\right) \le \frac{\sum_{k=1}^{n} (b_k - a_k)\phi\left(\frac{|f(b_k)-f(a_k)|}{b_k - a_k}\right)}{\Theta(S)} = \frac{\operatorname{Var}_\phi^R(f, S; [a,b])}{\Theta(S)} \le \frac{\operatorname{Var}_\phi^R(f; [a,b])}{\Theta(S)} .$$

Combining this with (3.73) and using the monotonicity of ϕ^{-1}, we obtain

$$\Gamma(f;S) = \Theta(S)\frac{\Gamma(f;S)}{\Theta(S)} \le \Theta(S)\phi^{-1}\left(\frac{\operatorname{Var}_\phi^R(f; [a,b])}{\Theta(S)}\right) \le \varepsilon .$$

This shows that f is absolutely continuous on $[a, b]$. By Theorem 3.18, f admits a representation

$$f(x) = \int_a^x g(t)\,dt \quad (a \le x \le b)$$

with a suitable function $g \in L_1([a, b])$ satisfying $f' = g$ a.e. on $[a, b]$. It remains to show that

$$\int_a^b \phi(|g(t)|)\,dt < \infty\,. \tag{3.74}$$

For $n \in \mathbb{N}$, consider the equidistant partition $P_n := \{t_{0,n}, t_{1,n}, \ldots, t_{n-1,n}, t_{n,n}\} \in \mathcal{P}([a, b])$ given by (3.62), and define functions $g_n : [a, b] \to \mathbb{R}$ as in the proof of Proposition 3.29. Then the sequence $(g_n)_n$ converges a.e. on $[a, b]$ to g, and therefore

$$\int_a^b \phi(|g(t)|)\,dt \le \sup\left\{\int_a^b \phi(|g_n(t)|)\,dt : n = 1, 2, 3, \ldots\right\},$$

by Fatou's theorem (Theorem 0.5). However,

$$\begin{aligned}
\int_a^b \phi(|g(t)|)\,dt &= \sum_{k=1}^n \int_{t_{k-1,n}}^{t_{k,n}} \phi\left(\frac{|f(t_{k,n}) - f(t_{k-1,n})|}{t_{k,n} - t_{k-1,n}}\right) dt \\
&= \sum_{k=1}^n \phi\left(\frac{|f(t_{k,n}) - f(t_{k-1,n})|}{t_{k,n} - t_{k-1,n}}\right)(t_{k,n} - t_{k-1,n}) \\
&= \operatorname{Var}_\phi^R(f, P_n; [a, b]) \le \operatorname{Var}_\phi^R(f; [a, b]),
\end{aligned}$$

not only proving (3.74), but also the estimate $\ge$ in (3.72).

Conversely, suppose now that $f \in AC([a, b])$ and $g := f'$ satisfies (3.74). Fix a partition $P = \{t_0, t_1, \ldots, t_n\} \in \mathcal{P}([a, b])$. Then

$$\phi\left(\frac{|f(t_k) - f(t_{k-1})|}{t_k - t_{k-1}}\right) = \phi\left(\frac{1}{t_k - t_{k-1}} \int_{t_{k-1}}^{t_k} g(t)\,dt\right).$$

Applying to the right-hand side the continuous Jensen inequality (2.19) for $\alpha(t) \equiv 1$, $u(t) := |g(t)|$ and $[c, d] := [t_{k-1}, t_k]$, we get the estimate

$$\phi\left(\frac{1}{t_k - t_{k-1}} \int_{t_{k-1}}^{t_k} g(t)\,dt\right) \le \frac{1}{t_k - t_{k-1}} \int_{t_{k-1}}^{t_k} \phi(|g(t)|)\,dt\,.$$

Multiplying by $t_k - t_{k-1}$ and summing up over $k = 1, 2, \ldots, n$ gives

$$\operatorname{Var}_\phi^R(f, P; [a, b]) = \sum_{k=1}^n \phi\left(\frac{|f(t_k) - f(t_{k-1})|}{t_k - t_{k-1}}\right)(t_k - t_{k-1}) \le \int_a^b \phi(|g(t)|)\,dt\,.$$

Since the last term in this estimate is independent of P, we have proved not only that $f \in RBV_\phi([a,b])$, but also the estimate $\le$ in (3.72). □

Clearly, for $\phi(t) = t^p$ with $1 < p < \infty$, we have $\varphi \in \infty_1$, and (3.72) reduces to (3.69). Thus, Theorem 3.36 contains Theorem 3.34 as a special case. Moreover, since Theorem 3.34 is false for $\phi(t) = t$, i.e. $RBV_\phi = BV$, we cannot drop the assumption $\phi \in \infty_1$ in Theorem 3.36.

3.6 Higher order Riesz-type variations

In this section, we use the higher order divided differences considered in Definition 2.71 to introduce and study higher order variations in Riesz's sense. The following Definitions 3.37 and 3.38 are parallel to Definitions 2.71 and 2.72.

Definition 3.37. Given a function $f : [a,b] \to \mathbb{R}$ and points $t_0, t_1, \dots, t_k \in [a,b]$, let the higher order divided differences be defined as in (2.139). Given a partition $P = \{t_0, t_1, \dots, t_m\} \in \mathcal{P}([a,b])$, $k \in \mathbb{N}$, and $p \in [1,\infty)$, we set

$$\mathrm{Var}^R_{k,p}(f, P; [a,b]) := \sum_{j=1}^{m-k+1} \frac{|f[t_j, \dots, t_{j+k-1}] - f[t_{j-1}, \dots, t_{j+k-2}]|^p}{|t_j - t_{j-1}|^{p-1}} \tag{3.75}$$

and call (3.75) the *(k, p)-variation* of f with respect to P (in Riesz's sense) on $[a,b]$. ■

Again, Definition 3.37 contains some notions of variation defined previously in the literature. In the special case $k = 1$ and $p = 1$, (3.75) reduces to Jordan's classical definition of $\mathrm{Var}(f, P; [a,b])$, while in the special case $k = 1$ and $p > 1$, (3.75) reduces to Riesz's variation $\mathrm{Var}^R_p(f, P; [a,b])$ which we studied in detail in Section 2.4. On the other hand, for arbitrary $k \in \mathbb{N}$ and $p = 1$, the $(k, 1)$-variation $\mathrm{Var}^R_{k,1}(f, P; [a,b])$ coincides with the $(k, 1)$-variation $\mathrm{Var}^W_{k,1}(f, P; [a,b])$ in Popoviciu's sense, see (2.142). However, for $k \in \mathbb{N}$ and $p > 1$, the expression (3.75) is of course different from the (k, p)-variation $\mathrm{Var}^W_{k,p}(f, P; [a,b])$ considered in (2.140) since we have the additional denominator $|t_j - t_{j-1}|^{p-1}$.

Definition 3.38. Given $k \in \mathbb{N}$ and $p \in [1,\infty)$, we call the (possibly infinite) number

$$\mathrm{Var}^R_{k,p}(f; [a,b]) := \sup\,\{\mathrm{Var}^R_{k,p}(f, P; [a,b]) : P \in \mathcal{P}([a,b])\}\,, \tag{3.76}$$

where the supremum is taken over all partitions of $[a,b]$, the *total (k, p)-variation* of f (in Riesz's sense) on $[a,b]$. The set

$$RBV_{k,p}([a,b]) := \{f \in B([a,b]) : \mathrm{Var}^R_{k,p}(f; [a,b]) < \infty\}$$

is called the *space of all functions of bounded (k, p)-variation* (in Riesz's sense) on $[a,b]$. ■

By what we observed before, we have the special cases

$$RBV_{1,p}([a,b]) = RBV_p([a,b]), \quad RBV_{k,1}([a,b]) = WBV_{k,1}([a,b]).$$

In rather the same way as we have done this for the space RBV_p in Proposition 2.51, one may show that the linear space $RBV_{k,p}$ with norm

$$\|f\|_{RBV_{k,p}} := \sum_{i=0}^{k-1} |f^{(i)}(a)| + \mathrm{Var}_{k,p}^R(f;[a,b])^{1/p} \tag{3.77}$$

is a Banach space. One may also show (Exercise 3.49) that $RBV_{k,p}([a,b])$ is "decreasing" with respect to $k \in \mathbb{N}$ and "increasing" with respect to $p \in [1,\infty)$.

The spaces $RBV_{k,p}([a,b])$ have particularly interesting properties for $p > 1$. As we have shown for $k = 1$ in Theorem 3.34, a function $f : [a,b] \to \mathbb{R}$ belongs to $RBV_{1,p}([a,b]) = RBV_p([a,b])$ for $1 < p < \infty$ if and only if $f \in AC([a,b])$ and $f' \in L_p([a,b])$; in this case

$$\mathrm{Var}_p^R(f;[a,b]) = \int_a^b |f'(t)|^p \, dt. \tag{3.78}$$

To formulate a parallel result for functions in $RBV_{k,p}([a,b])$, we denote by $AC^n([a,b])$ the linear space of all functions $f : [a,b] \to \mathbb{R}$ whose n-th derivative $f^{(n)}$ exists and is absolutely continuous on $[a,b]$, equipped with the norm[17]

$$\|f\|_{AC^n} = \sum_{j=0}^{n-1} |f^{(j)}(a)| + \|f^{(n)}\|_{AC} = \sum_{j=0}^{n} |f^{(j)}(a)| + \int_a^b |f^{(n+1)}(t)| \, dt. \tag{3.79}$$

The following is a generalization of Theorem 3.34 to higher order Riesz spaces.

Theorem 3.39. *Let $1 < p < \infty$ and $k \in \mathbb{N}$. Then a function f belongs to $RBV_{k,p}([a,b])$ if and only if $f \in AC^{k-1}([a,b])$ and $f^{(k)} \in L_p([a,b])$. Moreover, in this case, the equality*

$$\mathrm{Var}_{k,p}^R(f;[a,b]) = \frac{\|f^{(k)}\|_{L_p}^p}{((k-1)!)^p} = \int_a^b \left(\frac{|f^{(k)}(t)|}{(k-1)!} \right)^p dt \tag{3.80}$$

holds, where $\mathrm{Var}_{k,p}^R(f;[a,b])$ denotes the (k,p)-variation (3.76) of f in Riesz's sense.

Proof. For $k = 1$, the assertion of Theorem 3.39 is just Riesz's theorem (Theorem 3.34), so we may assume that $k \geq 2$. Following [229], we use a number of auxiliary results whose proofs are left to the reader as exercises.

Suppose first that $f : [a,b] \to \mathbb{R}$ belongs to $RBV_{k,p}([a,b])$. By Exercise 2.46, we know that then the derivative $f^{(k-1)}$ exists on $[a,b]$.

---

17 This is of course a special case of the general construction of X^n considered in (0.43). The special case $n = 1$ is considered in Exercise 3.53.

Let $P = \{t_0, t_1, \dots, t_m\} \in \mathcal{P}([a,b])$ be a partition of $[a,b]$. For any subinterval $[t_j, t_{j+1}]$, we define a partition $P_j \in \mathcal{P}([t_j, t_{j+1}])$ in the form

$$\begin{aligned} t_j = t_{j,1} < t_{j,2} < \dots < t_{j,k-1} < t_{j,k} = t_j + h < \dots \\ \dots < t_{j,k+1} = t_{j+1} - h < t_{j,k+2} < \dots < t_{j,2k} = t_{j+1}, \end{aligned} \tag{3.81}$$

where

$$h := \min \left\{ \frac{t_{j+1} - t_j}{2} : j = 0, 1, 2, \dots, m-1 \right\}.$$

Using the shortcut

$$F_{j,k}(P_j; h) := \frac{|f[t_{j+1} - h, t_{j,k+2}, \dots, t_{j,2k-1}] - f[t_j, t_{j,2}, \dots, t_{j,k-1}, t_j + h]|^p}{|t_{j+1} - t_j|^{p-1}},$$

by Exercise 3.48, the derivative $f^{(k-1)}$ is continuous, and so

$$\frac{|f^{(k-1)}(t_{j+1}) - f^{(k-1)}(t_j)|^p}{(t_{j+1} - t_j)^{p-1}} = \lim_{h \to 0+} \lim_{t_{j,k-1} \to t_j} \lim_{t_{j,k+2} \to t_{j+1}} F_{j,k}(P_j; h).$$

Consequently,

$$\begin{aligned} &\sum_{j=0}^{m} \frac{|f^{(k-1)}(t_{j+1}) - f^{(k-1)}(t_j)|^p}{|t_{j+1} - t_j|^{p-1}} \\ &\quad = \sum_{j=0}^{m} \lim_{h \to 0+} \lim_{t_{j,k-1} \to t_j} \lim_{t_{j,k+2} \to t_{j+1}} F_{j,k}(P_j; h) \le \mathrm{Var}^R_{k,p}(f; [a,b]). \end{aligned}$$

Thus, from Theorem 3.34, we conclude that $f^{(k-1)} \in AC([a,b])$ and $f^{(k)} \in L_p([a,b])$. Moreover,

$$\int_a^b \frac{|f^{(k)}(t)|^p}{((k-1)!)^p}\, dt \le \mathrm{Var}^R_{k,p}(f; [a,b]). \tag{3.82}$$

Conversely, suppose now that $f : [a,b] \to \mathbb{R}$ belongs to $AC^{k-1}([a,b])$ and $f^{(k)} \in L_p([a,b])$. Let P be a partition of $[a,b]$ of the form (3.81). Since $f^{(k-1)} \in AC([a,b])$ and $f^{(k)} \in L_p([a,b])$, by Exercise 2.49, we have

$$\begin{aligned} &\sum_{j=1}^{m} \frac{|f[t_{j,k+1}, \dots, t_{j,2k}] - f[t_{j,1}, \dots, t_{j,k}]|^p}{|t_{j,2k} - t_{j,1}|^{p-1}} \\ &\quad = \sum_{j=1}^{m} \frac{|f^{(k-1)}(\tau_{j,2k}) - f^{(k-1)}(\tau_{j-1,k})|^p}{((k-1)!)^p |t_{j,2k} - t_{j,1}|^{p-1}} \\ &\quad = \frac{1}{((k-1)!)^p} \sum_{j=1}^{m} \frac{1}{|t_{j,2k} - t_{j,1}|^{p-1}} \left| \int_{\tau_{j-1,k}}^{\tau_{j,2k}} f^{(k)}(\sigma)\, d\sigma \right|^p, \end{aligned}$$

where $\tau_{j,2k}$ belongs to the convex hull of $\{t_{j,k+1},\dots,t_{j,2k}\}$ and $\tau_{j-1,k}$ belongs to the convex hull of $\{t_{j,1},\dots,t_{j,k}\}$. By Hölder's inequality (0.108), we then get the estimates

$$\begin{aligned}
&\sum_{j=1}^{m} \frac{1}{|t_{j,2k}-t_{j,1}|^{p-1}} \left| \int_{\tau_{j-1,k}}^{\tau_{j,2k}} f^{(k)}(t)\,dt \right|^p \\
&\le \sum_{j=1}^{m} \frac{|\tau_{j,2k}-\tau_{j-1,k}|^{p-1}}{|t_{j,2k}-t_{j,1}|^{p-1}} \int_{\tau_{j-1,k}}^{\tau_{j,2k}} |f^{(k)}(t)|^p\,dt \\
&\le \sum_{j=1}^{m} \int_{\tau_{j-1,k}}^{\tau_{j,2k}} |f^{(k)}(t)|^p\,dt \le \sum_{j=1}^{m} \int_{t_{j,1}}^{t_{j,2k}} |f^{(k)}(t)|^p\,dt = \int_a^b |f^{(k)}(t)|^p\,dt\,.
\end{aligned}$$

This implies that $f \in RBV_{k,p}([a,b])$ and

$$\int_a^b \frac{|f^{(k)}(t)|^p}{((k-1)!)^p}\,dt \ge \mathrm{Var}^R_{k,p}(f;[a,b])\,. \tag{3.83}$$

Combining the estimates (3.82) and (3.83), we conclude that (3.80) is true, and so we are done. □

In the following Table 3.4 which is similar to Table 2.9, we recall the definition of higher order variations in Riesz's sense, together with corresponding integral representations.

Table 3.4. Higher order variations and Riesz spaces $RBV_{k,p}$.

$RBV_{k,p}([a,b])$	$p=1$	$1<p<\infty$
$k=1$	Jordan 1881	Riesz 1910
integral representation	$\mathrm{Var}^R_{1,1}(f)=\|f'\|_{L_1}$	$\mathrm{Var}^R_{1,p}(f)=\|f'\|_{L_p}$
of $f\in RBV_{1,p}\cap AC$	Varberg 1965	Riesz 1910
$k=2$	De la Vallée Poussin 1915	Merentes 1991
integral representation	$\mathrm{Var}^R_{2,1}(f)=\|f''\|_{L_1}$	$\mathrm{Var}^R_{2,p}(f)=\|f''\|^p_{L_p}$
of $f\in RBV_{2,p}\cap AC^1$	Russell 1974	Merentes 1992
$k\in\mathbb{N}$	Popoviciu 1934	Merentes/Sánchez 2012
integral representation	$\mathrm{Var}^R_{k,1}(f)=\frac{\|f^{(k)}\|_{L_1}}{(k-1)!}$	$\mathrm{Var}^R_{k,p}(f)=\frac{\|f^{(k)}\|^p_{L_p}}{((k-1)!)^p}$
of $f\in RBV_{k,p}\cap AC^{k-1}$	Russell 1978	Merentes/Sánchez 2012

In the same way as Medvedev has generalized the Riesz space $RBV_p([a,b])$, by replacing the power function $u\mapsto|u|^p$ by some general Young function ϕ (Definition 2.1), the higher order spaces $RBV_{k,p}([a,b])$ have been generalized as follows:

Definition 3.40. Given a partition $P = \{t_0, t_1, \dots, t_m\} \in \mathcal{P}([a,b])$, $k \in \mathbb{N}$, and a Young function $\phi : [0,\infty) \to [0,\infty)$, we set

$$\begin{aligned}&\mathrm{Var}^R_{k,\phi}(f, P; [a,b])\\ &:= \sum_{j=1}^{m-k+1} \phi\left(\frac{|f[t_j,\dots,t_{j+k-1}] - f[t_{j-1},\dots,t_{j+k-2}]|}{|t_j - t_{j-1}|}\right) |t_j - t_{j-1}|\end{aligned} \tag{3.84}$$

and call (3.84) the *(k,ϕ)-variation* of f with respect to P (in Riesz's sense) on $[a,b]$. Moreover, we call the (possibly infinite) number

$$\mathrm{Var}^R_{k,\phi}(f; [a,b]) := \sup\left\{\mathrm{Var}^R_{k,\phi}(f, P; [a,b]) : P \in \mathcal{P}([a,b])\right\}, \tag{3.85}$$

where the supremum is taken over all partitions of $[a,b]$, the *total (k,ϕ)-variation* of f (in Riesz's sense) on $[a,b]$. ■

Similarly, as in Definition 2.72, we now put

$$RBV_{k,\phi}([a,b]) := \{f \in B([a,b]) : \mathrm{Var}^R_{k,\phi}(cf; [a,b]) < \infty \text{ for some } c > 0\},$$

and

$$\|f\|_{RBV_{k,\phi}} := \sum_{j=0}^{k-1} |f^{(j)}(a)| + \inf\left\{\lambda > 0 : \mathrm{Var}^R_{k,\phi}(f/\lambda; [a,b]) \le 1\right\}. \tag{3.86}$$

As for $(RBV_{k,p}([a,b]), \|\cdot\|_{RBV_{k,p}})$, one may show that $(RBV_{k,\phi}([a,b]), \|\cdot\|_{RBV_{k,\phi}})$ is a Banach space.

Let us take a closer look at the two particular cases $k = 1$ and $k = 2$. For $k = 1$, the definitions (3.84)–(3.86) become

$$\mathrm{Var}^R_{1,\phi}(f, P; [a,b]) = \sum_{j=1}^{m} \phi\left(\frac{|f(t_j) - f(t_{j-1})|}{t_j - t_{j-1}}\right) |t_j - t_{j-1}|, \tag{3.87}$$

$$\mathrm{Var}^R_{1,\phi}(f; [a,b]) := \sup\left\{\mathrm{Var}^R_{1,\phi}(f, P; [a,b]) : P \in \mathcal{P}([a,b])\right\}, \tag{3.88}$$

$$RBV_{1,\phi}([a,b]) = \{f \in B([a,b]) : \mathrm{Var}^R_{1,\phi}(cf; [a,b]) < \infty \text{ for some } c > 0\},$$

and

$$\|f\|_{RBV_{1,\phi}} = |f(a)| + \inf\left\{\lambda > 0 : \mathrm{Var}^R_{1,\phi}(f/\lambda; [a,b]) \le 1\right\}, \tag{3.89}$$

respectively, which gives Medvedev's definition of the space $RBV_\phi([a,b])$. For $k = 2$ we get

$$\mathrm{Var}^R_{2,\phi}(f, P; [a,b]) = \sum_{j=1}^{m-1} \phi\left(\frac{|f[t_j, t_{j+1}] - f[t_{j-1}, t_j]|}{t_j - t_{j-1}}\right) |t_j - t_{j-1}|, \tag{3.90}$$

$$\mathrm{Var}^R_{2,\phi}(f; [a,b]) = \sup\left\{\mathrm{Var}^R_{2,\phi}(f, P; [a,b]) : P \in \mathcal{P}([a,b])\right\}, \tag{3.91}$$

$$RBV_{2,\phi}([a,b]) = \{f \in B([a,b]) : \mathrm{Var}^R_{2,\phi}(cf; [a,b]) < \infty \text{ for some } c > 0\}, \tag{3.92}$$

and

$$\|f\|_{RBV_{2,\phi}} = |f(a)| + |f'(a)| + \inf\left\{\lambda > 0 : \mathrm{Var}^R_{2,\phi}(f/\lambda; [a,b]) \le 1\right\}, \tag{3.93}$$

respectively. The space $(RBV_{2,\phi}([a,b]), \|\cdot\|_{RBV_{2,\phi}})$ in this form has been introduced and studied in [218], where also an analogue to Theorem 3.39 is proved.

3.7 Comments on Chapter 3

The basic facts about the interconnections between absolutely continuous functions and functions of bounded variation may be found in the books [76, 156, 182, 188, 238, 266, 311]. More sophisticated (and in part, surprising) properties of absolutely continuous functions are given in the survey papers [44, 45]. As Section 3.4 shows, absolutely continuous functions of one variable are intimately related to the concept of curve length. The analogue in higher dimensions is surface area; a beautiful discussion can be found in Goffman's survey [126].

Proposition 3.2 shows that the requirement $|g(x)| > 0$ for $g \in AC$ suffices to ensure that $1/g \in AC$, while from Exercises 1.1 and 1.2, it follows that $g \in BV$ has to be bounded away from zero to ensure[18] that $1/g \in BV$. The construction and properties of the Cantor set (and similar sets of positive measure) can be found in many standard textbooks on Real Analysis, e.g. [118], the important Vitali–Banach–Zaretskij theorem (under this or another name), for example, in [156, 182, 238]. The idea to extend the Cantor function (3.6) from the Cantor set (3.3) to the whole interval $[0, 1]$, preserving its monotonicity, is based on the following general principle: if $C \subset [a, b]$ is a subset with $a, b \in C$ and $f : C \to \mathbb{R}$ is increasing, then f extends to an increasing function $\hat{f} : [a, b] \to \mathbb{R}$. In fact, defining

$$\hat{f}(x) := \sup \{f(t) : t \in C \cap [a, x]\}, \tag{3.94}$$

one easily verifies that $\hat{f}$ is increasing and $\hat{f}(x) = f(x)$ for $x \in C$.

Our discussion of the class $AC_p([a, b])$ is taken from [183]. Table 3.2 shows that $AC_p \subseteq WBV_p$ for $p > 1$ (even $\subseteq C \cap WBV_p$ since all functions in AC_p are continuous). More precisely, in [185] the chain of inclusions (3.14) is proved. Roughly speaking, this means that the difference between the spaces $WBV_p([a, b])$ and $AC_p([a, b])$ is “not very large.” As pointed out before, however, all inclusions in (3.14) are strict.

We remark that Bruneau [65] defines p-absolutely continuous functions $f : [a, b] \to \mathbb{R}$ by requiring that for each $\varepsilon > 0$, there exists some $\delta > 0$ such that

$$\sum_{k=1}^{n} \mathrm{Var}_p^W (f; [a_k, b_k]) \leq \varepsilon$$

for any $S \in \Sigma([a, b])$ with $\Theta(S) \leq \delta$. As Proposition 3.14 shows, this coincides with Love's definition (our Definition 3.12) from [183]. Interestingly, such functions may be characterized differently: as the following theorem [65] shows, AC_p may be regarded as closure of $Lip_{1/p}$ in the WBV_p-norm (1.65).

Theorem 3.41. *A function $f : [a, b] \to \mathbb{R}$ is p-absolutely continuous if and only if there exists a sequence $(f_n)_n$ in $Lip_{1/p}([a, b])$ such that $\|f_n - f\|_{WBV_p} \to 0$ as $n \to \infty$.*

---

18 This is of course not surprising since BV function are bounded, but not necessarily continuous.

In [2], the author introduces and studies the four metrics

$$\begin{aligned} d_1(f,g) &:= \|f-g\|_{L_1} + |\operatorname{Var}(f;[a,b]) - \operatorname{Var}(g;[a,b])| \\ &= \int_a^b |f(t)-g(t)|\,dt + |\operatorname{Var}(f;[a,b]) - \operatorname{Var}(g;[a,b])|, \end{aligned} \tag{3.95}$$

$$\begin{aligned} d_2(f,g) &:= \|f-g\|_{L_1} + |L(\Gamma(f);[a,b]) - L(\Gamma(g);[a,b])| \\ &= \int_a^b |f(t)-g(t)|\,dt + |L(\Gamma(f);[a,b]) - L(\Gamma(g);[a,b])|, \end{aligned} \tag{3.96}$$

$$\begin{aligned} d_3(f,g) &:= \|f-g\|_\infty + |\operatorname{Var}(f;[a,b]) - \operatorname{Var}(g;[a,b])| \\ &= \sup_{a\le x\le b} |f(x)-g(x)| + |\operatorname{Var}(f;[a,b]) - \operatorname{Var}(g;[a,b])|, \end{aligned} \tag{3.97}$$

and

$$\begin{aligned} d_4(f,g) &:= \|f-g\|_\infty + |L(\Gamma(f);[a,b]) - L(\Gamma(g);[a,b])| \\ &= \sup_{a\le x\le b} |f(x)-g(x)| + |L(\Gamma(f);[a,b]) - L(\Gamma(g);[a,b])|, \end{aligned} \tag{3.98}$$

on the space $BV([a,b])$. Clearly, $d_1(f,g) \le d_3(f,g)$ and $d_2(f,g) \le d_4(f,g)$. Moreover, from (3.59), it follows that $d_1(f,g) \le d_2(f,g)$ and $d_3(f,g) \le d_4(f,g)$. We may summarize these inequalities in the form

$$d_1(f,g) \le \min\{d_2(f,g), d_3(f,g)\} \le \max\{d_2(f,g), d_3(f,g)\} \le d_4(f,g). \tag{3.99}$$

As is shown in [2], the topological properties of BV are quite different for each choice of these metrics. For example, equipped with either the metric (3.97) or (3.98), the spaces $BV([a,b])$ and $AC([a,b])$ are not complete and not even locally compact. On the other hand, in $BV([a,b])$ with the metric (3.95), every closed ball is compact.[19]

The following Luzin type result is mentioned in the papers [128, 129]:

Theorem 3.42 (Goffman–Liu). *A function $f : [a,b] \to \mathbb{R}$ is absolutely continuous if and only if for each $\varepsilon > 0$, we can find a continuously differentiable function $f_\varepsilon : [a,b] \to \mathbb{R}$ such that*

$$\lambda(\Delta(f,f_\varepsilon)) < \varepsilon, \qquad \int_{\Delta(f,f_\varepsilon)} |f'(t)|\,dt < \varepsilon, \qquad \int_{\Delta(f,f_\varepsilon)} |f_\varepsilon'(t)|\,dt < \varepsilon,$$

where $\Delta(f,g)$ is given by (0.47).

A comparison of Theorem 3.42 and Theorem 0.34 shows that the pair "measurable vs. continuous" in the Luzin theorem corresponds to the pair "AC vs. C^1" in the Goff-

19 This is a consequence of Helly's selection theorem (Theorem 1.11) and Lebesgue's convergence theorem (Theorem 0.4). Loosely speaking, it means that (BV, d_1) behaves rather like a finite dimensional space.

man–Liu theorem; moreover, for the latter, we obviously have to take into account the L_1-norm of the derivatives.

Theorem 3.42 suggests to equip the set $AC([a,b])$ with the metric

$$d(f,g) := \lambda(\Delta(f,g)) + \int\limits_{\Delta(f,g)} |f'(t)|\,dt + \int\limits_{\Delta(f,g)} |g'(t)|\,dt\,. \tag{3.100}$$

As far as we know, the metric space (AC, d) has not been studied yet in the literature.[20]

The results discussed in Section 3.3 may also be found in [76] or [156]. In the interesting survey article [309], the author cites the following proposition which he found "buried as an innocent problem" in Natanson's book [238]:

Proposition 3.43. *Let $f : [a,b] \to \mathbb{R}$ be differentiable with bounded derivative on a set $M \subseteq [a,b]$. Then*

$$\lambda^*(f(M)) \le \lambda^*(M) \sup_{x\in M} |f'(x)|\,, \tag{3.101}$$

where $\lambda^*(M)$ denotes the outer Lebesgue measure of M.

From Proposition 3.43, several results (among them, many of those presented in Section 3.3) may be derived rather easily. For example, if

$$Der_0(f) := \{x : a \le x \le b,\ f'(x) = 0\} \tag{3.102}$$

denotes the set of *critical points* of $f : [a,b] \to \mathbb{R}$, (3.101) implies that the set of *critical values*

$$f(Der_0(f)) = \{f(x) : a \le x \le b,\ f'(x) = 0\} \tag{3.103}$$

of f is always a nullset[21] and therefore "small." Of course, constant functions show that the set (3.102) may be very large. It is instructive to compare the sets (3.102) and (3.103) with the set of *extremal points*

$$Ext(f) := \{x : a \le x \le b,\ x \text{ is a local extremum for } f\} \tag{3.104}$$

and the set of *extremal values*

$$f(Ext(f)) = \{f(x) : a \le x \le b,\ x \text{ is a local extremum for } f\} \tag{3.105}$$

of f. Again, constant functions show that the set (3.104) may be very large, while the set (3.105) may be a singleton. Even more interesting is the characteristic function f of $[a,b]\cap\mathbb{Q}$ (Dirichlet function) which satisfies $Ext(f) = [a,b]$ and $f(Ext(f)) = \{0,1\}$. This function has a local extremum at *every* point, but is not constant on any subinterval.[22]

20 In [129], it is mentioned without proof that AC with the metric (3.100) is complete.

21 This result is known as *Sard's lemma* in the literature. We remark that this result is rather easy to prove for C^1-functions.

22 One may show that such a function must be discontinuous at every point, see Exercise 3.59.

In every calculus course, it is taught that $Ext(f) \subseteq Der_0(f)$ for differentiable functions f. Interestingly, the set (3.105) is even "smaller" than just a nullset, as the set (3.103). In fact, let $\hat{y} = f(\hat{x}) \in f(Ext(f))$ be such that f has a local maximum at $\hat{x}$, say, and choose an interval $I(\hat{y}) = [a(\hat{y}), b(\hat{y})]$ with rational endpoints such that $f(x) \leq f(\hat{x})$ for $a(\hat{x}) \leq x \leq b(\hat{x})$. Since the map $\hat{y} \mapsto I(\hat{y})$ is injective, and the system of all such intervals $I(\hat{y})$ is countable, we thus have proved[23] the following

Proposition 3.44. *For arbitrary functions $f : [a, b] \to \mathbb{R}$, the set* (3.105) *of extremal values of f is at most countable.*

A certain converse of Proposition 3.44 is true as well, see Exercise 3.58. The following example [12] illustrates the fact that although both sets (3.103) and (3.105) are "small," the gap between them may be considerable:

Example 3.45. Let C be the Cantor set (3.3), and let $g : [0, 1] \to \mathbb{R}$ and $f : [0, 1] \to \mathbb{R}$ be defined by

$$g(t) := \operatorname{dist}(t, C) = \min\{t - c : c \in C\}, \quad f(x) := \int_0^x g(t)\,dt\,.$$

Then $g \in Lip([0, 1])$ with $g(t) = 0$ on C and $g(t) > 0$ on $[0, 1] \setminus C$. Consequently, $f \in C^1([0, 1])$ is *strictly increasing*, and so $f(Ext(f)) = \{0, f(1)\}$. On the other hand, $f'(x) = 0$ on C, which means that $Der_0(f)$ contains the uncountable set C. However, the injectivity of f implies that then $f(Der_0(f))$ contains the uncountable set $f(C)$, and so is uncountable (although a nullset) itself. ♥

As we have seen in Example 3.15, there exist simple functions $f \in BV([a, b])$ which are everywhere differentiable on $[a, b]$ and have the property that f' is not Lebesgue integrable. Of course, as Theorem 3.18 shows, such a function cannot be absolutely continuous. A more dramatic refinement of this is an example, due to Katznelson and Stromberg[24] [157] of a function $f \in BV([a, b])$ with the following properties:
- f is everywhere differentiable on $[a, b]$ with $\|f'\|_\infty \leq 1$;
- both sets $\{x : a \leq x \leq b,\ f'(x) > 0\}$ and $\{x : a \leq x \leq b,\ f'(x) < 0\}$ are dense in $[a, b]$, so f is not monotone on any interval;
- the set $\{x : a \leq x \leq b,\ f'(x) = 0\}$ is also dense in $[a, b]$;
- f' is not Riemann integrable over any interval $[c, d] \subseteq [a, b]$.

---

23 The reader will have noticed that the proof of this result is similar to the proof of the countability of the set of discontinuities of a monotone function.

24 An explicit construction of this function is given in Example 13.2 of the book [268]. A more elementary example of a function $f \in BV([a, b])$ which is differentiable everywhere (with a bounded derivative) and such that f' vanishes on a dense subset of $[a, b]$ may be found in Example 13.3 of [268].

Jordan [153] has not only introduced the concept of bounded variation, but also studied functions of bounded variation. The connection between functions of bounded variation and rectifiable curves in the plane (Proposition 3.28 (c)) is also due to Jordan [154]. Theorem 3.25 which is taken from the paper [127] admits the following geometric interpretation. The family $\{g_h : 0 < h < 1\}$ defined in (3.44) converges, as $h \to 0+$, in $AC([0,1])$ to the boundary of a ball in $AC([0,1])$ centered at the origin if and only if g is (equivalent to) a function of bounded variation, and it converges to a point in $AC([0,1])$ if and only if g is (equivalent to) an absolutely continuous function.

The Riesz theorem (Theorem 3.34) and its generalization by Medvedev (Theorem 3.36) are of utmost importance in functional analysis and in the theory of differential equations. Roughly speaking, passing from the Riesz space RBV_p to the Medvedev space RBV_ϕ plays a similar role in applications to differential equations as passing from the Lebesgue space L_p to the Orlicz space L_ϕ in applications to integral equations. In [215], it is shown that Theorem 3.34 also holds for f replaced by

$$f^\vee(x) := \sup\left\{\frac{f(x+h) - f(x-h)}{2h} : 0 < h \le \min\{b-x, x-a\}\right\}$$

for $a < x < b$. We remark that the function $f^\vee$ occurs in the theory of maximal operators in Hardy's sense. More on integral representations of functions of bounded Riesz variation may be found in [283].

In view of their importance, let us summarize the main results of Section 3.3 in the following way, where we also consider the variation function (1.13).

Theorem 3.46. *The following four assertions on a function $f : [a,b] \to \mathbb{R}$ are equivalent.*
(a) *f is absolutely continuous;*
(b) *f' exists a.e., $f' \in L_1([a,b])$, and* (3.31) *holds;*
(c) *f' exists a.e., $f' \in L_1([a,b])$, and* (3.36) *holds;*
(d) *V_f is absolutely continuous.*

The implication (a) ⇒ (b) has been proved in Theorem 3.18, while the implication (b) ⇒ (c) has been discussed after the proof of Theorem 3.19. The implication (c) ⇒ (d) follows from (3.33), while the implication (d) ⇒ (a) is a consequence of the estimates

$$|f(x) - f(y)| \le \mathrm{Var}(f; [x,y]) \le V_f(y) - V_f(x)$$

which follow from (1.8) and (1.10).

The results proved in Section 3.3 imply some interesting decompositions. Writing as before $AC^o([a,b])$ for the space of all $f \in AC([a,b])$ satisfying $f(a) = 0$, we have the trivial decomposition

$$AC([a,b]) = \mathbb{R} \oplus AC^o([a,b]) \tag{3.106}$$

induced by the linear surjective isometry $f \mapsto f(a) + (f - f(a))$. Defining the integral operator J by

$$Jf(x) := \int_a^x f(t)\,dt\,, \tag{3.107}$$

Proposition 3.4 and Theorem 3.18 show that *J is a linear surjective isometry between $L_1([a,b])$ and $AC^o([a,b])$ with inverse $J^{-1}g = Dg = g'$*. Thus, we may replace (3.106) by the more interesting decomposition

$$AC([a,b]) = \mathbb{R} \oplus L_1([a,b]) \tag{3.108}$$

induced by the linear surjective isometry $f \mapsto f(a) + f'$. As a pleasant by-product, we get a very simple proof of the completeness of the space AC: since L_1 is complete and J is an isometry, $AC^o = J(L_1)$ is complete as well, and so AC is complete, by (3.106).

The higher order Riesz–Young–Medvedev spaces which we discussed in Section 3.6 have been studied in [226, 228, 229] and elsewhere. In [228], it is shown that $RBV_{k,\phi} \subseteq WBV_{k,1}$, with equality in case $\phi \notin \infty_1$, see (2.16). The important equality

$$\mathrm{Var}^R_{2,p}(f;[a,b]) = \int_a^b |f''(x)|^p\,dx \tag{3.109}$$

which holds in case $f' \in AC([a,b])$ (Table 3.4) and is analogous to (3.69) for $f \in AC([a,b])$ may be found in [220]. As far as we know, a rigorous proof of the general formula (3.80) appears in case $f^{(k-1)} \in AC([a,b])$ for the first time in [229]. In the paper [176], the authors also prove a parallel result to Theorem 3.39. To formulate this result, we again need the Orlicz class $\mathcal{L}_\phi([a,b])$ which we introduced in Definition 0.16.

Theorem 3.47. *Let ϕ be a Young function which satisfies condition ∞_1, see (2.16), and let $k \in \mathbb{N}$. Then a function f belongs to $RBV_{k,\phi}([a,b])$ if and only if $f \in AC^{k-1}([a,b])$ and $f^{(k)} \in \mathcal{L}_\phi([a,b])$. Moreover, in this case, the equality*

$$\mathrm{Var}^R_{k,\phi}(f;[a,b]) = \int_a^b \phi\left(\frac{|f^{(k)}(t)|}{(k-1)!}\right)dt \tag{3.110}$$

holds, where $\mathrm{Var}^R_{k,\phi}(f;[a,b])$ denotes the (k,ϕ)-variation (3.85) of f in Riesz's sense.

Relations between functions of (generalized) bounded variation and absolutely continuous functions have been also studied in a more general setting. For instance, [159] shows that a function $f \in \Lambda BV_\phi([a,b])$ (Definition 2.84) is absolutely continuous if and only if f' exists and satisfies (3.18). Since ΛBV_ϕ contains the spaces ΛBV and WBV_ϕ, this is a far reaching generalization of Theorem 3.18.

The Riesz–Medvedev theorem has a natural analogue for functions of two variables. However, formulating this analogue requires some preparation.

We start with the definition of the corresponding variation, imitating what we have done in Definition 1.42. Consider a Young function $\phi : [0,\infty) \to [0,\infty)$, a function, $f : [a,b] \times [c,d] \to \mathbb{R}$, and partitions $P = \{s_0, s_1, \dots, s_m\} \in \mathcal{P}([a,b])$ and $Q = \{t_0, t_1, \dots, t_n\} \in \mathcal{P}([c,d])$. We define three variations for f with respect to P and Q by

$$\mathrm{Var}_\phi^R(f(\cdot,c), P; [a,b]) := \sum_{i=1}^{m} \phi\left(\frac{|f(s_i,c) - f(s_{i-1},c)|}{s_i - s_{i-1}}\right)(s_i - s_{i-1}), \tag{3.111}$$

$$\mathrm{Var}_\phi^R(f(a,\cdot), Q; [c,d]) := \sum_{j=1}^{n} \phi\left(\frac{|f(a,t_j) - f(a,t_{j-1})|}{t_j - t_{j-1}}\right)(t_j - t_{j-1}), \tag{3.112}$$

and

$$\begin{aligned} &\mathrm{V}_{2,\phi}^R(f, P \times Q; [a,b] \times [c,d]) \\ &:= \sum_{i=1}^{m} \sum_{j=1}^{n} \phi\left(\frac{|\Delta_{i,j} f|}{(s_i - s_{i-1})(t_j - t_{j-1})}\right)(s_i - s_{i-1})(t_j - t_{j-1}), \end{aligned} \tag{3.113}$$

where we have used the shortcut

$$\Delta_{i,j} f := f(s_i,t_j) - f(s_{i-1},t_j) - f(s_i,t_{j-1}) + f(s_{i-1},t_{j-1}). \tag{3.114}$$

Of course, in case $\phi(t) = t$, these variations reduce to the three variations (1.76), (1.77), and (1.78). As in Section 1.4, we put

$$\mathrm{Var}_\phi^R(f(\cdot,c); [a,b]) := \sup\{\mathrm{Var}_\phi^R(f(\cdot,c), P; [a,b]) : P \in \mathcal{P}([a,b])\}, \tag{3.115}$$

$$\mathrm{Var}_\phi^R(f(a,\cdot); [c,d]) := \sup\{\mathrm{Var}_\phi^R(f(a,\cdot), Q; [c,d]) : Q \in \mathcal{P}([c,d])\}, \tag{3.116}$$

and

$$\begin{aligned} &\mathrm{V}_{2,\phi}^R(f; [a,b] \times [c,d]) \\ &:= \sup\{\mathrm{V}_{2,\phi}^R(f, P \times Q; [a,b] \times [c,d]) : P \in \mathcal{P}([a,b]), Q \in \mathcal{P}([c,d])\}, \end{aligned} \tag{3.117}$$

where all suprema are taken over the indicated partitions.

Definition 3.48. With the above notation, we call the (possibly infinite) number

$$\begin{aligned} &\mathrm{Var}_\phi^R(f; [a,b] \times [c,d]) \\ &:= \mathrm{Var}_\phi^R(f(\cdot,c); [a,b]) + \mathrm{Var}_\phi^R(f(a,\cdot); [c,d]) + \mathrm{V}_{2,\phi}^R(f; [a,b] \times [c,d]) \end{aligned} \tag{3.118}$$

the *total variation of f on $[a,b] \times [c,d]$ in the sense of Riesz–Medvedev*. In case $\mathrm{Var}_\phi^R(f; [a,b] \times [c,d]) < \infty$, we say that *$f$ has bounded Riesz–Medvedev variation on $[a,b] \times [c,d]$* and write $f \in RBV_\phi([a,b] \times [c,d])$. ∎

By what we have observed before, we get $RBV_\phi([a,b] \times [c,d]) = BV([a,b] \times [c,d])$ for the special choice $\phi(t) = t$. We point out, however, that not all properties of the one-dimensional case carry over to this space. For example, in Proposition 2.52 we have proved the inclusions

$$Lip([a,b]) \subseteq RBV_p([a,b]) \subseteq AC([a,b]) \tag{3.119}$$

for $p > 1$. It is a striking fact that the first inclusion in (3.119) is *not* true in the two-dimensional case, even for the simplest Young function $\phi(t) = t^p$, as the following example from [33] shows:

Example 3.49. On the unit square $[0, 1]\times[0, 1]$, consider for $n = 2, 3, 4, \dots$ the partitions

$$P_n := \{s_0, s_1, s_2, \dots, s_{2n-1}\}, \quad Q_n := \{0, t_1, t_2, \dots, t_{2n-1}\},$$

where

$$s_0 = t_0 := 0, \quad s_1 = t_1 := \frac{1}{n}, \quad s_2 = t_2 := \frac{2n-1}{2n(n-1)}, \quad s_3 = t_3 := \frac{1}{n-1}, \quad \dots$$
$$\dots, \quad s_{2n-3} = t_{2n-3} := \frac{1}{2}, \quad s_{2n-2} = t_{2n-2} := \frac{3}{4}, \quad s_{2n-1} = t_{2n-1} := 1.$$

Thus, $P_n \times Q_n$ is an equidistant lattice of constant mesh size $1/n(n-1)$. We define a function $f\colon [0, 1] \times [0, 1] \to \mathbb{R}$ by

$$f(x, y) := \begin{cases} 0 & \text{for } x = s_{2i-1} \text{ or } y = t_{2j-1}, \\ \frac{1}{k(k-1)} & \text{for } x = y = \frac{2k-1}{2k(k-1)}, \\ \text{linear} & \text{otherwise}. \end{cases}$$

Geometrically, the graph of f over the rectangle

$$R_{i,j} := \left[\frac{1}{i}, \frac{1}{i-1}\right] \times \left[\frac{1}{j}, \frac{1}{j-1}\right]$$

is the surface of a pyramid with vertex situated at the midpoint

$$P_{i,j} := \left(\frac{2i-1}{2i(i-1)}, \frac{2j-1}{2j(j-1)}, \frac{1}{i(i-1)}\right).$$

Evaluating the slope of any such pyramid, it is not hard to see that

$$|f(x, y) - f(\tilde{x}, \tilde{y})|^2 \le \frac{1}{4}(|x - \tilde{x}|^2 + |y - \tilde{y}|^2)\|, \tag{3.120}$$

which shows that f is Lipschitz continuous in the Euclidean norm on $[0, 1] \times [0, 1]$. On the other hand, even for the simple choice $\phi(t) = t^2$, the variation (3.113) of f is not finite, and so $f \notin RBV_\phi([0, 1] \times [0, 1])$.

To see this, we show that

$$\mathrm{V}^R_{2,\phi}(f, P_n \times Q_n; [0, 1] \times [0, 1]) \to \infty \quad (n \to \infty), \tag{3.121}$$

where P_n and Q_n are as before. Using the shortcut (3.114) and taking into account that $\phi(t) = t^2$, we get the chain of equalities

$$
\begin{aligned}
&\mathrm{V}^R_{2,\phi}(f, P_n \times Q_n; [0,1] \times [0,1]) \\
&= \sum_{i=1}^{2n-1} \sum_{j=1}^{2n-1} \phi\left(\frac{|\Delta_{i,j} f|}{(s_i - s_{i-1})(t_j - t_{j-1})} \right) (s_i - s_{i-1})(t_j - t_{j-1}) \\
&= \sum_{i=2}^{2n-1} \sum_{j=2}^{2n-1} \phi\left(\frac{\frac{1}{i(i-1)}}{\frac{1}{i(i-1)} \frac{1}{j(j-1)}} \right) \left(\frac{1}{i(i-1)} \frac{1}{j(j-1)} \right) \\
&= \sum_{i=2}^{2n-1} \sum_{j=2}^{2n-1} \frac{j^2(j-1)^2}{i(i-1)j(j-1)} = \sum_{i=2}^{2n-1} \sum_{j=2}^{2n-1} \frac{j(j-1)}{i(i-1)} .
\end{aligned}
$$

However, the second sum may be estimated for fixed $i \in \{2, 3, \ldots, 2n-1\}$ by

$$
\begin{aligned}
\sum_{j=2}^{2n-1} \frac{j(j-1)}{i(i-1)} &= \frac{(2n-1)(2n-2) + (2n-2)(2n-3) + \ldots + 3 \cdot 2 + 2 \cdot 1}{i(i-1)} \\
&\geq \frac{(2n-1) + (2n-2) + \ldots + 3 + 2 + 1}{i(i-1)} = \frac{n(2n-1)}{i(i-1)} ,
\end{aligned}
$$

which implies

$$
\begin{aligned}
\mathrm{V}^R_{2,\phi}(f, P_n \times Q_n; [0,1] \times [0,1]) &\geq \sum_{i=2}^{2n-1} \frac{n(2n-1)}{i(i-1)} \\
&= n(2n-1)\left(1 - \frac{1}{2n-1}\right) = 2n(n-1) .
\end{aligned}
$$

This proves (3.121), and hence $f \notin RBV_\phi([0,1] \times [0,1])$ as claimed. ♥

The reason for this counterexample is that the Lipschitz condition (3.120) is too weak to ensure that $f \in RBV_\phi$. In Proposition 3.2 of [33], it is shown that the combination of the three conditions

$$
\begin{aligned}
|f(s_i, t) - f(s_{i-1}, t)| &\leq L_1(t)|s_i - s_{i-1}| \quad (c \leq t \leq d) , \\
|f(s, t_j) - f(s, t_{j-1})| &\leq L_2(s)|t_j - t_{j-1}| \quad (a \leq s \leq b) ,
\end{aligned}
$$

and

$$
|\Delta_{ij} f| \leq L|(s_i - s_{i-1})(t_j - t_{j-1})|
$$

implies that

$$
\begin{aligned}
\mathrm{Var}^R_\phi(f(\cdot, c); [a,b]) &\leq \phi(L_1(c)) \sum_{i=1}^{m} |s_i - s_{i-1}| , \\
\mathrm{Var}^R_\phi(f(a, \cdot); [c,d]) &\leq \phi(L_2(a)) \sum_{j=1}^{n} |t_j - t_{j-1}| ,
\end{aligned}
$$

and

$$\mathrm{V}^R_{2,\phi}(f;[a,b]\times[c,d]) \le \phi(L)\sum_{i=1}^{m}\sum_{j=1}^{n}|(s_i-s_{i-1})(t_j-t_{j-1})|\,.$$

Consequently,

$$\mathrm{Var}^R_\phi(f;[a,b]\times[c,d]) \le \phi(L_1(c))(b-a)+\phi(L_2(a))(d-c)+\phi(L)(b-a)(d-c)\,,$$

and hence $f\in RBV_\phi([a,b]\times[c,d])$.

Now, we come to the definition of absolutely continuous functions of two variables. In [182], a definition is given which uses the concept of the "function of rectangles" and is rather technical. However, in [297], it is shown that this definition is equivalent to the following

Definition 3.50. We call a function $f:[a,b]\times[c,d]\to\mathbb{R}$ *absolutely continuous* and write $f\in AC([a,b]\times[c,d])$ if f admits a representation of the form

$$f(x,y)=f(a,c)+\int_a^x g_1(s)\,ds+\int_c^y g_2(t)\,dt+\int_a^x\int_c^y g(s,t)\,dt\,ds\,,$$

where $g_1\in L_1([a,b])$, $g_2\in L_1([c,d])$, and $g\in L_1([a,b]\times[c,d])$. ■

Definition 3.50 seems very natural since it is parallel to the fact that absolutely continuous functions of one variable are precisely primitives of L_1-functions, see Proposition 3.4 and Theorem 3.18. Another equivalent formulation is given in Exercise 3.62.

With this notion of absolute continuity, the following theorem which is perfectly analogous to Theorem 3.36 and involves the partial derivatives f_x, f_y and f_{xy} of f, has been proved in [33].

Theorem 3.51. *Let ϕ be a Young function which satisfies condition ∞_1, see* (2.16). *Then a function $f:[a,b]\times[c,d]\to\mathbb{R}$ belongs to $RBV_\phi([a,b]\times[c,d])$ if and only if $f\in AC([a,b]\times[c,d])$, $f_x(\cdot,c)\in\mathcal{L}_\phi([a,b])$, $f_y(a,\cdot)\in\mathcal{L}_\phi([c,d])$, and*[25] *$f_{xy}\in\mathcal{L}_\phi([a,b]\times[c,d])$. Moreover, in this case, the equality*

$$\begin{aligned}&\mathrm{Var}^R_\phi(f;[a,b]\times[c,d])\\&\quad=\int_a^b\phi(|f_x(x,c)|)\,dx+\int_c^d\phi(|f_y(a,y)|)\,dy+\int_a^b\int_c^d\phi(|f_{xy}(x,y)|)\,dy\,dx\end{aligned}$$

holds, where $\mathrm{Var}^R_\phi(f;[a,b]\times[c,d])$ *denotes the ϕ-variation* (3.118) *of f in the sense of Riesz and Medvedev.*

Absolutely continuous functions not only play a prominent role in the theory of the Lebesgue integral, but are also important in integration techniques like the substi-

25 Here, $f_{xy}\in\mathcal{L}_\phi([a,b]\times[c,d])$ means of course that $(\phi\circ f_{xy})\in L_1([a,b]\times[c,d])$.

tution formula[26] or the integration by parts formula. The latter is usually taught in courses on real functions in the following form:

Theorem 3.52. *For $f, g \in AC([a,b])$, the formula*

$$\int_a^b f'(x)g(x)\,dx = f(b)g(b) - f(a)g(a) - \int_a^b f(x)g'(x)\,dx \tag{3.122}$$

holds.

Note that both integrals in (3.122) make sense since the product of an L_1-function and an L_∞-function is integrable. However, one can weaken this result in several aspects. For example, in [74, 75], the following result is given:

Theorem 3.53. *Suppose that two functions $f, g : [a,b] \to \mathbb{R}$ have the property that their product fg is absolutely continuous, and the derivatives f' and g' exist a.e. on $[a,b]$. Moreover, assume that at least one of the functions $f'g$ or fg' is integrable. Then the formula* (3.122) *holds.*

In fact, under the hypotheses of Theorem 3.53, we know that the derivative $(fg)' = f'g + fg'$ exists a.e. on $[a,b]$ and is integrable. However, then *both* functions $f'g$ and fg' are integrable, and the assertion follows from the fundamental equality (3.31) for the Lebesgue integral.

To see that Theorem 3.53 is not only formally more general than Theorem 3.52, we consider the following

Example 3.54. Define $f, g : [0,1] \to \mathbb{R}$ by

$$f(x) := \begin{cases} \sin\frac{1}{x} & \text{for } 0 < x \le 1\,, \\ 0 & \text{for } x = 0\,, \end{cases} \qquad g(x) := \begin{cases} x^2\sin\frac{1}{x} & \text{for } 0 < x \le 1\,, \\ 0 & \text{for } x = 0\,. \end{cases}$$

In other words, in the notation of (0.86), we have $f = f_{0,-1}$ and $g = f_{2,-1}$. In Example 1.16, we have seen that the product function fg is Lipschitz continuous, and hence absolutely continuous. Moreover, the function

$$f(x)g'(x) := \begin{cases} 2x\sin^2\frac{1}{x} - \frac{1}{2}\sin\frac{2}{x} & \text{for } 0 < x \le 1\,, \\ 0 & \text{for } x = 0 \end{cases}$$

is certainly integrable, being the sum of a continuous and a bounded function. Thus, all hypotheses of Theorem 3.53 are met, and formula (3.109) holds for f and g. On the other hand, Theorem 3.52 does not apply since the function f is not continuous, let alone absolutely continuous on $[0,1]$. ♥

The phenomenon described in Example 3.54 is put into a more general framework in Exercise 3.64.

26 We will apply such formulas for change of variables in Section 5.3.

3.8 Exercises to Chapter 3

We state some exercises on the topics covered in this chapter; exercises marked with an asterisk * are more difficult.

Exercise 3.1. Show that $f \in AC([a,b])$ implies $|f| \in AC([a,b])$.

Exercise 3.2. Find a function $f \notin AC([0,1])$ such that $|f| \in AC([0,1])$. Compare this with Exercise 1.4.

Exercise 3.3. Suppose that $f \in C([a,b])$ and $|f| \in AC([a,b])$. Show that $f \in AC([a,b])$. Compare this with Exercise 1.5 and your example in Exercise 3.2.

Exercise 3.4. In the notation of Exercise 0.70, does $f, g \in AC([a,b])$ imply that $f \vee g, f \wedge g \in AC([a,b])$?

Exercise 3.5. If $f \in AC([a,b])$, show that $|f|^p \in AC([a,b])$ for $p \geq 1$. Is this also true for $0 < p < 1$?

Exercise 3.6. For which sequences $D = (d_n)_n$ is the zigzag function Z_D constructed in Definition 0.49 absolutely continuous on $[0,1]$?

Exercise 3.7. Let $f \in AC([a,b])$. Prove that for every $\varepsilon > 0$, there exists a $\delta > 0$ such that for all infinite collections $\{[a_n, b_n] : n \in \mathbb{N}\} \in \Sigma_\infty([a,b])$, the condition

$$\sum_{k=1}^{\infty} |b_k - a_k| \leq \delta$$

implies the condition

$$\sum_{k=1}^{\infty} |f(b_k) - f(a_k)| \leq \varepsilon .$$

Compare with Definition 3.1.

Exercise 3.8. Given $f : [a,b] \to \mathbb{R}$, suppose that for every $\varepsilon > 0$, there exists a $\delta > 0$ with the following property: for each collection $\{[a_1, b_1], \ldots, [a_n, b_n]\}$ of (not necessarily nonoverlapping!) subintervals of $[a,b]$, from

$$\sum_{k=1}^{n} |b_k - a_k| \leq \delta,$$

it follows that

$$\sum_{k=1}^{n} |f(b_k) - f(a_k)| \leq \varepsilon .$$

Prove that[27] $f \in Lip([a,b])$. Compare with Definition 3.1

27 Note that the converse is trivially true.

Exercise 3.9. Let C be the Cantor set (3.3), and let

$$[0,1] \setminus C = \bigcup_{n=1}^{\infty} (a_n, b_n)$$

be a representation of the complement of C as a countable union of open intervals (which is possible since $[0,1] \setminus C$ is open). Denote by $c_n := \frac{1}{2}(a_n + b_n)$ the corresponding midpoints. Given a positive sequence $(\delta_n)_n$ converging to zero, we define a function $f : [0,1] \to \mathbb{R}$ by

$$f(x) := \begin{cases} 0 & \text{for } x \in C, \\ \delta_n & \text{for } x = c_n \\ \delta_n \frac{x-a_n}{c_n-a_n} & \text{for } a_n < x < c_n, \\ \delta_n \frac{b_n-x}{b_n-c_n} & \text{for } c_n < x < b_n. \end{cases}$$

Prove that $f \in C([0,1]) \cap Lu([0,1])$, and that the total variation of f on $[0,1]$ is given by

$$\mathrm{Var}(f; [0,1]) = 2 \sum_{n=1}^{\infty} \delta_n .$$

Use this to construct, by a suitable choice of δ_n, examples of functions $f \in AC([0,1])$ and $f \in [C([0,1]) \cap Lu([0,1])] \setminus AC([0,1])$.

Exercise 3.10. Let $f : [a,b] \to \mathbb{R}$ be strictly increasing, and let

$$Der_{\infty}(f) := \{x : a \le x \le b,\ f'(x) = \infty\} .$$

Prove that $f \in AC([a,b])$ if and only if $f(Der_{\infty}(f))$ is a nullset. Illustrate this result by means of the strict Cantor function (3.7).

Exercise 3.11. Let $f : [a,b] \to [c,d]$ be a homeomorphism, and let $Der_0(f)$ be the set of critical points of f defined in (3.102). Using Exercise 3.10, prove that $f^{-1} \in AC([c,d])$ if and only if $f^{-1}(Der_0(f))$ is a nullset. Illustrate this result by means of the strict Cantor function (3.7).

Exercise 3.12. Prove that the Cantor function (3.6) has the integral

$$\int_0^1 \varphi(x)\, dx = \frac{1}{2} .$$

What is the integral of the strict Cantor function (3.7)?

Exercise 3.13. Let C be the Cantor set (3.3), and let $f : C \to \mathbb{R}$ be defined as follows. Given $x \in C$ with ternary representation (3.5), we put

$$f(x) := \sum_{k=1}^{\infty} x_k 3^{-k!} .$$

Prove the following statements.

(a) From $x, y \in C$ and $x - y > 3^{-n}$, it follows that $f(x) - f(y) \geq 3^{-n!}$.
(b) The map $f : C \to [0, 1]$ is injective, but not surjective.
(c) The map $f : C \to [0, 1]$ satisfies a Hölder condition

$$|f(x) - f(y)| \leq L|x - y|^{\alpha} \quad (x, y \in C)$$

for *any* $\alpha \in (0, 1)$.

Is f absolutely continuous?

Exercise 3.14. Prove that the Cantor function (3.6) belongs to $Lip_{\alpha}([0, 1])$ for $\alpha = \log 2/\log 3$, but does not belong to $Lip_{\beta}([0, 1])$ for any $\beta > \log 2/\log 3$.

Exercise 3.15. Construct two functions $g \in C^1([-1, 1])$ and $f \in AC([-1, 1]) \cap C^1([-1, 1] \setminus \{0\})$ such that $g \circ f \notin AC([-1, 1])$.

Exercise 3.16. Suppose that $f : [a, b] \to \mathbb{R}$ satisfies a Hölder condition

$$|f(x) - f(y)| \leq L|x - y|^{\alpha} \quad (a \leq x, y \leq b)$$

with some constant $L > 0$. Which condition on α then guarantees that $f \in AC([a, b])$?

Exercise 3.17*. Let $f : [a, b] \to \mathbb{R}$ be absolutely continuous, and let $g : [a, b] \to \mathbb{R}$ be increasing. Suppose that $\lambda(M_c(f)) = \lambda(M_c(g))$, where M_c is defined as in Exercise 0.1. Prove that g is then also absolutely continuous.

Exercise 3.18. Prove the inclusion $Lip([a, b]) \subseteq Lu([a, b])$, and give an example of a function $f \in Lu([0, 1]) \setminus Lip([0, 1])$.

Exercise 3.19. Is the inclusion $Lip_{\alpha}([a, b]) \subseteq Lu([a, b])$ also true for $\alpha < 1$? Compare with Exercise 3.8.

Exercise 3.20. Suppose that $f : [a, b] \to \mathbb{R}$ is *everywhere* differentiable on $[a, b]$. Prove that $f \in Lu([a, b])$. Compare with Exercise 3.18.

Exercise 3.21. Given $f, g \in Lu([a, b])$, show that $f \cdot g \in Lu([a, b])$. Is it also true that $f + g \in Lu([a, b])$?

Exercise 3.22. Is the equality

$$\bigcup_{p \geq 1} AC_p([a, b]) = C([a, b])$$

true? If not, construct a continuous function which does not belong to $AC_p([a, b])$ for any $p \geq 1$.

Exercise 3.23. Does the Cantor function (3.6) belong to any of the spaces $AC_p([0, 1])$?

Exercise 3.24. Given $p \geq 1$, determine the precise values of θ for which the special zigzag function Z_{θ} defined in (0.93) belongs to $AC_p([0, 1])$.

Exercise 3.25. Define $f\colon [0,1] \to \mathbb{R}$ by

$$f(x) := \begin{cases} \frac{x^{1/q}}{\log x} \sin^2 \frac{1}{x} & \text{for } 0 < x \le 1\,, \\ 0 & \text{for } x = 0\,. \end{cases}$$

Prove that $f \in AC_q([0,1])$, but $f \notin WBV_p([0,1])$ for $p < q$.

Exercise 3.26. Define $f\colon [0,1] \to \mathbb{R}$ by

$$f(x) := \sum_{k=1}^{\infty} \frac{\sin N^k x}{N^{k/q}}\,,$$

where $N \in \mathbb{N}$. Prove that for N sufficiently large, $f \in C([0,1]) \cap WBV_q([0,1])$, but $f \notin AC_q([0,1])$.

Exercise 3.27. Given a bounded function $f\colon [a,b] \to \mathbb{R}$, suppose that for every $\varepsilon > 0$, there exists a function $g \in AC_p([a,b])$ such that

$$\inf \left\{ \sum_{j=1}^{m} \operatorname{Var}_p^W (f - g; [t_{j-1}, t_j]) : \{t_0, t_1, \ldots, t_m\} \in \mathcal{P}([a,b]) \right\} \le \varepsilon\,,$$

see (3.16). Prove that then also $f \in AC_p([a,b])$.

Exercise 3.28. Are the two norms (3.42) and (3.43) equivalent on $AC([a,b])$? Are they equivalent to the norm $\|f\|_{AC} := |f(a)| + \|f'\|_{L_1}$?

Exercise 3.29. Show that the space $AC([a,b])$ with any of the norms (3.42) or (3.43) is separable. Compare this with Exercise 1.49.

Exercise 3.30. Use Theorem 3.20 to show that $Lip([a,b])$ is dense in $AC([a,b])$ with respect to the norm (3.42).

Exercise 3.31. Suppose that $f \in C([a,b])$ has the property that the set $S := \{x : a \le x \le b,\ f'(x) \text{ does not exist}\}$ is countable, and $f'(x) \equiv 0$ on $[a,b] \setminus S$. Prove that f is constant. Is the same true if S is an uncountable nullset?

Exercise 3.32. Let $f \in AC([a,b])$, and let $N \subset [a,b]$ be a nullset such that f' exists on $[a,b] \setminus N$. Suppose that the restriction $f'|_{[a,b]\setminus N}$ of f' to $[a,b] \setminus N$ satisfies a Lipschitz condition with Lipschitz constant L. Prove that f' then exists on the whole interval $[a,b]$ (as a one-sided derivative at the boundary points a and b) and satisfies a Lipschitz condition with Lipschitz constant L.

Exercise 3.33. Is the inequality (3.32) in Theorem 3.19 always strict for $f \in BV([a,b]) \setminus AC([a,b])$?

Exercise 3.34. Illustrate Theorem 3.25 by means of the Cantor function $\varphi \in BV([0,1]) \setminus AC([0,1])$ defined in (3.6).

Exercise 3.35. Carry out the details in the last part of Example 3.15.

Exercise 3.36. Carry out the details in the last part of Example 3.23.

Exercise 3.37. Illustrate Theorem 3.25 by means of the function $g = \varphi'$ from Example 3.16.

Exercise 3.38*. Construct a function $g \in L_1([-1, 1])$ such that $g(0) = 0$,

$$\frac{1}{h}\int_0^h |g(t)|\,dt \to \infty \quad (h \to 0)\,,$$

but the function $f : [-1, 1] \to \mathbb{R}$ defined by

$$f(x) := \int_{-1}^{x} g(t)\,dt$$

is differentiable at 0 with $f'(0) = 0$.

Exercise 3.39. Let $f : [a, b] \to \mathbb{R}$ be increasing, and let (3.38) be the decomposition of f as a sum of an absolutely continuous function $f_{\rm ac}$ and a singular function $f_{\rm sg}$. Show that both $f_{\rm ac}$ and $f_{\rm sg}$ are then also increasing.

Exercise 3.40. We define a sequence of “sawtooth functions” as follows. For $0 \le x \le 1$, we first put $f_1(x) := \frac{1}{2} - |x - \frac{1}{2}|$. For $n = 2, 3, 4, \ldots$, we then define f_n on $[0, 2^{-n+1}]$ by $f_n(x) := 2^{-n} - |x - 2^{-n}|$ and extend f_n periodically to the whole interval $[0, 1]$. Prove the following properties of the sequence $(f_n)_n$.

(a) The sequence $(f_n)_n$ converges uniformly on $[0, 1]$ to $f(x) \equiv 0$.
(b) All functions f_n have the same graph length $L(\Gamma(f_n); [0, 1]) = \sqrt{2}$.
(c) The convergence $L(\Gamma(f_n)) \to L(\Gamma(f))$ does *not* hold as $n \to \infty$.

Also, calculate the total variations $\mathrm{Var}(f_n; [0, 1])$ and $\mathrm{Var}(f; [0, 1])$ and comment on the result.

Exercise 3.41. Let $f : [0, 1] \to [0, 1]$ be monotonically increasing with $f(0) = 0$ and $f(1) = 1$. Show that the graph length of f satisfies $\sqrt{2} \le L(\Gamma(f)) \le 2$. Moreover, prove that $L(\Gamma(f)) = \sqrt{2}$ if and only if $f(x) = x$, and $L(\Gamma(f)) = 2$ if and only if $f'(x) \equiv 0$ a.e. on $[0, 1]$.

Exercise 3.42. Calculate $L(\Gamma(f))$ for $f(x) = x^2$ on $[0, 1]$.

Exercise 3.43. Let $f : [a, b] \to \mathbb{R}$ be differentiable with (Riemann) integrable derivative f' on $[a, b]$. Prove that then (3.60) holds, and compare with Example 3.15.

Exercise 3.44. For $f \in BV([a, b])$, let V_f be the variation function (1.13) of f and L_f be the length function (3.57) of f. Prove that $L_f \in BV([a, b])$ and

$$\mathrm{Var}(L_f; [a, b]) = L(\Gamma(V_f); [a, b])\,,$$

i.e. the total variation of the length function is the length of the graph of the variation function.

Exercise 3.45. Is the inequality (3.64) in Theorem 3.31 always strict for $f \in BV([a,b]) \setminus AC([a,b])$?

Exercise 3.46. Define $g : [0,1] \to \mathbb{R}$ by

$$g(x) := \begin{cases} \operatorname{ent}\left(\frac{1}{\operatorname{ent}(1/x)}\right) & \text{for } 0 < x \le 1\,, \\ 0 & \text{for } x = 0\,, \end{cases}$$

where $\operatorname{ent}(\xi)$ denotes the integer part of ξ. Show that $g \in BV([0,1]) \setminus AC([0,1])$ and calculate $\operatorname{Var}(g; [0,1])$ and $L(\Gamma(g); [0,1])$.

Exercise 3.47. Determine all $p \in [1,\infty)$ for which the function from Definition 0.47 belongs to $RBV_p([0,1])$.

Exercise 3.48*. Let $1 < p < \infty$, $k \ge 2$, and $f \in RBV_{k,p}([a,b])$, see Definition 3.38. Prove that the derivative $f^{(k-2)}$ belongs then to $RBV_{2,1}([a,b])$. Conclude that the right and left derivative $f_+^{(k-1)}$ and $f_-^{(k-1)}$ exist on $[a,b]$ and are, respectively, right and left continuous.

Exercise 3.49. Prove the (strict) inclusions

$$RBV_{k,p}([a,b]) \supset RBV_{k+1,p}([a,b]) \quad (k = 1,2,3,\ldots;\ 1 \le p < \infty)$$

and

$$RBV_{k,p}([a,b]) \subset RBV_{k,q}([a,b]) \quad (k = 1,2,3,\ldots;\ 1 \le p < q < \infty)$$

and compare with Proposition 2.73.

Exercise 3.50. Prove the continuous imbedding

$$RBV_{k+1,p}([a,b]) \hookrightarrow RBV_{k,p}([a,b]) \quad (k \in \mathbb{N}, 1 \le p < \infty)\,,$$

where $RBV_{k,p}([a,b])$ is the space defined by the norm (3.77), and calculate the sharp imbedding constant $c(RBV_{k+1}, RBV_k)$.

Exercise 3.51. Let ϕ and ψ be two Young functions which satisfy condition ∞_1 and

$$\psi(t) \le \phi(ct) \quad (t \ge T)$$

for some constants $c > 0$ and $T > 0$. Prove the continuous imbedding

$$RBV_\phi([a,b]) \hookrightarrow RBV_\psi([a,b])\,,$$

where $RBV_\phi([a,b])$ is the space defined by the norm (2.99),

Exercise 3.52. Prove the continuous imbedding

$$RBV_{k,\phi}([a,b]) \hookrightarrow RBV_{k,1}([a,b]) \quad (k \in \mathbb{N}),$$

where $RBV_{k,\phi}([a,b])$ is the space defined by the norm (3.86), and ϕ is a Young function satisfying condition ∞_1.

Exercise 3.53. The set $AC^1([a,b])$ consists of all differentiable functions with an absolutely continuous derivative on $[a,b]$. Show that $AC^1([a,b])$, equipped with the norm (0.44), is a Banach space. Also, show that $AC^1 \hookrightarrow Lip$ and calculate the sharp imbedding constant $c(AC^1, Lip)$, see (0.36).

Exercise 3.54. The space $AC([a,b])$ can be imbedded in $AC(\mathbb{R})$ by extending each $f \in AC([a,b])$ to be constant outside $[a,b]$. Prove that a closed bounded set $M \subset AC([a,b])$ is compact in $(AC([a,b]), \|\cdot\|_{AC})$ if and only if both

$$\lim_{n\to\infty} \sup_{f\in M} \operatorname{Var}(f; \mathbb{R} \setminus (-n,n)) = 0$$

and

$$\lim_{\tau\to 0} \sup_{f\in M} \|f - f_\tau\|_{AC} = 0,$$

where f_τ denotes the shift $f_\tau(t) := f(t+\tau)$ of f.

Exercise 3.55. Using Exercise 3.33 (but not general facts from functional analysis), show that the closed unit ball in $(AC([a,b]), \|\cdot\|_{AC})$ is not compact.

Exercise 3.56. Find examples of functions $f, g \in BV([0,1])$ for which the first and last inequalities in (3.99) are strict.

Exercise 3.57. Calculate $d_k(\varphi,\psi)$ $(k = 1,2,3,4)$ for the four metrics (3.95)–(3.98), where φ is the Cantor function (3.6) and ψ is the strict Cantor function (3.7), and check the estimates (3.99).

Exercise 3.58. Given a countable set $M = \{y_1, y_2, y_3, \ldots\}$, define $f: [0,1] \to \mathbb{R}$ by

$$f(x) := \begin{cases} y_k & \text{for } \frac{1}{k} - \frac{1}{2k(k+1)} \le x \le \frac{1}{k} + \frac{1}{2k(k+1)}, \\ \text{linear} & \text{otherwise}. \end{cases}$$

Prove that $f \in C([0,1])$ and $f(Ext(f)) = M$, see (3.105).

Exercise 3.59. Suppose that $f: [a,b] \to \mathbb{R}$ satisfies $Ext(f) = [a,b]$ and $f(Ext(f)) = \{c,d\}$ for $c \neq d$, see (3.104) and (3.105). Prove that f is discontinuous everywhere on $[a,b]$.

Exercise 3.60. Suppose that $f \in C([a,b])$ satisfies $Ext(f) = [a,b]$, see (3.104). Does it follow that f is constant?

Exercise 3.61. For $0 < \alpha < 1$, let $f_\alpha : [0,1] \times [0,1] \to \mathbb{R}$ be defined by $f_\alpha(x,y) := (x-y)^\alpha$. Use Theorem 3.51 to determine all $p \geq 1$ such that $f_\alpha \in RBV_p([0,1] \times [0,1])$. Also, calculate $\|f_\alpha\|_{RBV_p}$ in this case, and compare with Example 3.35.

Exercise 3.62. Prove that $f \in AC([a,b] \times [c,d])$, see Definition 3.50, if and only if $f(a,\cdot) \in AC([c,d])$, $f(\cdot,y) \in AC([a,b])$ for every $y \in [c,d]$, $f_x(x,\cdot) \in AC([c,d])$ for almost every $x \in [a,b]$, and $f_{xy} \in L_1([a,b] \times [c,d])$.

Exercise 3.63*. Given $g \in L_1([a,b])$, define f as in (3.20) and put

$$E_c(g) := \left\{x : a < x < b, \limsup_{h \to 0} \left|\frac{g(x+h) - g(x)}{h}\right| > c\right\} \quad (c > 0)\,.$$

Prove that

$$\lambda(E_c(g)) \leq \frac{4}{c}\|g\|_{L_1}$$

and compare with Exercise 3.10.

Exercise 3.64. For $\alpha, \beta, \gamma, \delta \in \mathbb{R}$, define $f, g : [0,1] \to \mathbb{R}$ by

$$f(x) := \begin{cases} x^\alpha \sin x^\beta & \text{for } 0 < x \leq 1\,, \\ 0 & \text{for } x = 0\,, \end{cases} \qquad g(x) := \begin{cases} x^\gamma \sin x^\delta & \text{for } 0 < x \leq 1\,, \\ 0 & \text{for } x = 0\,. \end{cases}$$

Using Table 2.4 in Chapter 2 to determine all values of α, β, γ, and δ for which Theorem 3.53 applies to f and g, but Theorem 3.52 does not.

Exercise 3.65. Formulate and prove an analogue to Theorem 3.46 for $f \in Lip([a,b])$, and illustrate this by means of the function g from Example 3.45.

Exercise 3.66. A continuous function $f : [a,b] \to \mathbb{R}$ is called *piecewise linear* if there exists a partition $P = \{t_0, t_1, \ldots, t_m\} \in \mathcal{P}([a,b])$ such that f is affine on each interval $[t_{j-1}, t_j]$. Prove that every piecewise linear function is absolutely continuous and calculate its norm (3.42).

Exercise 3.67. Denote by $PL([a,b])$ the set of all piecewise linear functions on $[a,b]$, see Exercise 3.66. Show that $PL([a,b]) = J(S([a,b]))$, where $S([a,b])$ denotes the set of step functions introduced in Section 0.4, and J denotes the linear integral operator (3.107).

Exercise 3.68. Use the result of Exercise 3.67 and the fact that $S([a,b])$ is dense in $L_1([a,b])$ to prove that $PL([a,b])$ is dense in $AC([a,b])$ with respect to the norm (3.43).

4 Riemann–Stieltjes integrals

By means of the classical Riemann–Stieltjes integral, it is possible to construct an isometry between the dual space of the Chebyshev space $C([a, b])$ and a subspace of (suitably regularized) functions in $BV([a, b])$. Similarly, an analogous isometry makes it possible to identify the dual space of the Lebesgue space $L_p([a, b])$ $(1 < p < \infty)$ with the space $RBV_{p/(p-1)}([a, b])$, where $RBV_p([a, b])$ denotes the space of functions of bounded p-variation in Riesz's sense introduced in Chapter 2. Here, one does not need regularizations since all functions in $RBV_p([a, b])$ are continuous for $p > 1$. One may consider the same isometry also for the case $p = 1$ which leads to the space $Lip([a, b])$ of Lipschitz continuous functions on $[a, b]$, but not for the case $p = \infty$ which has to be treated differently. In the last section, we show how to extend the Riemann–Stieltjes integral to the case when the "integrator" does not necessarily belong to the classical space $BV([a, b])$, but to the larger space $\Phi BV([a, b])$ of functions of bounded Schramm variation introduced in Section 2.3.

4.1 Classical RS-integrals

In this section, we start with the definition of the classical Riemann–Stieltjes integral and discuss some of its properties.

To begin with, let $\alpha : [a, b] \to \mathbb{R}$ be monotonically increasing, and let $f : [a, b] \to \mathbb{R}$ be bounded. Given a partition $P = \{t_0, t_1, \dots, t_m\} \in \mathcal{P}([a, b])$, consider the numbers

$$M_j := \sup\{f(x) : t_{j-1} \le x \le t_j\}, \quad m_j := \inf\{f(x) : t_{j-1} \le x \le t_j\} \tag{4.1}$$

for $j = 1, 2, \dots, m$. As usual, we will call the real number

$$L_\alpha(f, P; [a, b]) := \sum_{j=1}^{m} m_j(\alpha(t_j) - \alpha(t_{j-1})), \tag{4.2}$$

the *lower Riemann–Stieltjes sum* (or *lower RS-sum*) of f with respect to P and α, and the real number

$$U_\alpha(f, P; [a, b]) := \sum_{j=1}^{m} M_j(\alpha(t_j) - \alpha(t_{j-1})), \tag{4.3}$$

the *upper Riemann–Stieltjes sum* (or *upper RS-sum*) of f with respect to P and α. A useful monotonicity property of these sums is given in the following

Proposition 4.1. *For $P, Q \in \mathcal{P}([a, b])$ with $P \subseteq Q$, the estimates*

$$L_\alpha(f, P; [a, b]) \le L_\alpha(f, Q; [a, b]) \le U_\alpha(f, Q; [a, b]) \le U_\alpha(f, P; [a, b]) \tag{4.4}$$

hold.

Proof. Let $P = \{t_0, t_1, \ldots, t_m\} \in \mathcal{P}([a,b])$, and let $Q = \{t_0, \ldots, t_{i-1}, t_*, t_i, \ldots, t_m\}$ be a partition consisting of P and one additional point $t_* \in (t_{i-1}, t_i)$. Then we have

$$L_\alpha(f, Q; [a,b]) = \sum_{j=1}^{i-1} m_j(\alpha(t_j) - \alpha(t_{j-1})) + m_{i-1}^*(\alpha(t_*) - \alpha(t_{i-1})) + m_i^*(\alpha(t_i) - \alpha(t_*)) + \sum_{j=i+1}^{m} m_j(\alpha(t_j) - \alpha(t_{j-1})),$$

where

$$m_{i-1}^* := \inf\{f(x) : t_{i-1} \le x \le t_*\} \quad m_i^* := \inf\{f(x) : t_* \le x \le t_i\}.$$

Obviously, $m_{i-1}^* \ge m_i$ and $m_i^* \ge m_i$, and hence

$$L_\alpha(f, P; [a,b]) \le L_\alpha(f, P \cup \{t_*\}; [a,b]).$$

By applying induction on the number of added points, we derive the first estimate in (4.4). The third estimate is proved in the same way, while the second estimate is of course trivial. □

Definition 4.2. We say that a function $f \in B([a,b])$ is *Riemann–Stieltjes integrable* (or *RS-integrable*, for short) with respect to α on $[a,b]$ if for each $\varepsilon > 0$, there exists a partition $P \in \mathcal{P}([a,b])$ such that

$$U_\alpha(f, P; [a,b]) - L_\alpha(f, P; [a,b]) \le \varepsilon. \tag{4.5}$$

In this case, we write $f \in RS_\alpha([a,b])$. ■

Thus, the condition $f \in RS_\alpha([a,b])$ means that the equality

$$\inf\{U_\alpha(f, P; [a,b]) : P \in \mathcal{P}([a,b])\} = \sup\{L_\alpha(f, P; [a,b]) : P \in \mathcal{P}([a,b])\} \tag{4.6}$$

holds. In this case, we denote the common value in (4.6) by

$$I = \int_a^b f(x)\, d\alpha(x) = \int_a^b f\, d\alpha \tag{4.7}$$

and call it the *Riemann–Stieltjes integral* (or *RS-integral*, for short) of f with respect to α on $[a,b]$. The function f in (4.7) is often called the *integrand* and the function α the *integrator*. Obviously, in case $\alpha(x) = x$, Definition 4.2 gives nothing other than Riemann integrability, and (4.7) is the classical Riemann integral of f. In this case, we simply write[1] $f \in RS([a,b])$. We point out, however, that, in general, the function α need not even be continuous.

In the next proposition, we collect some useful properties of the RS-integral.

1 Of course, we had better write $R([a,b])$ instead of $RS([a,b])$ for Riemann integrable functions; however, this could be confused with the space of regular functions which we introduced in Chapter 0.

Proposition 4.3. *Let $\alpha, \beta : [a,b] \to \mathbb{R}$ be monotonically increasing, $f, g \in RS_\alpha([a,b])$, and $\mu \in \mathbb{R}$. Then the following is true.*

(a) *The integral (4.7) is additive with respect to the integrand, i.e.*[2]

$$\int_a^b (f+g)\,d\alpha = \int_a^b f\,d\alpha + \int_a^b g\,d\alpha\,. \tag{4.8}$$

(b) *The integral (4.7) is additive with respect to the integrator, i.e.*[3]

$$\int_a^b f\,d(\alpha+\beta) = \int_a^b f\,d\alpha + \int_a^b f\,d\beta\,. \tag{4.9}$$

(c) *The integral (4.7) is homogeneous with respect to the integrand, i.e.*

$$\int_a^b (\mu f)\,d\alpha - \mu \int_a^b f\,d\alpha \tag{4.10}$$

for $\mu \in \mathbb{R}$.

(d) *The integral (4.7) is homogeneous with respect to the integrator, i.e.*

$$\int_a^b f\,d(\mu\alpha) = \mu \int_a^b f\,d\alpha(x) \tag{4.11}$$

for $\mu > 0$.

(e) *The integral (4.7) is monotone with respect to the integrand, which means that $f(x) \le g(x)$ on $[a,b]$ implies*

$$\int_a^b f\,d\alpha \le \int_a^b g\,d\alpha\,. \tag{4.12}$$

In particular,

$$m(f)[\alpha(b)-\alpha(a)] \le \int_a^b f\,d\alpha \le M(f)[\alpha(b)-\alpha(a)]\,, \tag{4.13}$$

where $m(f)$ is defined in (0.61) and $M(f)$ in (0.62).

2 The statement (a) is meant in the following sense: from $f, g \in RS_\alpha([a,b])$, it follows that also $f+g \in RS_\alpha([a,b])$, and (4.8) holds.

3 The statement (b) is meant in the following sense: from $f \in RS_\alpha([a,b]) \cap RS_\beta([a,b])$, it follows that also $f \in RS_{\alpha+\beta}([a,b])$, and (4.9) holds.

Proof. In the sequel, we drop the interval $[a, b]$ in (4.2) and (4.3) and the argument x in (4.7) so as not to overburden the notation.

(a) For every $P \in \mathcal{P}([a, b])$, we have

$$L_\alpha(f, P) + L_\alpha(g, P) \le L_\alpha(f + g, P) \le U_\alpha(f + g, P) \le U_\alpha(f, P) + U_\alpha(g, P) . \tag{4.14}$$

From $f, g \in RS_\alpha([a, b])$, it follows that for each $\varepsilon > 0$, we can find partitions $P_f, P_g \in \mathcal{P}([a, b])$ such that

$$U_\alpha(f, P_f) - L_\alpha(f, P_f) \le \varepsilon, \quad U_\alpha(g, P_g) - L_\alpha(g, P_g) \le \varepsilon .$$

Adding side by side of the above inequalities, we get

$$U_\alpha(f, P_f) + U_\alpha(g, P_g) - L_\alpha(f, P_f) - L_\alpha(g, P_g) \le 2\varepsilon .$$

Taking now into account (4.14) and Proposition 4.1, we obtain

$$\begin{aligned} L_\alpha(f, P_f) + L_\alpha(g, P_g) &\le L_\alpha(f + g, P_f \cup P_g) \\ &\le U_\alpha(f + g, P_f \cup P_g) \le U_\alpha(f, P_f) + U_\alpha(g, P_g) . \end{aligned}$$

Consequently, if we denote by P the common refinement $P_f \cup P_g$ of P_f and P_g, we conclude that

$$U_\alpha(f + g, P) - L_\alpha(f + g, P) \le 2\varepsilon ,$$

which shows that $f + g \in RS_\alpha([a, b])$. For the same partition P, we further have

$$U_\alpha(f, P) \le \int_a^b f\, d\alpha + \varepsilon, \quad U_\alpha(g, P) \le \int_a^b g\, d\alpha + \varepsilon .$$

Consequently, (4.14) implies that

$$\int_a^b (f + g)\, d\alpha \le U_\alpha(f + g, P) \le \int_a^b f\, d\alpha + \int_a^b g\, d\alpha + 2\varepsilon$$

and so

$$\int_a^b (f + g)\, d\alpha \le \int_a^b f\, d\alpha + \int_a^b g\, d\alpha$$

since $\varepsilon > 0$ was arbitrary. Replacing f by $-f$ and g by $-g$, we obtain the reverse estimate, and so we have proved (4.8). The assertion (b) is proved similarly, building on the inequalities

$$L_\alpha(f, P) + L_\beta(f, P) \le L_{\alpha+\beta}(f, P) \le U_{\alpha+\beta}(f, P) \le U_\alpha(f, P) + U_\beta(f, P)$$

for every $P \in \mathcal{P}([a, b])$.

The assertion (c) follows from the equalities

$$L_\alpha(\mu f, P) = \mu L_\alpha(f, P), \quad U_\alpha(\mu f, P) = \mu U_\alpha(f, P),$$

while the assertion (d) follows from the equalities

$$L_{\mu\alpha}(f, P) = \mu L_\alpha(f, P), \quad U_{\mu\alpha}(f, P) = \mu U_\alpha(f, P).$$

To prove (e), it suffices to observe that $f(x) \leq g(x)$ on $[a, b]$ clearly implies both $L_\alpha(f, P) \leq L_\alpha(g, P)$ and $U_\alpha(f, P) \leq U_\alpha(g, P)$. The estimate (4.13) is a consequence of (4.12). □

Before discussing several classes of RS-integrable functions, we want to extend Definition 4.2 to functions α of bounded variation. Proposition 4.3 (b) suggests how to do this: given $\alpha \in BV([a, b])$, consider any decomposition $\alpha = \beta - \gamma$ of α as the difference of two monotonically increasing functions $\beta, \gamma : [a, b] \to \mathbb{R}$ (for example, the Jordan decomposition described in Theorem 1.5). Then we define the RS-integral of $f \in RS_\beta([a, b]) \cap RS_\gamma([a, b])$ in the natural way by

$$\int_a^b f\,d\alpha := \int_a^b f\,d\beta - \int_a^b f\,d\gamma. \tag{4.15}$$

Proposition 4.3 (b) guarantees that this definition does not depend on the choice of β and γ.

The Riemann integral may be introduced not only by means of upper and lower sums, but also with so-called Riemann sums in which the numbers M_j and m_j in (4.1) are replaced by the value of f at an arbitrary intermediate points. The same is possible for the Riemann–Stieltjes integral.

Definition 4.4. Let $\alpha : [a, b] \to \mathbb{R}$ be monotonically increasing, and let $f : [a, b] \to \mathbb{R}$ be bounded. Given a partition $P = \{t_0, t_1, \ldots, t_m\} \in \mathcal{P}([a, b])$, choose a set $\Pi = \{\tau_1, \tau_2, \ldots, \tau_m\}$ of points satisfying

$$a = t_0 \leq \tau_1 \leq t_1 \leq \ldots \leq t_{m-1} \leq \tau_m \leq t_m = b. \tag{4.16}$$

Then we will call the sum

$$S_\alpha(f, P, \Pi; [a, b]) := \sum_{j=1}^{m} f(\tau_j)(\alpha(t_j) - \alpha(t_{j-1})) \tag{4.17}$$

the *Riemann–Stieltjes sum* (or *RS-sum*, for short) in what follows. ■

By definition, the notation

$$\lim_{\mu(P)\to 0} S_\alpha(f, P, \Pi; [a, b]) =: A, \tag{4.18}$$

where $\mu(P)$ denotes the mesh size (1.2) of P, then has the following meaning: for any $\varepsilon > 0$, there exists a $\delta > 0$ such that

$$|S_\alpha(f, P, \Pi; [a,b]) - A| \leq \varepsilon \tag{4.19}$$

for any partition P satisfying $\mu(P) \leq \delta$ and any set Π of intermediate points[4] satisfying (4.16).

Proposition 4.5. *If the limit* (4.18) *exists, then* $f \in RS_\alpha([a,b])$ *and*

$$\lim_{\mu(P)\to 0} S_\alpha(f, P, \Pi; [a,b]) = \int_a^b f(x)\, d\alpha(x)\,, \tag{4.20}$$

i.e. the RS-integral of f with respect to α coincides with this limit.

Proof. Denote the limit as in (4.18) by A, and let $\varepsilon > 0$. By definition, we find a $\delta > 0$ such that

$$A - \varepsilon \leq S_\alpha(f, P, \Pi; [a,b]) \leq A + \varepsilon \tag{4.21}$$

for any partition P satisfying $\mu(P) \leq \delta$. If we let the intermediate points τ_j vary over the intervals $[t_{j-1}, t_j]$ $(j = 1, 2, \ldots, m)$ and take the corresponding infimum and supremum over $S_\alpha(f, P, \Pi; [a,b])$, from (4.21), we get

$$A - \varepsilon \leq L_\alpha(f, P; [a,b]) \leq U_\alpha(f, P; [a,b]) \leq A + \varepsilon\,.$$

By Definition 4.2, this means nothing more than $f \in RS_\alpha([a,b])$ and

$$\int_a^b f(x)\, d\alpha(x) = A$$

since the real number A in (4.18) is uniquely determined. □

We observe that the converse implication of the statement of Proposition 4.5 is also true: for every $f \in RS_\alpha([a,b])$, the limit (4.18) exists and coincides with the RS-integral of f with respect to α. Indeed, this is a simple consequence of the inequalities

$$L_\alpha(f, P; [a,b]) \leq S_\alpha(f, P, \Pi; [a,b]) \leq U_\alpha(f, P; [a,b])$$

which holds for increasing α and all sets $\Pi = \{\tau_1, \tau_2, \ldots, \tau_m\}$ satisfying (4.16).

Now, we state some useful criteria for the existence of an RS-integral. The following criterion is a consequence of Proposition 4.5.

---

4 Observe that $\mu(P) \to 0$ implies $\mu(\Pi) \to 0$, and so the relation (4.18) is actually independent of Π. For this reason, we will sometimes write $S_\alpha(f, P; [a,b])$ instead of $S_\alpha(f, P, \Pi; [a,b])$ whenever we consider the limit $\mu(P) \to 0$ in the sequel.

Corollary 4.6. *Let $\alpha : [a,b] \to \mathbb{R}$ be monotonically increasing. Then $f \in RS_\alpha([a,b])$ if and only if for each $\varepsilon > 0$, there exists $\delta > 0$ such that for arbitrary partitions $P_1, P_2 \in \mathcal{P}([a,b])$ with $\mu(P_i) \leq \delta$, we have*

$$|S_\alpha(f, P_1, \Pi_1; [a,b]) - S_\alpha(f, P_2, \Pi_2; [a,b])| \leq \varepsilon, \tag{4.22}$$

where Π_1 and Π_2 are arbitrary sets of intermediate points of P_1 and P_2, respectively.

Proof. The statement follows immediately from the Cauchy criterion for the existence of the limit (4.18). □

We may derive from Corollary 4.6 another useful property of the RS-integral which is often used in explicit calculations.

Proposition 4.7. *From $f \in RS_\alpha([a,b])$, it follows that $f \in RS_\alpha([a,c])$ and $f \in RS_\alpha([c,b])$ for each $c \in (a,b)$; moreover, the equality*

$$\int_a^b f(x)\,d\alpha(x) = \int_a^c f(x)\,d\alpha(x) + \int_c^b f(x)\,d\alpha(x) \tag{4.23}$$

holds in this case.

Proof. Given $\varepsilon > 0$, by Corollary 4.6, we find $\delta > 0$ such that (4.22) holds for arbitrary partitions $P_1, P_2 \in \mathcal{P}([a,b])$ with $\mu(P_i) \leq \delta$ and arbitrary sets Π_1 and Π_2 of intermediate points of P_1 and P_2, respectively.[5]

Fix $P_1', P_2' \in \mathcal{P}([a,c])$ with $\mu(P_1') \leq \delta$ and $\mu(P_2') \leq \delta$, and corresponding sets Π_1' and Π_2' of intermediate points. Moreover, choose any $P'' \in \mathcal{P}([c,b])$ with $\mu(P'') \leq \delta$ and a corresponding set Π'' of intermediate points. Then $P_i := P_i' \cup P'' \in \mathcal{P}([a,b])$ satisfies $\mu(P_i) \leq \delta$ $(i = 1, 2)$, and thus by applying the above condition, we get

$$|S_\alpha(f, P_1', \Pi_1'; [a,c]) - S_\alpha(f, P_2', \Pi_2'; [a,c])| = |S_\alpha(f, P_1, \Pi_1; [a,b]) - S_\alpha(f, P_2, \Pi_2; [a,b])| \leq \varepsilon,$$

where $\Pi_i := \Pi_i' \cup \Pi''$ $(i = 1, 2)$. From this, we conclude that $f \in RS_\alpha([a,c])$.

The assertion $f \in RS_\alpha([c,b])$ is proved analogously by considering partitions $P' \in \mathcal{P}([a,c])$ and $P_1'', P_2'' \in \mathcal{P}([c,b])$. It remains to prove (4.23). However, any pair of partitions $P' \in \mathcal{P}([a,c])$ and $P'' \in \mathcal{P}([c,b])$ gives rise to a partition $P := P' \cup P'' \in \mathcal{P}([a,b])$ which satisfies $\mu(P) = \max\{\mu(P'), \mu(P'')\}$. For the corresponding RS-sums (involving sets Π', Π'', and $\Pi := \Pi' \cup \Pi''$) and RS-integrals, we then get

$$S_\alpha(f, P, \Pi; [a,b]) = S_\alpha(f, P', \Pi'; [a,c]) + S_\alpha(f, P'', \Pi''; [c,b])$$

and, passing to the limit $\mu(P) \to 0$,

$$\int_a^b f(x)\,d\alpha(x) = \int_a^c f(x)\,d\alpha(x) + \int_c^b f(x)\,d\alpha(x),$$

and so we are done. □

5 To be precise, we have to assume here that α is monotonically increasing because Corollary 4.6 holds only for such α. However, for $\alpha \in BV([a,b])$, the assertion may be obtained by means of the usual Jordan decomposition.

The reader might ask why Proposition 4.7 is formulated only in one direction. Recall that for the classical Riemann integral, the other direction is also true: $f \in RS([a,c])$ and $f \in RS([c,b])$ implies $f \in RS([a,b])$ with[6]

$$\int_a^b f(x)\,dx = \int_a^c f(x)\,dx + \int_c^b f(x)\,dx\,.$$

The reason for the missing analogue in Proposition 4.7 is simply that it is not true, as the following example shows.

Example 4.8. Let $f : [-1,1] \to \mathbb{R}$ and $\alpha : [-1,1] \to \mathbb{R}$ be defined by $f := \chi_{(0,1]}$ and $\alpha := \chi_{[-1,0]}$, i.e.

$$f(x) = \begin{cases} 0 & \text{for } -1 \le x \le 0\,, \\ 1 & \text{for } 0 < x \le 1\,, \end{cases} \qquad \alpha(x) = \begin{cases} 1 & \text{for } -1 \le x \le 0\,, \\ 0 & \text{for } 0 < x \le 1\,. \end{cases}$$

Clearly, $f \in B([-1,1])$ and $\alpha \in BV([-1,1])$. A straightforward calculation shows that $f \in RS_\alpha([-1,0]) \cap RS_\alpha([0,1])$ with

$$\int_{-1}^0 f(x)\,d\alpha(x) = \int_0^1 f(x)\,d\alpha(x) = 0\,. \tag{4.24}$$

On the other hand, both functions f and α are discontinuous at 0; as we shall later see (Theorem 4.14), this implies that $f \notin RS_\alpha([-1,1])$. ♥

The following is a refinement of Example 4.8, insofar as the function f is even *unbounded* near its (unique) point of discontinuity.

Example 4.9. Let $f : [-1,1] \to \mathbb{R}$ and $\alpha : [-1,1] \to \mathbb{R}$ be defined by

$$f(x) := \begin{cases} 0 & \text{for } -1 \le x \le 0\,, \\ \frac{1}{x} & \text{for } 0 < x \le 1\,, \end{cases} \qquad \alpha(x) := \begin{cases} x & \text{for } -1 \le x \le 0\,, \\ 0 & \text{for } 0 < x \le 1\,. \end{cases}$$

Observe that we even have $\alpha \in BV([-1,1]) \cap C([-1,1])$ in this example. As before, a straightforward calculation shows that $f \in RS_\alpha([-1,0]) \cap RS_\alpha([0,1])$ with (4.24). On the other hand, for any partition $P \in \mathcal{P}([-1,1])$ which does not contain 0, the upper sum $U_\alpha(f,P;[-1,1])$ does not exist since f is unbounded on $(0,1]$, and so $f \notin RS_\alpha([-1,1])$. ♥

The following proposition gives an important criterion for the existence of Riemann–Stieltjes integrals.

6 Here, the point c may also lie outside the interval $[a,b]$ if we adopt the usual convention that the Riemann integral over $[b,a]$ is the negative of the Riemann integral over $[a,b]$.

Proposition 4.10. *Let $f \in B([a,b])$, and let $\alpha : [a,b] \to \mathbb{R}$ be monotonically increasing. Then $f \in RS_\alpha([a,b])$ if and only if for each $\varepsilon > 0$, there exists $\delta > 0$ such that* (4.5) *holds for arbitrary partitions $P = \{t_0, t_1, \dots, t_m\} \in \mathcal{P}([a,b])$ with $\mu(P) \le \delta$.*

Proof. Assume first that $f \in RS_\alpha([a,b])$, let $\varepsilon > 0$, and choose $\delta > 0$ such that the assertion of Corollary 4.6 holds for this δ. Fix $P = \{t_0, t_1, \dots, t_m\} \in \mathcal{P}([a,b])$ with $\mu(P) \le \delta$ as well as two sets $\Pi_1 = \{\tau_1^1, \tau_2^1, \dots, \tau_m^1\}$ and $\Pi_2 = \{\tau_1^2, \tau_2^2, \dots, \tau_m^2\}$ of intermediate points corresponding to $P_1 = P_2 = P$. Applying Corollary 4.6, we then obtain

$$\begin{aligned} &\left| \sum_{j=1}^m f(\tau_j^1)(\alpha(t_j) - \alpha(t_{j-1})) - \sum_{j=1}^m f(\tau_j^2)(\alpha(t_j) - \alpha(t_{j-1})) \right| \\ &\quad \le |S_\alpha(f, P, \Pi_1; [a,b]) - S_\alpha(f, P, \Pi_2; [a,b])| \le \frac{\varepsilon}{2}\,. \end{aligned} \tag{4.25}$$

For each $i \in \{1, 2, \dots, m\}$, we choose the points $\tau_i^1, \tau_i^2 \in [t_{i-1}, t_i]$ in such a way that

$$|f(\tau_i^2) - f(\tau_i^1)| = f(\tau_i^2) - f(\tau_i^1) \ge M_i - m_i - \frac{\varepsilon}{2m(\alpha(b) - \alpha(a))}\,,$$

where M_i and m_i are defined as in (4.1) and we may assume without loss of generality that $\alpha(a) < \alpha(b)$. Then from (4.25), we obtain

$$\begin{aligned} U_\alpha(f, P; [a,b]) - L_\alpha(f, P; [a,b]) &= \sum_{i=1}^m (M_i - m_i)(\alpha(t_i) - \alpha(t_{i-1})) \\ \le \sum_{i=1}^m \left(f(\tau_i^2) - f(\tau_i^1) + \frac{\varepsilon}{2m(\alpha(b) - \alpha(a))} \right) (\alpha(t_i) - \alpha(t_{i-1})) \\ = \sum_{i=1}^m \left(f(\tau_i^2) - f(\tau_i^1) \right) (\alpha(t_i) - \alpha(t_{i-1})) + \frac{\varepsilon}{2} \sum_{i=1}^m \frac{\alpha(t_i) - \alpha(t_{i-1})}{m(\alpha(b) - \alpha(a))} \le \varepsilon\,. \end{aligned}$$

Thus, we have verified that (4.5) holds for every $f \in RS_\alpha([a,b])$. The converse implication is evident. □

The following theorem describes two important classes of RS-integrable functions and is analogous to a well-known result for the Riemann integral.

Theorem 4.11.
(a) *For $\alpha \in BV([a,b])$, every continuous function $f : [a,b] \to \mathbb{R}$ is RS-integrable with respect to α.*
(b) *For $\alpha \in BV([a,b]) \cap C([a,b])$, every function $f : [a,b] \to \mathbb{R}$ of bounded variation is RS-integrable with respect to α, and the equality* (4.20) *holds.*

Proof. By what we have observed above, we may assume without loss of generality that α is increasing and nonconstant.

(a) Being continuous on the compact interval $[a,b]$, the function f is uniformly continuous. So, given $\varepsilon > 0$, we find $\delta > 0$ such that $|x - y| \le \delta$ implies $|f(x) - f(y)| \le \varepsilon$ for all $x, y \in [a,b]$.

Let $P = \{t_0, t_1, \dots, t_m\} \in \mathcal{P}([a,b])$ be any partition satisfying $\mu(P) \le \delta$, and let M_j and m_j be defined by (4.1). Then

$$U_\alpha(f,P;[a,b]) - L_\alpha(f,P;[a,b]) \le (\alpha(b)-\alpha(a)) \sum_{j=1}^{m} (M_j - m_j) \le (\alpha(b)-\alpha(a))\varepsilon$$

which shows that $f \in RS_\alpha([a,b])$.

(b) For each $m \in \mathbb{N}$, we consider a partition $P_m := \{t_0, t_1, \dots, t_m\}$ with the property that

$$\alpha(t_j) - \alpha(t_{j-1}) = \frac{\alpha(b)-\alpha(a)}{m} \quad (j = 1, 2, \dots, m)$$

which is possible by the continuity of α. Suppose that f is monotonically increasing. Then we have $M_j = f(t_j)$ and $m_j = f(t_{j-1})$ in (4.1), and hence

$$\begin{aligned} U_\alpha(f,P;[a,b]) - L_\alpha(f,P;[a,b]) &= \frac{\alpha(b)-\alpha(a)}{m} \sum_{j=1}^{m} [f(t_j) - f(t_{j-1})] \\ &= \frac{\alpha(b)-\alpha(a)}{m} [f(b) - f(a)] \le \varepsilon \end{aligned} \tag{4.26}$$

if we choose m sufficiently large, which shows that $f \in RS_\alpha([a,b])$. Now, since

$$L_\alpha(f,P;[a,b]) \le S_\alpha(f,P,\Pi;[a,b]) \le U_\alpha(f,P;[a,b])$$

for every partition $P \in \mathcal{P}([a,b])$, the estimate (4.26) shows that (4.20) is also true. For general $\alpha \in BV([a,b])$, we use again a decomposition argument. □

Theorem 4.11 contains the well-known result that both continuous and monotone functions are Riemann integrable. Exercise 4.20 shows that the equality (4.20) may fail if we drop the continuity assumption on α in Theorem 4.11 (b).

We point out that RS-integrals may also be defined for much more general functions than f continuous and $\alpha \in BV([a,b])$, or f monotone and $\alpha \in BV([a,b]) \cap C([a,b])$. Both functions f and α may even be *unbounded*; however, in this case, they must “interact” in a certain sense in order to “compensate” the unboundedness. Here is an example.

Example 4.12. Let $f : [0,3] \to \mathbb{R}$ and $\alpha : [0,3] \to \mathbb{R}$ be defined by

$$f(x) := \begin{cases} \frac{1}{x-1} & \text{for } 0 \le x < 1, \\ 0 & \text{for } 1 \le x \le 3, \end{cases} \qquad \alpha(x) := \begin{cases} 0 & \text{for } 0 \le x \le 2, \\ \frac{1}{x-2} & \text{for } 2 < x \le 3. \end{cases}$$

A straightforward calculation shows that $U_\alpha(f,P;[0,3]) = L_\alpha(f,P;[0,3]) = 0$ for all $P \in \mathcal{P}([0,3])$ satisfying $\mu(P) < 1/3$, and so $f \in RS_\alpha([0,3])$. ♥

In view of Theorem 4.11 and Example 4.12, one might ask if (and how) discontinuities of f could be “compensated” by α, or vice versa, to ensure that $f \in RS_\alpha([a,b])$. For example, one may show that “extremely nonintegrable” functions like the Dirichlet function become RS-integrable only for an extremely regular choice of α (Exercise 4.17). Likewise, the following proposition shows that a discontinuity of α at some point must be "compensated" by the continuity of f at this point. For the proof, we introduce some notation.

Definition 4.13. Let $M \subseteq \mathbb{R}$, $c \in M$, and f be a function which is defined on some neighborhood of c. Then we call the limit

$$\operatorname{osc}(f;c) := \lim_{\delta\to 0+} \operatorname{osc}(f;[c-\delta,c+\delta])\,, \tag{4.27}$$

where $\operatorname{osc}(f;A)$ denotes the oscillation (1.12), the *local oscillation* of f at c. ■

Observe that the condition (4.5) for a partition $P = \{t_0, t_1, \dots, t_m\} \in \mathcal{P}([a,b])$ may be written in the form

$$\sum_{j=1}^{m} \operatorname{osc}(f;[t_{j-1},t_j])(\alpha(t_j)-\alpha(t_{j-1}) \le \varepsilon\,. \tag{4.28}$$

This condition will be crucial in the following

Theorem 4.14. *Let $f \in RS_\alpha([a,b])$ for some $\alpha \in BV([a,b])$, and suppose that α is discontinuous at some point $c \in (a,b)$. Then f must be continuous at c.*

Proof. Suppose that both f and α are discontinuous at c. This means, in particular, that $\operatorname{osc}(f;c) > 0$, where $\operatorname{osc}(f;c) > 0$ denotes the local oscillation (4.27) of f at c. Concerning the function α, we distinguish different kinds of discontinuity.

Suppose first that α has a discontinuity of first kind (jump) at c, which means that $\alpha(c-) \neq \alpha(c+)$, or a discontinuity of a second kind at c, which means that $\alpha(c-)$ or $\alpha(c+)$ does not exist. Then we find $\eta > 0$ such that for any $\delta > 0$, there exist points $\sigma \in [a,c)$ and $\tau \in (c,b]$ with $0 < \tau - \sigma < \delta$ and $|\alpha(\tau) - \alpha(\sigma)| \ge \eta$. Now, we fix a partition $P = \{t_0, t_1, \dots, t_m\} \in \mathcal{P}([a,b])$ with the property that $\mu(P) < \delta$ and $t_{i-1} = \sigma$ and $t_i = \tau$ for some $i \in \{1, 2, \dots, m\}$.

Furthermore, we choose sets $\Xi := \{\xi_1, \xi_2, \dots, \xi_m\}$ and $H := \{\eta_1, \eta_2, \dots, \eta_m\}$ of points in $[a,b]$ which satisfy $\xi_j = \eta_j \in [t_{j-1}, t_j]$ for $j \neq i$ and $\xi_i, \eta_i \in [t_{i-1}, t_i] = [\sigma, \tau]$. In addition, we may require that

$$|f(\xi_i) - f(\eta_i)| \ge \frac{1}{2} \operatorname{osc}(f;c)$$

with $\operatorname{osc}(f;c)$ as above. By construction, we have

$$\left|S_\alpha(f,P,\Xi;[a,b]) - S_\alpha(f,P,H;[a,b])\right| = |\alpha(\tau)-\alpha(\sigma)|\,|f(\xi_i)-f(\eta_i)| \ge \frac{\eta}{2}\operatorname{osc}(f;c) > 0$$

for the corresponding Riemann–Stieltjes sums. Since $\delta > 0$ was arbitrary, Corollary 4.6 implies that $f \notin RS_\alpha([a,b])$, contradicting our hypothesis.

Suppose now that α has a removable discontinuity at c, which means that $\alpha(c-) = \alpha(c+) \neq \alpha(c)$. Then we repeat the above reasoning, but now take $\sigma := c$ if f is right continuous at c, or $\tau := c$ if f is left continuous at c. We then get the same contradiction as before, and the proof is complete. □

Observe that the above reasoning is "symmetric" in f and α. This leads to a refined version of Theorem 4.14 which reads as follows.

Theorem 4.15. *Let $f \in RS_\alpha([a, b])$ for some $\alpha \in BV([a, b])$. Then, at each point of the interval $[a, b]$, at least one of the functions f and α is continuous.*

Proposition 4.3 (a) and (c) and equality (4.15) show that the set $RS_\alpha([a, b])$ is a linear space. In the following Proposition 4.16, we show that $RS_\alpha([a, b])$ is also an algebra and prove some other useful properties.

Proposition 4.16. *Let $\alpha \in BV([a, b])$. Then the following is true.*
(a) *If $f \in RS_\alpha([a, b])$ with $f([a, b]) \subseteq [c, d]$ and $h \in C([c, d])$, then $h \circ f \in RS_\alpha([a, b])$.*
(b) *If $f, g \in RS_\alpha([a, b])$, then $fg \in RS_\alpha([a, b])$.*
(c) *If $f \in RS_\alpha([a, b])$, then $|f| \in RS_\alpha([a, b])$.*

Proof. (a) Let $\varepsilon > 0$. Since h is uniformly continuous on $[c, d]$, we may find $\delta \in (0, \varepsilon)$ such that $|h(u) - h(v)| \leq \varepsilon$ for all $u, v \in [c, d]$ satisfying $|u - v| \leq \delta$.

By assumption, we find a partition $P = \{t_0, t_1, \ldots, t_m\} \in \mathcal{P}([a, b])$ such that

$$U_\alpha(f, P; [a, b]) - L_\alpha(f, P; [a, b]) \leq \delta^2 . \tag{4.29}$$

We denote by M_j and m_j the same numbers as in (4.1), and by K_j and k_j the corresponding numbers for $h \circ f$, i.e.

$$K_j := \sup \{h(f(x)) : t_{j-1} \leq x \leq t_j\} \tag{4.30}$$

and

$$k_j := \inf \{h(f(x)) : t_{j-1} \leq x \leq t_j\} . \tag{4.31}$$

We split the index set $\{t_0, t_1, \ldots, t_m\}$ into two disjoint parts I and J by requiring that $j \in I$ if $M_j - m_j \leq \delta$ and $j \in J$ if $M_j - m_j > \delta$.

For $j \in I$, we have $K_j - k_j \leq \varepsilon$, by our choice of δ, and so

$$\sum_{j \in I} (K_j - k_j)(\alpha(t_j) - \alpha(t_{j-1})) \leq \varepsilon \sum_{j \in I} (\alpha(t_j) - \alpha(t_{j-1})) \leq \varepsilon \operatorname{Var}(\alpha; [a, b]) . \tag{4.32}$$

On the other hand, for $j \in J$, we have $K_j - k_j \leq 2\|h\|_C$. However, (4.29) implies that

$$\begin{aligned} \delta \sum_{j \in J} (\alpha(t_j) - \alpha(t_{j-1})) &< \sum_{j \in J} (M_j - m_j)(\alpha(t_j) - \alpha(t_{j-1})) \\ &\leq U_\alpha(f, P; [a, b]) - L_\alpha(f, P; [a, b]) \leq \delta^2 , \end{aligned}$$

and hence

$$\begin{aligned} \sum_{j \in J} (K_j - k_j)(\alpha(t_j) - \alpha(t_{j-1})) &\leq 2\|h\|_C \sum_{j \in J} (\alpha(t_j) - \alpha(t_{j-1})) \\ &\leq 2\|h\|_C \delta \leq 2\|h\|_C \varepsilon . \end{aligned} \tag{4.33}$$

Combining (4.32) and (4.33), we conclude that

$$U_\alpha(h \circ f, P; [a,b]) - L_\alpha(h \circ f, P; [a,b]) \le \left[\mathrm{Var}(\alpha; [a,b]) + 2\|h\|_C\right] \varepsilon ,$$

and so $h \circ f \in RS_\alpha([a,b])$ as claimed.

(b) Observing that

$$fg = \frac{1}{4}\left[(f+g)^2 - (f-g)^2\right] ,$$

choosing $h(u) := u^2$, and using Proposition 4.3 (a) and (c), the assertion follows.

(c) The statement follows from (a) for the choice $h(u) = |u|$. □

Again, Proposition 4.16 contains a well-known result for the Riemann integral in case $\alpha(x) = x$. The next important theorem shows that in case $\alpha \in C^1([a,b])$ the RS-integral (4.7) may always be reduced to the familiar Riemann integral.

Theorem 4.17. *Let $\alpha \in C^1([a,b])$ and $f \in RS([a,b])$. Then $f \in RS_\alpha([a,b])$ and*

$$\int_a^b f(t)\, d\alpha(t) = \int_a^b f(t)\dot\alpha(t)\, dt , \tag{4.34}$$

where $\dot\alpha$ denotes the derivative of α with respect to t, and the integral on the right-hand side of (4.34) is the usual Riemann integral.

Proof. Observe first that $f\dot\alpha \in RS([a,b])$ by the well-known properties of the Riemann integral. Let $\varepsilon > 0$. Then we may find $\delta_1 > 0$ such that

$$\left|\sum_{j=1}^m f(\tau_j)\dot\alpha(\tau_j)\left(t_j - t_{j-1}\right) - \int_a^b f(x)\dot\alpha(x)\, dx\right| \le \varepsilon \tag{4.35}$$

for any partition $P = \{t_0, t_1, \dots, t_m\} \in \mathcal{P}([a,b])$ with $\mu(P) \le \delta_1$ and any choice of intermediate points $\tau_j \in [t_{j-1}, t_j]$. Likewise, we may find $\delta_2 > 0$ such that

$$\left|\sum_{j=1}^m \dot\alpha(\tau_j)\left(t_j - t_{j-1}\right) - \int_a^b \dot\alpha(x)\, dx\right| \le \varepsilon$$

for any partition $P = \{t_0, t_1, \dots, t_m\} \in \mathcal{P}([a,b])$ with $\mu(P) \le \delta_2$ and any choice of intermediate points $\tau_j \in [t_{j-1}, t_j]$. If $\sigma_j \in [t_{j-1}, t_j]$ is another intermediate point, we have

$$\sum_{j=1}^m |\dot\alpha(\tau_j) - \dot\alpha(\sigma_j)|\left(t_j - t_{j-1}\right) \le 2\varepsilon \tag{4.36}$$

whenever $\mu(P) \le \delta_2$ because $\dot\alpha$ is continuous. Now, we fix any partition $P = \{t_0, t_1, \dots, t_m\} \in \mathcal{P}([a,b])$ satisfying $\mu(P) \le \delta := \min\{\delta_1, \delta_2\}$. By the mean value theorem, we find points $\sigma_j \in [t_{j-1}, t_j]$ such that

$$\alpha(t_j) - \alpha(t_{j-1}) = \dot\alpha(\sigma_j)\left(t_j - t_{j-1}\right) .$$

Consequently,

$$\sum_{j=1}^{m} f(\tau_j)\left[\alpha(t_j)-\alpha(t_{j-1})\right] \\ = \sum_{j=1}^{m} f(\tau_j)\dot{\alpha}(\sigma_j)\left(t_j-t_{j-1}\right)+\sum_{j=1}^{m} f(\tau_j)\left[\dot{\alpha}(\tau_j)-\dot{\alpha}(\sigma_j)\right]\left(t_j-t_{j-1}\right). \tag{4.37}$$

However, (4.35) and (4.36) imply that

$$\left|\sum_{j=1}^{m} f(\tau_j)\left[\alpha(t_j)-\alpha(t_{j-1})\right]-\int_a^b f(x)\dot{\alpha}(x)\,dx\right| \le (2\|f\|_\infty+1)\varepsilon.$$

This shows that

$$\lim_{\mu(P)\to 0}\sum_{j=1}^{m} f(\tau_j)\left[\alpha(t_j)-\alpha(t_{j-1})\right]=\lim_{\mu(P)\to 0} S_{\dot{\alpha}}(f,P;[a,b])=\int_a^b f(x)\dot{\alpha}(x)\,dx,$$

by Proposition 4.5, which proves the assertion. □

An extension of Theorem 4.17 from functions $\alpha \in C^1([a,b])$ to functions $\alpha \in AC([a,b])$ is given in Theorem 4.43 in Section 4.5.

In Proposition 4.10, we have given a necessary and sufficient condition for the RS-integrability of a function f with respect to some increasing integrator α. The following theorem from [182] gives a more complete picture and is valid for general functions $\alpha \in BV([a,b])$.

Theorem 4.18. *Let $f \in B([a,b])$ and $\alpha \in BV([a,b])$. Then the following three statements are equivalent.*

(a) $f \in RS_\alpha([a,b])$.

(b) *For each $\varepsilon > 0$, there exists $\delta > 0$ such that for any partition $P = \{t_0, t_1, \dots, t_m\} \in \mathcal{P}([a,b])$ with $\mu(P) \le \delta$, we have*

$$\sum_{j=1}^{m} \operatorname{osc}(f;[t_{j-1},t_j])|\alpha(t_j)-\alpha(t_{j-1})| \le \varepsilon, \tag{4.38}$$

where $\operatorname{osc}(f;A)$ *denotes the oscillation* (1.12) of f on $A \subseteq [a,b]$.

(c) *For each $\varepsilon > 0$, there exists $\delta > 0$ such that for any partition $P = \{t_0, t_1, \dots, t_m\} \in \mathcal{P}([a,b])$ with $\mu(P) \le \delta$, we have*

$$\sum_{j=1}^{m} \operatorname{osc}(f;[t_{j-1},t_j]) \operatorname{Var}(\alpha;[t_{j-1},t_j]) \le \varepsilon. \tag{4.39}$$

Proof. We show first that (a) implies (b). Let $f \in RS_\alpha([a,b])$. Since $\alpha \in BV([a,b])$, according to Theorem 1.5, we can represent α as the difference $\alpha = \beta - \gamma$ of two monotonically increasing functions $\beta, \gamma : [a,b] \to \mathbb{R}$, and so (4.15) holds. Given $\varepsilon > 0$, by Proposition 4.10, we can find $\delta > 0$ such that for an arbitrary partition $P = \{t_1, t_2, \dots, t_m\} \in$

$\mathcal{P}([a,b])$ with $\mu(P) \le \delta$, we have

$$\sum_{j=1}^{m} \operatorname{osc}(f;[t_{j-1},t_j])(\beta(t_j) - \beta(t_{j-1})) \le \frac{\varepsilon}{2}$$

and

$$\sum_{j=1}^{m} \operatorname{osc}(f;[t_{j-1},t_j])(\gamma(t_j) - \gamma(t_{j-1})) \le \frac{\varepsilon}{2}.$$

Keeping in mind the above estimates, we get

$$\begin{aligned}
&\sum_{j=1}^{m} \operatorname{osc}(f;[t_{j-1},t_j])|\alpha(t_j) - \alpha(t_{j-1})| \\
&\quad = \sum_{j=1}^{m} \operatorname{osc}(f;[t_{j-1},t_j])|\beta(t_j) - \gamma(t_j) - \beta(t_{j-1}) + \gamma(t_{j-1})| \\
&\quad = \sum_{j=1}^{m} \operatorname{osc}(f;[t_{j-1},t_j])|[\beta(t_j) - \beta(t_{j-1})] \quad [\gamma(t_j) - \gamma(t_{j-1})]| \\
&\quad \le \sum_{j=1}^{m} \operatorname{osc}(f;[t_{j-1},t_j])[|\beta(t_j) - \beta(t_{j-1})| + |\gamma(t_j) - \gamma(t_{j-1})|] \\
&\quad = \sum_{j=1}^{m} \operatorname{osc}(f;[t_{j-1},t_j])(\beta(t_j) - \beta(t_{j-1})) \\
&\qquad + \sum_{j=1}^{m} \operatorname{osc}(f;[t_{j-1},t_j])(\gamma(t_j) - \gamma(t_{j-1})) \le \varepsilon\,.
\end{aligned}$$

This shows that condition (b) holds, and so we have proved the first implication.

Now, we prove that (b) implies (c). We first show[7] that under condition (b), for each point $x \in [a,b]$, at least one of the functions f or α is continuous at x. Given $\varepsilon > 0$, choose $\delta > 0$ according to condition (b). Fix a point $x_0 \in [a,b]$ and assume, for example, that α is discontinuous at x_0. Then we have

$$\lim_{x \to x_0+} \alpha(x) \ne \alpha(x_0)$$

or

$$\lim_{x \to x_0-} \alpha(x) \ne \lim_{x \to x_0+} \alpha(x)\,.$$

Further, take a number $h > 0$ such that $h < \min\{\delta, b - x_0\}$ and consider an arbitrary partition $P = \{t_0, t_1, \dots, t_m\} \in \mathcal{P}([a,b])$ with $\mu(P) \le \delta$. We add to the partition P the two points x_h and $x_0 + h$, where $x_h = x_0$ or $x_h = x_0 - h$, and denote by P' the extended partition $P \cup \{x_h, x_0 + h\}$. Obviously, $\mu(P') \le \delta$ and, in view of (4.38), we have

$$|f(x_0 + h) - f(x_0)||\alpha(x_0 + h) - \alpha(x_h)| \le \varepsilon\,.$$

7 Note that we cannot apply Theorem 4.14 here since we do not suppose that $f \in RS_\alpha([a,b])$.

Hence, letting $h \to 0$, we get $f(x_0+h) \to f(x_0)$ which means that f is right continuous at x_0. Similarly, we can prove the left continuity of f at x_0, and so we conclude that the function f is continuous at x_0.

Repeating the above reasoning under the assumption that the function f is discontinuous at some point x_0, we may prove that the function α is continuous at x_0. So, we proved the assertion.

Fix $\varepsilon > 0$. Choose $\delta_0 > 0$ according to condition (b) corresponding to the number $\varepsilon/3$. Let $M := \operatorname{osc}(f; [a, b])$, and let $P = \{z_0, z_1, \dots, z_k\} \in \mathcal{P}([a, b])$ be a partition satisfying[8]

$$\operatorname{Var}(\alpha, [a, b]) - \frac{\varepsilon}{3M} \le \sum_{j=1}^{k} |\alpha(z_j) - \alpha(z_{j-1})|. \tag{4.40}$$

Obviously, if we take any partition $P' \supseteq P$, then (4.40) also holds for P'. As we have seen, condition (b) implies that at every point of the interval $[a, b]$, at least one of the functions f or α is continuous. In view of Proposition 1.7, this implies that the same is true for at least one of the functions f or V_α, where V_α denotes the variation function (1.13) of α. Since these functions are bounded, we can find $\delta_1 > 0$ such that

$$\operatorname{osc}(f; [x', x]) \operatorname{Var}(\alpha; [x', x]) \le \frac{\varepsilon}{3M} \tag{4.41}$$

provided that $|x - x'| \le \delta_1$ and $x' < z_j < x$ for $j = 1, 2, \dots, k$.

Next, let $P_1 = \{x_0, x_1, \dots, x_n\} \in \mathcal{P}([a, b])$ be such that $\mu(P_1) \le \delta := \min\{\delta_0, \delta_1\}$. Obviously, inequality (4.41) holds for the partition P_1 with ε replaced by $\varepsilon/3$. Extending the partition P_1 by including those points z_j which are distinct from points of P_1, we obtain a new partition $P_2 = \{x'_0, x'_1, \dots, x'_p\}$ satisfying

$$\operatorname{Var}(\alpha; [a, b]) - \frac{\varepsilon}{3M} \le \sum_{j=1}^{p} |\alpha(x'_j) - \alpha(x'_{j-1})|.$$

Consequently, we get

$$\sum_{j=1}^{p} [\operatorname{Var}(\alpha; [x'_{j-1}, x'_j]) - |\alpha(x'_j) - \alpha(x'_{j-1})|] \le \frac{\varepsilon}{3M}.$$

This implies

$$\begin{aligned} S_0 &:= \sum_{j=1}^{p} \operatorname{osc}(f; [x'_{j-1}, x'_j]) \operatorname{Var}(\alpha; [x'_{j-1}, x'_j]) \\ &\le \sum_{j=1}^{p} \operatorname{osc}(f; [x'_{j-1}, x'_j]) \{\operatorname{Var}(\alpha; [x'_{j-1}, x'_j]) - |\alpha(x'_j) - \alpha(x'_{j-1})|\} \\ &\quad + \sum_{j=1}^{p} \operatorname{osc}(f; [x'_{j-1}, x'_j]) |\alpha(x'_j) - \alpha(x'_{j-1})| \le \frac{2}{3}\varepsilon. \end{aligned}$$

8 Of course, without loss of generality, we may assume that f is not constant.

Finally, we notice that we can write

$$S := \sum_{j=1}^{n} \operatorname{osc}(f; [x_{j-1}, x_j]) \operatorname{Var}(\alpha; [x_{j-1}, x_j]) = S' + S'' ,$$

where S'' denotes the sum of those terms for which there exists k such that $x_{j-1} < z_k < x_j$, while S' denotes the remainder sum of S, being a partial sum of S_0. Obviously, we have $S' \leq S_0 \leq 2\varepsilon/3$. Furthermore, from (4.41), we know that $S'' \leq \varepsilon/3$, and so $S \leq \varepsilon$ as claimed.

It remains to show that (c) implies (a). To this end, we represent the function α again in the form $\alpha = \beta - \gamma$, where[9]

$$\beta(x) = \frac{1}{2}[V_\alpha(x) + \alpha(x)] , \quad \gamma(x) = \frac{1}{2}[V_\alpha(x) - \alpha(x)] ,$$

and V_α denotes the variation function (1.13) of α. Clearly,

$$\operatorname{Var}(\alpha; [x, y]) = \operatorname{Var}(\beta; [x, y]) + \operatorname{Var}(\gamma; [x, y]) \tag{4.42}$$

for an arbitrary interval $[x, y] \subseteq [a, b]$. Fix $\varepsilon > 0$ and choose $\delta > 0$ according to condition (c). Then, in view of (4.39), for any partition $P = \{t_0, t_1, \dots, t_m\} \in \mathcal{P}([a, b])$ with $\mu(P) \leq \delta$, we have

$$\sum_{j=1}^{m} \operatorname{osc}(f; [t_{j-1}, t_j]) \operatorname{Var}(\alpha; [t_{j-1}, t_j]) \leq \varepsilon .$$

From (4.42), we then obtain the inequality

$$\sum_{j=1}^{m} \operatorname{osc}(f; [t_{j-1}, t_j])\{\operatorname{Var}(\beta; [t_{j-1}, t_j]) + \operatorname{Var}(\gamma; [t_{j-1}, t_j])\} \leq \varepsilon ,$$

and hence

$$\sum_{j=1}^{m} \operatorname{osc}(f; [t_{j-1}, t_j])(\beta(t_j) - \beta(t_{j-1})) = \sum_{j=1}^{m} \operatorname{osc}(f; [t_{j-1}, t_j]) \operatorname{Var}(\beta; [t_{j-1}, t_j]) \leq \varepsilon$$

and

$$\sum_{j=1}^{m} \operatorname{osc}(f; [t_{j-1}, t_j])(\gamma(t_j) - \gamma(t_{j-1})) = \sum_{j=1}^{m} \operatorname{osc}(f; [t_{j-1}, t_j]) \operatorname{Var}(\gamma; [t_{j-1}, t_j]) \leq \varepsilon .$$

From these estimates and Proposition 4.10, we deduce that both $f \in RS_\beta([a, b])$ and $f \in RS_\gamma([a, b])$, and so by the definition, the RS-integral for BV-functions, that $f \in RS_\alpha([a, b])$. This completes the proof. □

9 This is the Jordan decomposition involving "slowly increasing" functions, see Exercise 1.27.

Observe that the implication (c) $\Rightarrow$ (b) in Theorem 4.18 is trivial by (1.8). However, the nontrivial implication (b) $\Rightarrow$ (c) in our proof shows that the difference is not essential if $\mu(P)$ is "small enough."

Theorem 4.18 has a useful consequence which relates RS-integrability of a function f with respect to a function α and that with respect to its variation function V_α.

Corollary 4.19. *Let $f \in B([a,b])$ and $\alpha \in BV([a,b])$. Then $f \in RS_\alpha([a,b])$ if and only if $f \in RS_{V_\alpha}([a,b])$, where V_α denotes the variation function* (1.13) *of α.*

Proof. Being increasing, the function V_α has bounded variation on $[a,b]$. Now, $f \in RS_\alpha([a,b])$ implies (4.38), and so also (4.39), by Theorem 4.18. However, the equality

$$\mathrm{Var}(\alpha; [t_{j-1}, t_j]) = V_\alpha(t_j) - V_\alpha(t_{j-1}) \tag{4.43}$$

shows that then $f \in RS_{V_\alpha}([a,b])$, again by Theorem 4.18. Since this argument is symmetric in α and V_α, we have proved the desired equivalence. □

In view of Corollary 4.19, the question arises if there is any relation between the RS-integral of a function f with respect to V_α and its RS-integral with respect to its parent function α. The following theorem contains such a relation which gives a useful estimate for RS-integrals with interesting applications to integral equations of fractional order [44]. We did not find the proof of this estimate in the literature,[10] and so we give a short proof here.

Theorem 4.20. *For $f \in RS_\alpha([a,b])$, the estimate*

$$\left| \int_a^b f(x)\, d\alpha(x) \right| \le \int_a^b |f(x)|\, dV_\alpha(x) \tag{4.44}$$

holds.

Proof. From Corollary 4.19, we know that $f \in RS_{V_\alpha}([a,b])$, and so the integral appearing on the right-hand side of (4.44) is well-defined.

Take an arbitrary partition $\{t_0, t_1, \dots, t_m\} \in \mathcal{P}([a,b])$, and choose a set of intermediate points $\Pi = \{\tau_1, \tau_2, \dots, \tau_m\}$ for the partition P satisfying (4.16). Then

$$\begin{aligned} \left| \sum_{j=1}^m f(\tau_j)(\alpha(t_j) - \alpha(t_{j-1})) \right| &\le \sum_{j=1}^m |f(\tau_j)|\, |\alpha(t_j) - \alpha(t_{j-1})| \\ &\le \sum_{j=1}^m |f(\tau_j)|\, \mathrm{Var}(\alpha; [t_{j-1}, t_j]) = \sum_{j=1}^m |f(\tau_j)|(V_\alpha(t_j) - V_\alpha(t_{j-1}))\,, \end{aligned}$$

where we have used (1.8). Since the above inequality is satisfied for arbitrary partitions $P \in \mathcal{P}([a,b])$ and arbitrary sets Π of intermediate points, in view of Proposition 4.5 and the continuity of the function $u \mapsto |u|$, we obtain (4.44). □

10 In some textbooks and monographs, e.g. [237], this estimate is mentioned without proof.

We point out that the estimate (4.44) is not true if we also put $d\alpha(x)$, rather than $dV_\alpha(x)$, on the right-hand side, see Exercise 4.42. For another proof of the important Theorem 4.20, see Exercise 4.44.

In the next theorem, we prove a fundamental estimate for the RS-integral and deduce two natural convergence theorems.

Theorem 4.21. *The RS-integral has the following properties.*

(a) *The fundamental estimate*

$$\left|\int_a^b f(t)\,d\alpha(t)\right| \leq \sup_{a\leq x\leq b} |f(x)|\,\mathrm{Var}(\alpha;[a,b]) \tag{4.45}$$

holds for all $\alpha \in BV([a,b])$ and $f \in RS_\alpha([a,b])$.

(b) *If $(f_n)_n$ is a sequence of bounded functions $f_n : [a,b] \to \mathbb{R}$ which satisfies $\|f_n - f\|_\infty \to 0$ as $n \to \infty$, then*

$$\lim_{n\to\infty} \int_a^b f_n(x)\,d\alpha(x) = \int_a^b f(x)\,d\alpha(x) \tag{4.46}$$

for all $\alpha \in BV([a,b])$.

(c) *If $(\alpha_n)_n$ is a sequence of functions $\alpha_n : [a,b] \to \mathbb{R}$ of bounded variation which satisfies $\mathrm{Var}(\alpha_n - \alpha;[a,b]) \to 0$ as $n \to \infty$, then*

$$\lim_{n\to\infty} \int_a^b f(x)\,d\alpha_n(x) = \int_a^b f(x)\,d\alpha(x) \tag{4.47}$$

for all $f \in B([a,b])$.

Proof. As before, we may assume without loss of generality that α is monotonically increasing. By V_α, we denote the variation function (1.13) of α.

(a) In view of Theorem 4.20 and Proposition 4.3 (e), we obtain

$$\begin{aligned}\left|\int_a^b f(x)\,d\alpha(x)\right| &\leq \int_a^b |f(x)|\,dV_\alpha(x)\\ &\leq \|f\|_\infty \int_a^b dV_\alpha(x) = \sup_{a\leq x\leq b} |f(x)|\,\mathrm{Var}(\alpha;[a,b])\end{aligned}$$

as claimed.

(b) Using again Theorem 4.20 and Proposition 4.3 (e), we get

$$\left|\int_a^b f_n(x)\,d\alpha(x) - \int_a^b f(x)\,d\alpha(x)\right| \leq \int_a^b |f_n(x) - f(x)|\,dV_\alpha(x) \leq \|f_n - f\|_\infty\,\mathrm{Var}(\alpha;[a,b])\,,$$

which proves the assertion.

(c) Applying (4.44) with α replaced by $\alpha_n - \alpha$, we obtain

$$\left|\int_a^b f(x)\,d\alpha_n(x) - \int_a^b f(x)\,d\alpha(x)\right| = \left|\int_a^b f(x)\,d(\alpha_n(x) - \alpha(x))\right|$$

$$\leq \int_a^b |f(x)|\,dV_{\alpha_n-\alpha}(x) \leq \|f\|_\infty \operatorname{Var}(\alpha_n - \alpha; [a,b]),$$

and the assertion follows. □

Theorem 4.21 shows that the map

$$(f,\alpha) \mapsto \int_a^b f(x)\,d\alpha(x)$$

is a continuous bilinear form on the product $C([a,b]) \times BV([a,b])$. This bilinear form will play an important role in the following two sections.

Of course, the convergence results contained in Theorem 4.21 (b) and (c) are not too surprising since they refer to the "natural" norms on $B([a,b])$ (uniform convergence) and $BV([a,b])$ (convergence in variation). Recall that the estimate

$$|\alpha_n(x) - \alpha(x)| \leq \operatorname{Var}(\alpha_n - \alpha; [a,b]) \quad (a \leq x \leq b)$$

implies that under the hypotheses of Theorem 4.21 (c), the sequence $(\alpha_n)_n$ also converges uniformly on $[a,b]$ to α. The following example shows that (4.47) may fail if we *only* assume uniform convergence of $(\alpha_n)_n$ in $BV([a,b]) \cap C([a,b])$ on $[a,b]$ to some function $\alpha \in BV([a,b]) \cap C([a,b])$.

Example 4.22. Let $f : [0,2\pi] \to \mathbb{R}$ and $\alpha_n : [0,2\pi] \to \mathbb{R}$ be defined by

$$f(x) := \sum_{m=1}^{\infty} \frac{\cos m^6 x}{m^2}, \quad \alpha_n(x) := \frac{\sin nx}{\sqrt{n}} \quad (n = 1,2,3\ldots). \tag{4.48}$$

Since the series in (4.48) converges uniformly on $[0,2\pi]$, the function f is continuous; moreover, the sequence $(\alpha_n)_n$ obviously converges uniformly on $[0,2\pi]$ to zero.

Since $\alpha_n \in C^1([0,2\pi])$, we may use Theorem 4.17 to calculate the RS-integral of f with respect to α_n and obtain

$$I_n := \int_0^{2\pi} f(x)\,d\alpha_n(x) = \int_0^{2\pi} f(x)\dot\alpha_n(x)\,dx = \sqrt{n}\int_0^{2\pi} f(x)\cos nx\,dx.$$

Putting the definition (4.48) of f in the integral and observing that

$$\int_0^{2\pi} \cos mx \cos nx\,dx = \begin{cases} \pi & \text{if } m = n, \\ 0 & \text{if } m \neq n, \end{cases}$$

we end up with

$$I_n = \begin{cases} \sqrt[6]{n} \int_0^{2\pi} \cos^2 nx\,dx = \pi\sqrt[6]{n} & \text{if } n = p^6 \text{for some } p \in \mathbb{N}, \\ 0 & \text{otherwise}. \end{cases}$$

This shows that the sequence of integrals $(I_n)_n$ is unbounded, and so it cannot be convergent, although the sequence $(\alpha_n)_n$ converges uniformly on $\mathbb{R}$ to 0. ♥

The crucial point in Example 4.22 is of course that we have chosen the "wrong type of convergence" for the sequence $(\alpha_n)_n$. From Theorem 4.21 (c), it follows that $(\alpha_n)_n$ cannot converge in the norm of $BV([0, 2\pi])$ to 0. Indeed, consider the partition

$$P_n := \{0, 2\pi\} \cup \{s_0, s_1, \ldots, s_n\} \cup \{t_0, t_1, \ldots, t_n\} \in \mathcal{P}([0, 2\pi])$$

with points

$$s_k := \frac{(1+4k)\pi}{2n}, \quad t_k := \frac{(3+4k)\pi}{2n} \quad (k = 1, 2, \ldots, n).$$

From $\alpha_n(s_k) = 1/\sqrt{k}$ and $\alpha_n(t_k) = -1/\sqrt{k}$, it follows then that

$$\mathrm{Var}(\alpha_n; [0, 2\pi]) \geq \mathrm{Var}(\alpha_n, P_n; [0, 2\pi]) \geq \sum_{k=1}^{n} \frac{2}{\sqrt{k}} \to \infty \quad (n \to \infty),$$

which shows that the sequence $(\alpha_n)_n$ is unbounded in the norm of $BV([0, 2\pi])$.

In some cases, the fundamental estimate (4.45) in Theorem 4.21 may be sharpened. For example, the following estimates are useful in the theory of Fourier series:

Proposition 4.23. *Let $\alpha \in BV([0, 2\pi])$. Then the following is true.*

(a) *The estimate*

$$\left| \int_0^{2\pi} \alpha(x) \cos nx\,dx \right| \leq \frac{1}{n} \mathrm{Var}(\alpha; [0, 2\pi]) \tag{4.49}$$

holds for all $n \in \mathbb{N}$; moreover, there exists a function $\tilde{\alpha} \in BV([0, 2\pi])$ such that

$$\left| \int_0^{2\pi} \tilde{\alpha}(x) \cos nx\,dx \right| = \frac{1}{n} \mathrm{Var}(\tilde{\alpha}; [0, 2\pi]) \tag{4.50}$$

for infinitely many $n \in \mathbb{N}$.

(b) *The estimate*

$$\left| \int_0^{2\pi} \alpha(x) \sin nx\,dx \right| \leq \frac{2}{n} \mathrm{Var}(\alpha; [0, 2\pi]) \tag{4.51}$$

holds for all $n \in \mathbb{N}$; moreover, there exists a function $\hat{\alpha} \in BV([0, 2\pi])$ such that

$$\left| \int_0^{2\pi} \hat{\alpha}(x) \sin nx\,dx \right| = \frac{2}{n} \mathrm{Var}(\hat{\alpha}; [0, 2\pi]) \tag{4.52}$$

for infinitely many $n \in \mathbb{N}$.

Proof. Suppose first that $\alpha : [0, 2\pi] \to \mathbb{R}$ is increasing. Since

$$\int_0^{2\pi} \cos nx\, dx = 0 \quad (n = 1, 2, 3, \ldots),$$

we may suppose that $\alpha(0) = 0$ and so $\mathrm{Var}(\alpha; [0, 2\pi]) = \alpha(2\pi)$. From the third mean value theorem for the Riemann integral (Exercise 4.25), we get

$$\left| \int_0^{2\pi} \alpha(x) \cos nx\, dx \right| \leq \alpha(2\pi) \sup_{0 \leq c \leq 2\pi} \left| \int_c^{2\pi} \cos nx\, dx \right| \leq \frac{1}{n} \mathrm{Var}(\alpha; [0, 2\pi]). \tag{4.53}$$

In the general case $\alpha \in BV([0, 2\pi])$, we may find increasing functions β and γ such that $\alpha = \beta - \gamma$ and $\mathrm{Var}(\alpha; [0, 2\pi]) = \mathrm{Var}(\beta; [0, 2\pi]) + \mathrm{Var}(\gamma; [0, 2\pi])$, see Exercise 1.27. Then we obtain

$$\begin{aligned} \left| \int_0^{2\pi} \alpha(x) \cos nx\, dx \right| &\leq \left| \int_0^{2\pi} \beta(x) \cos nx\, dx \right| + \left| \int_0^{2\pi} \gamma(x) \cos nx\, dx \right| \\ &\leq \frac{1}{n} \mathrm{Var}(\beta; [0, 2\pi]) + \frac{1}{n} \mathrm{Var}(\gamma; [0, 2\pi]) = \frac{1}{n} \mathrm{Var}(\alpha; [0, 2\pi]) \end{aligned}$$

as claimed. Choosing, in particular, $\tilde{\alpha}(x) := \chi_{[0,\pi/2]}(x)$ for $0 \leq x \leq 2\pi$, we have $\mathrm{Var}(\tilde{\alpha}; [0, 2\pi]) = 1$ and

$$\left| \int_0^{2\pi} \tilde{\alpha}(x) \cos nx\, dx \right| = \left| \int_0^{\pi/2} \cos nx\, dx \right| = \frac{1}{n} \left| \sin \frac{n\pi}{2} \right| = \frac{1}{n}$$

for all odd $n \in \mathbb{N}$, and so we have proved (a).

The proof of (b) is very similar. We start again with an increasing function and replace (4.53) by

$$\left| \int_0^{2\pi} \alpha(x) \sin nx\, dx \right| \leq \alpha(2\pi) \sup_{0 \leq c \leq 2\pi} \left| \int_c^{2\pi} \sin nx\, dx \right| \leq \frac{2}{n} \mathrm{Var}(\alpha; [0, 2\pi]). \tag{4.54}$$

The remaining part using the Jordan decomposition of $\alpha \in BV([0, 2\pi])$ is precisely the same as in (a). To prove the last assertion, we choose now $\hat{\alpha}(x) := \chi_{[0,\pi]}(x)$ for $0 \leq x \leq 2\pi$ and get $\mathrm{Var}(\hat{\alpha}; [0, 2\pi]) = 1$ and

$$\left| \int_0^{2\pi} \hat{\alpha}(x) \sin nx\, dx \right| = \left| \int_0^{\pi} \sin nx\, dx \right| = \frac{1}{n} |1 - \cos n\pi| = \frac{2}{n}$$

for all odd $n \in \mathbb{N}$. □

Now, we present two useful results on RS-integrals which for the Riemann integral (i.e. $\alpha(x) = x$) reduce to well-known integration techniques.

Proposition 4.24 (integration by parts). *Let $f \in BV([a,b]) \cap C([a,b])$ and $\alpha \in BV([a,b])$. Then the equality*

$$\int_a^b f(x)\,d\alpha(x) = f(b)\alpha(b) - f(a)\alpha(a) - \int_a^b \alpha(x)\,df(x) \tag{4.55}$$

holds.

Proof. Let $P = \{t_0, t_1, \dots, t_m\} \in \mathcal{P}([a,b])$ and $\Pi = \{\tau_0, \tau_1, \dots, \tau_m, \tau_{m+1}\}$ be a set of intermediate points satisfying $\tau_0 = a$, $\tau_{m+1} = b$, and (4.16). For the RS-sums (4.17), we then obtain[11]

$$\begin{aligned} S_\alpha(f, P, \Pi; [a,b]) &= \sum_{j=1}^{m} f(\tau_j)(\alpha(t_j) - \alpha(t_{j-1})) \\ &= f(\tau_{m+1})\alpha(t_m) - f(\tau_0)\alpha(t_0) - \sum_{k=1}^{m+1} \alpha(t_{k-1})(f(\tau_k) - f(\tau_{k-1})) \\ &= f(b)\alpha(b) - f(a)\alpha(a) - S_f(\alpha, \Pi, P; [a,b]) . \end{aligned}$$

Now, $\mu(P) \to 0$ implies $\mu(\Pi) \to 0$, and Proposition 4.5 shows that in this case

$$\lim_{\mu(P)\to 0} S_\alpha(f, P, \Pi; [a,b]) = \int_a^b f(x)\,d\alpha(x)$$

and

$$\lim_{\mu(\Pi)\to 0} S_f(\alpha, \Pi, P; [a,b]) = \int_a^b \alpha(x)\,df(x)$$

which gives (4.55). □

Of course, in case $\alpha(x) = x$, equality (4.55) is nothing else but the classical integration by parts formula. In the proof of Proposition 4.24, we have used the hypotheses $f \in BV([a,b]) \cap C([a,b])$ and $\alpha \in BV([a,b])$ for technical reasons. If $\alpha \in BV([a,b])$, but f is only continuous, we may use formula (4.55) to *define* the RS-integral on the right-hand side. This gives yet another possibility of extending the RS-integral to larger classes of functions.

The next proposition extends a well-known mean value theorem for the Riemann integral, see Exercise 4.23.

---

11 Here, we consider first $P = \{t_0, t_1, \dots, t_m\}$ as a partition of $[a,b]$ and $\Pi = \{\tau_0, \tau_1, \dots, \tau_m, \tau_{m+1}\}$ as a set of intermediate points, and afterwards vice versa. This is possible since $\tau_{k-1} \le t_{k-1} \le \tau_k$ for $k = 1, 2, \dots, m+1$.

Proposition 4.25 (mean value theorem). *Let $f \in C([a,b])$, and let $\alpha : [a,b] \to \mathbb{R}$ be monotonically increasing. Then there exists $\xi \in [a,b]$ such that*

$$\int_a^b f(x)\,d\alpha(x) = f(\xi)(\alpha(b) - \alpha(a))\,. \tag{4.56}$$

Proof. Defining $m(f)$ as in (0.61) and $M(f)$ as in (0.62), from the estimate (4.13), we conclude that we may choose $\eta \in [m(f), M(f)]$ with

$$\int_a^b f(x)\,d\alpha(x) = \eta(\alpha(b) - \alpha(a))\,.$$

Since f is continuous, by the intermediate value theorem, we may find $\xi \in [a,b]$ satisfying $f(\xi) = \eta$ which proves the assertion. □

Proposition 4.25 is usually called the *first mean value theorem* for the RS-integral. Another mean value theorem may be found in Exercise 4.22.

We close this section with an application of Proposition 4.24 to so-called generalized trapezoid inequalities for functions of bounded variation. Recall that the Dragomir trapezoid inequality [105] for functions $\alpha \in BV([a,b])$ states that

$$\left| \int_a^b \alpha(t)\,dt - [\alpha(b) - \alpha(a)]\frac{b-a}{2} \right| \leq \frac{1}{2}(b-a)\operatorname{Var}(\alpha; [a,b])\,. \tag{4.57}$$

More generally, in [77], the authors consider the function

$$J_\alpha(x) := \int_a^b \alpha(t)\,dt - \alpha(a)(x-a) - \alpha(b)(b-x) \quad (a \leq x \leq b)\,. \tag{4.58}$$

Then the following estimate holds for the function J_α.

Proposition 4.26. *For $\alpha \in BV([a,b])$, the function* (4.58) *satisfies the estimate*

$$|J_\alpha(x)| \leq \max\{x-a, b-x\}\operatorname{Var}(\alpha; [a,b]). \tag{4.59}$$

Proof. We claim that

$$J_\alpha(x) = \int_a^b (x-t)\,d\alpha(t) \quad (a \leq x \leq b)\,. \tag{4.60}$$

Indeed, applying (4.55) for fixed $x \in [a,b]$ to the function $f(t) := x - t$ and integrating with respect to t, we obtain, by Theorem 4.17 ,

$$\begin{aligned}\int_a^b (x-t)\,d\alpha(t) &= f(b)\alpha(b) - f(a)\alpha(a) - \int_a^b \alpha(t)\,d(x-t)\\ &= (x-b)\alpha(b) - (x-a)\alpha(a) + \int_a^b \alpha(t)\,dt = J_\alpha(x)\,.\end{aligned}$$

Applying (4.45) to the RS-integral in (4.60) yields

$$\left|\int_a^b (x-t)\,d\alpha(t)\right| \le \sup_{a\le t\le b} |x-t|\,\mathrm{Var}(\alpha;[a,b]) = \max\{x-a, b-x\}\,\mathrm{Var}(\alpha;[a,b])$$

which proves the assertion. □

Putting, in particular, $x := (b+a)/2$ in (4.59), we get precisely the trapezoidal formula (4.57). Moreover, applying (4.59) to the special choice $\alpha := \chi_{(a,b)}$ gives $\|\alpha\|_{L_1} = b-a$ and $\mathrm{Var}(\alpha;[a,b]) = 2$, which shows that the constant $1/2$ on the right-hand side of (4.57) is sharp.

4.2 Bounded variation and duality

One of the numerous useful properties of the Riemann–Stieltjes integral is that it allows us to rather easily describe the dual space of several well-known Banach spaces.

Consider first the Banach space $X = C([a,b])$ of all continuous functions $f : [a,b] \to \mathbb{R}$ with the usual maximum norm (0.45). We want to associate with each $\ell \in X^*$ a unique $\alpha \in BV([a,b])$ in such a way that that X^* becomes isomorphic to a subspace of $BV([a,b])$, and such that ℓ and α are related by the formula

$$\langle f,\ell\rangle = \langle f,\ell_\alpha\rangle = \int_a^b f(t)\,d\alpha(t) \tag{4.61}$$

which we already encountered after the proof of Theorem 4.21. Here, we have adopted, as in Section 0.2, the notation $\langle f,\ell\rangle$ for $\ell(f)$ to put evidence on the duality between X and X^*.

First, we recall the concept of a normalized function of bounded variation, see Definition 1.2. A function $\alpha \in BV([a,b])$ is *normalized* if $\alpha(a) = 0$ (i.e. α belongs to the subspace $BV^o([a,b])$) and, in addition,

$$\alpha(x_0) = \alpha(x_0+) = \lim_{x\to x_0+} \alpha(x) \tag{4.62}$$

for $a \le x_0 < b$. In this case, we write $\alpha \in NBV([a,b])$. For further use, we now discuss a procedure to associate to each BV-function a normalized BV-function.

Definition 4.27. We associate to each $\alpha \in BV([a,b])$ a function $\alpha^\#$ given by

$$\alpha^\#(x) := \begin{cases} 0 & \text{for } x = a, \\ \alpha(x+) - \alpha(a) & \text{for } a < x < b, \\ \alpha(b) - \alpha(a) & \text{for } x = b, \end{cases} \tag{4.63}$$

and call $\alpha^\#$ the (right) *normalization* of α. ■

The name “normalization” is of course motivated by the fact that always $\alpha^\# \in NBV([a,b])$, and so $\alpha \mapsto \alpha^\#$ defines a map from $BV([a,b])$ into $NBV([a,b])$. Geometrically, this map has two effects on α: it shifts the graph of α vertically to fit the condition $\alpha^\#(a) = 0$, and it “fills the holes in the graph from the right” at any point where α has a jump. In the next Proposition 4.28, we will summarize some interesting analytical properties of this map. To this end, we define an equivalence relation $\approx$ on $BV([a,b])$ by writing $\alpha \approx \beta$ if $\ell_\alpha = \ell_\beta$ in (4.61), i.e.

$$\int_a^b f(t)\,d\alpha(t) = \int_a^b f(t)\,d\beta(t)$$

for all $f \in C([a,b])$.

Proposition 4.28. *The normalization* (4.63) *has the following properties.*

(a) *We have* $\alpha^\# \approx \alpha$, *i.e. every function* $\alpha \in BV([a,b])$ *is equivalent to its normalization.*
(b) *Each equivalence class with respect to* $\approx$ *contains exactly one normalized function.*
(c) *The estimate*

$$\mathrm{Var}(\alpha^\#; [a,b]) \le \mathrm{Var}(\alpha; [a,b]) \tag{4.64}$$

holds.

Proof. It is not hard to see that $\beta \approx 0$ if and only if

$$\beta(a) = \beta(b) = \beta(x+) = \beta(x-) \quad (a < x < b)\,. \tag{4.65}$$

Applying this to the function $\beta = \alpha^\# - \alpha$, we get

$$\beta(a) = \alpha^\#(a) - \alpha(a) = -\alpha(a) = \alpha^\#(b) - \alpha(b) = \beta(b)$$

and

$$\beta(x\pm) = \alpha^\#(x\pm) - \alpha(a) - \alpha(x\pm) = -\alpha(a)$$

for $a < x < b$, and so (a) is true. The statement (b) is of course an immediate consequence of (a).

To prove (c), let $\varepsilon > 0$, fix a partition $P = \{t_0, t_1, \dots, t_m\} \in \mathcal{P}([a,b])$, and choose points $\tau_0 := a$, $\tau_m := b$, and $\tau_j \in (t_j, t_{j+1})$ $(j = 1, 2, \dots, m-1)$ at which α is continuous and which are so close to t_j that

$$|\alpha(t_j+) - \alpha(\tau_j)| < \frac{\varepsilon}{2m}\,.$$

Then

$$\sum_{j=1}^m |\alpha^\#(t_j) - \alpha^\#(t_{j-1})| \le \sum_{j=1}^m |\alpha(\tau_j) - \alpha(\tau_{j-1})| + \varepsilon \le \mathrm{Var}(\alpha) + \varepsilon\,,$$

and so $\mathrm{Var}(\alpha^\#) \le \mathrm{Var}(\alpha) + \varepsilon$. Since $\varepsilon > 0$ was arbitrary, we have proved that $\mathrm{Var}(\alpha^\#) \le \mathrm{Var}(\alpha)$. □

Proposition 4.28 shows, in particular, that the map $\alpha \mapsto \alpha^\#$ is a bounded operator between $BV^o([a,b])$ and $NBV([a,b])$ which satisfies $\|\alpha^\#\|_{BV} \le \|\alpha\|_{BV}$. As Exercise 4.47 shows, however, this map is not an isometry.

The question arises regarding how to characterize the relation $\approx$ more explicitly, i.e. without referring to the integral (4.61). The following proposition gives a complete characterization which we formulate in terms of the set

$$\Delta(\alpha,\beta) = \Delta(\alpha,\beta;[a,b]) := \{x : a \le x \le b,\ \alpha(x) \neq \beta(x)\} \tag{4.66}$$

which we already introduced in (0.47):

Proposition 4.29. *For $\alpha, \beta \in BV([a,b])$, we have $\alpha \approx \beta$ if and only if $\Delta(\alpha,\beta)$ is at most countable and contained in (a,b).*

Proof. Suppose first that $\Delta(\alpha,\beta) = \{x_1, x_2, x_3, \ldots\} \subset (a,b)$, and let $\gamma := \alpha - \beta$ and $f \in C([a,b])$. Then the real numbers

$$a_n := \alpha(x_n) - \beta(x_n) \quad (n = 1, 2, 3, \ldots)$$

satisfy

$$\gamma(t) = \sum_{n=1}^{\infty} a_n \left(\chi_{[x_n,b]}(t) - \chi_{(x_n,b]}(t)\right), \quad \sum_{n=1}^{\infty} |a_n| < \infty$$

by construction. Applying Exercise 4.27 to this sequence $(a_n)_n$ (and $b_n := -a_n$), we conclude that

$$\int_a^b f(x)\,d\gamma(x) = \sum_{n=1}^{\infty} (a_n - a_n) f(x_n) = 0\,,$$

which implies $\gamma \approx 0$, and thus $\alpha \approx \beta$.

Conversely, suppose now that $\gamma \approx 0$ for some $\gamma \in BV([a,b])$ which means that

$$\int_a^b f(x)\,d\gamma(x) = 0$$

for all $f \in C([a,b])$. We claim that γ is “almost constant” in the sense that the set $\{x : a \le x \le b,\ \gamma(x) \neq \gamma(a)\}$ is at most countable.

Considering the function $f(x) \equiv 1$, we see that $\gamma(b) = \gamma(a)$. Since γ has bounded variation, the set $D(\gamma)$ of all discontinuity points of γ in $[a,b]$ is at most countable. We show that $\gamma(x) = \gamma(a)$ for all $x \in (a,b)\backslash D(\gamma)$. In fact, given such a point x, choose $n_0 \in \mathbb{N}$ so large that $n_0(b-x) > 1$. For $n \ge n_0$, we define continuous functions $f_n : [a,b] \to \mathbb{R}$ by

$$f_n(t) := \begin{cases} 1 & \text{for } a \le t \le x\,, \\ 0 & \text{for } x + \frac{1}{n} \le t \le b\,, \\ \text{linear} & \text{for } x \le t \le x + \frac{1}{n}\,. \end{cases}$$

By construction, then we have

$$0 = \int_a^b f_n(t)\,d\gamma(t) = \int_a^x d\gamma(t) + \int_x^{x+1/n} f_n(t)\,d\gamma(t)$$
$$\le \gamma(x) - \gamma(a) + \max_{x\le t\le x+1/n} |f_n(t)|\,\mathrm{Var}(\gamma;[x,x+1/n])$$
$$= \gamma(x) - \gamma(a) + \mathrm{Var}(\gamma;[x,x+1/n])$$

by Theorem 4.21 (a). However,

$$\mathrm{Var}(\gamma;[x,x+1/n]) \to 0 \quad (n\to\infty),$$

and so $\gamma(x) = \gamma(a)$ for $x \in (a,b)\setminus D(\gamma)$ as claimed. □

It is very easy to find an example which shows that the countability assumption on $\Delta(\alpha,\beta)$ in Proposition 4.29 cannot be dropped. The following example shows that the requirement $\Delta(\alpha,\beta)\subseteq(a,b)$ cannot be dropped either.

Example 4.30. On $[a,b] = [0,1]$, let $f(x)\equiv 1$, $\alpha(x)\equiv 0$, and $\beta(x) = \chi_{\{1\}}(x)$. Then $\Delta(\alpha,\beta) = \{1\}$ is finite, but

$$\int_0^1 f(x)\,d\alpha(x) = 0 \ne 1 = \int_0^1 f(x)\,d\beta(x),$$

i.e. $\alpha \not\sim \beta$. ♥

Now, we are in a position to prove the announced duality theorem between the spaces $C([a,b])$ and $NBV([a,b])$.

Theorem 4.31. *The dual space $C([a,b])^*$ of the space $C([a,b])$ with the usual norm*

$$\|f\|_C = \max_{a\le t\le b} |f(t)| \tag{4.67}$$

may be identified with the space $NBV([a,b])$. More precisely, the map

$$\Phi : NBV([a,b]) \to C([a,b])^*, \quad \Phi(\alpha) := \ell_\alpha,$$

with ℓ_α as in (4.61) for $f\in C([a,b])$ and $\alpha\in NBV([a,b])$, is a linear surjective isometry.

Proof. The fundamental estimate (4.45) shows that

$$|\langle f,\ell_\alpha\rangle| = \left|\int_a^b f(t)\,d\alpha(t)\right| \le \|f\|_C\|\alpha\|_{BV}$$

for $f\in C([a,b])$ and $\alpha\in BV^o([a,b])$), and so (4.61) indeed defines a bounded linear functional ℓ_α on X with $\|\ell_\alpha\|_{C^*}\le\|\alpha\|_{BV}$. This means that Φ is well-defined and satisfies $\|\Phi(\alpha)\|_{C^*}\le\|\alpha\|_{BV}$.

The interesting and nontrivial part is of course to show that Φ is surjective. So, let ℓ be an arbitrary bounded linear functional on $C([a,b])$; by the Hahn–Banach theorem (Theorem 0.21), we may extend ℓ to a bounded linear functional (which we also denote by ℓ) on the space $B([a,b])$ with norm (0.39). We have to show that there exists a function $\alpha \in NBV([a,b]))$ such that $\|\ell\|_{C^*} = \mathrm{Var}(\alpha)$ and $\ell = \ell_\alpha$. For $a < s \leq b$, we put $z_s := \chi_{[a,s]}$, i.e.

$$z_s(t) := \begin{cases} 1 & \text{if } a \leq t \leq s\,, \\ 0 & \text{if } s < t \leq b\,. \end{cases}$$

Obviously, $z_s \in B([a,b])$ with $\|z_s\|_\infty \leq 1$. In addition, for $s = a$, we put $z_a(t) \equiv 0$. The function $\alpha : [a,b] \to \mathbb{R}$ defined by $\alpha(s) := \langle z_s, \ell\rangle$ satisfies both $\alpha(a) = \langle z_a, \ell\rangle = 0$ and (4.62); we claim that $\alpha \in BV([a,b]))$ and

$$\|\alpha\|_{BV} = \mathrm{Var}(\alpha) \leq \|\ell\|_{C^*}\,. \tag{4.68}$$

To see this, fix a partition $P = \{t_0, t_1, \ldots, t_m\} \in \mathcal{P}([a,b])$. We use the shortcut

$$\epsilon_j := sgn\left[\alpha(t_j) - \alpha(t_{j-1})\right] \quad (j = 1, 2, \ldots, m)$$

and consider the step function

$$w(t) := \begin{cases} \epsilon_1 & \text{for } a \leq t \leq t_1\,, \\ \epsilon_j & \text{for } t_{j-1} < t \leq t_j \end{cases}$$

for $j = 2, 3, \ldots, m$. Since

$$w(t) = \sum_{j=1}^m \epsilon_j \left[z_{t_j}(t) - z_{t_{j-1}}(t)\right],$$

applying the linear functional ℓ to w, we obtain

$$\begin{aligned} \langle w, \ell\rangle &= \sum_{j=1}^m \epsilon_j \left[\langle z_{t_j}, \ell\rangle - \langle z_{t_{j-1}}, \ell\rangle\right] \\ &= \sum_{j=1}^m \epsilon_j \left[\alpha(t_j) - \alpha(t_{j-1})\right] = \sum_{j=1}^m \left|\alpha(t_j) - \alpha(t_{j-1})\right|\,, \end{aligned}$$

by our definition of ϵ_j and α. However, $|\langle w, \ell\rangle| \leq \|\ell\|_{C^*}$ since $\|w\|_\infty \leq 1$, and so

$$\sum_{j=1}^m \left|\alpha(t_j) - \alpha(t_{j-1})\right| \leq \|\ell\|_{C^*}\,. \tag{4.69}$$

Since the right-hand side of (4.69) is independent of the partition P, we see that, in fact, $\alpha \in BV([a,b]))$ and $\mathrm{Var}(\alpha) \leq \|\ell\|_{C^*}$ as claimed.

It remains to show that $\ell = \ell_\alpha$. With the same partition P as before and $f \in C([a,b])$, we define $z_P : [a,b] \to \mathbb{R}$ by

$$z_P(t) := \sum_{j=1}^m f(t_{j-1}) \left[z_{t_j}(t) - z_{t_{j-1}}(t)\right].$$

A comparison with (4.17) shows that we may consider the value of ℓ at z_P as a Riemann–Stieltjes sum

$$\langle z_P, \ell\rangle = \sum_{j=1}^{m} f(t_{j-1})\left[\alpha(t_j) - \alpha(t_{j-1})\right] = S_\alpha(f, P, \Pi; [a, b]) \tag{4.70}$$

where[12] $\Pi = P \setminus \{t_m\} = \{t_0, t_1, \dots, t_{m-1}\}$. Moreover ,

$$|z_P(t) - f(t)| = \begin{cases} |f(a) - f(t)| & \text{for } a \le t \le t_1\,, \\ |f(t_{j-1}) - f(t)| & \text{for } t_{j-1} < t \le t_j \end{cases}$$

for $j = 2, 3, \dots, m$. Since f is uniformly continuous on $[a, b]$, we have $\|z_P - f\|_\infty \to 0$, and so $\langle z_P, \ell\rangle \to \langle f, \ell\rangle$, as $\mu(P) \to 0$. However, we also know from Proposition 4.5 that

$$\langle f, \ell\rangle = \lim_{\mu(P)\to 0} S_\alpha(f, P, \Pi; [a, b]) = \int_a^b f(t)\, d\alpha(t)\,. \tag{4.71}$$

Thus, we have proved that $\ell = \ell_\alpha = \Phi(\alpha)$ which means that Φ is surjective.

Thus far, we have not made any assertion about the *uniqueness* of α, i.e. the injectivity of Φ, and it is only here that we need the normalization (4.63) of α.

In the first part of the proof, we have seen that

$$\|\Phi(\alpha)\|_{C^*} = \|\ell_\alpha\|_{C^*} \le \|\alpha\|_{BV}$$

if ℓ_α is given by (4.61). Since $\alpha \approx \alpha^\#$, by Proposition 4.28 (a), the integral in (4.61) does not change if we pass from the integrator α to the integrator $\alpha^\#$. So, we also have

$$\|\Phi(\alpha)\|_{C^*} = \|\ell_\alpha\|_{C^*} \le \|\alpha^\#\|_{BV}\,.$$

On the other hand, in Proposition 4.28 (c), we have proved that

$$\|\alpha^\#\|_{BV} = \mathrm{Var}(\alpha^\#) \le \mathrm{Var}(\alpha) \le \|\ell_\alpha\|_{C^*} = \|\Phi(\alpha)\|_{C^*}\,.$$

Thus, the map Φ is an isometry, hence injective, and the proof is complete. □

It follows from the definition of the norm of a bounded linear functional (and our proof of Theorem 4.31 shows this as well) that

$$\mathrm{Var}(\alpha; [a, b]) = \sup\left\{\int_a^b f(x)\, d\alpha(x) : f \in C([a, b]),\ \|f\|_C \le 1\right\} \tag{4.72}$$

for all $\alpha \in NBV([a, b])$. Conversely, the Hahn–Banach theorem (Corollary 0.22 (c)) implies that

$$\|f\|_C = \sup\left\{\int_a^b f(x)\, d\alpha(x) : \alpha \in NBV([a, b]), \mathrm{Var}(\alpha; [a, b]) \le 1\right\} \tag{4.73}$$

for all $f \in C([a, b])$. This may also be proved quite easily by a direct computation:

12 More precisely, we take $\Pi = \{\tau_1, \tau_2, \dots, \tau_m\}$ with $\tau_j := t_{j-1}$; this set Π obviously satisfies (4.16).

Proposition 4.32. *For any $f \in C([a,b])$, the equality* (4.73) *holds.*

Proof. Denoting the right-hand side of (4.73) by S, from (4.45), we immediately get the estimate $\|f\|_C \geq S$. To prove the reverse estimate, choose $\xi \in [a,b]$ such that $\|f\|_C = |f(\xi)|$. If $\xi < b$, we define $\alpha : [a,b] \to \mathbb{R}$ by

$$\alpha(x) := sgn\, f(\xi)\chi_{(\xi,b]}(x) = \begin{cases} 1 & \text{if } f(\xi) > 0 \text{ and } \xi < x \leq b, \\ -1 & \text{if } f(\xi) < 0 \text{ and } \xi < x \leq b, \\ 0 & \text{if } f(\xi) = 0 \text{ or } a \leq x \leq \xi. \end{cases}$$

Clearly, $\alpha \in NBV([a,b])$ and $\mathrm{Var}(\alpha;[a,b]) \leq 1$. Moreover,

$$S \geq \int_a^b f(x)\,d\alpha(x) = |f(\xi)| = \|f\|_C.$$

In case $\xi = b$, an analogous reasoning shows that $S \geq |f(b-t)|$ for any $t \in (0, b-a]$, and the assertion follows from the continuity of f. □

4.3 Bounded p-variation and duality

Now, we are going to prove a parallel result to Theorem 4.31 by replacing $BV([a,b])$ by the spaces $RBV_p([a,b])$ for $1 < p < \infty$, see Definition 2.50. This result once more emphasizes the importance of these spaces. Recall that $RBV_p^o([a,b])$ denotes the subspace of all $f \in RBV_p([a,b])$ satisfying $f(a) = 0$, see (3.70).

Theorem 4.33. *For $1 < p < \infty$, the dual space $L_p([a,b])^*$ of the space $L_p([a,b])$ with the usual norm*

$$\|f\|_{L_p} = \left(\int_a^b |f(t)|^p\,dt \right)^{1/p} \tag{4.74}$$

may be identified with the space $RBV_{p'}^o([a,b])$, where $p' = p/(p-1)$. More precisely, the map

$$\Phi : RBV_{p'}^o([a,b]) \to L_p([a,b])^*, \quad \Phi(\alpha) := \ell_\alpha,$$

with ℓ_α as in (4.61) for $f \in L_p([a,b])$ and $\alpha \in RBV_{p'}^o([a,b])$, is a linear surjective isometry.

Proof. The proof is quite similar to that of the preceding Theorem 4.31, so we point out only the differences. One technical advantage is that we need not normalize α now since functions from $RBV_{p'}([a,b])$ are continuous for $p' > 1$.

Given $f \in L_p([a,b])$, $\alpha \in RBV_{p'}^o([a,b])$, and a partition $P = \{t_0, t_1, \ldots, t_m\} \in \mathcal{P}([a,b])$, we define $S_\alpha(f, P, \Pi; [a,b])$ as in (4.70). By Hölder's inequality (0.108), we then get

$$
\begin{aligned}
|S_\alpha(f,P,\Pi;[a,b])| &= \sum_{j=1}^m |f(t_{j-1}|\,|\alpha(t_j)-\alpha(t_{j-1})| \\
&\le \left(\sum_{j=1}^m \frac{|\alpha(t_j)-\alpha(t_{j-1})|^{p'}}{|t_j-t_{j-1}|^{p'-1}}\right)^{1/p'} \left(\sum_{j=1}^m |f(t_{j-1})|^p|t_j-t_{j-1}|\right)^{1/p} \qquad (4.75)\\
&\le \mathrm{Var}^R_{p'}(\alpha)^{1/p'}\left(\sum_{j=1}^m |f(t_{j-1})|^p|t_j-t_{j-1}|\right)^{1/p}.
\end{aligned}
$$

Now, if f is continuous on $[a,b]$, then the limit (4.71) exists, and (4.75) together with $\alpha(a)=0$ shows that

$$
\left|\int_a^b f(t)\,d\alpha(t)\right| \le \mathrm{Var}^R_{p'}(\alpha)^{1/p'}\left(\int_a^b |f(t)|^p\,dt\right)^{1/p} = \|f\|_{L_p}\|\alpha\|_{RBV_{p'}}. \qquad (4.76)
$$

Since $C([a,b])$ is dense in $L_p([a,b])$ with respect to the norm (4.74), there exists a unique extension of the integral (4.61) to the entire space $L_p([a,b])$, and we use the same notation for this extension. Thus, we have proved that the map Φ is well-defined and satisfies $\|\Phi(\alpha)\|_{L_p^*} \le \|\alpha\|_{RBV_{p'}}$.

To show that Φ is surjective, we proceed as in the proof of Theorem 4.31. Thus, given an arbitrary bounded linear functional ℓ on $L_p([a,b])$, we have to find a $\alpha \in RBV_{p'}([a,b]))$ such that

$$
\alpha(a)=0, \quad \|\ell\|_{L_p^*} = \mathrm{Var}^R_{p'}(\alpha)^{1/p'}, \quad \ell=\ell_\alpha.
$$

Defining z_s as before, we have $z_s \in L_p([a,b])$ with $\|z_s\|_{L_p} = s^{1/p}$. Putting again $\alpha(s) := \langle z_s,\ell\rangle$ and fixing a partition $P=\{t_0,t_1,\dots,t_m\}\in\mathcal{P}([a,b])$, we obtain

$$
\begin{aligned}
|\alpha(t_j)-\alpha(t_{j-1})| &= |\langle z_{t_j},\ell\rangle - \langle z_{t_{j-1}},\ell\rangle| \le \|\ell\|_{L_p^*}\|z_{t_j}-z_{t_{j-1}}\|_{L_p} \\
&= \|\ell\|_{L_p^*}\left(\int_a^b \left|\chi_{[a,t_j]}(t)-\chi_{[a,t_{j-1}]}(t)\right|^p dt\right)^{1/p} = \|\ell\|_{L_p^*}|t_j-t_{j-1}|,
\end{aligned}
$$

and hence

$$
\sum_{j=1}^m \frac{|\alpha(t_j)-\alpha(t_{j-1})|^{p'}}{|t_j-t_{j-1}|^{p'-1}} \le \|\ell\|^{p'}_{L_p^*}\sum_{j=1}^m |t_j-t_{j-1}|^{p'/p-1/(p-1)} = \|\ell\|^{p'}_{L_p^*}
$$

since

$$
\frac{p'}{p}-\frac{1}{p-1} = \frac{p}{p(p-1)}-\frac{1}{p-1} = 0.
$$

This shows that $\alpha \in RBV_{p'}([a,b]))$ and $\|\alpha\|_{RBV_{p'}} \leq \|\ell\|_{L_p^*}$. Now, we prove that $\|\alpha\|_{RBV_{p'}} = \|\ell\|_{L_p^*}$. In fact, all functions of the form

$$f(t) := \sum_{j=1}^{m} \xi_j \chi_{[t_{j-1},t_j)} \quad (\xi_1, \xi_2, \ldots, \xi_m \in \mathbb{R}) \tag{4.77}$$

belong to $L_p([a,b])$ and satisfy

$$\|f\|_{L_p} = \left(\sum_{j=1}^{m} |\xi_j|^p \, |t_j - t_{j-1}| \right)^{1/p} .$$

Applying the functional ℓ to such a function and using (0.108) yields

$$\begin{aligned} |\langle f, \ell \rangle| &\leq \sum_{j=1}^{m} |\xi_j| \, |\alpha(t_j) - \alpha(t_{j-1})| \\ &\leq \left(\sum_{j=1}^{m} \frac{|\alpha(t_j) - \alpha(t_{j-1})|^{p'}}{|t_j - t_{j-1}|^{p'-1}} \right)^{1/p'} \left(\sum_{j=1}^{m} |\xi_j|^p \, |t_j - t_{j-1}|^{(p'-1)p/p'} \right)^{1/p} , \end{aligned}$$

so $|\langle f, \ell \rangle| \leq \mathrm{Var}_{p'}^R(\alpha)^{1/p'} \|f\|_{L_p}$. Since the set of functions of the form (4.77) is dense in $L_p([a,b])$, we conclude that $\|\ell\|_{L_p^*} \leq \mathrm{Var}_{p'}^R(\alpha)^{1/p'} = \|\alpha\|_{RBV_{p'}}$.

It remains to prove (4.61). Choosing $f = z_s$ in (4.71) gives

$$\lim_{\mu(P) \to 0} S_\alpha(z_s, P, \Pi; [a,b]) = \int_a^b z_s(t) \, d\alpha(t) = \langle z_s, \ell_\alpha \rangle ,$$

where we may consider, without loss of generality, only partitions P which contain the point s. For such partitions, we have, by (4.70),

$$S_\alpha(z_s, P, \Pi; [a,b]) = \sum_{j=1}^{m} \left[\alpha(t_j) - \alpha(t_{j-1}) \right] \chi_{[a,s]}(t_{j-1}) = \alpha(s) - \alpha(a) = \alpha(s) ,$$

and hence, after passing to the limit as $\mu(P) \to 0$,

$$\int_a^b z_s(t) \, d\alpha(t) = \alpha(s) .$$

Applying this to functions f of the form (4.77), we get

$$\langle f, \ell \rangle = \int_a^b f(t) \, d\alpha(t) = \sum_{j=1}^{m} \left[\alpha(t_j) - \alpha(t_{j-1}) \right] \xi_j .$$

Again, since the set of functions of the form (4.77) is dense in $L_p([a,b])$, we have proved (4.61) for all $f \in L_p([a,b])$. Thus, we may put $\Phi(\alpha) = \ell$, and the proof of the surjectivity of Φ is complete. The injectivity of Φ follows from the isometry property $\|\Phi(\alpha)\|_{L_p^*} = \|\alpha\|_{RBV_{p'}}$. □

Theorem 4.33 refers to the case $p > 1$. One could expect that a parallel result holds in case $p = 1$, where $RBV_\infty^o([a,b]))$ has to be interpreted as space $Lip^o([a,b])$ of all Lipschitz continuous functions $\alpha : [a,b] \to \mathbb{R}$ with $\alpha(a) = 0$. In fact, for $f \in L_1([a,b])$, we obtain the estimate

$$\left| \int_a^b f(t)\, d\alpha(t) \right| \leq \sup_{x \neq y} \frac{|\alpha(x) - \alpha(y)|}{|x-y|} \int_a^b |f(t)|\, dt = \|f\|_{L_1} \|\alpha\|_{Lip}, \tag{4.78}$$

which is the precise analogue to (4.76). The remaining part of the proof carries over; we summarize with the following

Theorem 4.34. *The dual space $L_1([a,b])^*$ of the space $L_1([a,b])$ with norm*

$$\|f\|_{L_1} = \int_a^b |f(t)|\, dt \tag{4.79}$$

may be identified with the space $RBV_\infty^o([a,b]) = Lip^o([a,b])$ of Lipschitz continuous functions $\alpha : [a,b] \to \mathbb{R}$ satisfying $\alpha(a) = 0$. More precisely, the map

$$\Phi : Lip^o([a,b]) \to L_1([a,b])^*, \quad \Phi(\alpha) := \ell_\alpha,$$

with ℓ_α as in (4.61) for $f \in L_1([a,b])$ and $\alpha \in Lip^o([a,b]))$, is a linear surjective isometry.

The Theorems 4.33 and 4.34 are somewhat unsatisfactory, insofar as they connect the Riesz space $RBV_{p'}$ to the Lebesgue space L_p which is quite different in nature. However, the Riesz theorem (Theorem 3.34) allows us to establish a more "intrinsic" duality inside the scale of RBV_p-spaces.

Indeed, from equality (3.69), it follows that for $1 < p < \infty$, the differential operator $Df := f'$ defines a linear surjective isometry between the spaces $(RBV_p^o, \|\cdot\|_{RBV_p})$ and $(L_p, \|\cdot\|_{L_p})$. So, we may combine this with the duality result established in Theorem 4.33 to get the following more satisfactory version:

Theorem 4.35. *For $1 < p < \infty$, the dual space $RBV_p^o([a,b])^*$ of the space $RBV_p^o([a,b])$ with norm (2.90) may be identified with the space $RBV_{p'}^o([a,b])$, where $p' = p/(p-1)$. More precisely, the map*

$$\Psi : RBV_{p'}^o([a,b]) \to RBV_p^o([a,b])^*, \quad \Psi(\alpha) := \ell_\alpha,$$

with

$$\langle f, \ell_\alpha \rangle := \int_a^b f'(t)\, d\alpha(t)$$

for $f \in RBV_p^o([a,b])$ and $\alpha \in RBV_{p'}^o([a,b])$, is a linear surjective isometry.

4.4 Nonclassical RS-integrals

The classical Riemann–Stieltjes integral has been generalized in various directions. Some of these generalizations aim at avoiding the nonexistence of the integral (4.7) due to the presence of common points of discontinuity of f and α, see Theorem 4.15. We will mention some results of this type in the next section.

Other extensions of the RS-integral (4.7) consist of replacing the class $BV([a,b])$ with larger classes related to higher order variations $\mathrm{Var}_k^W(\alpha;[a,b])$ (Section 2.7), or the Waterman class $\Lambda BV([a,b])$ discussed in Section 2.2, or the Schramm class $\Phi BV([a,b])$ discussed in Section 2.3. We will describe the corresponding procedure now for the space $\Phi BV([a,b])$; unfortunately, this requires a great deal of technical definitions and results.

Our main theorem in this section asserts that under suitable growth conditions on two Schramm sequences $\Phi = (\phi_n)_n$ and $\Psi = (\psi_n)_n$, the integral (4.7) exists for $f \in \Phi BV([a,b]) \cap C([a,b])$ and $\alpha \in \Psi BV([a,b])$, for a precise formulation, see Theorem 4.40 below. The whole material of this section is taken from Schramm's survey paper [286].

Recall that $\Phi = (\phi_n)_n$ is called a Schramm sequence if each $\phi_n : [0,\infty) \to [0,\infty)$ is a convex Young function and $\phi_{n+1}(x) \le \phi_n(x)$ for all $n \in \mathbb{N}$. If, in addition,

$$\sum_{n=1}^{\infty} \phi_n(x) = \infty \quad (x > 0)\,, \tag{4.80}$$

then $\Phi = (\phi_n)_n$ is called a divergent Schramm sequence. The space $\Phi BV([a,b])$ consists, by definition, of all functions $f : [a,b] \to \mathbb{R}$ satisfying $\mathrm{Var}_\Phi(cf) < \infty$ for some $c > 0$, where

$$\begin{aligned}\mathrm{Var}_\Phi(f) &= \mathrm{Var}_\Phi(f;[a,b])\\ &= \sup\left\{\sum_{n=1}^{\infty} \phi_n(|f(b_n) - f(a_n)|) : \{[a_n,b_n] : n \in \mathbb{N}\} \in \Sigma_\infty([a,b])\right\}\end{aligned} \tag{4.81}$$

and the supremum in (4.81) is taken over all infinite collections in $\Sigma_\infty([a,b])$, see Section 1.2. This space, equipped with the norm

$$\|f\|_\Phi = |f(a)| + \inf\{\lambda > 0 : \mathrm{Var}_\Phi(f/\lambda) \le 1\}\,, \tag{4.82}$$

is a Banach space; in particular, the infimum in (4.82) is a norm on the subspace $\Phi BV^o([a,b])$ of all $f \in \Phi BV([a,b])$ satisfying $f(a) = 0$.

To introduce a Riemann–Stieltjes integral (4.7) for α belonging to $\Phi BV([a,b])$, we need a series of technical lemmas. We begin with the following

Lemma 4.36. *Let $\Phi = (\phi_n)_n$ and $\Psi = (\psi_n)_n$ be two Schramm sequences satisfying*

$$\sum_{k=1}^{\infty} \phi_k^{-1}(1/k)\psi_k^{-1}(1/k) < \infty\,. \tag{4.83}$$

Then the following holds.

(a) *If A and B are any nonnegative real numbers, then*

$$\sum_{k=1}^{\infty} \phi_k^{-1}(A/k)\psi_k^{-1}(B/k) < \infty\,. \tag{4.84}$$

(b) *There exists a convex Young function* $\Gamma : [0,\infty) \to [0,\infty)$ *satisfying* $\Gamma(x) = o(x)$ *as* $x \to 0$ *and*

$$\sum_{k=1}^{\infty} (\Gamma \circ \phi_k)^{-1}(1/k)(\Gamma \circ \psi_k)^{-1}(1/k) < \infty\,. \tag{4.85}$$

Proof. (a) Choose $m \in \mathbb{N}$ such that $m \geq A$ and $m \geq B$. For $jm \leq k < (j+1)m$, we then have

$$\begin{aligned} &\sum_{k=1}^{\infty} \phi_k^{-1}(A/k)\psi_k^{-1}(B/k) \\ &\quad \leq \sum_{k=1}^{m-1} \phi_k^{-1}(A/k)\psi_k^{-1}(B/k) + m\sum_{j=1}^{\infty} \phi_j^{-1}(1/j)\psi_j^{-1}(1/j) < \infty \end{aligned} \tag{4.86}$$

since both ϕ_k^{-1} and ψ_k^{-1} are increasing.

(b) From (a) it follows that

$$\sum_{k=1}^{\infty} \phi_k^{-1}(3n/k)\psi_k^{-1}(3n/k) < \infty \quad (n = 1,2,3,\dots)\,.$$

Thus, we may choose a positive increasing sequence $(k_n)_n$ of natural numbers so that

$$k_{n+1} > \left(1 + \frac{1}{n}\right)k_n, \quad \sum_{k=k_n}^{\infty} \phi_k^{-1}(3n/k)\psi_k^{-1}(3n/k) < \frac{1}{n^2} \quad (n = 1,2,3,\dots)\,.$$

We define a function $\gamma : [0,\infty) \to [0,\infty)$ by

$$\gamma(t) := \begin{cases} 0 & \text{for } t = 0\,, \\ \frac{1}{n} & \text{for } \frac{1}{k_{n+1}} < t \leq \frac{1}{k_n}\,, \\ 1 + t & \text{for } t > \frac{1}{k_1}\,. \end{cases}$$

Then γ is positive and increasing, and $\gamma(t) \to 0$ as $t \to 0$. Consequently, the function $\Gamma : [0,\infty) \to [0,\infty)$ defined by

$$\Gamma(x) := \int_0^x \gamma(t)\,dt \quad (x \geq 0)$$

is increasing and convex and satisfies $\Gamma(0) = 0$ and $\Gamma(x) = o(x)$ as $x \to 0$. We claim that Γ has the required properties. If $nk_n < k < nk_{n+1}$, we have

$$\Gamma\left(\frac{3n}{k}\right) = \int_0^{3n/k} \gamma(t)\,dt \geq \int_{n/k}^{2n/k} \gamma(t)\,dt \geq \frac{n}{k}\gamma\left(\frac{n}{k}\right) \geq \frac{1}{k}\,,$$

while for $nk_{n+1} \leq k < (n+1)k_{n+1}$, we have

$$\Gamma\left(\frac{3n}{k}\right) = \int_0^{3n/k} \gamma(t)\,dt \geq \int_{(n+1)/k}^{(2n+1)/k} \gamma(t)\,dt \geq \frac{n}{k}\gamma\left(\frac{n+1}{k}\right) = \frac{1}{k}\,.$$

Consequently, applying Γ^{-1} to the first and last term, we deduce that

$$\Gamma^{-1}\left(\frac{1}{k}\right) \leq \frac{3n}{k} \qquad (nk_{n+1} \leq k < (n+1)k_{n+1})\,.$$

Combining these estimates, we arrive at the chain of inequalities

$$\begin{aligned}
&\sum_{k=1}^{\infty}(\Gamma \circ \phi_k)^{-1}(1/k)(\Gamma \circ \psi_k)^{-1}(1/k) \\
&\quad = \sum_{k=1}^{k_1-1} \phi_k^{-1}(\Gamma^{-1}(1/k))\psi_k^{-1}(\Gamma^{-1}(1/k)) \\
&\qquad + \sum_{n=1}^{\infty} \sum_{k=nk_n}^{(n+1)k_{n+1}-1} \phi_k^{-1}(\Gamma^{-1}(1/k))\psi_k^{-1}(\Gamma^{-1}(1/k)) \\
&\quad \leq \sum_{k=1}^{k_1-1} \phi_k^{-1}(\Gamma^{-1}(1/k))\psi_k^{-1}(\Gamma^{-1}(1/k)) + \sum_{n=1}^{\infty} \sum_{k=nk_n}^{(n+1)k_{n+1}-1} \phi_k^{-1}(3n/k)\psi_k^{-1}(3n/k) \\
&\quad \leq \sum_{k=1}^{k_1-1} \phi_k^{-1}(\Gamma^{-1}(1/k))\psi_k^{-1}(\Gamma^{-1}(1/k)) + \sum_{n=1}^{\infty} \sum_{k=k_n}^{\infty} \phi_k^{-1}(3n/k)\psi_k^{-1}(3n/k) \\
&\quad \leq \sum_{k=1}^{k_1-1} \phi_k^{-1}(\Gamma^{-1}(1/k))\psi_k^{-1}(\Gamma^{-1}(1/k)) + \sum_{n=1}^{\infty} \frac{1}{n^2} < \infty\,,
\end{aligned}$$

and so we have proved (b). □

Observe that the condition (4.83) is trivially satisfied if only finitely many functions ϕ_k or ψ_k are different from zero. This shows that Lemma 4.36 covers the spaces BV, WBV_p, and WBV_ϕ.

Also, note that $\Gamma(x) = o(x)$ implies $(\Gamma \circ \phi_n)(x) = o(\phi_n(x))$ as $x \to 0$, so $\Phi BV([a,b]) \subseteq \Gamma\Phi BV([a,b])$. For a Schramm sequence $\Phi = (\phi_n)_n$ and Γ as in Lemma 4.36 (b), we define

$$\chi_n(\Gamma, \Phi; x) := \begin{cases} \frac{\Gamma(\phi_n(x))}{\phi_n(x)} & \text{for } x > 0\,, \\ 0 & \text{for } x = 0\,. \end{cases} \tag{4.87}$$

Then $\chi_n(\Gamma, \Phi; x) \leq \chi_{n+1}(\Gamma, \Phi; x)$ for $x \geq 0$ and $\chi_n(\Gamma, \Phi; x) \to 0$ as $x \to 0$. Moreover, denoting by $\operatorname{osc}(f; [a,b])$ the oscillation (1.12) of f on $[a,b]$, from the trivial estimate

$\phi_1(\mathrm{osc}(f;[a,b])) \le \mathrm{Var}_\Phi(f)$, we get

$$\begin{aligned}\mathrm{Var}_{\Gamma\circ\Phi}(f) &= \sup\left\{\sum_{n=1}^{\infty}\Gamma(\phi_n(|f(b_n)-f(a_n)|))\right\}\\ &= \sup\left\{\sum_{n=1}^{\infty}\chi_n(\Gamma,\Phi;|f(b_n)-f(a_n)|)\phi_n(|f(b_n)-f(a_n)|)\right\}\\ &\le \chi_1(\Gamma,\Phi;\mathrm{osc}(f:[a,b]))\,\mathrm{Var}_\Phi(f)\end{aligned} \tag{4.88}$$

where both suprema are taken over all collections $\{[a_n,b_n] : n \in \mathbb{N}\} \in \Sigma_\infty([a,b])$. This again shows that $\Phi BV([a,b]) \subseteq \Gamma\Phi BV([a,b])$.

We now introduce some notation which will simplify our computations. Given a Schramm sequence $\Phi = (\phi_n)_n$ and a sequence $A = (a_n)_n$ of real numbers, we use the shortcut

$$\rho_n(\Phi;A) := \sum_{k=1}^{n}\phi_k(|a_k|) \quad (n = 1,2,3,\ldots). \tag{4.89}$$

Moreover, we denote by $\mathfrak{I}$ the family of all strictly increasing sequences $I = (i_n)_n$ starting from $i_0 \ge 0$. We associate to each pair of sequences $A = (a_n)_n$ and $I = (i_n)_n \in \mathfrak{I}$ another sequence $I(A)$ with terms

$$I(A) := \left(\sum_{k=i_0+1}^{i_1} a_k, \sum_{k=i_1+1}^{i_2} a_k, \ldots, \sum_{k=i_{n-1}+1}^{i_n} a_k, \ldots\right). \tag{4.90}$$

Loosely speaking, the sequence $I(A)$ is derived from the sequence A by "adding its terms sectionwise" according to the sections prescribe by the index set I. For example, if I is the set of prime numbers, then

$$I(A) = (a_1 + a_2, a_3, a_4 + a_5, a_6 + a_7, a_8 + a_9 + a_{10} + a_{11}, a_{12} + a_{13}, \ldots).$$

We also put

$$\rho_n^*(\Phi;A) := \sup\{\rho_n(\Phi;I(A)) : I \in \mathfrak{I}\}, \tag{4.91}$$

where the supremum in (4.91) is taken over all sequences $I \in \mathfrak{I}$. Using this notation, we now have the following

Lemma 4.37. *Let $A = (a_n)_n$ and $B = (b_n)_n$ be real sequences, and let $\Phi = (\phi_n)_n$ and $\Psi = (\psi_n)_n$ be two Schramm sequences. Then the following is true.*

(a) *The estimate*

$$\sqrt[n]{|a_1a_2\cdots a_n|} \le \phi_n^{-1}\left(\frac{\rho_n(\Phi;I(A))}{n}\right) \tag{4.92}$$

holds for all $n \in \mathbb{N}$.

(b) *For every $n \in \mathbb{N}$, there exists a $k_0 \in \mathbb{N}$, $1 \le k_0 \le n$, such that*

$$|a_{k_0}b_{k_0}| \le \phi_n^{-1}\left(\frac{\rho_n(\Phi;I(A))}{n}\right)\psi_n^{-1}\left(\frac{\rho_n(\Psi;I(B))}{n}\right). \tag{4.93}$$

Proof. By the arithmetic-geometric mean inequality and the monotonicity of ϕ_n, we have

$$\phi_n\left(\sqrt[n]{|a_1 a_2 \cdots a_n|}\right) \le \phi_n\left(\frac{1}{n}\left(|a_1| + |a_2| + \ldots + |a_n|\right)\right).$$

Moreover, from Jensen's inequality (2.18), it follows that

$$\phi_n\left(\frac{1}{n}\sum_{k=1}^{n}|a_k|\right) \le \frac{1}{n}\sum_{k=1}^{n}\phi_k(|a_k|) = \frac{1}{n}\rho_n(\Phi; I(A))$$

and applying ϕ_n^{-1}, we obtain (a).

To prove (b), we choose $k_0 \in \{1, 2, \ldots, n\}$ such that $|a_{k_0} b_{k_0}| = \min\{|a_k b_k| : k = 1, 2, \ldots, n\}$. Then

$$\begin{aligned} |a_{k_0} b_{k_0}| &\le \left|\prod_{k=1}^{n} a_k b_k\right|^{1/n} = \left|\prod_{k=1}^{n} a_k\right|^{1/n} \left|\prod_{k=1}^{n} b_k\right|^{1/n} \\ &\le \phi_n^{-1}\left(\frac{\rho_n(\Phi; I(A))}{n}\right) \psi_n^{-1}\left(\frac{\rho_n(\Psi; I(B))}{n}\right), \end{aligned} \tag{4.94}$$

where in the last inequality we have used (a). □

Before we state the main result of this section, we need another two auxiliary propositions. We start with the following refinement of Lemma 4.37 (b).

Proposition 4.38. *Under the hypotheses of Lemma 4.37, the estimate*

$$\begin{aligned} \left|\sum_{k=1}^{n}\sum_{i=1}^{k} a_j b_k\right| &\le \phi_1^{-1}\left(\rho_1^*(\Phi; I(A))\right) \psi_1^{-1}\left(\rho_1^*(\Psi; I(B))\right) \\ &\quad + \sum_{k=1}^{n-1} \phi_k^{-1}\left(\frac{\rho_k^*(\Phi; I(A))}{k}\right) \psi_k^{-1}\left(\frac{\rho_k^*(\Psi; I(B))}{k}\right) \end{aligned} \tag{4.95}$$

is true.[13]

Proof. We prove (4.95) by induction over n. For $n = 1$, we have

$$|a_1 b_1| = \phi_1^{-1}(\phi_1(|a_1|))\psi_1^{-1}(\psi_1(|b_1|)) \le \phi_1^{-1}\left(\rho_1^*(\Phi; I(A))\right) \psi_1^{-1}\left(\rho_1^*(\Psi; I(B))\right).$$

Suppose that $n \ge 2$ and (4.95) holds true for $n - 1$. Defining a shifted sequence $A' = (a_n')_n$ by $a_k' := a_{k+1}$, by Lemma 4.37 (b), we find a $k_0 \in \{1, 2, \ldots, n-1\}$ such that

$$|a_{k_0+1} b_{k_0}| \le \phi_{n-1}^{-1}\left(\frac{\rho_{n-1}(\Phi; I(A'))}{n-1}\right) \psi_{n-1}^{-1}\left(\frac{\rho_{n-1}(\Psi; I(B))}{n-1}\right).$$

From

$$\rho_{n-1}(\Phi; I(A')) \le \rho_{n-1}^*(\Phi; I(A')) \le \rho_{n-1}(\Phi; I(A))$$

13 In case $n = 1$, we take the last sum in (4.95) to be 0.

and

$$\rho_{n-1}(\Psi; I(B)) \le \rho_{n-1}^*(\Psi; I(B)),$$

it follows that

$$|a_{k_0+1}b_{k_0}| \le \phi_{n-1}^{-1}\left(\frac{\rho_{n-1}^*(\Phi; I(A))}{n-1}\right)\psi_{n-1}^{-1}\left(\frac{\rho_{n-1}^*(\Psi; I(B))}{n-1}\right). \tag{4.96}$$

To reduce our claim to the induction hypothesis, we define two sequences $C = (c_n)_n$ and $D = (d_n)_n$ by

$$c_k := \begin{cases} a_k & \text{for } k < k_0\,, \\ a_{k_0} + a_{k_0+1} & \text{for } k = k_0\,, \\ a_{k+1} & \text{for } k > k_0\,, \end{cases}$$

and

$$d_k := \begin{cases} b_k & \text{for } k < k_0\,, \\ b_{k_0} + b_{k_0+1} & \text{for } k = k_0\,, \\ b_{k+1} & \text{for } k > k_0\,, \end{cases}$$

respectively, and note that $\rho_n^*(\Phi; I(C)) \le \rho_n^*(\Phi; I(A))$ and $\rho_n^*(\Psi; I(D)) \le \rho_n^*(\Psi; I(B))$. A straightforward calculation shows that

$$\sum_{k=1}^{n-1}\sum_{i=1}^{k} c_j d_k = a_{k_0+1}b_{k_0} + \sum_{k=1}^{n}\sum_{i=1}^{k} a_j b_k\,.$$

Thus, combining this with (4.96), we obtain

$$\begin{aligned}\left|\sum_{k=1}^{n}\sum_{i=1}^{k} a_j b_k\right| &\le |a_{k_0+1}b_{k_0}| + \left|\sum_{k=1}^{n-1}\sum_{i=1}^{k} c_j d_k\right| \\ &\le \phi_1^{-1}(\rho_1^*(\Phi; I(C)))\psi_1^{-1}(\rho_1^*(\Psi; I(D))) \\ &\quad + \sum_{k=1}^{n-2}\phi_k^{-1}\left(\frac{\rho_k^*(\Phi; I(C))}{k}\right)\psi_k^{-1}\left(\frac{\rho_k^*(\Psi; I(D))}{k}\right) \\ &\quad + \phi_{n-1}^{-1}\left(\frac{\rho_{n-1}^*(\Phi; I(A))}{n-1}\right)\psi_{n-1}^{-1}\left(\frac{\rho_{n-1}^*(\Psi; I(B))}{n-1}\right) \\ &\le \phi_1^{-1}(\rho_1^*(\Phi; I(A)))\psi_1^{-1}(\rho_1^*(\Psi; I(B))) \\ &\quad + \sum_{k=1}^{n-1}\phi_k^{-1}\left(\frac{\rho_k^*(\Phi; I(A))}{k}\right)\psi_k^{-1}\left(\frac{\rho_k^*(\Psi; I(B))}{k}\right)\end{aligned}$$

which is (4.95) for arbitrary n. □

Now, we examine Riemann–Stieltjes sums. As in Definition 4.4, we use the notation (4.17), where $P = \{t_0, t_1, \dots, t_m\} \in \mathcal{P}([a,b])$ is a partition of $[a,b]$ and $\Pi = \{\tau_1, \tau_2, \dots, \tau_m\}$ is a set of intermediate points satisfying $t_{j-1} \le \tau_j \le t_j$ for $j = 1, 2, \dots, m$. In the following proposition, we give an upper estimate for (4.17) in terms of finitely many elements of two Schramm sequences whose number depends on the length m of the partition P.

Proposition 4.39. *Let $\Phi = (\phi_n)_n$ and $\Psi = (\psi_n)_n$ be two Schramm sequences, and fix sets $P = \{t_0, t_1, \dots, t_m\} \in \mathcal{P}([a,b])$ and $\Pi = \{\tau_1, \tau_2, \dots, \tau_m\}$ as above. Suppose that $f \in \Phi BV^o([a,b])$ and $\alpha \in \Psi BV([a,b])$. Then the estimate*

$$S_\alpha(f, P, \Pi; [a,b]) \le 2\sum_{j=1}^{m-1} \phi_j^{-1}\left(\frac{\mathrm{Var}_\Phi(f)}{j}\right)\psi_j^{-1}\left(\frac{\mathrm{Var}_\Psi(\alpha)}{j}\right) \tag{4.97}$$

is true, where $S_\alpha(f, P, \Pi; [a,b])$ is defined by (4.17).

Proof. We prove the assertion by making special choices for the intermediate points in Π. If we take $\tau_0 := a$, we get

$$\begin{aligned} |S_\alpha(f, P, \Pi; [a,b])| &= \left|\sum_{j=1}^{m} f(\tau_j)(\alpha(t_j) - \alpha(t_{j-1}))\right| \\ &= \left|\sum_{j=1}^{m}\sum_{i=1}^{j}(f(\tau_i) - f(\tau_{i-1}))(\alpha(t_j) - \alpha(t_{j-1})\right| . \end{aligned} \tag{4.98}$$

Now, we define a sequence $A = (a_n)_n$ by $a_n := f(\tau_n) - f(\tau_{n-1})$ for $n \le m$ and $a_n := 0$ otherwise, and another sequence $B = (b_n)_n$ by $b_n := \alpha(t_n) - \alpha(t_{n-1})$ for $n \le m$ and $b_n := 0$ otherwise. Then from (4.98), we have

$$\begin{aligned} |S_\alpha(f, P, \Pi; [a,b])| &= \left|\sum_{j=1}^{m}\sum_{i=1}^{j} a_j b_j\right| \\ &\le \phi_1^{-1}(\rho_1^*(\Phi; I(A)))\,\psi_1^{-1}(\rho_1^*(\Psi; I(B))) \\ &\quad + \sum_{j=1}^{m-1}\phi_j^{-1}\left(\frac{\rho_j^*(\Phi; I(A))}{j}\right)\psi_j^{-1}\left(\frac{\rho_j^*(\Psi; I(B))}{j}\right) \\ &\le \phi_1^{-1}(\mathrm{Var}_\Phi(f))\,\psi_1^{-1}(\mathrm{Var}_\Psi(\alpha)) \\ &\quad + \sum_{j=1}^{m-1}\phi_j^{-1}\left(\frac{\mathrm{Var}_\Phi(f)}{j}\right)\psi_j^{-1}\left(\frac{\mathrm{Var}_\Psi(\alpha)}{j}\right) \\ &\le 2\sum_{j=1}^{m-1}\phi_j^{-1}\left(\frac{\mathrm{Var}_\Phi(f)}{j}\right)\psi_j^{-1}\left(\frac{\mathrm{Var}_\Psi(\alpha)}{j}\right) \end{aligned}$$

which is (4.97). □

Now, we are in a position to state our main result on the existence of the RS integral (4.7) for α in a larger class than just $BV([a,b])$.

Theorem 4.40. *Let $\Phi = (\phi_n)_n$ and $\Psi = (\psi_n)_n$ be two Schramm sequences satisfying (4.83). If $f \in \Phi BV^o([a,b]) \cap C([a,b])$ and $\alpha \in \Psi BV([a,b])$, then the Riemann–Stieltjes integral (4.7) exists.*

Proof. Let $P = \{x_0, x_1, \dots, x_m\} \in \mathcal{P}([a,b])$ and $Q = \{y_0, y_1, \dots, y_n\} \in \mathcal{P}([a,b])$ be partitions of $[a,b]$, and let $\Xi = \{\xi_1, \xi_2, \dots, \xi_m\}$ and $H = \{\eta_1, \eta_2, \dots, \eta_n\}$, respectively, be sets of corresponding intermediate points, i.e.

$$x_0 \le \xi_1 \le x_1 \le \ldots \le x_{m-1} \le \xi_m \le x_m, \quad y_0 \le \eta_1 \le y_1 \le \ldots \le y_{n-1} \le \eta_n \le y_n .$$

We define two step functions $g, h : [a,b] \to \mathbb{R}$ by

$$g(t) := \begin{cases} 0 & \text{for } t = a, \\ f(\xi_j) & \text{for } x_{i-1} < t \le x_j \end{cases}$$

for $i = 1, 2, \ldots, m$ and

$$h(t) := \begin{cases} 0 & \text{for } t = a, \\ f(\eta_j) & \text{for } y_{j-1} < t \le y_j \end{cases}$$

for $j = 1, 2, \ldots, n$. These functions satisfy

$$S_\alpha(g, P, \Xi; [a,b]) = \sum_{i=1}^{m} g(\xi_j)(\alpha(x_j) - \alpha(x_{i-1})) = \sum_{i=1}^{m+n} g(t_j)(\alpha(t_j) - \alpha(t_{i-1}))$$

and

$$S_\alpha(h, Q, H; [a,b]) = \sum_{j=1}^{n} h(\eta_j)(\alpha(y_j) - \alpha(y_{j-1})) = \sum_{j=1}^{m+n} h(t_j)(\alpha(t_j) - \alpha(t_{j-1})),$$

respectively. Consequently,

$$S_\alpha(g, P, \Xi; [a,b]) - S_\alpha(h, Q, H; [a,b]) = 2\sum_{k=1}^{m+n} \frac{1}{2}\left(g(t_k) - h(t_k)\right)\left(\alpha(t_k) - \alpha(t_{k-1})\right), \quad (4.99)$$

the factors 2 and 1/2 being for later convenience. By $R = P \cup Q = \{t_0, t_1, \ldots, t_{m+n}\}$, we denote the common refinement of the partitions P and Q. Then (4.99) may be rewritten in the form

$$S_\alpha(g, P, \Xi; [a,b]) - S_\alpha(h, Q, H; [a,b]) = 2S_\alpha(d, R, R; [a,b]),$$

where the function $d := \frac{1}{2}(g - h)$ belongs to $\Phi BV^o([a,b]) \subseteq \Gamma\Phi BV^o([a,b])$. Now, we apply Proposition 4.39 to the Schramm sequences $\Gamma \circ \Phi$ and $\Gamma \circ \Psi$, with Γ as in Lemma 4.36 (b), and to f replaced by d. As a result, we obtain the estimate

$$\begin{aligned} &\left|S_\alpha(g, P, \Xi; [a,b]) - S_\alpha(h, Q, H; [a,b])\right| \\ &\qquad \le 4\sum_{k=1}^{m+n} (\Gamma \circ \phi_k)^{-1}\left(\frac{\mathrm{Var}_{\Gamma\circ\Phi}(h)}{k}\right)(\Gamma \circ \psi_k)^{-1}\left(\frac{\mathrm{Var}_{\Gamma\circ\Psi}(\alpha)}{k}\right). \end{aligned}$$

We further get, with χ_n given by (4.87) and $\mathrm{osc}(f; [a,b])$ given by (1.12),

$$\begin{aligned} \mathrm{Var}_{\Gamma\circ\Phi}(d; [a,b]) &\le \chi_1(\Phi, \mathrm{osc}(d; [a,b]))\,\mathrm{Var}_\Phi(d; [a,b]) \\ &< \frac{1}{2}\chi_1(\Phi, \mathrm{osc}(d; [a,b]))\left(\mathrm{Var}_\Phi(g; [a,b]) + \mathrm{Var}_\Phi(h; [a,b])\right) \\ &\le \chi_1(\Phi, \mathrm{osc}(d; [a,b]))\,\mathrm{Var}_\Phi(f; [a,b]) \\ &\le \chi_1(\Phi, \mathrm{osc}((g - f)/2; [a,b])) + \mathrm{osc}((h - f)/2; [a,b]))\,\mathrm{Var}_\Phi(f; [a,b]). \end{aligned}$$

Let $\varepsilon > 0$. Since f is uniformly continuous on $[a, b]$, we may find $\delta > 0$ such that $\text{osc}((g - f)/2; [a, b]) + \text{osc}((h - f)/2; [a, b]) \leq \varepsilon$ for $\mu(P) \leq \delta$ and $\mu(Q) \leq \delta$. Then

$$\begin{aligned} &|S_\alpha(g, P, \Xi; [a, b]) - S_\alpha(h, Q, H; [a, b])| \\ &\qquad \leq \sum_{k=1}^{\infty} (\Gamma \circ \phi_k)^{-1} \left(\frac{\chi_1(\Phi; \varepsilon) \operatorname{Var}_\Phi(f)}{k} \right) (\Gamma \circ \psi_k)^{-1} \left(\frac{\operatorname{Var}_{\Gamma \circ \Psi(f)}(\alpha)}{k} \right). \end{aligned} \tag{4.100}$$

Since $\chi_1(\Phi; \varepsilon) \to 0$ as $\varepsilon \to 0$, we may make the right-hand side of (4.100) as small as we wish by choosing ε sufficiently small. The proof is complete. □

When applying Theorem 4.40 to specific Schramm sequences $\Phi = (\phi_n)_n$ and $\Psi = (\psi_n)_n$, the crucial condition to be verified is (4.83). As mentioned before, this condition trivially holds if only finitely many terms ϕ_n and ψ_n are different from zero. For example, in the most important example of the space $BV([a, b])$, this is true, and so we regain Theorem 4.11 as a special case of Theorem 4.40. A more interesting case, where the series (4.83) is infinite, refers to the Waterman space $\Lambda BV([a, b])$ introduced in Definition 2.15.

Let $\Lambda = (\lambda_n)_n$ and $M = (\mu_n)_n$ be two Waterman sequences in the sense of Definition 2.15. Then the corresponding spaces $\Lambda BV([a, b])$ and $MBV([a, b])$, equipped with the norms (2.30), may be interpreted as Schramm spaces $\Phi BV([a, b])$ and $\Psi BV([a, b])$ by choosing $\phi_n(t) = \lambda_n t$ and $\psi_n(t) = \mu_n t$. Consequently, the crucial condition (4.83) holds if and only if

$$\sum_{k=1}^{\infty} \frac{1}{k^2 \lambda_k \mu_k} < \infty. \tag{4.101}$$

Therefore, as a corollary of Theorem 4.40, we obtain the following

Theorem 4.41. *Let $\Lambda = (\lambda_n)_n$ and $M = (\mu_n)_n$ be two Waterman sequences satisfying* (4.101). *Then the Riemann–Stieltjes integral* (4.7) *exists for $f \in \Lambda BV^o([a, b]) \cap C([a, b])$ and $\alpha \in MBV([a, b])$. Moreover, the estimate*

$$\left| \int_a^b f(x)\, d\alpha(x) \right| \leq 2\|f\|_{\Lambda BV} \|\alpha\|_{\Lambda BV} \sum_{k=1}^{\infty} \frac{1}{k^2 \lambda_k \mu_k} \tag{4.102}$$

holds, where $\|\cdot\|_{\Lambda BV}$ denotes the norm (2.30).

We illustrate Theorem 4.41 by a simple example where condition (4.101) may be easily verified.

Example 4.42. Choosing $\lambda_n := n^{-p}$ and $\mu_n := n^{-q}$ for $0 < p, q \leq 1$, we get the Waterman spaces $\Lambda_p BV([a, b])$ and $\Lambda_q BV([a, b])$ which we introduced in Definition 2.29. For this choice of λ_n and μ_n, condition (4.101) reads

$$\sum_{k=1}^{\infty} \frac{1}{k^{2-p-q}} = \zeta(2 - p - q, 0) < \infty,$$

where we have used the shortcut (0.17), and this is true precisely for $p + q < 1$. So, in this case, we get the estimate

$$\left| \int_a^b f(x)\, d\alpha(x) \right| \le 2\zeta(2 - p - q, 0)\, \|f\|_{\Lambda_p BV} \|\alpha\|_{\Lambda BV_q}$$

for any pair of functions $f \in \Lambda_p BV([a,b])$ and $\alpha \in \Lambda_q BV([a,b])$. ♥

4.5 Comments on Chapter 4

Riemann–Stieltjes integrals are treated in many calculus textbooks, e.g. [76, 139, 146–148, 156, 181, 186, 238]. In Section 4.1, we followed the presentation in [269] and [182], and in the other section, as in Chapter 3, in Carothers' beautiful book [76].

The particularly important Theorem 4.11 shows that $\alpha \in BV([a,b])$ implies $C([a,b]) \subseteq RS_\alpha([a,b])$, while $\alpha \in BV([a,b]) \cap C([a,b])$ implies $BV([a,b]) \subseteq RS_\alpha([a,b])$. In particular, continuous functions and functions of bounded variation are Riemann integrable. When passing from increasing integrators to BV-integrators, we used for $\alpha \in BV([a,b])$ the decomposition $\alpha = \beta - \gamma$ and the fact that

$$RS_\beta([a,b]) \cap RS_\gamma([a,b]) \subseteq RS_{\beta-\gamma}([a,b]) = RS_\alpha([a,b]), \tag{4.103}$$

by Proposition 4.3 (b). We would of course like to have equality in (4.103), i.e.

$$RS_\beta([a,b]) \cap RS_\gamma([a,b]) = RS_\alpha([a,b]) \tag{4.104}$$

because this would truly reduce the study of BV-integrators to the case of increasing integrators. Unfortunately, (4.104) is generally not true for just any splitting $\alpha = \beta - \gamma$. For example, in case $\beta(x) = \gamma(x) = x$, we have $\alpha(x) \equiv 0$, and so the right-hand side of (4.103) is much larger than just $RS([a,b])$. However, one can show that (4.104) *is* true for the canonical Jordan decomposition from Theorem 1.5, see Exercise 4.41.

Theorems 4.14 and 4.15 are crucial for the existence of the Riemann–Stieltjes integral. Given a partition $P \in \mathcal{P}([a,b])$ and using the notation (4.2) and (4.3), we have the estimate

$$U_\alpha(f, P; [a,b]) - L_\alpha(f, P; [a,b]) \ge \operatorname{osc}(f; x) \operatorname{osc}(\alpha; x) \quad (x \notin P),$$

where $\operatorname{osc}(f; c)$ denotes the local oscillation (4.27). Consequently, in order for $f \in RS_\alpha([a,b])$, we must have $\operatorname{osc}(f; x) \operatorname{osc}(\alpha; x) = 0$ for "most" values of x, in accordance with Theorem 4.15.

We point out that the convergence relation (4.47) holds if we only suppose that $(\alpha_n)_n$ is a bounded sequence in $BV([a,b])$ which converges pointwise to α. This fact is known as *Helly's first theorem*. Example 4.22 is taken from Rudin's book [269].

Theorem 4.17 gives conditions under which a Riemann–Stieltjes integral reduces to a Riemann integral. What about Lebesgue integrals? We take this one step further and now consider the right integral in (4.34) as a Lebesgue integral.

Theorem 4.43. *Let $\alpha \in AC([a,b])$ and $f \in C([a,b])$. Then $f \in RS_\alpha([a,b])$ and*

$$\int_a^b f(t)\,d\alpha(t) = \int_a^b f(t)\dot{\alpha}(t)\,dt, \tag{4.105}$$

where $\dot{\alpha}$ denotes the derivative of α with respect to t a.e. on $[a,b]$, and the integral on the right-hand side of (4.105) *is the Lebesgue integral.*

Proof. Given a partition $P = \{t_0, t_1, \ldots, t_m\}$ and a set $\Pi = \{\tau_1, \tau_2, \ldots, \tau_m\}$ of intermediate points satisfying (4.16), consider the Riemann–Stieltjes sum (4.17). Since α is absolutely continuous, we may write

$$\alpha(t_j) - \alpha(t_{j-1}) = \int_{t_{j-1}}^{t_j} \dot{\alpha}(t)\,dt,$$

and hence

$$S_\alpha(f, P, \Pi; [a,b]) = \sum_{j=1}^m f(\tau_j)[\alpha(t_j) - \alpha(t_{j-1})] = \sum_{j=1}^m \int_{t_{j-1}}^{t_j} \dot{\alpha}(t)\,dt\,.$$

Consequently,

$$\begin{aligned}\left|S_\alpha(f, P.\Pi; [a,b]) - \int_a^b f(t)\dot{\alpha}(t)\,dt\right| &= \left|\sum_{j=1}^m \int_{t_{j-1}}^{t_j} \left[f(\tau_j) - f(t)\right] \dot{\alpha}(t)\,dt\right| \\ &\le \sum_{j=1}^m \operatorname{osc}(f; [t_{j-1}, t_j]) \int_{t_{j-1}}^{t_j} |\dot{\alpha}(t)|\,dt \\ &\le \|\dot{\alpha}\|_{L_1} \max\{\operatorname{osc}(f; [t_{j-1}, t_j]) : j = 1, 2, \ldots, m\}\,.\end{aligned}$$

Since f is continuous, by assumption, we have $\operatorname{osc}(f; [t_{j-1}, t_j]) \to 0$ as $\mu(P) \to 0$, and so

$$\lim_{\mu(P)\to 0} S_\alpha(f, P, \Pi; [a,b]) = \int_a^b f(t)\dot{\alpha}(t)\,dt$$

which proves the assertion. □

As an immediate consequence of Theorem 4.43, we get the following result which is parallel to Proposition 4.24:

Proposition 4.44 (integration by parts). *Let $f \in AC([a,b])$ and $\alpha \in AC([a,b])$. Then the equality*

$$\int_a^b f(x)\,d\alpha(x) = f(b)\alpha(b) - f(a)\alpha(a) - \int_a^b \alpha(x)\,df(x) \tag{4.106}$$

holds.

One of the most important properties of the Riemann–Stieltjes integral is that of providing the right tool for the duality theory exposed in Sections 4.2 and 4.3. Our proof of Theorem 4.31 is purely "functional-analytic," being based on the Hahn–Banach theorem, and has the advantage of carrying over with only minor technical changes to other function spaces. However, it is interesting to note that there exist other proofs which are purely "analytic" and essentially use special properties of the space $C([a,b])$. We give here a proof taken from the book [76] of the surjectivity of the map Φ which is the crucial part of Theorem 4.31. Without loss of generality, let $[a,b] = [0,1]$. The proof in [76] uses the *Bernstein polynomials* $p_{n,k} : [0,1] \to \mathbb{R}$ $(n = 1,2,3,\dots)$ defined by

$$p_{k,n}(x) := \binom{n}{k} x^k (1-x)^{n-k} \quad (k = 0,1,2,\dots,n)\,.$$

It is well known that for any function $f \in C([0,1])$, the sequence $(B_n(f))_n$ given by

$$B_n(f) := \sum_{k=0}^{n} f(k/n)p_{k,n}$$

converges *uniformly* on $[0,1]$ to f as $n \to \infty$. We use this fact to now give an alternative proof of the following part of Theorem 4.31:

Theorem 4.45. *The map*

$$\Phi : BV([a,b]) \to C([a,b])^*\,, \quad \Phi(\alpha) := \ell_\alpha\,,$$

with ℓ_α as in (4.61) is onto.

Proof. Let ℓ be an arbitrary bounded linear functional on $C([0,1])$. Then for any $f \in C([0,1])$, we have

$$\ell(B_n(f)) = \sum_{k=0}^{n} f(k/n)\ell(p_{k,n}) \to \ell(f) \quad (n \to \infty)$$

by what we have observed before. In particular, observe that the numbers $\ell(p_{k,n})$ $(n = 1,2,3,\dots; k = 0,1,2,\dots,n)$ do not depend on f.

We define a sequence $(\alpha_n)_n$ of functions $\alpha_n \in BV([0,1])$ by[14]

$$\alpha_n(x) := \begin{cases} 0 & \text{for } x = 0\,, \\ \ell(p_{n,1}) & \text{for } 0 < x < \frac{1}{n}\,, \\ \ell(p_{n,k}) & \text{for } \frac{k}{n} \le x < \frac{k+1}{n}\,, \\ \ell(p_{n,n}) & \text{for } x = 1\,. \end{cases}$$

14 The assertion $\alpha_n \in BV([0,1])$ (actually, $\alpha_n \in NBV([0,1])$) follows from the fact that α_n is a step function with a jump of size $\ell(p_{n,k})$ at the finitely many points k/n $(k = 0,1,\dots,n)$.

By construction, we then have

$$\int_0^1 f(x)\,d\alpha_n(x) = \sum_{k=0}^{n} f(k/n)\ell(p_{k,n})\,. \tag{4.107}$$

We use Helly's selection principle (Theorem 1.11) to show that the sequence $(\alpha_n)_n$ contains a subsequence which converges pointwise on $[0,1]$ to some $\alpha \in BV([0,1])$. To this end, we have to show that the sequence $(\alpha_n)_n$ is bounded in the norm (1.16).

Recall that the sequences (4.107) satisfy

$$\sum_{k=0}^{n} p_{k,n}(x) = \sum_{k=0}^{n} |p_{k,n}(x)| = 1 \quad (0 \le x \le 1)\,.$$

Consequently,

$$\begin{aligned}\mathrm{Var}(\alpha_n) &= \sum_{k=0}^{n} |\ell(p_{k,n})| - \left|\sum_{k=0}^{n} \pm\ell(p_{k,n})\right| = \left|\ell\left(\sum_{k=0}^{n} \pm p_{k,n}\right)\right| \\ &\le \|\ell\|_{C^*} \left\|\sum_{k=0}^{n} \pm p_{k,n}\right\|_C \le \|\ell\|_{C^*} \left\|\sum_{k=0}^{n} |p_{k,n}|\right\|_C = \|\ell\|_{C^*}\,.\end{aligned}$$

By Helly's theorem, we know, passing if necessary to a subsequence, that $\alpha_n(x) \to \alpha(x)$ for all $x \in [0,1]$, where $\alpha \in BV([0,1])$, and even $\mathrm{Var}(\alpha_n - \alpha; [0,1]) \to 0$ as $n \to \infty$. From Theorem 4.21 (c) we conclude that (4.47) holds true, and combining this with (4.107) yields

$$\int_0^1 f(x)\,d\alpha(x) = \lim_{n\to\infty} \int_0^1 f(x)\,d\alpha_n(x) = \lim_{n\to\infty} \sum_{k=0}^{n} f(k/n)\ell(p_{k,n}) = \ell(f)\,.$$

However, this means precisely that $\ell = \ell_\alpha = \Phi(\alpha)$, and so we are done. □

An integral of Riemann–Stieltjes type for integrands $f \in \kappa BV([a,b]) \cap C([a,b])$ and integrators $\alpha \in \kappa BV([a,b])$, including duality theory, has been introduced and discussed by Korenblum in [164].

Riemann–Stieltjes integrals have also been defined with respect to functions of higher order variation, e.g. by Russell [272, 275, 277]. We describe such a definition in the simplest possible situation.

Let $f : [a,b] \to \mathbb{R}$ be bounded, and let $\alpha \in WBV_{2,1}([a,b])$, which means that the second order variation of α in the sense of De la Vallée Poussin (Definition 2.72) is finite. Fix a partition $P = \{t_0, t_1, \ldots, t_m\} \in \mathcal{P}([a,b])$ and a set $\Pi = \{\tau_1, \tau_2, \ldots, \tau_m\}$ satisfying (4.16). In analogy to (4.17), we call the sum

$$S_\alpha^2(f, P, \Pi; [a,b]) := \sum_{j=1}^{m-1} f(\tau_j)(\alpha[t_j, t_{j+1}] - \alpha[t_{j-1}, t_j])\,,$$

where $\alpha[s,t]$ denotes the divided difference (2.139), a *second order Riemann–Stieltjes sum*. Russell [272] then defines[15] a second order RS-integral as follows.

Definition 4.46. Suppose that the limit

$$\lim_{\mu(P)\to 0} S^2_\alpha(f,P,\Pi;[a,b]) =: A$$

exists in the following sense: for any $\varepsilon > 0$, there exists a $\delta > 0$ such that

$$|S^2_\alpha(f,P,\Pi;[a,b]) - A| \le \varepsilon$$

for any partition P satisfying $\mu(P) \le \delta$ and any set Π of intermediate points satisfying (4.16). Then we write $f \in RS^2_\alpha([a,b])$ and call the number

$$A = \int_a^b f(x)\frac{d^2\alpha(x)}{dx}$$

the *second order Riemann–Stieltjes integral* of f with respect to α. ■

In the papers [272, 275, 277], one can find some natural properties of this integral as well as a connection with the usual RS-integral (4.7). Suppose, for example, that $f \in C([a,b])$ and $\beta \in BV([a,b]) \cap C([a,b])$, and let

$$\alpha(x) := \int_a^x \beta(t)\,dt \quad (a \le x \le b).$$

Then $\alpha \in WBV_{2,1}([a,b])$ and, as one should expect,

$$\int_a^b f(x)\frac{d^2\alpha(x)}{dx} = \int_a^b f(x)d\beta(x).$$

Riemann–Stieltjes integrals have been extended in many different directions. We confine ourselves to mention the paper [248] where the author defines generalized RS-integrals with respect to charges[16] and studies the linear space of all functions which are integrable with respect to every charge. Related questions are discussed in [150].

15 In fact, Russell's definition is much more general, involving another increasing function u and using terms like u-bounded variation, u-derivative, u-convexity etc.; our definition corresponds to the special case $u(x) = x$.

16 In the terminology of [248], a *charge* is an additive continuous function on bounded subsets of BV.

4.6 Exercises to Chapter 4

We state now some exercises on the topics covered in this chapter; exercises marked with an asterisk * are more difficult.

Exercise 4.1. Calculate the integral

$$\int_{-\pi}^{\pi} (x+2)\, d\alpha(x)\,,$$

where $\alpha(x) = e^x[\chi_{(0,1]} - \chi_{[-1,0)}]$.

Exercise 4.2. Calculate the integral

$$\int_{0}^{\pi} (x-1)\, d\alpha(x)\,,$$

where $\alpha(x) = \cos x\, sgn\, x$.

Exercise 4.3. Calculate the integral

$$\int_{0}^{1} x\, d\varphi(x)\,,$$

where φ denotes the Cantor function (3.6) on $[0, 1]$.

Exercise 4.4. Calculate the integral

$$\int_{0}^{1} x^3\, d\varphi(x)\,,$$

where φ denotes the Cantor function (3.6) on $[0, 1]$.

Exercise 4.5. Show that

$$\int_{0}^{1} \varphi(1-x)\, d\varphi(x) = \frac{1}{2}\,,$$

where φ denotes the Cantor function (3.6) on $[0, 1]$.

Exercise 4.6. Give an example of functions $f \in B([a, b])$ and $\alpha \in BV([a, b])$ such that $f^2 \in RS_\alpha([a, b])$ and $|f| \in RS_\alpha([a, b])$, but $f \notin RS_\alpha([a, b])$.

Exercise 4.7. Prove that $f \in RS_\alpha([a, b])$ implies $f^3 \in RS_\alpha([a, b])$. Is the converse also true?

Exercise 4.8. Find functions $f \in RS_\alpha([a, b])$ with $f([a, b]) \subseteq [c, d]$ and $g \in RS_\alpha([c, d])$ such that $g \circ f \notin RS_\alpha([a, b])$.

Exercise 4.9. Find functions $f \in C([a,b])$ with $f([a,b]) \subseteq [c,d]$ and $g \in RS_\alpha([c,d])$ such that $g \circ f \notin RS_\alpha([a,b])$.

Exercise 4.10. Let $f : [-1,1] \to \mathbb{R}$ be bounded. We define three functions $\alpha_i \in BV([-1,1])$ $(i = 1,2,3)$ by

$$\alpha_1(t) := \begin{cases} 0 & \text{for } -1 \le t \le 0\,, \\ 2 & \text{for } 0 < t \le 1\,, \end{cases} \qquad \alpha_2(t) := \begin{cases} 0 & \text{for } -1 \le t < 0\,, \\ 2 & \text{for } 0 \le t \le 1\,, \end{cases}$$

and

$$\alpha_3(t) := \alpha_1(t) + \alpha_2(t) = \begin{cases} 0 & \text{for } -1 \le t < 0\,, \\ 1 & \text{for } t = 0\,, \\ 2 & \text{for } 0 < t \le 1\,. \end{cases}$$

(a) Show that $f \in RS_{\alpha_1}([-1,1])$ if and only if f is right continuous at zero, i.e. $f(0+) = f(0)$; in this case, we have

$$\int_{-1}^{1} f(x)\,d\alpha_1(x) = f(0)\,.$$

(b) Formulate and prove an analogous result for α_2.
(c) Show that $f \in RS_{\alpha_3}([-1,1])$ if and only if f is continuous at zero.
(d) Deduce that if f is continuous at zero, the equality

$$\int_{-1}^{1} f(x)\,d\alpha_1(x) = \int_{-1}^{1} f(x)\,d\alpha_2(x) = \int_{-1}^{1} f(x)\,d\alpha_3(x) = f(0)$$

is true.

Exercise 4.11. Let $(x_n)_n$ be a sequence of pairwise different real numbers $x_n \in (0,1)$, and let $(c_n)_n$ be a positive real sequence such that

$$\sum_{n=1}^{\infty} c_n < \infty\,.$$

Define $\alpha : [-1,1] \to \mathbb{R}$ by

$$\alpha(x) := \sum_{n=1}^{\infty} c_n \alpha_1(x - x_n)\,,$$

where α_1 is defined as in the preceding Exercise 4.10. Show that $f \in RS_\alpha([-1,1])$ for any $f \in C([-1,1])$ and

$$\int_{-1}^{1} f(x)\,d\alpha(x) = \sum_{n=1}^{\infty} c_n f(x_n)\,.$$

Exercise 4.12. Show that

$$\int_0^3 x\,d\alpha(x) = \frac{3}{2},$$

where $\alpha(x) := x - \text{ent}\,x$.

Exercise 4.13. Construct functions $f \in C([-1,1])$ and $\alpha \in BV([-1,1])$ such that $f\chi_{[0,1]} \notin RS_\alpha([-1,1])$.

Exercise 4.14. Let $c \in (a,b)$ and $\alpha := \chi_{[c,b]}$, and let $f : [a,b] \to \mathbb{R}$ be some function. Prove that $f \in RS_\alpha([a,b])$ if and only if f is continuous at c. What is the value of the RS-integral

$$I = \int_a^b f(x)\,d\alpha(x)$$

in this case?

Exercise 4.15. For $\sigma > 0$ and $\tau \in \mathbb{R}$, let $f_\sigma : [0,1] \to \mathbb{R}$ and $\alpha_\tau : [0,1] \to \mathbb{R}$ be defined by

$$f_\sigma(x) := \begin{cases} x^\sigma & \text{for } 0 < x \le 1, \\ 0 & \text{for } x = 0, \end{cases} \qquad \alpha_\tau(x) := \begin{cases} x^\tau & \text{for } 0 < x \le 1, \\ 0 & \text{for } x = 0. \end{cases}$$

Show that $f_\sigma \in RS_{\alpha_\tau}([0,1])$ if and only if $\tau \ge 0$, and calculate the RS-integral of f_σ with respect to α_τ in this case.

Exercise 4.16. For $\sigma \in \mathbb{R}$, let $f_\sigma : [0,1] \to \mathbb{R}$ and $\alpha : [0,1] \to \mathbb{R}$ be defined by

$$f_\sigma(x) := \begin{cases} x^\sigma & \text{for } 0 < x \le 1, \\ 0 & \text{for } x = 0, \end{cases} \qquad \alpha(x) := \begin{cases} \cos\frac{1}{x} & \text{for } 0 < x \le 1, \\ 0 & \text{for } x = 0. \end{cases}$$

Show that $f_\sigma \in RS_\alpha([0,1])$ if and only if $\sigma > 0$, and calculate the RS-integral of f_σ with respect to α in this case.

Exercise 4.17. Let $f = \chi_{[0,1]\cap\mathbb{Q}}$ be the Dirichlet function, and let $\alpha \in BV([0,1])$. Show that $f \in RS_\alpha([0,1])$ if and only if α is constant.

Exercise 4.18. Let $\alpha \in BV([a,b])$, and suppose that $f \in B([a,b])$ belongs to both $RS_\alpha([a,c])$ and $RS_\alpha([c,b])$ for some $c \in (a,b)$. Assume that α is continuous at c. Prove that then $f \in RS_\alpha([a,b])$ with (4.23).

Exercise 4.19. Let $f, g \in C([a,b])$ and $\alpha \in BV([a,b])$, and define $\beta : [a,b] \to \mathbb{R}$ by

$$\beta(x) := \int_a^x f(t)\,d\alpha(t) \quad (a \le x \le b).$$

Show that

$$\int_a^b g(t)\,d\beta(t) = \int_a^b f(t)g(t)\,d\alpha(t)$$

and illustrate this by means of a nontrivial example.

Exercise 4.20. Using the notation of Definition 4.4 and Exercise 4.10, show that $\alpha_2 \in RS_{\alpha_1}([-1,1])$, although the limit

$$\lim_{\mu(P)\to 0} S_{\alpha_1}(\alpha_2, P; [-1,1])$$

does not exist. Why does this not contradict Theorem 4.11 (b)?

Exercise 4.21. Give an example of functions $\alpha \in BV([a,b])$ and $f \in C([a,b]) \setminus BV([a,b])$ such that one may define the RS-integral

$$\int_a^b \alpha(x)\, df(x)$$

building on the equality (4.55) in Proposition 4.24.

Exercise 4.22. Let $\alpha \in BV([a,b]) \cap C([a,b])$, and let $f : [a,b] \to \mathbb{R}$ be monotonically increasing. Prove that there exists $\xi \in [a,b]$ such that

$$\int_a^b f(x)\, d\alpha(x) = f(a)(\alpha(\xi) - \alpha(a)) + f(b)(\alpha(b) - \alpha(\xi))\,.$$

Exercise 4.23. Given $f \in C([a,b])$, prove that there exists a point $\xi \in [a,b]$ such that

$$\int_a^b f(x)\, dx = f(\xi)(b-a)\,.$$

Exercise 4.24. Let $f, g \in C([a,b])$ with $g(x) \geq 0$ on $[a,b]$. Prove that there exists a point $\xi \in [a,b]$ such that

$$\int_a^b f(x)g(x)\, dx = f(\xi) \int_a^b g(x)\, dx\,.$$

Show that this result contains that of Exercise 4.23 as a special case, and that one cannot drop the assumption $g(x) \geq 0$.

Exercise 4.25. Let $f \in C^1([a,b])$ and $g \in C([a,b])$. Prove that there exists a point $c \in (a,b)$ such that

$$\int_a^b f(x)g(x)\, dx = f(a) \int_a^c g(x)\, dx + f(b) \int_c^b g(x)\, dx\,.$$

Exercise 4.26. Let $x_0 \in [a,b]$ be fixed. Construct three functions $f : [a,b] \to \mathbb{R}$, $\alpha : [a,b] \to \mathbb{R}$, and $\beta : [a,b] \to \mathbb{R}$ such that $\Delta(\alpha,\beta) = \{x_0\}$ (see (4.66)) and $f \in RS_\alpha([a,b]) \setminus RS_\beta([a,b])$.

Exercise 4.27. Given a sequence $(x_n)_n$ of points $x_n \in [a, b]$, define $\gamma : [a, b] \to \mathbb{R}$ by

$$\gamma(t) := \sum_{n=1}^{\infty} \left(a_n \chi_{[x_n, b]}(t) + b_n \chi_{(x_n, b]}(t) \right),$$

where $(a_n)_n$ and $(b_n)_n$ are two real sequence satisfying

$$\sum_{n=1}^{\infty} |a_n| < \infty, \quad \sum_{n=1}^{\infty} |b_n| < \infty.$$

(a) Show that $\gamma \in BV([a, b])$.
(b) If $f : [a, b] \to \mathbb{R}$ is a bounded function which is continuous at each point x_n, prove that $f \in RS_\gamma([a, b])$ and

$$\int_a^b f(x)\, d\gamma(x) = \sum_{n=1}^{\infty} (a_n + b_n) f(x_n).$$

Exercise 4.28. Let $\alpha_n : [0, 1] \to \mathbb{R}$ be defined by

$$\alpha_n(x) := \frac{\operatorname{ent}(n^x)}{n} \quad (n = 1, 2, 3, \ldots),$$

where ent(ξ) denotes the integer part of ξ. Show that $\alpha_n \in BV([0, 1])$ and that the sequence $(\alpha_n)_n$ converges in the norm (1.12) to some $\alpha \in BV([0, 1])$. Given $f \in C([0, 1])$, use Theorem 4.21 (c) to show that

$$\lim_{n \to \infty} \frac{1}{n} \sum_{k=2}^{n} f(\log k / \log n) = f(1).$$

Illustrate this result for the example $f(x) = x^2$.

Exercise 4.29. Let $\alpha \in BV([0, 2\pi])$ with $\alpha(0) = \alpha(2\pi)$. Show that then the estimate

$$\left| \int_0^{2\pi} \alpha(x) \sin nx\, dx \right| \leq \frac{1}{n} \operatorname{Var}(\alpha; [0, 2\pi])$$

holds for all $n \in \mathbb{N}$. Compare with Proposition 4.23 (b).

Exercise 4.30. Show that there exists a function $\hat{\alpha} \in BV([0, 2\pi])$ such that $\hat{\alpha}(0) = \hat{\alpha}(2\pi)$ and

$$\left| \int_0^{2\pi} \hat{\alpha}(x) \sin nx\, dx \right| = \frac{1}{n} \operatorname{Var}(\hat{\alpha}; [0, 2\pi])$$

for infinitely many $n \in \mathbb{N}$. Compare this with Proposition 4.23 (b) and Exercise 4.29.

Exercise 4.31. Let $f, \alpha : [a, b] \to \mathbb{R}$ be continuous and let α be in addition strictly increasing. Define $F : [a, b] \to \mathbb{R}$ by

$$F(x) := \int_a^x f(t)\, d\alpha(t) \quad (a \leq x \leq b).$$

The so-called *α-derivative* of F is then defined by

$$\frac{dF}{d\alpha}(x) := \lim_{h\to 0} \frac{F(x+h) - F(x)}{\alpha(x+h) - \alpha(x)}.$$

Prove that

$$\frac{dF}{d\alpha}(x) = f(x)$$

for all $x \in [a,b]$.

Exercise 4.32. Let $f \in BV([a,b]) \cap C([a,b])$. Prove that

$$\int_a^b f(x)\,df(x) = \frac{f^2(b) - f^2(a)}{2}.$$

Use this result to solve again Exercise 4.5.

Exercise 4.33. Let $f \in C([1,\infty))$, and let $\alpha(t) := \mathrm{ent}(t)$ denote the integer part of $t \geq 1$. Compute the RS-integral

$$F(x) := \int_1^x f(t)d\alpha(t) \quad (x \geq 1)$$

for $x \in \mathbb{N}$ and $x \notin \mathbb{N}$.

Exercise 4.34. Suppose that $\alpha \in BV([a,b])$ has the property that the space $RS_\alpha([a,b])$ contains all step functions. Prove that $\alpha \in C([a,b])$.

Exercise 4.35. Show that

$$\bigcap_{\alpha\nearrow} RS_\alpha([a,b]) = C([a,b]),$$

where the intersection is taken over all *increasing* functions $\alpha : [a,b] \to \mathbb{R}$.

Exercise 4.36. If $\alpha \in BV([a,b]) \cap C([a,b])$, show that the value of the RS-integral (4.7) does not depend on the values of f at any *finite* number of points. Is this still true if we change "finite" to "countable?" Explain.

Exercise 4.37. Let $\alpha \in BV([a,b])$, $f \in B([a,b])$, $P \in \mathcal{P}([a,b])$, and Π a set of intermediate points. Prove that

$$|S_\alpha(f, P, \Pi; [a,b])| \leq \|f\|_\infty \operatorname{Var}(\alpha, P; [a,b]),$$

where $S_\alpha(f, P, \Pi; [a,b])$ denotes the RS-sum (4.17), $\|f\|_\infty$ the norm (0.39) of f, and $\operatorname{Var}(\alpha, P; [a,b])$ the variation (1.3) of α.

Exercise 4.38. Let $f \in C^1([1,\infty))$, and let α be defined as in Exercise 4.33. Using the integration by parts formula (4.55), prove that

$$\sum_{k=1}^n f(k) = \alpha(n)f(n) - \int_1^n f'(t)\alpha(t)\,dt$$

and

$$\sum_{k=1}^{2n}(-1)^k f(k) = \int_1^{2n} f'(t)(\alpha(t) - 2\alpha(t/2)\,dt$$

for any $n \in \mathbb{N}$.

Exercise 4.39*. Suppose that $\alpha : [a,b] \to \mathbb{R}$ is right-continuous and increasing. Given $\varepsilon > 0$ and a subinterval $[c,d] \subset [a,b]$, construct a continuous function $f : [a,b] \to [0,1]$ such that

$$\int_a^b f\,d\alpha \geq \alpha(d) - \alpha(c) - \varepsilon\,.$$

Can you also take $\varepsilon = 0$?

Exercise 4.40*. Suppose that $\alpha \in BV([a,b])$ is right-continuous. Given $\varepsilon > 0$ and a partition $P \in \mathcal{P}([a,b])$, construct a continuous function $f : [a,b] \to [0,1]$ such that

$$\int_a^b f\,d\alpha \geq \mathrm{Var}(\alpha, P; [a,b]) - \varepsilon\,.$$

Use this to again prove the equality (4.72).

Exercise 4.41. Let $\alpha \in BV([a,b])$, $\beta(x) := V_\alpha(x) = \mathrm{Var}(\alpha; [a,x])$, and $\gamma(x) := \beta(x) - \alpha(x)$. Show that $RS_\alpha([a,b]) = RS_\beta([a,b]) \cap RS_\gamma([a,b])$.

Exercise 4.42. Find functions $\alpha \in BV([a,b])$ and $f \in RS_\alpha([a,b])$ such that

$$\left|\int_a^b f(x)\,d\alpha(x)\right| > \int_a^b |f(x)|\,d\alpha(x)\,.$$

Exercise 4.43. Let $\tau : [a,b] \to [c,d]$ be strictly increasing, continuous, and surjective (and hence a homeomorphism). Given $g \in RS_\alpha([c,d])$, show that $f := g \circ \tau \in RS_\beta([a,b])$, where $\beta := \alpha \circ \tau$. Moreover, prove that

$$\int_a^b f\,d\beta = \int_c^d g\,d\alpha$$

in this case.

Exercise 4.44. Let p_α and n_α denote the positive respectively negative variation (Theorem 1.6) of $\alpha \in BV([a,b])$, so

$$\alpha(x) = p_\alpha(x) - n_\alpha(x) + \alpha(a)\,, \quad V_\alpha(x) = p_\alpha(x) + n_\alpha(x)\,.$$

Show that

$$\left|\int_a^b f\,d\alpha\right| \leq \int_a^b |f|\,dp_\alpha + \int_a^b |f|\,dn_\alpha\,.$$

Use this fact to again prove the estimate (4.44).

Exercise 4.45. Calculate the regularization $f^\#$ for the function f from Example 1.4.

Exercise 4.46. Given $\alpha \in BV([a,b])$ and $f \in C([a,b])$, show that

$$\left| \int_a^b \alpha \, df \right| \leq \|\alpha\|_{BV} \|f - f(b)\|_\infty ,$$

where $\|\cdot\|_\infty$ denotes the norm (0.39).

Exercise 4.47. Find an example of a function $\alpha \in BV^o([0,1])$ for which the inequality (4.64) is strict.

Exercise 4.48. Discuss Example 0.24 from the viewpoint of Theorem 4.31.

Exercise 4.49. For $1 < p < \infty$ and $p' = p/(p-1)$, prove directly (i.e. without making use of the abstract Theorem 0.23) that

$$\|f\|_{L_p} = \sup \left\{ \int_a^b f(x) \, d\alpha(x) : \alpha \in RBV^o_{p'}([a,b]), \mathrm{Var}^R_{p'}(\alpha; [a,b]) \leq 1 \right\}$$

for all $f \in L_p([a,b])$, and compare with Proposition 4.32.

Exercise 4.50. For $0 < \tau < 1$, let $\alpha_\tau : [0,1] \to \mathbb{R}$ be defined by $\alpha_\tau(x) = x^\tau$ as in Example 2.78 and Example 3.35. Illustrate the abstract Theorems 4.33 and 4.35 for $\alpha = \alpha_\tau$.

5 Nonlinear composition operators

Given a function $h : \mathbb{R} \to \mathbb{R}$, in this chapter, we study the nonlinear composition operator C_h defined by $C_h f := h \circ f$ for f belonging to several spaces of functions of (generalized) bounded variation. It turns out that the typical condition, both necessary and sufficient, for C_h to map such a space into itself is a local Lipschitz condition on h. Afterwards, we briefly consider sufficient conditions on h under which the corresponding operator C_h is bounded and/or continuous in norm; such conditions are often close to be also necessary. In view of applications, Lipschitz conditions for the operator C_h are of particular interest; in the last two sections, we will show that a global Lipschitz condition for C_h often leads to a strong degeneracy for h, while a local Lipschitz condition for C_h is fulfilled for sufficiently large classes of nonlinear functions h. Apart from our main object of study, the space $BV([a, b])$, we also consider other function spaces which frequently arise in applications, such as spaces of Lipschitz, Hölder, or absolutely continuous functions.

5.1 The composition operator problem

Let X be some space of functions $f : [a, b] \to \mathbb{R}$, and let $h : \mathbb{R} \to \mathbb{R}$ be a fixed function. Under appropriate hypotheses on h, we may then define a nonlinear operator C_h on X by putting[1]

$$C_h f(x) := h(f(x)) \quad (a \leq x \leq b) . \tag{5.1}$$

This operator is usually called the (autonomous) *composition operator* generated by h. More generally, if $h : [a, b] \times \mathbb{R} \to \mathbb{R}$ is a fixed function of two variables, one may also consider the (nonautonomous) *superposition operator*

$$S_h f(x) := h(x, f(x)) \quad (a \leq x \leq b) . \tag{5.2}$$

It turns out that the behavior of the superposition operator (5.2) is far more complicated than that of the composition operator (5.1). We postpone the study of the nonautonomous operator (5.2) to the next Chapter 6; in this chapter, we confine ourselves to the autonomous operator (5.1).

The most important problem related to the operator (5.1) reads as follows:

– *Given a function class X, find conditions on the function h, possibly both necessary and sufficient, under which the operator C_h generated by h maps the class X into itself.*

---

1 Sometimes, this operator is denoted by H, especially in the Russian literature, where it is often called *Nemytskij operator*. Recall that H is the Russian letter for the Roman letter N; this fact allowed Hercule Poirot in Agatha Christie's "Murder on the Orient Express" to discover the role of Princess Natalia Dragomiroff in the story.

This problem is sometimes referred to as the *composition operator problem* (or COP, for short) in the literature. Here, we will use the notation

$$COP(X) := \{h : C_h(X) \subseteq X\} = \{h : h \circ f \in X \text{ for all } f \in X\}. \tag{5.3}$$

The problem of determining the set $COP(X)$ for given X is sometimes almost trivial, and sometimes highly nontrivial. For example, it is easy to see that $COP(C) = C$, which means that the operator (5.1) maps the space $C([a, b])$ into itself if and only if[2] the corresponding function h is continuous on $\mathbb{R}$. This means, in particular, that the outer function h may (and has to) be taken from the same class C as the inner function f to end up again in the class C. As we shall see later, for many function spaces X, the class $COP(X)$ is essentially different from X itself.

In what follows, we will be interested in the problem of describing $COP(X)$ for the various function classes X which we introduced and studied in previous chapters. To begin with, let us make a trivial, though useful remark. Whenever the identity $f(x) = x$ belongs to the class X (and this is true for most function spaces), we automatically have $COP(X) \subseteq X$. The difficult part is of course to see whether or not this inclusion is strict, or we have equality [3] $COP(X) = X$.

Before starting our discussion, we make a general remark. Denoting for the moment a space of functions $f : [a, b] \to \mathbb{R}$ by $X([a, b])$, to emphasize the dependence on the domain of definition $[a, b]$, we introduce a special notion.

Definition 5.1. We call a function space X *COP-invariant* if, whenever the operator (5.1) maps the space $X([a, b])$ into itself and is bounded in the norm of $X([a, b])$ for some interval $[a, b]$, it also maps the space $X([c, d])$ into itself and is bounded in the norm of $X([c, d])$ for any other interval $[c, d]$. ■

The COP-invariance of a space X justifies dropping the interval in the notation $COP(X)$. The usual way to prove COP-invariance is by considering the strictly increasing affine bijection

$$\ell(t) := \frac{b-a}{d-c}(t-c) + a \quad (c \le t \le d) \tag{5.4}$$

between $[c, d]$ and $[a, b]$ with inverse

$$\ell^{-1}(s) = \frac{d-c}{b-a}(s-a) + c \quad (a \le s \le b), \tag{5.5}$$

---

2 The inclusion $C \subseteq COP(C)$ is an easy exercise in every first-year calculus course, while the inclusion $COP(C) \subseteq C$ uses the Tietze–Urysohn extension lemma, see Theorem 0.33.

3 To be precise, such an equality is not quite correct since X consists of functions defined on $[a, b]$, while $COP(X)$ consists of functions defined on the real line. The notation $COP(X) \subseteq X$ actually means $COP(X([a, b])) \subseteq X_{\mathrm{loc}}((-\infty, \infty))$, i.e. h belongs to the class X over the real line if and only if $h|_{[a,b]} \in X$. This is the reason why we have to *localize* the function h, as the formulation of Theorem 5.9 and similar theorems shows.

which we already considered in (0.80) and (0.81), and to show that it respects the structure of X. This leads to the following simple lemma which relates the COP-invariance of a function space X with its shift-invariance introduced in Definition 0.45.

Lemma 5.2. *Every shift-invariant function space X is COP-invariant.*

Proof. The proof is almost evident. The shift invariance of X means that, for ℓ given by (5.4), the operator $Lf := f \circ \ell$ is a linear isomorphism between $X([a,b])$ and $X([c,d])$ with inverse $L^{-1}g = g \circ \ell^{-1}$. So, if $C_h : X([a,b]) \to X([a,b])$ is bounded, then $L \circ C_h \circ L^{-1} : X([c,d]) \to X([c,d])$ is bounded as well, and vice versa. This shows that X is COP-invariant. □

We illustrate Definition 5.1 by means of four important examples.

Example 5.3. We show that both the Wiener space $X = WBV_p$ and the Riesz space $X = RBV_p$ are COP-invariant for $p \geq 1$. To this end, by Lemma 5.2, it suffices to show that X is shift-invariant.

The function $\ell : [c,d] \to [a,b]$ defined by (5.4) is an affine homeomorphism with inverse (5.5) which satisfies $\ell(c) = a$ and $\ell(d) = b$. Thus, $\ell : \mathcal{P}([c,d]) \to \mathcal{P}([a,b])$ with

$$\ell(P) = \ell(\{t_0, t_1, \ldots, t_m\}) = \{\ell(t_0), \ell(t_1), \ldots, \ell(t_m)\}$$

defines a 1-1 correspondence between all partitions of $[c,d]$ and all partitions of $[a,b]$ since ℓ is strictly increasing. Consequently, for $f \in WBV_p([a,b])$, we obtain

$$\mathrm{Var}_p^W(f, \ell(P); [a,b]) = \sum_{j=1}^{m} |f(\ell(t_j)) - f(\ell(t_{j-1}))|^p = \mathrm{Var}_p^W(f \circ \ell, P; [c,d]) .$$

Passing to the supremum with respect to $P \in \mathcal{P}([c,d])$ and $\ell(P) \in \mathcal{P}([a,b])$, we conclude that

$$\mathrm{Var}_p^W(f; [a,b]) = \mathrm{Var}_p^W(f \circ \ell; [c,d]) , \tag{5.6}$$

and so we have proved the shift-invariance of the space $X = WBV_p$.

For the proof of the shift-invariance of the space $X = RBV_p$, the equality (5.6) has to be replaced with

$$\mathrm{Var}_p^R(f; [a,b]) = \left(\frac{d-c}{b-a}\right)^{p-1} \mathrm{Var}_p^R(f \circ \ell; [c,d]) . \tag{5.7}$$

The remaining part of the proof is the same. ♥

Observe that our reasoning in Example 5.3 gives even a more precise result than just COP-invariance: from (5.6) and $\ell(c) = a$, it follows that the map $f \mapsto f \circ \ell$ is an *isometry* between $WBV_p([a,b])$ and $WBV_p([c,d])$ for $p \geq 1$ (in particular, between $BV([a,b])$ and $BV([c,d])$). However, (5.7) shows that in case $p > 1$, the map $f \mapsto f \circ \ell$ is not an isometry, but only an isomorphism between $RBV_p([a,b])$ and $RBV_p([c,d])$.

Example 5.4. Let us now show that the space $X = AC$ of absolutely continuous functions is shift-invariant, and hence COP-invariant as well. Similarly, as before, we use the fact that the map ℓ from (5.4) defines, for all $T \in \Sigma([c,d])$, through the equality

$$\begin{aligned}\ell(T) &= \ell(\{[c_1,d_1],[c_2,d_2],\dots,[c_n,d_n]\})\\ &= \{[\ell(c_1),\ell(d_1)],[\ell(c_2),\ell(d_2)],\dots,[\ell(c_n),\ell(d_n)]\}\end{aligned}$$

a 1-1 correspondence between $\Sigma([c,d])$ and $\Sigma([a,b])$. So, for $f \in AC([a,b])$, we have

$$\Gamma(f;\ell(T)) = \sum_{k=1}^{n} |f(\ell(d_k)) - f(\ell(c_k))| = \Gamma(f \circ \ell; T)\,, \tag{5.8}$$

where we use the notation (3.1). Given $\varepsilon > 0$, choose $\eta > 0$ such that $\Gamma(f \circ \ell; T) \le \varepsilon$ for all collections $T = \{[c_1,d_1],[c_2,d_2],\dots,[c_n,d_n]\} \in \Sigma([c,d])$ satisfying $\Theta(T) \le \eta$. Then putting $\delta := \frac{b-a}{d-c}\eta$ and $S := \ell(T)$, we see that

$$\Theta(S) = \Theta(\ell(T)) = \sum_{k=1}^{n} |\ell(d_k) - \ell(c_k)| = \frac{b-a}{d-c}\Theta(T)\,.$$

Consequently, $\Theta(S) \le \delta$ implies $\Theta(T) \le \eta$, and hence $\Gamma(f;S) \le \varepsilon$, by (5.8). This shows that $f \circ \ell \in AC([c,d])$, and so the space AC is shift-invariant. ♥

Example 5.4 admits a natural generalization which is given in Exercise 5.1. The following example generalizes Example 5.3.

Example 5.5. Let ϕ be a Young function which satisfies the condition ∞_1, see (2.16). We claim that the space $X = RBV_\phi$ is COP-invariant. We could do this by imitating the reasoning of Example 5.3, but this would require to replace (5.7) by a suitable equality (or estimate) for $\mathrm{Var}_\phi^R(f;[a,b])$ and $\mathrm{Var}_\phi^R(f \circ \ell;[c,d])$. Instead, we may prove the COP-invariance of $X = RBV_\phi$ directly by using Medvedev's theorem.

Given $h : \mathbb{R} \to \mathbb{R}$, suppose that the operator (5.1) maps the space $RBV_\phi([a,b])$ into itself. We use the fact that the function $\ell : [c,d] \to [a,b]$ defined by (5.4) is an affine diffeomorphism with constant derivative $\ell'(t) \equiv \frac{b-a}{d-c} > 0$ and inverse (5.5).

Fix $g \in RBV_\phi([c,d])$. By Medvedev's theorem (Theorem 3.36), we know that g is absolutely continuous on $[c,d]$ and satisfies

$$\int_c^d \phi(\beta|g'(t)|)\,dt < \infty$$

for some $\beta > 0$. By what we have seen in Example 5.4, the function $f := g \circ \ell^{-1}$ then belongs to $AC([a,b])$. Moreover, putting $\alpha := \frac{b-a}{d-c}\beta$ and using the substitution $s = \ell(t)$, we obtain

$$\int_a^b \phi(\alpha|f'(s)|)\,ds = \frac{b-a}{d-c}\int_c^d \phi(\beta|g'(t)|)\,dt < \infty\,.$$

Again, from Medvedev's theorem, it follows that $f \in RBV_\phi([a,b])$, and so $C_h f = h \circ f \in RBV_\phi([a,b])$, by assumption. So, $C_h g = h \circ f \circ \ell \in RBV_\phi([c,d])$. ♥

Example 5.6. Finally, we show that the space $X = \Lambda BV$ of functions of bounded Waterman variation (Definition 2.15) is shift-invariant, and hence COP-invariant.

In the same way as in Example 5.4, the affine homeomorphism ℓ maps $\Sigma_\infty([c,d])$ into $\Sigma_\infty([a,b])$ by means of the formula

$$\begin{aligned}\ell(T_\infty) &= \ell(\{[c_1,d_1],[c_2,d_2],[c_3,d_3],\dots\})\\ &= \{[\ell(c_1),\ell(d_1)],[\ell(c_2),\ell(d_2)],[\ell(c_3),\ell(d_3)],\dots\}.\end{aligned}$$

So, for $f \in \Lambda BV([a,b])$ and $T_\infty \in \Sigma_\infty([c,d])$, we have

$$\mathrm{Var}_\Lambda(f,\ell(T_\infty);[a,b]) = \sum_{k=1}^{\infty} \lambda_k |f(\ell(d_k)) - f(\ell(c_k))| = \mathrm{Var}_\Lambda(f\circ\ell, T_\infty;[c,d]).$$

Passing to the supremum with respect to $T_\infty \in \Sigma_\infty([c,d])$ and $\ell(T_\infty) \in \Sigma_\infty([a,b])$, we conclude that

$$\mathrm{Var}_\Lambda(f;[a,b]) = \mathrm{Var}_\Lambda(f\circ\ell;[c,d]),$$

and so we have proved the shift-invariance of the space $X = WBV_p$. ♥

Again, our proof shows that the map $f \mapsto f\circ\ell$ is an *isometry* between $\Lambda BV([a,b])$ and $\Lambda BV([c,d])$.

Before discussing the COP for the operator (5.1), in some spaces, we state a preliminary result which is of independent interest [251].

Proposition 5.7. *Suppose that the operator* (5.1) *maps the Waterman space* $\Lambda BV([a,b])$ *into itself. Then the generating function* $h:\mathbb{R}\to\mathbb{R}$ *is continuous.*

Proof. Let $\Lambda = (\lambda_n)_n$ be an arbitrary Waterman sequence, see Definition 2.15. Without loss of generality, we may assume that $[a,b] = [0,1]$, $h(0) = 0$, and h is discontinuous at 0. Choose a positive decreasing sequence $(u_n)_n$ which converges to 0 and satisfies

$$h(u_n) \ge 1, \quad \sum_{n=1}^{\infty} u_n < 1.$$

Let

$$a_n := \frac{2n+1}{2n(n+1)}, \quad b_n := \frac{1}{n},$$

and consider the collection $S_\infty := \{[a_n,b_n] : n \in \mathbb{N}\} \in \Sigma_\infty([0,1])$. We define peak functions $f:[0,1]\to\mathbb{R}$ by

$$f(t) := \begin{cases} 0 & \text{for } t = 0 \text{ or } t = b_n,\ n = 1,2,3,\dots,\\ u_n & \text{for } t = a_n,\ n = 1,2,3,\dots,\\ \text{linear} & \text{otherwise}.\end{cases}$$

Note that then $f \in BV([0,1])$, and hence $f \in \Lambda BV([0,1])$, by (2.33). On the other hand,

$$|(h\circ f)(b_n) - (h\circ f)(a_n)| = (h\circ f)(a_n) = h(u_n) \ge 1, \tag{5.9}$$

and therefore

$$\mathrm{Var}_\Lambda(f, S_\infty; [0,1]) = \sum_{n=1}^{\infty} \lambda_n |h(f(b_n)) - h(f(a_n))| = \sum_{n=1}^{\infty} \lambda_n h(u_n) \geq \sum_{n=1}^{\infty} \lambda_n = \infty,$$

by (5.9). This shows that $C_h f = h \circ f$ does not belong to $\Lambda BV([0,1])$, and so the proof is complete. □

We point out that Proposition 5.7 also holds for other spaces of functions of (generalized) bounded variation, see Exercise 5.8.

Now, we discuss the COP for the operator (5.1) in the function spaces BV, WBV_p, RBV_p, AC, Lip, and Lip_α. If X denotes one of these spaces, we certainly have $COP(X) \subseteq X_{\mathrm{loc}}(\mathbb{R})$ since the identity $f(x) = x$ belongs to X. It is a remarkable fact that except for $X = Lip$, this inclusion is *strict* for all the other spaces X:

Example 5.8. Let $h : \mathbb{R} \to \mathbb{R}$ be the "seagull function" defined by

$$h(u) := \min\left\{\sqrt{|u|}, 1\right\} = \begin{cases} \sqrt{|u|} & \text{for } |u| < 1, \\ 1 & \text{for } |u| \geq 1; \end{cases} \tag{5.10}$$

we will consider this function over and over for constructing counterexamples in what follows. Clearly, it suffices to consider h on the interval $[0,1]$. Of course, $h \in Lip_{1/2}([0,1])$, and so $h \in WBV_2([0,1])$, by (1.68). Moreover, our calculations in Example 3.35 show that $h \in RBV_p([0,1])$ for $1 \leq p < 2$ and

$$\mathrm{Var}_p^R(h; [0,1]) = \frac{1}{2^{p-1}(2-p)} \qquad (1 \leq p < 2).$$

So, from (2.93), we know that h is also absolutely continuous.[4] Of course, h is *not* locally Lipschitz continuous on $\mathbb{R}$.

We show now that the composition operator C_h generated by (5.10) does not map any of the spaces $X \in \{BV, WBV_p, RBV_p, AC, Lip, Lip_\alpha\}$ into itself. The function $f : [0,1] \to \mathbb{R}$ defined by

$$f(x) := \begin{cases} x^2 \sin^2 \frac{1}{x} & \text{for } 0 < x \leq 1, \\ 0 & \text{for } x = 0 \end{cases} \tag{5.11}$$

belongs to $Lip([0,1])$ since f is differentiable with bounded derivative. On the other hand, the composed function $C_h f = h \circ f$ does not even belong to $BV([0,1])$ since it is basically the function from Example 1.8. So, this example shows simultaneously that $C_h(Lip) \not\subseteq Lip$, $C_h(RBV_p) \not\subseteq RBV_p$, $C_h(AC) \not\subseteq AC$, and $C_h(BV) \not\subseteq BV$.

4 The absolute continuity of h may also be deduced from Theorem 3.9 since h has the Luzin property and is continuous and monotone.

The simple function $f(x) := x^\alpha$ which belongs to $Lip_\alpha([0,1])$ is mapped by C_h into the function $C_h f(x) = x^{\alpha/2}$ which does not belong to $Lip_\alpha([0,1])$. To show that $C_h(WBV_p) \nsubseteq WBV_p$ for general p, we use the special zigzag function Z_θ defined in (0.93). According to Table 2.4, we have $Z_\theta \in WBV_p([0,1])$ if and only if $p\theta > 1$. So, if we choose $1 < p\theta \le 2$, then $Z_\theta \in WBV_p([0,1])$, but $C_h Z_\theta = h \circ Z_\theta \notin WBV_p([0,1])$.

Note that all of these examples also show that the requirement $h \in Lip_\alpha(\mathbb{R})$ is *not sufficient* for ensuring that C_h maps any of the spaces BV, WBV_p, RBV_p, AC, Lip, or Lip_α into itself. ♥

Interestingly, Example 5.8 also shows that the monotonicity requirement on f is important in Proposition 2.79. In fact, the function f in (5.11) belongs to RBV_p for any $p \ge 1$, and the function h in (5.10) belongs to RBV_q for $q < 2$. If Proposition 2.79 would apply in this case, we could choose in (2.154) any $r \ge 1$ satisfying

$$1 - \frac{1}{r} < 1 \cdot \left(1 - \frac{1}{2}\right) = \frac{1}{2},$$

i.e. $r < 2$ to guarantee that $h \circ f \in RBV_r$. However, we have seen above that $h \circ f$ does not even belong to $BV = RBV_1$.

In terms of (5.3), we may summarize the contents of Example 5.8 in the set of inequalities

$$COP(BV) \neq BV_{\mathrm{loc}}(\mathbb{R}), \quad COP(WBV_p) \neq WBV_{p,\mathrm{loc}}(\mathbb{R}),$$
$$COP(RBV_p) \neq RBV_{p,\mathrm{loc}}(\mathbb{R}), \quad COP(AC) \neq AC_{\mathrm{loc}}(\mathbb{R}),$$
$$COP(Lip_\alpha) \neq Lip_{\alpha,\mathrm{loc}}(\mathbb{R}) \quad (0 < \alpha < 1).$$

The reason for the bad behavior of the operator C_h in Example 5.8 will become clear in a moment: it is the *lack of Lipschitz continuity* of the seagull function h. In fact, we shall show now in a series of theorems that, for $p \ge 1$ and $0 < \alpha \le 1$,

$$\begin{aligned} COP(BV) &= COP(WBV_p) = COP(RBV_p) \\ &= COP(AC) = COP(Lip_\alpha) = Lip_{\mathrm{loc}}(\mathbb{R}). \end{aligned} \tag{5.12}$$

Throughout the following, the local Lipschitz condition

$$|h(u) - h(v)| \le k(r)|u - v| \quad (u, v \in \mathbb{R},\ |u|, |v| \le r) \tag{5.13}$$

will play a crucial role. Occasionally, we will also need the analogous condition

$$|h'(u) - h'(v)| \le k_1(r)|u - v| \quad (u, v \in \mathbb{R},\ |u|, |v| \le r) \tag{5.14}$$

for the derivative of h (if it exists, of course). Together with the characteristics $k(r)$ and $k_1(r)$, the characteristics

$$\tilde{k}(r) := \sup_{|u| \le r} |h(u)| \quad (r > 0) \tag{5.15}$$

and

$$\tilde{k}_1(r) := \sup_{|u|\le r} |h'(u)| \quad (r > 0) \tag{5.16}$$

will also play a prominent role. The mean value theorem shows that these characteristics are related by the simple estimates

$$\tilde{k}(r) \le k(r)r + |h(0)|, \quad \tilde{k}_1(r) \le k_1(r)r + |h'(0)|\,. \tag{5.17}$$

We start with the classical function space $BV([a,b])$. The COP for this space was completely solved by Josephy [155] who proved the following.

Theorem 5.9. *The operator* (5.1) *maps the space* $BV([a,b])$ *into itself if and only if the corresponding function* h *is locally Lipschitz on* $\mathbb{R}$*, i.e. for each* $r > 0$*, there exists* $k(r) > 0$ *such that* (5.13) *holds.*

Proof. Although we will prove a more general result below (Theorem 5.10), we reproduce here Josephy's original proof.

Suppose first that h satisfies (5.13) for each $r > 0$. By the COP-invariance of the space BV (Example 5.3), we may assume without loss of generality that $[a,b] = [0,1]$. For $f \in BV([0,1])$ and $P \in \mathcal{P}([0,1])$ we have then

$$\mathrm{Var}(C_h f, P; [0,1]) \le k(\|f\|_\infty)\,\mathrm{Var}(f, P; [0,1])\,, \tag{5.18}$$

where $\|\cdot\|_\infty$ denotes the norm (0.39), and so $C_h f \in BV([0,1])$ with

$$\|C_h f\|_{\mathrm{BV}} \le |h(f(0))| + k(\|f\|_\infty)\|f\|_{\mathrm{BV}}\,. \tag{5.19}$$

The nontrivial part is of course to show that the hypothesis $C_h(BV) \subseteq BV$ implies the local Lipschitz condition (5.13). Suppose that $h \notin Lip_{\mathrm{loc}}(\mathbb{R})$; without loss of generality, we may assume that $h \notin Lip([0,1])$. Then we may find sequences $(u_n)_n$ and $(v_n)_n$ in $[0,1]$ such that

$$|h(u_n) - h(v_n)| > (n^2 + n)\delta_n\,, \tag{5.20}$$

where $\delta_n := |u_n - v_n|$. Since the identity $f(t) = t$ belongs to $BV([0,1])$, we know that $h \in BV([0,1])$ and, in particular, h is bounded on $[0,1]$, say $|h(u)| \le 1/2$ for $0 \le u \le 1$. So, from (5.20), we get

$$(n^2 + n)\delta_n < |h(u_n) - h(v_n)| \le 1\,,$$

and hence

$$\delta_n < \frac{1}{n^2 + n} \quad (n = 1, 2, 3, \dots)\,. \tag{5.21}$$

Passing to appropriate subsequences if necessary, we may suppose that $(u_n)_n$ and $(v_n)_n$ converge to some point $u^* \in [0,1]$ satisfying

$$|u_n - u^*| < \frac{1}{(n+1)^2} \quad (n = 1, 2, 3, \dots)\,.$$

Now, we define a function $f : [0,1] \to \mathbb{R}$ in the following way. Let $f(0) := 0$, $f(1) := v_1$, and define f on $I_n := \left[\frac{1}{n+1}, \frac{1}{n}\right]$ by

$$f(x) := \begin{cases} u_n & \text{if } x = \frac{1}{n+1} + k\delta_n \text{ for some } k \in \mathbb{N}, \\ v_n & \text{otherwise on } I_n . \end{cases}$$

We claim that $f \in BV([0,1])$, but $C_h f \notin BV([0,1])$, which proves the assertion. To show this, consider the natural number

$$m_n := \operatorname{ent}\left(\frac{1}{(n^2+n)\delta_n}\right) \quad (n = 1, 2, 3, \dots), \tag{5.22}$$

where ent ξ denotes the integer part of $\xi \in \mathbb{R}$. By (5.21), we then have

$$\frac{1}{2(n^2+n)\delta_n} \le m_n < \frac{1}{(n^2+n)\delta_n} .$$

Since f is a step function on I_n, the variation of f on the interval I_n may be estimated from above by

$$\begin{aligned} \operatorname{Var}(f; I_n) &\le 2m_n\delta_n + |u_{n-1} - u^*| + |u^* - u_n| + \delta_n \\ &\le \frac{2}{n^2+n} + \frac{1}{n^2} + \frac{1}{(n+1)^2} + \frac{1}{n^2+n} \le \frac{5}{n^2} . \end{aligned}$$

Consequently,

$$\operatorname{Var}(f; [0,1]) \le 5 \sum_{n=1}^{\infty} \frac{1}{n^2} < \infty ,$$

and so $f \in BV([0,1])$. To see that $C_h f \notin BV([0,1])$, consider the special partition $P_n := \{t_0, t_1, \dots, t_{2m_n}\} \cup \left\{\frac{1}{n}\right\} \in \mathcal{P}(I_n)$, where

$$t_j := \frac{1}{n+1} + \frac{j}{2}\delta_n \quad (j = 0, 1, \dots, 2m_n) ,$$

with m_n given by (5.22). Then

$$C_h f(t_j) = h(f(t_j)) = \begin{cases} h(u_n) & \text{for } j \text{ even} , \\ h(v_n) & \text{for } j \text{ odd} , \end{cases}$$

and so

$$\operatorname{Var}(C_h f, P_n; I_n) \ge 2m_n|h(u_n) - h(v_n)| \ge \frac{2}{2(n^2+n)\delta_n}(n^2+n)\delta_n = 2 .$$

We conclude that $\operatorname{Var}(C_h f; [0,1]) = \infty$, and the proof is complete. □

The assertion of Theorem 5.9 is really remarkable: it shows that we can guarantee that $h \circ f \in BV$ for $f \in BV$ only for locally Lipschitz outer functions h. Even more surprising,

Theorem 1.28 shows that the same is true if we start with the much smaller class of monotonically increasing inner functions f.

Clearly, one cannot expect that the function h is *globally* Lipschitz on $\mathbb{R}$ if the corresponding composition operator maps $BV([a,b])$ into itself. For instance, the map $h(u) := u^2$ is not globally Lipschitz, but the corresponding operator C_h clearly maps $BV([0,1])$ into itself, by Theorem 5.9.

Observe that the order of composition in Theorem 5.9 is important: if $f : [0,1] \to [0,1]$ has bounded variation and $h : [0,1] \to \mathbb{R}$ is Lipschitz continuous, then $h \circ f$ has bounded variation. Conversely, the functions from Example 5.8 show that $f \in Lip$ and $h \in BV$ does *not* imply that $h \circ f \in BV$.

Next, we consider the COP for the spaces Lip, AC and RBV_p simultaneously in the following somewhat surprising generalization of Theorem 5.9.

Theorem 5.10. *The following six statements about the composition operator* (5.1) *and the generating function $h : \mathbb{R} \to \mathbb{R}$ are equivalent.*
(a) *The operator* (5.1) *maps $Lip([a,b])$ into $BV([a,b])$.*
(b) *The operator* (5.1) *maps $Lip([a,b])$ into $Lip([a,b])$.*
(c) *The operator* (5.1) *maps $RBV_p([a,b])$ into $RBV_p([a,b])$.*
(d) *The operator* (5.1) *maps $AC([a,b])$ into $AC([a,b])$.*
(e) *The operator* (5.1) *maps $BV([a,b])$ into $BV([a,b])$.*
(f) *The function h satisfies* (5.13).

Proof. The inclusions

$$Lip([a,b]) \subseteq RBV_p([a,b]) \subseteq AC([a,b]) \subseteq BV([a,b]) \tag{5.23}$$

show that any of the statements (b), (c), (d) or (e) implies (a). So, we first show that (a) implies (f).

Suppose that h does not satisfy a local Lipschitz condition, which means that there is some $r > 0$ for which (5.13) does not hold for any constant $k(r) > 0$. In particular, for any natural number n, we cannot have an estimate of the form

$$|h(u) - h(v)| \le c|u - v| \quad (u, v \in [-r, r],\ |u - v| \le r/n)$$

since otherwise (5.13) would hold with $k(r) = nc$. Consequently, there exist $u_k < v_k$ such that

$$\delta_k := v_k - u_k < \frac{1}{k^2}, \quad |h(v_k) - h(u_k)| > k^2|v_k - u_k| \quad (k = 1, 2, \ldots). \tag{5.24}$$

Passing, if necessary, to a subsequence, we can assume without loss of generality that there exists $u_\infty \in [-r, r]$ such that $|u_k - u_\infty| \le 1/2k^2$ for all k, and so $|u_k - u_{k+1}| \le 1/k^2$. Take $n_k := \text{ent}\,(1/k^2\delta_k)$, i.e. n_k is the unique natural number satisfying

$$\frac{1}{k^2\delta_k} \le n_k < \frac{1}{k^2\delta_k} + 1 \quad (k = 1, 2, 3, \ldots). \tag{5.25}$$

Then
$$\delta_k n_k \leq \frac{1}{k^2} + \delta_k < \frac{2}{k^2} \quad (k = 1, 2, 3, \ldots).$$
Consequently, the strictly increasing sequence $(t_k)_k$ defined recursively by
$$t_1 := 0, \quad t_{k+1} := t_k + |u_k - u_{k+1}| + 2n_k\delta_k$$
is actually bounded by
$$T := \sum_{k=1}^{\infty} (|u_k - u_{k+1}| + 2n_k\delta_k) \leq \sum_{k=1}^{\infty} \frac{5}{k^2} < \infty.$$
Now, we define a function $f : [0, T] \to \mathbb{R}$ by
$$f(x) := \begin{cases} u_k & \text{if } x = t_k + 2m\delta_k \text{ for } m \in \{0, \ldots, n_k\}, \\ v_k & \text{if } x = t_k + (2m-1)\delta_k \text{ for } m \in \{1, \ldots, n_k\}, \\ \text{linear} & \text{if } t_k + (m-1)\delta_k \leq x \leq t_k + m\delta_k \text{ for } m \in \{1, \ldots, 2n_k\}, \\ \text{linear} & \text{if } t_k + 2n_k\delta_k \leq x \leq t_{k+1}, \\ u_\infty & \text{if } x = T. \end{cases}$$
Note that $|u_k - v_k| = \delta_k$ and $|u_k - u_{k+1}| \leq |t_{k+1} - (t_k + 2n_k\delta_k)|$, and so f is actually Lipschitz continuous on $[0, T)$ with Lipschitz constant $L \leq 1$. Consequently, $f|_{[0,T)}$ has a unique continuous extension to a Lipschitz continuous function with Lipschitz constant L on $[0, T]$, and since $u_k \to u_\infty$, actually f itself is this extension.

Assume now by contradiction that the operator (5.1) maps $Lip([a, b])$ into $BV([a, b])$. Applying the result of Example 5.3 with $[c, d] = [0, u_\infty]$, this would imply that the function $C_h f = h \circ f$ belongs to $BV([0, u_\infty])$. On the other hand, by construction of f, (5.24) and (5.25), for any $m \in \mathbb{N}$, we have
$$\mathrm{Var}(h \circ f; [0, u_\infty]) \geq \sum_{k=1}^{m} 2n_k|h(v_k) - h(u_k)| > \sum_{k=1}^{m} 2n_k k^2\delta_k \geq 2m,$$
which is a contradiction. This proves that statement (f) of Theorem 5.10 follows from (a), and so also any of the other statements (b), (c), (d), or (e).

Conversely, we show now that (f) implies each of the statements (b), (c), (d) or (e), and so also statement (a). Assume that h satisfies the local Lipschitz condition (5.13). Then C_h maps the space $BV([a, b])$ into itself, by Theorem 5.9, and so statement (d) is fulfilled. It remains to show that C_h maps each of the spaces $X = Lip([a, b])$ and $X = AC([a, b])$ into itself. Indeed, for any function $f \in X$, there is some r with $|f(t)| \leq r$ for all $t \in [a, b]$, and thus $C_h f = h \circ f$ satisfies
$$|C_h f(s) - C_h f(t)| = |h(f(s)) - h(f(t))| \leq k(r)|f(s) - f(t)| \quad (s, t \in [a, b]),$$
and so also
$$\sum_{k=1}^{n} |C_h f(b_k) - C_h f(a_k)| \leq k(r) \sum_{k=1}^{n} |f(b_k) - f(a_k)|$$

for every finite collection $S = \{[a_1, b_1], \dots, [a_n, b_n]\} \in \Sigma([a, b])$. These estimates show that if f is Lipschitz continuous or absolutely continuous, then $C_h f$ also has the respective property. The proof is complete. □

Note that Theorem 5.10 covers not only the spaces occurring in the conditions (a)–(e), but any other "intermediate" space. We will prove this in the more general setting of the space $RBV_\phi([a, b])$ in Theorem 5.13 below. We also remark that Theorem 5.10 solves Exercise 3.5, see Exercise 5.7.

One might wonder why the space $WBV_p([0, 1])$ is not covered by Theorem 5.10. In fact, one may show that (5.13) is also equivalent to the inclusion $C_h(WBV_p) \subseteq WBV_p$ (see Theorem 5.12 below); this was recently proved in [15]. However, it is *not* true that (5.13) is equivalent to the inclusion $C_h(Lip) \subseteq WBV_p$ which by (1.68) in case $p > 1$ would be still weaker than (a):

Example 5.11. Consider again the operator C_h generated by the seagull function (5.10). For any partition $\{t_0, t_1, \dots, t_m\} \in \mathcal{P}([0, 1])$, the estimate

$$\mathrm{Var}_2^W(C_h f, P; [0, 1]) = \sum_{j=1}^{m} \left| h(f(t_j)) - h(f(t_{j-1})) \right|^2$$

$$= \sum_{j=1}^{m} \left| \sqrt{|f(t_j)|} - \sqrt{|f(t_{j-1})|} \right|^2 \le \sum_{j=1}^{m} \left| f(t_j) - f(t_{j-1}) \right|$$

shows that the operator C_h maps $BV([0, 1])$ into $WBV_2([0, 1])$, and so also $Lip([0, 1])$ into $WBV_2([0, 1])$. However, the function (5.10) certainly does not satisfy (5.13). ♥

Of course, the function (5.10) belongs to the Hölder space $Lip_{1/2}([0, 1])$. So, it is not surprising that in order to include the family of spaces $WBV_p([0, 1])$, we have to replace (5.13) by the local Hölder condition

$$|h(u) - h(v)| \le k(r)|u - v|^\beta \quad (u \in \mathbb{R},\ |u|, |v| \le r) \tag{5.26}$$

for some fixed $\beta \in (0, 1]$. We then obtain the following result which is parallel to Theorem 5.10.

Theorem 5.12. *Any of the following four equivalent conditions on the operator* (5.1) *implies condition* (5.26) *on the corresponding function h, where* $0 < \beta \le 1$*:*
(a) *The operator* (5.1) *maps* $WBV_p([a, b])$ *into* $WBV_{p/\beta}([a, b])$.
(b) *The operator* (5.1) *maps* $WBV_p([a, b])$ *into* $WBV_q([a, b])$ *for any* $q \ge p/\beta$.
(c) *The operator* (5.1) *maps* $BV([a, b])$ *into* $WBV_{1/\beta}([a, b])$.
(d) *The operator* (5.1) *maps* $Lip_\alpha([a, b])$ *into* $WBV_{1/\alpha\beta}([a, b])$.

Proof. The proof is similar to that of Theorem 5.10, and so we only sketch the idea. Since condition (a) implies (b), (b) implies (c), and (c) implies (d), by (1.68) and (1.72), we only have to prove that (d) implies (5.26). So, if we assume that (5.26) is false, we may find sequences $(u_k)_k$ and $(v_k)_k$ such that

$$|u_k - v_k| \le \frac{1}{k^2}, \quad |h(u_k) - h(v_k)| > k^2 |u_k - v_k|^\beta \quad (k = 1, 2, \dots),$$

and then construct a function $f \in Lip_\alpha([a,b])$ such that $h \circ f \notin WBV_{1/\alpha\beta}([a,b])$ precisely in the same way as in the proof of Theorem 5.10. □

Of course, in case $\beta = 1$, condition (5.26) reduces to condition (5.13), and we may recover from Theorem 5.12 some parts of Theorem 5.10. Thus, (a), (b) and (c) in Theorem 5.12 then all reduce (for $p = 1$) to (e) in Theorem 5.10, while (d) in Theorem 5.12 becomes (a) in Theorem 5.10 if, in addition, $\alpha = 1$. For $\beta = 1$, we also get from (a) the equivalence of (5.13) and the inclusion $C_h(WBV_p) \subseteq WBV_p$. Note that, by Theorem 5.12, the composition operator C_h generated by the seagull function (5.10) maps WBV_p into WBV_{2p} and Lip_α into $WBV_{2/\alpha}$.

Now, we give the complete solution of the COP for the spaces RBV_ϕ, ΛBV, and WBV_ϕ which is parallel to Theorem 5.9. Theorem 5.13 has been proved in [227], while Theorems 5.14 and 5.15 have been proved in [251].

Theorem 5.13. *The operator* (5.1) *maps the space* $RBV_\phi([a,b])$ *into itself if and only if the corresponding function h satisfies* (5.13).

Proof. In case $\phi \notin \infty_1$, we have $RBV_\phi = BV$, by Proposition 2.57, and the statement follows from Theorem 5.9. So, let us assume that $\phi \in \infty_1$. By Example 5.5, we may suppose without loss of generality that $[a,b] = [0,1]$. First, let $h : \mathbb{R} \to \mathbb{R}$ be locally Lipschitz on $\mathbb{R}$, and let $f \in RBV_\phi([0,1])$. Then f is bounded on $[0,1]$ and $\mathrm{Var}_\phi^R(\lambda f; [0,1]) < \infty$ for some $\lambda > 0$ (Definition 2.55). Considering (5.13) for $r := \|f\|_\infty$, for any partition $P = \{t_0, t_1, \ldots, t_m\} \in \mathcal{P}([0,1])$, we obtain the estimate

$$\begin{aligned}&\sum_{j=1}^m \phi\left(\frac{\lambda|h(f(t_j)) - h(f(t_{j-1}))|}{k(\|f\|_\infty)|t_j - t_{j-1}|}\right)|t_j - t_{j-1}|\\ &\leq \sum_{j=1}^m \phi\left(\frac{\lambda|f(t_j) - f(t_{j-1})|}{|t_j - t_{j-1}|}\right)|t_j - t_{j-1}| \leq \mathrm{Var}_\phi^R(\lambda f; [0,1]) < \infty\,.\end{aligned}$$

This shows that for $\mu := \lambda/k(\|f\|_\infty)$, we get $\mathrm{Var}_\phi^R(\mu C_h f; [0,1]) < \infty$, and hence $C_h f \in RBV_\phi[0,1]$ as claimed.

The converse implication may be proved quite easily as a corollary of Theorem 5.10. On the one hand, for $f \in Lip([a,b])$ with Lipschitz constant $L > 0$ and any partition $P = \{t_0, t_1, \ldots, t_m\} \in \mathcal{P}([a,b])$, we have

$$\begin{aligned}\mathrm{Var}_\phi(f, P) &= \sum_{j=1}^m \phi\left(\frac{|f(t_j) - f(t_{j-1})|}{t_j - t_{j-1}}\right)|t_j - t_{j-1}|\\ &\leq \sum_{j=1}^m \phi(L)|t_j - t_{j-1}| = \phi(L)(b-a)\,,\end{aligned}$$

which shows that $f \in RBV_\phi([a,b])$, and hence $Lip([a,b]) \subseteq RBV_\phi([a,b])$. On the other hand, from Proposition 2.56, it follows that the inclusion $RBV_\phi([a,b]) \subseteq BV([a,b])$ is also true. So, the assertion follows from Theorem 5.10. □

Now, we discuss the COP for the space ΛBV of functions of bounded Waterman variation. Recall that a Waterman sequence is a decreasing sequence $\Lambda = (\lambda_n)_n$ converging to zero and satisfying (2.24). The corresponding function space ΛBV has been introduced in Definition 2.15.

Theorem 5.14. *The operator* (5.1) *maps the space* $\Lambda BV([a,b])$ *into itself if and only if the corresponding function* h *satisfies* (5.13).

Proof. The sufficiency of (5.13) is simple to prove. In fact, if h satisfies (5.13) and $f \in \Lambda BV([a,b])$ with $\|f\|_{\Lambda BV} \le r$ is fixed, then for each $S_\infty = \{[a_n, b_n] : n \in \mathbb{N}\} \in \Sigma_\infty([a,b])$, we have

$$\begin{aligned} \mathrm{Var}_\Lambda(h \circ f, S_\infty; [a,b]) &= \sum_{k=1}^{\infty} \lambda_k |h(f(b_k)) - h(f(a_k))| \\ &\le k(r) \sum_{k=1}^{\infty} \lambda_k |f(b_k) - f(a_k)| = k(r)\, \mathrm{Var}_\Lambda(f, S_\infty; [a,b]), \end{aligned} \tag{5.27}$$

which shows that $f \in \Lambda BV([a,b])$ implies $C_h f = h \circ f \in \Lambda BV([a,b])$.

To prove the necessity of (5.13), suppose that h does not satisfy a local Lipschitz condition. Without loss of generality, we may assume that there is a sequence $(J_n)_n$ of disjoint intervals $J_n = [p_n, q_n]$ with $p_n \to 0$ and $q_n \to 0$ as $n \to \infty$, as well as an unbounded real sequence $(c_n)_n$, with $c_1 > 1$, such that

$$\sum_{n=1}^{\infty} \frac{1}{c_n} < \infty, \quad |h(q_n) - h(p_n)| \ge c_n |q_n - p_n|. \tag{5.28}$$

Since $q_n - p_n \to 0$ and h is continuous, by what we have observed before, we see that $h(q_n) - h(p_n) \to 0$ as well, and so $c_n(q_n - p_n) \to 0$, by (5.28). Moreover, we may choose the intervals J_n in such a way that the sequence $(c_n(q_n - p_n))_n$ is monotonically decreasing.

Building on the sequences $(J_n)_n$ and $(c_n)_n$, we now construct a function $f \in \Lambda BV([0,1])$ such that $h \circ f \notin \Lambda BV([0,1])$. Fix a Waterman sequence $\Lambda = (\lambda_n)_n$, where without loss of generality, $\lambda_1 \le 1$. The function $L : [0,\infty) \to [0,\infty)$ defined by

$$L(x) := \begin{cases} 0 & \text{for } x = 0, \\ \lambda[1,n] & \text{for } x = n \in \mathbb{N}, \\ \text{linear} & \text{otherwise}, \end{cases} \tag{5.29}$$

where we used the shortcut (2.42), is strictly increasing and piecewise linear with $L(0) = 0$ and $L(x) \to \infty$ as $x \to \infty$, and hence a bijection on $[0,\infty)$. Denoting

$$k_n := \mathrm{ent}\, L^{-1}\left(\frac{1}{c_n(q_n - p_n)} + 1\right) \quad (n = 1, 2, 3, \dots), \tag{5.30}$$

where ent ξ is the integer part of ξ, the monotonicity of L implies that

$$\lim_{n\to\infty} k_n = \infty, \quad \lambda[1, k_n] = L(k_n) > \frac{1}{c_n(q_n - p_n)},$$

and so $|h(q_n) - h(p_n)|\lambda[1, k_n] > 1$, by (5.28). Choose $C > 1$ such that

$$\frac{1}{c_n} < (q_n - p_n)\lambda[1, k_n] < \frac{C}{c_n}. \tag{5.31}$$

For this C, we then have

$$\sum_{n=1}^{\infty}(q_n - p_n)\lambda[1, k_n] < C\sum_{n=1}^{\infty}\frac{1}{c_n} < \infty. \tag{5.32}$$

For each $n \in \mathbb{N}$, let $\{I_{n,1}, I_{n,2}, \dots, I_{n,k_n}\}$ be a finite collection of nonoverlapping intervals contained in $(2^{-n}, 2^{-(n-1)})$ such that $I_{n,m}$ is situated to the left of $I_{n,m+1}$ for each m. Now, we define $f : [0, 1] \to \mathbb{R}$ in the following way. First, we put $f(0) = f(1) := 0$. Next, for each $n \in \mathbb{N}$, we define $f : [0, 1] \to \mathbb{R}$ as an increasing linear map of $I_{n,m}$ $(m = 1, 2, \dots, k_n)$ onto J_n. Finally, we define f to be linear and continuous on the remaining components of $[0, 1]$. We claim that the function f constructed in this way belongs to $\Lambda BV([0, 1])$.

To prove this, we use Proposition 2.31. According to that proposition, it suffices to show that $L \circ I_f \in L_1(\mathbb{R})$, where L is the function (5.29) and I_f denotes the Banach indicatrix of f, see Definition 0.38. In fact, from the definition of L in (5.29), it follows immediately that we may take $\mu_f(y) := \lambda[1, I_f(y)]$, where we use the notation of Proposition 2.31. Now, (5.32) shows that

$$\begin{aligned}\int_{-\infty}^{\infty}\mu_f(y)\,dy &\le q_1\lambda[1, 2] + \sum_{n=1}^{\infty}(q_n - p_n)\lambda[1, 2k_n]\\ &\le q_1\lambda[1, 2] + 2\sum_{n=1}^{\infty}(q_n - p_n)\lambda[1, k_n] < \infty,\end{aligned}$$

and hence $\mu_f \in L_1(\mathbb{R})$ and so $f \in \Lambda BV([0, 1])$.

It remains to show that $C_h f = h \circ f \notin \Lambda BV([0, 1])$. To prove this, we write $I_{n,m} = [a_{n,m}, b_{n,m}]$ and note that

$$\begin{aligned}\mathrm{Var}_\Lambda\left(C_h f; [2^{-n}, 2^{-(n-1)}]\right) &\ge \sum_{m=1}^{k_n}\lambda_m\left|h(f(b_{n,m})) - h(f(a_{n,m}))\right|\\ &= |h(q_n) - h(p_n)|\lambda[1, k_n] > 1,\end{aligned} \tag{5.33}$$

as we have seen above. If $C_h f$ were in $\Lambda BV([0, 1])$, then the right continuity of $C_h f$ at 0 would imply, by Exercise 2.14, that

$$\lim_{n\to\infty}\mathrm{Var}_\Lambda\left(C_h f; [2^{-n}, 2^{-(n-1)}]\right) = 0,$$

contradicting (5.33). Consequently, $C_h f$ does not belong to $\Lambda BV([0, 1])$, and the proof is complete. □

Now, we are ready to discuss the COP for the space WBV_ϕ of all functions of bounded Wiener–Young variation, at least for a restricted class of Young functions ϕ. Recall that the δ_2-condition was given in (2.4), condition ∞_1 in (2.16), and condition 0_1 in (2.17).

Theorem 5.15. *Let ϕ be a Young function which satisfies a δ_2-condition and condition ∞_1, but not condition 0_1. Then the operator* (5.1) *maps the space $WBV_\phi([a,b])$ into itself if and only if the corresponding function h satisfies* (5.13).

Proof. The proof is very similar to that of Theorem 5.14, and so we only sketch the general idea and the differences. Again, we may suppose without loss of generality that $[a,b] = [0,1]$.

The fact that $f \in WBV_\phi([0,1])$ implies $h \circ f \in WBV_\phi([0,1])$ for h satisfying (5.13) follows again from a simple calculation. To prove the necessity of (5.13), suppose that h does not satisfy a local Lipschitz condition, and define the sequence $(J_n)_n$ of disjoint intervals $J_n = [p_n, q_n]$ and the unbounded real sequence $(c_n)_n$ satisfying (5.28) as in the proof of Theorem 5.14. However, the numbers k_n from (5.30) will now be defined in a different way.

Define a function $P : [0,\infty) \to [0,\infty)$ by

$$P(x) := \begin{cases} 0 & \text{for } x = 0\,, \\ \frac{1}{\phi(c_n(q_n - p_n))} & \text{for } x = n \in \mathbb{N}\,, \\ \text{linear} & \text{otherwise}\,. \end{cases}$$

By construction, P is then an increasing homeomorphism from $[0,\infty)$ onto $[0,\infty)$. Instead of (5.30), we now put

$$k_n := \operatorname{ent}(P(n) + 1) \quad (n = 1,2,3,\ldots)\,. \tag{5.34}$$

In particular, $k_n\phi(c_n(q_n - p_n)) > 1$ for all $n \in \mathbb{N}$. Now, let f be defined as in the proof of Theorem 5.14, but with this new definition of the numbers k_n. We claim that the function f constructed in this way belongs to $WBV_\phi([0,1])$, but $C_h f = h \circ f$ does not.

To prove the first assertion, let $P = \{t_0, t_1, \ldots, t_m\} \in \mathcal{P}([0,1])$ be an arbitrary partition of $[0,1]$; we have to find an upper bound for

$$\operatorname{Var}_\phi(f, P; [0,1]) = \sum_{j=1}^{m} \phi(|f(t_j) - f(t_{j-1})|) \tag{5.35}$$

which does not depend on P. For the extreme indices $j = 1$ and $j = m$, we have

$$\phi(|f(t_1) - f(0)|) = \phi(|f(t_1)|) \le \phi(q_1)$$

and

$$\phi(|f(1) - f(t_{m-1})|) = \phi(|f(t_{j-1})|) \le \phi(q_1)\,,$$

respectively. For the remaining indices $j \in \{2,3,\ldots,m-1\}$, we distinguish two cases.

Suppose first that there exists $n \in \mathbb{N}$ such that $f(t_j) \in J_n$ and $f(t_{j-1}) \in J_n$. Writing again $I_{n,m} = [a_{n,m}, b_{n,m}]$, we observe that $a_{n,1} \le t_j, t_{j-1} \le b_{n,k_n}$. Clearly, $\phi(|f(t_j) -$

$f(t_{j-1})|) \leq \phi(q_n - p_n)$. Note that increasing the number of partition points along intervals of monotonicity does not increase the sum (5.35), and so the contribution of all such terms is bounded by $2k_n\phi(q_n - p_n)$.

Now, suppose that there is no $n \in \mathbb{N}$ such that $f(t_j) \in J_n$ and $f(t_{j-1}) \in J_n$. Then there exists a smallest index k satisfying $f(t_j) \leq q_k$. Suppose that there are r such intervals having this same smallest index k satisfying the above.[5] If

$$[t_{j_0}, t_{j_0+1}], [t_{j_0+1}, t_{j_0+2}], \ldots, [t_{j_0+r-1}, t_{j_0+r}]$$

are these intervals, we have, by the convexity of ϕ,

$$\sum_{l=1}^{r} \phi\left(|f(t_{j_0+l}) - f(t_{j_0+l-1})|\right) \leq \phi\left(|f(t_{j_0+r}) - f(t_{j_0})|\right) \leq \phi(q_j)\,.$$

By combining all of the obtained information, we see that (5.35) may be estimated from above by

$$\sum_{j=1}^{m} \phi(|f(t_j) - f(t_{j-1})|) \leq 2\phi(q_1) + 2\sum_{n=1}^{\infty} k_n\phi(c_n(q_n - p_n)) + \sum_{n=1}^{\infty} \phi(q_n) \tag{5.36}$$

which is obviously independent of the choice of the partition P. Observe that q_n was chosen so that the last series in (5.36) converges. The other series in (5.36) may be estimated by

$$\begin{aligned}
\sum_{n=1}^{\infty} k_n\phi(c_n(q_n - p_n)) &= \sum_{n=1}^{\infty} \operatorname{ent}(P(n) + 1)\phi(c_n(q_n - p_n)) \\
&\leq \sum_{n=1}^{\infty} \left[\frac{1}{\phi(c_n(q_n - p_n))} + 1\right] \phi(q_n - p_n) \\
&\leq \sum_{n=1}^{\infty} \left[\frac{\phi(q_n - p_n)}{c_n phi(q_n - p_n)} + \phi(q_n - p_n)\right] \\
&= \sum_{n=1}^{\infty} \frac{1}{c_n} + \sum_{n=1}^{\infty} \phi(q_n - p_n) \leq \sum_{n=1}^{\infty} \frac{1}{c_n} + \sum_{n=1}^{\infty} \phi(q_n) < \infty
\end{aligned}$$

where we have used the fact that $\phi(c_n(q_n - p_n)) \geq c_n\phi(q_n - p_n)$ since ϕ is convex and $c_n \geq 1$. So, we have proved that $f \in WBV_\phi([0,1])$.

It remains to show that $C_h f = h \circ f \notin WBV_\phi([0,1])$. Writing $I_{n,m} = [a_{n,m}, b_{n,m}]$ and $J_n = [p_n, q_n]$ as before and taking k_n as in (5.34), we get

$$\begin{aligned}
\sum_{j=1}^{k_n} \phi\left(|C_h f(b_{n,m}) - C_h f(a_{n,m})|\right) &= \sum_{j=1}^{k_n} \phi\left(|h(q_n) - h(p_n))|\right) \\
&= k_n\phi\left(|h(q_n) - h(p_n)|\right) \geq k_n\phi\left(c_n(q_n - p_n)\right) \\
&\geq \frac{\phi\left(c_n(q_n - p_n)\right)}{\phi\left(c_n(q_n - p_n)\right)} = 1\,.
\end{aligned}$$

5 Observe that this situation can arise at most once for each index k.

Thus, for this collection of intervals $I_{n,m} = [a_{n,m}, b_{n,m}]$, we have

$$\sum_{n=1}^{\infty}\sum_{j=1}^{k_n} \phi(|C_h f(b_{n,m}) - C_h f(a_{n,m})|) \geq \sum_{n=1}^{\infty} 1 = \infty$$

which shows that $C_h = h \circ f$ does not belong to $WBV_\phi([0,1])$. □

The functions $\phi(t) = t^p$ or $\phi(t) = (t+1)\log(t+1)$ may serve as examples of Young functions which satisfy all the hypotheses of Theorem 5.15.

We may summarize the contents of Theorems 5.13, 5.14 and 5.15 as equalities

$$COP(RBV_\phi) = COP(\Lambda BV) = COP(WBV_\phi) = Lip_{\mathrm{loc}}(\mathbb{R}) \tag{5.37}$$

which is completely analogous to (5.12).

Again, our favorite function (5.10) may be used to show that the requirement $h \in X_{\mathrm{loc}}(\mathbb{R})$ for $X \in \{RBV_\phi, \Lambda BV, WBV_\phi\}$ does not imply $C_h(X) \subseteq X$, i.e. $h \in COP(X)$. We have already proved this assertion in Example 5.8 for the space RBV_p, and in Example 5.11 for the space WBV_p. So, it remains to show this for the Waterman space ΛBV; we state this as

Example 5.16. Consider the Waterman space $\Lambda_q BV([0,1])$ generated by the sequence $\lambda_n := n^{-q}$ for $0 < q \leq 1$, see Definition 2.29, and let $h : \mathbb{R} \to \mathbb{R}$ be the seagull function (5.10). Then $h \in \Lambda_q BV([0,1])$ for *any* $q > 0$ because $h \in BV([0,1])$, see (2.33).

In Section 2.2, we have shown that the special zigzag function Z_θ defined in (0.93) belongs to $\Lambda_q BV([0,1])$ if and only if $\theta + q > 1$; in this case,

$$\mathrm{Var}_{\Lambda_q}(Z_\theta; [0,1]) = \sum_{n=1}^{\infty} \frac{1}{n^{\theta+q}} < \infty .$$

So, if we choose $1 - q < \theta \leq 2(1-q)$, then[6] $Z_\theta \in \Lambda_q BV([0,1])$, but $C_h Z_\theta = h \circ Z_\theta \notin \Lambda_q BV([0,1])$. ♥

Our examples show again that the requirement $h \in Lip_{\alpha,loc}(\mathbb{R})$ is not sufficient for the operator C_h to map any of the spaces RBV_ϕ, ΛBV or WBV_ϕ into itself.

One might ask why the proofs of Theorems 5.14 and 5.15 are so complicated and, in particular, why we could not use Theorem 5.10 to prove them, as we did for Theorem 5.13. The point is that for applying Theorem 5.10, the spaces $\Lambda BV([a,b])$ and $WBV_\phi([a,b])$ should *be contained* in $BV([a,b])$; however, as the relations (2.23) and (2.28) show, they *contain* $BV([a,b])$ as a (usually, strict) subspace.

In the following table, where we use the shortcut $[a,b] = I$ and $[c,d] = J$, we make a comparison of different conditions of $h : J \to \mathbb{R}$, under which the corresponding composition operator C_h maps all functions from some space $X = X(I)$ whose range is contained in the domain J of h, into the same space $X = X(I)$.

6 To be precise, we do not have $h \circ Z_\theta = Z_{\theta/2}$ because this function is not piecewise linear; however, the statement $h \circ Z_\theta \notin \Lambda_q BV([0,1])$ follows from a direct calculation.

Table 5.1. Asymmetry in compositions of functions.

$f \in BV(I),\ h \in BV(J)$	$\not\Rightarrow$	$h \circ f \in BV(I)$	(Example 5.8)
$f \in BV(I),\ h \in Lip_\alpha(J)$	$\not\Rightarrow$	$h \circ f \in BV(I)$	(Example 5.8)
$f \in BV(I),\ h \in Lip(J)$	$\Leftrightarrow$	$h \circ f \in BV(I)$	(Theorem 5.9)
$f \in RBV_p(I),\ h \in RBV_p(J)$	$\not\Rightarrow$	$h \circ f \in RBV_p(I)$	(Example 5.8)
$f \in RBV_p(I),\ h \in Lip_\alpha(J)$	$\not\Rightarrow$	$h \circ f \in RBV_p(I)$	(Example 5.8)
$f \in RBV_p(I),\ h \in Lip(J)$	$\Leftrightarrow$	$h \circ f \in RBV_p(I)$	(Theorem 5.10)
$f \in RBV_\phi(I),\ h \in RBV_\phi(J)$	$\not\Rightarrow$	$h \circ f \in RBV_\phi(I)$	(Example 5.8)
$f \in RBV_\phi(I),\ h \in Lip_\alpha(J)$	$\not\Rightarrow$	$h \circ f \in RBV_\phi(I)$	(Example 5.8)
$f \in RBV_\phi(I),\ h \in Lip(J)$	$\Leftrightarrow$	$h \circ f \in RBV_\phi(I)$	(Theorem 5.13)
$f \in \Lambda BV(I),\ h \in \Lambda BV(J)$	$\not\Rightarrow$	$h \circ f \in \Lambda BV(I)$	(Example 5.16)
$f \in \Lambda BV(I),\ h \in Lip_\alpha(J)$	$\not\Rightarrow$	$h \circ f \in \Lambda BV(I)$	(Example 5.16)
$f \in \Lambda BV(I),\ h \in Lip(J)$	$\Leftrightarrow$	$h \circ f \in \Lambda BV(I)$	(Theorem 5.14)
$f \in WBV_p(I),\ h \in WBV_p(J)$	$\not\Rightarrow$	$h \circ f \in WBV_p(I)$	(Example 5.8)
$f \in WBV_p(I),\ h \in Lip_\alpha(J)$	$\not\Rightarrow$	$h \circ f \in WBV_p(I)$	(Example 5.8)
$f \in WBV_p(I),\ h \in Lip(J)$	$\Leftrightarrow$	$h \circ f \in WBV_p(I)$	(Theorem 5.12)
$f \in WBV_\phi(I),\ h \in WBV_\phi(J)$	$\not\Rightarrow$	$h \circ f \in WBV_\phi(I)$	(Example 5.8)
$f \in WBV_\phi(I),\ h \in Lip_\alpha(J)$	$\not\Rightarrow$	$h \circ f \in WBV_\phi(I)$	(Example 5.8)
$f \in WBV_\phi(I),\ h \in Lip(J)$	$\Leftrightarrow$	$h \circ f \in WBV_\phi(I)$	(Theorem 5.15)
$f \in AC(I),\ h \in AC(J)$	$\not\Rightarrow$	$h \circ f \in AC(I)$	(Example 5.8)
$f \in AC(I),\ h \in Lip_\alpha(J)$	$\not\Rightarrow$	$h \circ f \in AC(I)$	(Example 5.8)
$f \in AC(I),\ h \in Lip(J)$	$\Leftrightarrow$	$h \circ f \in AC(I)$	(Theorem 5.10)
$f \in Lip_\alpha(I),\ h \in Lip_\alpha(J)$	$\not\Rightarrow$	$h \circ f \in Lip_\alpha(I)$	(Example 5.8)
$f \in Lip_\alpha(I),\ h \in Lip(J)$	$\Leftrightarrow$	$h \circ f \in Lip_\alpha(I)$	(Theorem 5.24)

This table exhibits a certain *asymmetry* in such conditions insofar as the requirement $h \in X$ is *never* sufficient for guaranteeing that $C_h(X) \subseteq X$. Roughly speaking, this table shows that *local Lipschitz continuity* of h is the right condition, while *local Hölder continuity* does not suffice. The important point in Table 5.1 is of course that the crucial condition $h \in Lip([c, d])$ in every row where a theorem is cited is also *necessary* for the operator C_h to map the underlying space into itself. Thus, the equivalence arrow $\Leftrightarrow$ in the third row, say, means that $h \circ f \in BV([a, b])$ *for all* functions $f \in BV([a, b])$ satisfying $f([a, b]) \subseteq [c, d]$ *if and only if* $h \in Lip([c, d])$, and similarly for the other seven equivalence arrows.

We remark that we did not prove the result mentioned in the last row of Table 5.1 for the Hölder space Lip_α; for technical reasons, we postpone the proof to Theorem 5.24 below. However, we have already shown in Example 5.8 that Lip_α is not stable under composition.

Let us point out another important aspect of the crucial condition $h \in Lip_{loc}(\mathbb{R})$. We know that every function $f \in BV$ (and therefore also every function which belongs to Lip, RBV_p, or AC) is differentiable almost everywhere. So, the question arises as to whether or not the classical *chain rule*

$$(h \circ f)'(x) = h'(f(x))f'(x) \tag{5.38}$$

holds under the hypotheses of Theorem 5.12, and in which sense it has to be interpreted. The following theorem answers this question for the space AC.

Theorem 5.17. *Let $f \in AC([a,b])$ with $f([a,b]) \subseteq [c,d]$, and let $h \in Lip([c,d])$. Then (5.38) holds for almost all $x \in [a,b]$, where $h'(f(x))f'(x)$ is interpreted to be zero whenever $f'(x) = 0$, even if h is not differentiable at $f(x)$.*

Proof. Since $f \in AC([a,b])$ and $g := h \circ f \in AC([a,b])$, by Theorem 5.10, there exists a nullset $N \subset [a,b]$ such that both $f'(x)$ and $g'(x)$ exist for $x \in [a,b] \setminus N$. Fix $x_0 \in [a,b] \setminus N$; we distinguish two cases.

Suppose first that $f'(x_0) = 0$. Given $\varepsilon > 0$, we find then some $h > 0$ such that $|f(x) - f(x_0)| \leq \varepsilon |x - x_0|$ for $x \in (x_0 - h, x_0 + h) \cap [a,b]$. It follows that

$$|g(x) - g(x_0)| \leq lip(h)|f(x) - f(x_0)| \leq lip(h)\varepsilon|x - x_0| \quad (|x - x_0| < h),$$

and hence $g'(x_0) = 0 = h'(f(x_0))f'(x_0)$ as claimed.

Now, suppose that $f'(x_0) \neq 0$. For $f(x) \neq f(x_0)$, we then have

$$\frac{g(x) - g(x_0)}{x - x_0} = \frac{h(f(x)) - h(f(x_0))}{f(x) - f(x_0)} \frac{f(x) - f(x_0)}{x - x_0}. \tag{5.39}$$

Since both $f'(x_0) \neq 0$ and $g'(x_0)$ exist, by assumption, we may pass to the limit for $x \to x_0$ in (5.39) and obtain

$$\lim_{x \to x_0} \frac{g(x) - g(x_0)}{f(x) - f(x_0)} = \frac{g'(x_0)}{f'(x_0)}. \tag{5.40}$$

Denoting the right-hand side of (5.40) by ρ, for $\varepsilon > 0$, there exists some $\delta > 0$ such that

$$\rho - \varepsilon < \frac{h(f(x)) - h(f(x_0))}{f(x) - f(x_0)} < \rho + \varepsilon \quad (|x - x_0| < \delta). \tag{5.41}$$

Moreover, from $f'(x_0) \neq 0$, it follows that f maps a neighborhood of x_0 onto a neighborhood of $f(x_0)$. More precisely, there exists some $\eta > 0$ such that any $y \in (f(x_0) - \eta, f(x_0) + \eta) \cap [c,d]$ may be expressed as $y = f(x)$ for some suitable $x \in (x_0 - \delta, x_0 + \delta) \cap [a,b]$. Putting this into (5.41) yields

$$\rho - \varepsilon < \frac{h(y) - h(f(x_0))}{y - f(x_0)} < \rho + \varepsilon \quad (|y - f(x_0)| < \eta).$$

This shows that $g'(x_0)$ exists and $g'(x_0) = h'(f(x_0))f'(x_0) = \rho$ as desired. □

Theorem 5.17 allows us to prove the following formula for *change of variables* for the Lebesgue integral; another result of this type may be found in Exercise 5.24.

Theorem 5.18 (change of variables). *Let $f \in AC([a,b])$ with $f([a,b]) \subseteq [c,d]$, and let $g \in L_\infty([c,d])$. Then $(g \circ f)f' \in L_1([a,b])$, and*

$$\int_{f(\alpha)}^{f(\beta)} g(y)\,dy = \int_{\alpha}^{\beta} g(f(t))f'(t)\,dt \tag{5.42}$$

for every interval $[\alpha,\beta] \subset [a,b]$.

Proof. Defining

$$h(u) := \int_{c}^{u} g(y)\,dy \quad (c \le u \le d),$$

from Theorem 3.20, we know that $h \in Lip([c,d])$ with $lip(h;[c,d]) = \|g\|_{L_\infty}$; moreover, $h'(u) = g(u)$ a.e. on $[c,d]$. Theorem 5.10 implies that $h \circ f \in AC([a,b])$, and Theorem 5.17 implies that (5.38) holds a.e. on $[a,b]$. Fix $\alpha, \beta \in [a,b]$ with $\alpha < \beta$. From Theorem 3.18, we conclude that

$$h(f(\beta)) - h(f(\alpha)) = \int_{f(\alpha)}^{f(\beta)} h'(y)\,dy = \int_{f(\alpha)}^{f(\beta)} g(y)\,dy.$$

On the other hand,

$$h(f(\beta)) - h(f(\alpha)) = \int_{\alpha}^{\beta} (h \circ f)'(t)\,dt = \int_{\alpha}^{\beta} g(f(t))f'(t)\,dt,$$

by (5.38), which proves the assertion. □

The following example shows that Theorem 5.18 becomes false if we replace the assumption $f \in AC$ by the weaker assumption $f \in BV$.

Example 5.19. Let $f = \varphi$ be the Cantor function (3.6) and $g(y) := y$, and let $[\alpha,\beta] = [a,b] = [0,1]$. Then the left-hand side of (5.42) is 1, while the right-hand side is 0. ♥

5.2 Boundedness and continuity

All results in the preceding section refer to the autonomous composition operator (5.1). In case of the nonautonomous superposition operator (5.2), the situation becomes more complicated, as we will see in the next chapter. As a rule, even finding criteria on $h : [a,b] \times \mathbb{R} \to \mathbb{R}$ under which the operator (5.2) maps a given space X into itself is a highly nontrivial problem.

In this section, we are interested in conditions on a function $h : \mathbb{R} \to \mathbb{R}$, possibly both necessary and sufficient, under which the composition operator (5.1) is bounded or continuous[7] between two given function spaces.

There are, however, some function spaces in which the operator (5.1) is really "well-behaved." For example, in the space C of continuous functions and in the Lebesgue space L_p, one knows simple conditions, both necessary and sufficient, on h under which the operator C_h maps the space $C([a, b])$ into itself, or the space $L_p([a, b])$ into the space $L_q([a, b])$. In addition, in all these spaces, we get both boundedness and continuity of the operator C_h "for free:"

Theorem 5.20. *The operator* (5.1) *maps the space* $C([a, b])$ *into itself if and only if the function h is continuous on* $\mathbb{R}$. *In this case, the operator* (5.1) *is automatically bounded and continuous in the norm* (0.45).

Theorem 5.21. *For* $1 \le p, q < \infty$, *the operator* (5.1) *maps the space* $L_p([a, b])$ *into the space* $L_q([a, b])$ *if and only if the function h satisfies the growth condition*

$$|h(u)| \le \alpha + \beta |u|^{p/q} \quad (u \in \mathbb{R}) \tag{5.43}$$

for some constants $\alpha, \beta > 0$. *In this case, the operator* (5.1) *is automatically bounded and continuous in the norm* (0.11).

The proof of Theorem 5.20 is a simple consequence of the Tietze–Urysohn extension theorem (Theorem 0.33), while the proof of Theorem 5.21 relies on a result of Krasnosel'skij, see [165–168] or [170].

Observe that the requirements on h in Theorems 5.20 and 5.21 are quite different: in Theorem 5.20, we have to impose a simple analytic property (continuity of h on $\mathbb{R}$), while in Theorem 5.21, we have to impose a growth condition (polynomial growth of h of degree at most p/q for large values of $u \in \mathbb{R}$). In the special case $p = q$, the estimate (5.43) shows that h has to be of sublinear growth for large values of u.

Before passing to other spaces of continuous functions, let us briefly consider the COP for the most general space we have introduced in Chapter 0, namely, the space $R([a, b])$ of regular functions. First, we show, by means of a counterexample, that the regularity of h on the real axis is too weak to imply that the operator C_h maps the space $R([a, b])$ into itself.

Example 5.22. Let $A \subset [0, 1]$ be an uncountable Cantor set of positive measure (see, e.g. [76] or [118]), and define $f : [0, 1] \to \mathbb{R}$ by

$$f(x) := \operatorname{dist}(x, A) = \inf \{|x - a| : a \in A\}.$$

7 Boundedness means that the operator maps bounded sets into bounded sets. We point out that in contrast to linear operators, a nonlinear operator like (5.1) or (5.2) may be bounded but discontinuous, or continuous but unbounded. This is one of the reasons for the numerous difficulties which we encounter when dealing with these harmless looking operators.

Clearly, f is (Lipschitz) continuous, and therefore regular. Moreover, the function $h := \chi_{\{0\}}$ is certainly regular on $\mathbb{R}$. However, the composition $C_h f = h \circ f = \chi_A$ is not regular, by Proposition 0.35, since it is discontinuous on the uncountable set A. ♥

Example 5.22 shows even more than we claimed: continuity of h on the whole real axis, *except for one point*, does not even imply the very weak condition $C_h(Lip) \subseteq R$! It is somewhat surprising that the "right" condition for the inclusion $C_h(R) \subseteq R$ is the same as for the inclusion $C_h(C) \subseteq C$:

Theorem 5.23. *The operator* (5.1) *maps the space* $R([a,b])$ *into itself if and only if the function* h *is continuous on* $\mathbb{R}$. *In this case, the operator* (5.1) *is automatically bounded and continuous in the norm* (0.39).

Proof. Suppose first that $h : \mathbb{R} \to \mathbb{R}$ is continuous, and let $f \in R([a,b])$. By Theorem 0.36, we find a strictly increasing function $\tau : [a,b] \to [a,b]$ and a continuous function $g : [a,b] \to \mathbb{R}$ such that $f = g \circ \tau$. However, then the function $C_h f = (h \circ g) \circ \tau$ has an analogous representation, and again from Theorem 0.36, it follows that $C_h f$ is regular.

Conversely, suppose now that $h : \mathbb{R} \to \mathbb{R}$ is discontinuous at some point $u_0 \in \mathbb{R}$, where without loss of generality[8] $h(u_0-) < h(u_0+)$, see (0.54). Choose $\varepsilon > 0$ and $\delta > 0$ such that $h(u) > h(u_0-) + \varepsilon$ for $u_0 < u \leq u_0 + \delta$, and define $f : [a,b] \to \mathbb{R}$ by

$$f(x) := h(u_0) + \min\{\text{dist}(x, A), \delta\} \quad (a \leq x \leq b),$$

where $A \subset [a,b]$ is an uncountable Cantor set as in Example 5.22. Then $C_h f = h \circ f$ is discontinuous at each point of A, and so it cannot belong to $R([a,b])$.

We still have to show that the operator C_h is bounded and continuous in the norm (0.39). Fix $r > 0$ and $f \in R([a,b])$ with $\|f\|_\infty \leq r$. Then $f(x) \in [-r, r]$ for $a \leq x \leq b$, and so $h(f(x)) \in [-R, R]$ for $a \leq x \leq b$ and some $R > 0$ since h is continuous on the compact interval $[-r, r]$. Similarly, the uniform continuity of h on any compact interval implies that C_h is continuous at every point $f \in R([a,b])$ in the norm (0.39). □

For $0 < \alpha \leq 1$, we consider now the Banach space $Lip_\alpha([a,b])$ of all Hölder continuous (in particular, Lipschitz continuous for $\alpha = 1$) functions on $[a,b]$, equipped with the usual norm (0.71) or the equivalent norm (0.77). Interestingly, in this space, we do not have a completely analogous result to Theorem 5.20 or Theorem 5.23:

Theorem 5.24. *The operator* (5.1) *maps the space* $Lip_\alpha([a,b])$ *into itself if and only if the function* h *satisfies* (5.13). *In this case, the operator* (5.1) *is automatically bounded in the norm* (0.71).

8 Here, we use the fact that the assumption $C_h(R) \subseteq R$ implies that h is regular on the real line, and so both unilateral limits $h(u_0-)$ and $h(u_0+)$ exist.

Proof. First of all, we note that we cannot use Theorem 5.10 here in case $0 < \alpha < 1$ since $C_h(Lip_\alpha) \subseteq Lip_\alpha$ implies $C_h(Lip) \subseteq Lip_\alpha$, but not $C_h(Lip) \subseteq BV$, as Example 1.23 shows.

The sufficiency of (5.13) for $C_h(Lip_\alpha) \subseteq Lip_\alpha$ is trivial; it is the necessity which requires a subtle construction. We already proved the necessity of (5.13) for $C_h(Lip) \subseteq Lip$ in Theorem 5.10. So, let $0 < \alpha < 1$, and assume that $h \notin Lip_{\mathrm{loc}}(\mathbb{R})$. As in the proof of Theorem 5.9, we can then find a constant $r_0 > 0$ and two convergent sequences $(u_n)_n$ and $(v_n)_n$ in $[-r_0, r_0]$ such that $u_n \neq v_n$ and

$$|h(u_n) - h(v_n)| > n|u_n - v_n| \quad (n = 1, 2, 3, \ldots). \tag{5.44}$$

Passing to subsequences, if necessary, we may assume that both $u_n \to u^*$ and $v_n \to u^*$ as $n \to \infty$. Consider the sequence of intervals

$$I_n := [u^* - \gamma_{n+1}, u^* + \gamma_{n+1}], \quad \gamma_n := \frac{b-a}{2^{n_0+n+2}},$$

where $n_0 \in \mathbb{N}$ is so large that $\gamma_n < 1$ for all $n \in \mathbb{N}$. Then for every n, we find k_n such that $u_{k_n}, v_{k_n} \in I_n$ which implies that

$$\delta_n := |u_{k_n} - v_{k_n}|^{1/\alpha} < 2^{1/\alpha}\gamma_{n+1}^{1/\alpha} = \gamma_n^{1/\alpha}.$$

This generates another sequence $(\delta_n)_n$ with $0 < \delta_n < \gamma_n < 1$. We define sequences $(s_n)_n$ and $(t_n)_n$ in $[a, b]$ by

$$s_1 := a + \frac{b-a}{4}, \qquad s_n := s_1 + \gamma_1 + \delta_1 + \ldots + \gamma_{n-1} + \delta_{n-1}, \qquad t_n := s_n + \delta_n.$$

Obviously, $\gamma_m < s_n - s_m$ for $n > m$. Since $0 < \tau \leq 1$ implies $\tau \leq \tau^\alpha$ and $\tau > 1$ implies $\tau^\alpha > 1$, we find $\gamma_m < (s_n - s_m)^\alpha$ for $n > m$, and similarly $\gamma_m < (s_n - t_m)^\alpha$ and $\gamma_m < (t_n - t_m)^\alpha$ for $n > m$. On the set $M := \{s_1, s_2, s_3, \ldots\} \cup \{t_1, t_2, t_3, \ldots\}$, we define a function f by

$$f(x) := \begin{cases} u_{k_n} & \text{for } x = s_n, \\ v_{k_n} & \text{for } x = t_n. \end{cases}$$

A straightforward calculation shows that

$$|f(s) - f(t)| \leq |s - t|^\alpha \quad (s, t \in M).$$

Applying now the McShane extension to f (Theorem 0.42), we may extend f to a function $\overline{f} \in Lip_\alpha([a, b])$ with Hölder constant 1. By assumption, the function $\overline{g} := h \circ \overline{f}$ belongs then also to $Lip_\alpha([a, b])$. So, there exists some $\overline{L} > 0$ satisfying

$$|\overline{g}(x) - \overline{g}(y)| \leq \overline{L}|x - y|^\alpha \quad (a \leq x, y \leq b). \tag{5.45}$$

Putting, in particular, $x := s_n$ and $y := t_n$ in (5.45), we end up with

$$|\overline{g}(x) - \overline{g}(y)| = |h(u_{k_n}) - h(v_{k_n})| \leq \overline{L}|s_n - t_n|^\alpha = \overline{L}\delta_n^\alpha = \overline{L}|u_{k_n} - v_{k_n}|.$$

Comparing this with (5.44) yields $\overline{L} \geq k_n$ for all $n \in \mathbb{N}$, a contradiction. So, we have proved that h satisfies (5.13).

The proof of the automatic boundedness of C_h is easy. In fact, $\|f\|_{Lip_\alpha} \leq r$ implies $\|f\|_C \leq r$, and hence

$$\frac{|C_h f(x) - C_h f(y)|}{|x-y|^\alpha} \leq k(r)\frac{|f(x) - f(y)|}{|x-y|^\alpha} \leq k(r) lip_\alpha(f) \leq k(r)r\,,$$

with $k(r)$ as in (5.13). So, $lip_\alpha(C_h f) \leq k(r) lip_\alpha(f)$ which shows that C_h is bounded in the norm (0.71). □

The reader may have noticed that in contrast to Theorems 5.20 and 5.21, we did not claim automatic continuity of the operator (5.1) in the space $Lip_\alpha([a,b])$. In fact, the operator C_h may map this space into itself without being continuous:

Example 5.25. We give this counterexample for the case $\alpha = 1$, a similar example in case $0 < \alpha < 1$ may be easily constructed. Let $h : \mathbb{R} \to \mathbb{R}$ be defined by

$$h(u) := \min\{|u|, 1\}\,. \tag{5.46}$$

Then the corresponding operator C_h maps $Lip_\alpha([0,1])$ into itself and is bounded, by Theorem 5.24. However, C_h is not continuous in the norm (0.71). To see this, consider the functions $f(t) := t$ and $f_n(t) := t + 1/n$ $(n = 1, 2, 3, \ldots)$. Clearly,

$$\|f_n - f\|_{Lip_\alpha} = |f_n(0) - f(0)| = \frac{1}{n} \to 0 \quad (n \to \infty)\,.$$

On the other hand, we have

$$C_h f_n(t) = \begin{cases} t + \frac{1}{n} & \text{for } 0 \leq t \leq \tau_n\,, \\ 1 & \text{for } \tau_n < t \leq 1\,, \end{cases}$$

where $\tau_n := 1 - 1/n$, and hence

$$lip(C_h f_n - C_h f) \geq \frac{|h(f_n(\tau_n)) - h(f_n(1)) - h(f(\tau_n)) - h(f(1))}{1 - \tau_n} = \frac{1 - \tau_n}{1 - \tau_n} = 1$$

which shows that $\|C_h f_n - C_h f\|_{Lip} \nrightarrow 0$ as $n \to \infty$. ♥

One might ask which additional property of the function h is "missing" in Theorem 5.24 to also ensure the continuity of the operator C_h. Surprisingly, this question has a very natural answer: in [125], it was shown that C_h is continuous in $Lip_\alpha([a,b])$ if and only if $h \in C^1(\mathbb{R})$. We give a slightly more general result which states that this is even equivalent to the uniform continuity of C_h on bounded subsets.[9]

---

9 Recall that an operator $A : X \to Y$ between two normed spaces $(X, \|\cdot\|_X)$ and $(Y, \|\cdot\|_Y)$ is called *uniformly continuous* if for each $\varepsilon > 0$, one can find a $\delta > 0$ such that $\|f - g\|_X \leq \delta$ implies $\|Af - Ag\|_Y \leq \varepsilon$. If this holds for $\|f\|_X, \|g\|_X \leq r$, and δ is allowed to depend on r, A is called *uniformly continuous on bounded sets*. We will study these properties in detail in the next chapter for the superposition operator (5.2).

Theorem 5.26. *The following three statements for $C_h : Lip_\alpha([a,b]) \to Lip_\alpha([a,b])$ are equivalent:*
(a) *The function h belongs to* $C^1(\mathbb{R})$.
(b) *The operator* (5.1) *is uniformly continuous on bounded subsets in the norm* (0.71).
(c) *The operator* (5.1) *is continuous in the norm* (0.71).

Proof. To prove that (a) implies (b), suppose first that $h \in C^1(\mathbb{R})$. Since this clearly implies (5.13), we know that C_h maps $Lip_\alpha([a,b])$ into itself and is bounded in the norm (0.71). To show that C_h is uniformly continuous on bounded sets, let $f, g \in Lip_\alpha([a,b])$ satisfy $\|f\|_{Lip_\alpha} \leq r$ and $\|g\|_{Lip_\alpha} \leq r$, and fix $\varepsilon > 0$. From $h \in C^1(\mathbb{R})$, it follows that we find $\delta \in (0,\varepsilon)$ such that $\|f - g\|_C \leq \delta$ implies $\|C_h f - C_h g\|_C \leq \varepsilon$, and $|u - v| \leq \delta$ implies $|h'(u) - h'(v)| \leq \varepsilon$ for $u, v \in [-r, r]$. So, we have, in particular, $|C_h f(a) - C_h g(a)| \leq \varepsilon$, and it remains to estimate $lip_\alpha(h \circ f - h \circ g)$.

Using the Lagrange formula

$$h(u) - h(v) = (u - v) \int_0^1 h'[v + \tau(u - v)]\, d\tau \quad (u, v \in \mathbb{R}),$$

for $s \neq t$, we obtain

$$\begin{aligned}
&\frac{|C_h f(s) - C_h f(t) - C_h g(s) + C_h g(t)|}{|s-t|^\alpha} \\
&\quad \leq lip_\alpha(g) \int_0^1 \left|h'[g(s) + \tau(g(t) - g(s))] - h'[f(s) + \tau(f(t) - f(s))]\right| d\tau \\
&\qquad + lip_\alpha(g - f) \int_0^1 \left|h'[f(s) + \tau(f(t) - f(s))]\right| d\tau\,.
\end{aligned} \tag{5.47}$$

Now, from $|f(s) - g(s)| \leq \delta$ and $|f(t) - g(t)| \leq \delta$, we get

$$\begin{aligned}
&\left|g(s) + \tau(g(t) - g(s)) - f(s) + \tau(f(t) - f(s))\right| \\
&\qquad\qquad \leq (1 + \tau)|f(s) - g(s)| + \tau|f(t) - g(t)| \leq \delta\,,
\end{aligned}$$

and so the first integrand in (5.47) may be estimated by

$$\left|h'[g(s) + \tau(g(t) - g(s))] - h'[f(s) + \tau(f(t) - f(s))]\right| \leq \varepsilon\,.$$

On the other hand, the second integrand remains bounded by

$$\left|h'[f(s) + \tau(f(t) - f(s))]\right| \leq \tilde{k}_1(r)$$

with $\tilde{k}_1(r)$ given by (5.16). Consequently, from $lip_\alpha(f - g) \leq \delta$, we conclude that

$$\frac{|C_h f(s) - C_h f(t) - C_h g(s) + C_h g(t)|}{|s-t|^\alpha} \leq lip_\alpha(g)\varepsilon + \tilde{k}_1(r)\delta \leq (lip_\alpha(g) + \tilde{k}_1(r))\varepsilon\,,$$

and so we have proved (b).

The fact that (b) implies (c) is of course trivial; so it remains to prove that (c) implies (a). Suppose that the operator (5.1) maps $Lip_\alpha([a,b])$ into itself and is both bounded and continuous. Since h then satisfies (5.13), by Theorem 5.24, both limits

$$L_-(u) := \liminf_{h\to 0} \frac{h(u+h)-h(u)}{h}, \quad L_+(u) := \limsup_{h\to 0} \frac{h(u+h)-h(u)}{h}$$

exist and are finite for any $u \in \mathbb{R}$. Moreover, the set $N := \{u \in \mathbb{R} : L_-(u) < L_+(u)\}$ is a nullset; we claim that $N = \emptyset$, and so $h'(u) = L_-(u) = L_+(u)$ exists for *all* $u \in \mathbb{R}$.

Suppose that we can find $u_0 \in N$ and choose two decreasing positive[10] sequences $(h_n)_n$ and $(k_n)_n$, both converging to zero, such that

$$L_-(u_0) = \lim_{n\to\infty} \frac{h(u_0+h_n)-h(u_0)}{h_n}, \quad L_+(u_0) = \lim_{n\to\infty} \frac{h(u_0+k_n)-h(u_0)}{k_n}.$$

Since N is a nullset, we may find another sequence $(\delta_n)_n$ converging to zero such that $h'(u_0+\delta_n)$ exists for all $n \in \mathbb{N}$. Define functions $f_n : [a,b] \to \mathbb{R}$ by

$$f_n(x) := (x-a)^\alpha + u_0 + \delta_n \quad (a \le x \le b,\ n = 1,2,3,\dots).$$

Clearly, all functions f_n belong to $Lip_\alpha([a,b])$, and the sequence $(f_n)_n$ converges in the norm of $Lip_\alpha([a,b])$ to the function $f_0(x) = (x-a)^\alpha + u_0$ which also belongs to $Lip_\alpha([a,b])$. Now, we use our continuity assumption on the operator C_h; as a consequence, we know that $\|C_h f_n - C_h f_0\|_{Lip_\alpha} \to 0$ as $n \to \infty$. In particular, this implies that we may find $n_0 \in \mathbb{N}$ such that

$$lip_\alpha(h \circ f_n - h \circ f_0) \le \varepsilon := \frac{L_+(u_0) - L_-(u_0)}{4}$$

for $n \ge n_0$. By definition of the functions f_n and f_0, this means that

$$|h((s-a)^\alpha + u_0 + \delta_n) - h((t-a)^\alpha + u_0 + \delta_n) - h((s-a)^\alpha + u_0) + h((t-a)^\alpha + u_0)| \\ \le \varepsilon |s-t|^\alpha$$

for all $s,t \in [a,b]$ and $n \ge n_0$. Substituting now $t := a$ and $s := a + h_m^{1/\alpha}$, we get

$$\big|h(u_0+\delta_n+h_m) - h(u_0+\delta_n) - h(u_0+h_m) - h(u_0)\big| \le \varepsilon |h_m| \tag{5.48}$$

for $n \ge n_0$ and m sufficiently large. Dividing (5.48) by $|h_m|$ and letting $m \to \infty$ yields

$$|h'(u_0+\delta_n) - L_-(u_0)| \le \varepsilon \quad (n \ge n_0). \tag{5.49}$$

10 The assumption $h_n > 0$ and $k_n > 0$ may be made without loss of generality; otherwise, the following construction has to be modified accordingly.

The same argument with h_m replaced by k_m shows that

$$|h'(u_0 + \delta_n) - L_+(u_0)| \leq \varepsilon \quad (n \geq n_0)\,. \tag{5.50}$$

Finally, combining (5.49) and (5.50), we obtain

$$\begin{aligned} &L_+(u_0) - L_-(u_0) \\ &\quad \leq |L_+(u_0) - h'(u_0 + \delta_n)| + |h'(u_0 + \delta_n) - L_-(u_0)| \leq 2\varepsilon = \frac{1}{2}\,[L_+(u_0) - L_-(u_0)] \end{aligned}$$

for $n \geq n_0$, a contradiction. So, our assumption $N \neq \emptyset$ was false, and h' exists on the whole real line. Moreover, since (5.49) and (5.50) can be verified for any $u_0 \in \mathbb{R}$, any $\varepsilon > 0$, and any sequence $(\delta_n)_n$ converging to zero, we may also conclude that h' is continuous at every point, i.e. $h \in C^1(\mathbb{R})$. □

Theorem 5.26 explains why the function (5.46) in Example 5.25 has to be chosen Lipschitz continuous, but not continuously differentiable at each point.

Now, we turn to our main object of study, namely, the space BV and its various generalizations. The following result shows that also for these spaces, we get boundedness of C_h as a "fringe benefit."

Theorem 5.27. *Let $1 \leq p < \infty$, and let X be any of the spaces $AC([a, b])$, $BV([a, b])$, $WBV_p([a, b])$, or $RBV_p([a, b])$. Suppose that the operator (5.1) maps the space X into itself. Then this operator is automatically bounded in the norm of X.*

Proof. We know that the hypothesis $C_h(X) \subseteq X$ implies that h satisfies (5.13). For $f \in X$ with $\|f\|_X \leq r$, we certainly have $|f(a)| \leq r$, so (5.13) shows that $|h(f(a))| \leq |h(0)| + k(r)r$. Moreover, for any such f, we have $|f(x)| \leq r$ for $a \leq x \leq b$, hence $\|f\|_\infty \leq r$.

Given any partition $P = \{t_0, t_1, \dots, t_m\} \in \mathcal{P}([a, b])$, by (5.13), we have, for $p \geq 1$,

$$\sum_{j=1}^{m} |h(f(t_j)) - h(f(t_{j-1}))|^p \leq k(r)^p \sum_{j=1}^{m} |f(t_j) - f(t_{j-1})|^p\,,$$

which implies that $\mathrm{Var}_p^W(h \circ f) \leq k(r)^p \,\mathrm{Var}_p^W(f)$; so C_h is bounded in the spaces $BV([a, b])$, $WBV_p([a, b])$, and $AC([a, b])$. The analogous estimates

$$\sum_{j=1}^{m} \frac{|h(f(t_j)) - h(f(t_{j-1}))|^p}{(t_j - t_{j-1})^{p-1}} \leq k(r)^p \sum_{j=1}^{m} \frac{|f(t_j) - f(t_{j-1})|^p}{(t_j - t_{j-1})^{p-1}}$$

shows that C_h is bounded in the space $RBV_p([a, b])$ as well. □

We point out that Theorem 5.27 does not assert continuity of the operator C_h in any of the spaces covered by this theorem. In fact, continuity conditions which are both necessary and sufficient seem to be unknown in these spaces. We give just one sufficient condition on the function h which, in case of the spaces BV and AC, guarantees the *uniform* continuity of C_h on bounded subsets.

Proposition 5.28. *Suppose that $h \in Lip^1_{loc}(\mathbb{R})$. Then the operator* (5.1) *is uniformly continuous on bounded subsets of $BV([a, b])$ and $AC([a, b])$.*

We do not give the proof here since we are going to prove a more general result in Theorem 5.50 below. The following counterexample shows that the condition $h \in Lip_{loc}(\mathbb{R})$ is too weak to guarantee uniform continuity of C_h on bounded subsets of BV or AC.

Example 5.29. Let $h : \mathbb{R} \to \mathbb{R}$ be defined by $h(u) := |u|$. Since $h \in Lip_{loc}(\mathbb{R})$ (even $h \in Lip(\mathbb{R})$), the corresponding operator C_h maps each of the spaces $BV([0, 2\pi])$ and $AC([0, 2\pi])$ into itself and is bounded, by Theorem 5.10.

Consider the sequences $(f_n)_n$ and $(g_n)_n$ defined by

$$f_n(x) := \frac{\cos nx}{n}, \quad g_n(x) := f_n(x) + \frac{1}{n}.$$

Clearly, $f_n \in AC([0, 2\pi])$ with $\mathrm{Var}(f_n; [0, 2\pi]) = 4$, so both sequences $(f_n)_n$ and $(g_n)_n$ are bounded in the norm (1.16). Moreover, $\|f_n - g_n\|_{BV} \to 0$ as $n \to \infty$.

For $n \in \mathbb{N}$, consider the partition $P_n := \{t_0, t_1, \ldots, t_{4n}\} \in \mathcal{P}([0, 2\pi])$ defined by

$$t_0 := 0, t_1 := \frac{\pi}{2n}, \ldots, t_j := \frac{j\pi}{2n}, \ldots, t_{4n} := 2\pi.$$

An easy calculation then shows that

$$f_n(t_{j-1}) = -\frac{1}{n}, \quad f_n(t_j) = g_n(t_{j-1}) = 0, \quad g_n(t_j) = \frac{1}{n}.$$

Consequently,

$$\begin{aligned}&|C_h f_n(t_j) - C_h g_n(t_j) - C_h f_n(t_{j-1}) + C_h g_n(t_{j-1})|\\ &= \big||f_n(t_j)| - |g_n(t_j)| - |f_n(t_{j-1})| + |g_n(t_{j-1})|\big| = \frac{2}{n},\end{aligned}$$

which implies that

$$\begin{aligned}\mathrm{Var}(C_h f_n - C_h g_n; [0, 2\pi]) &\geq \mathrm{Var}(C_h f_n - C_h g_n, P_n; [0, 2\pi])\\ &= \sum_{j=1}^{4n} |C_h f_n(t_j) - C_h g_n(t_j) - C_h f_n(t_{j-1}) + C_h g_n(t_{j-1})| \geq 8,\end{aligned}$$

and so $\|C_h f_n - C_h g_n\|_{BV} \not\to 0$ as $n \to \infty$. ♥

Let us summarize the boundedness and continuity behavior of the composition operator (5.1) discussed so far in a series of tables. Although these tables are not very exciting in the case of the autonomous operator (5.1), we state them here in order to make a comparison with the nonautonomous operator (5.2) in Chapter 6 (Tables 6.1–6.4).

Of course, the situation is most satisfactory in Tables 5.2–5.4, where all statements are equivalent. On the other hand, the Tables 5.6–5.8 do not contain conditions on h, both necessary and sufficient, under which the operator C_h is continuous in the norm of the corresponding space. As far as we know, such continuity criteria are not known. Of course, one may try to find conditions which are just sufficient (Ex-

Table 5.2. The operator C_h in $L_p([a,b])$.

C_h bounded in L_p	$\Leftrightarrow$	$C_h(L_p) \subseteq L_p$	$\Leftrightarrow$	C_h continuous in L_p
		$\Updownarrow$		
		$\lvert h(u)\rvert \le \alpha + \beta\lvert u\rvert$		

Table 5.3. The operator C_h in $R([a,b])$.

C_h bounded in R	$\Leftrightarrow$	$C_h(R) \subseteq R$	$\Leftrightarrow$	C_h continuous in R
		$\Updownarrow$		
		$h \in C(\mathbb{R})$		

Table 5.4. The operator C_h in $C([a,b])$.

C_h bounded in C	$\Leftrightarrow$	$C_h(C) \subseteq C$	$\Leftrightarrow$	C_h continuous in C
		$\Updownarrow$		
		$h \in C(\mathbb{R})$		

Table 5.5. The operator C_h in $Lip_\alpha([a,b])$ $(0 < \alpha \le 1)$.

C_h bounded in Lip_α	$\Leftrightarrow$	$C_h(Lip_\alpha) \subseteq Lip_\alpha$	$\Leftarrow$	C_h continuous in Lip_α
		$\Updownarrow$		$\Updownarrow$
		$h \in Lip_{\mathrm{loc}}(\mathbb{R})$	$\Leftarrow$	$h \in C^1(\mathbb{R})$

Table 5.6. The operator C_h in $BV([a,b])$.

C_h bounded in BV	$\Leftrightarrow$	$C_h(BV) \subseteq BV$	$\Leftarrow$	C_h continuous in BV
		$\Updownarrow$		$\Uparrow$
		$h \in Lip_{\mathrm{loc}}(\mathbb{R})$	$\Leftarrow$	$h \in Lip^1_{\mathrm{loc}}(\mathbb{R})$

Table 5.7. The operator C_h in $AC([a,b])$.

C_h bounded in AC	$\Leftrightarrow$	$C_h(AC) \subseteq AC$	$\Leftarrow$	C_h continuous in AC
		$\Updownarrow$		$\Uparrow$
		$h \in Lip_{\mathrm{loc}}(\mathbb{R})$	$\Leftarrow$	$h \in Lip^1_{\mathrm{loc}}(\mathbb{R})$

Table 5.8. The operator C_h in $RBV_p([a,b])$ $(p > 1)$.

C_h bounded in RBV_p	$\Leftrightarrow$	$C_h(RBV_p) \subseteq RBV_p$	$\Leftarrow$	C_h continuous in RBV_p
		$\Updownarrow$		
		$h \in Lip_{\mathrm{loc}}(\mathbb{R})$		

ercises 5.9 and 5.10). Conditions for the uniform continuity or uniform boundedness of C_h on bounded sets will be studied more systematically in Section 6.4 in the next chapter.

5.3 Spaces of differentiable functions

Recall that given a space X of functions $f : [a, b] \to \mathbb{R}$ with norm $\|\cdot\|_X$, by X^1, we denote the space of all primitives of functions in X, i.e. $X^1 := \{f : f' \in X\}$, equipped with the natural norm

$$\|f\|_{X^1} := |f(a)| + \|f'\|_X , \tag{5.51}$$

see Definition 0.32. In this section, we shall study the composition operator (5.1) in such spaces for $X \in \{C, AC, BV, WBV_p, RBV_p, Lip\}$. By (5.51), the corresponding norms on these spaces are

$$\|f\|_{C^1} = |f(a)| + \max_{a \leq x \leq b} |f'(x)| , \tag{5.52}$$

$$\|f\|_{AC^1} = |f(a)| + |f'(a)| + \int_a^b |f''(t)|\, dt , \tag{5.53}$$

$$\|f\|_{WBV_p^1} = |f(a)| + |f'(a)| + \mathrm{Var}_p^W(f'; [a, b])^{1/p} , \tag{5.54}$$

$$\|f\|_{RBV_p^1} = |f(a)| + |f'(a)| + \mathrm{Var}_p^R(f'; [a, b])^{1/p} , \tag{5.55}$$

and

$$\|f\|_{Lip^1} = |f(a)| + |f'(a)| + lip(f'; [a, b]) , \tag{5.56}$$

respectively. Taking, in particular, $p = 1$ in (5.54) or (5.55), we get the space BV^1 with norm

$$\|f\|_{BV^1} = |f(a)| + |f'(a)| + \mathrm{Var}(f'; [a, b]) . \tag{5.57}$$

Interestingly, it turns out that the operator (5.1) does not behave in the space X^1 in the same way as in the corresponding parent space X. We illustrate this by means of the composition operator problem described in Section 5.1: when analyzing the set (5.3) for given X, one may establish, loosely speaking, the following "golden rule" which applies quite frequently:

- If not all functions in X are differentiable, then $COP(X) = Lip_{\mathrm{loc}}(\mathbb{R})$.
- If all functions in X are differentiable, then $COP(X) = X_{\mathrm{loc}}(\mathbb{R})$.

In other words, in the first case, the operator (5.1) maps X into itself if and only if the corresponding function h satisfies (5.13), and so the COP has an *extrinsic and universal solution*. We have proved this above for $X = BV$ in Theorem 5.9, for $X = RBV_p$, $X = Lip$ and $X = AC$ in Theorem 5.10, for $X = WBV_p$ in Theorem 5.12, for $X = RBV_\phi$ in

Theorem 5.13, for $X = \Lambda BV$ in Theorem 5.14, for $X = WBV_\phi$ in Theorem 5.15, and for $X = Lip_\alpha$ in Theorem 5.24.

On the other hand, in the second case, the operator (5.1) maps X into itself if and only if the corresponding function h belongs (locally) to the same class, which is therefore an *algebra* with respect to composition, and so the COP has an *intrinsic and individual solution*, as we will show now.

As an example of the "golden rule," we start with the simplest case $X = C$, i.e. with the space $X^1 = C^1([a,b])$ with norm (5.52).

Theorem 5.30. *The operator* (5.1) *maps the space* $C^1([a,b])$ *into itself if and only if the function h is continuously differentiable on* $\mathbb{R}$. *In this case, the operator* (5.1) *is automatically bounded and continuous in the norm* (5.52).

Proof. The proof is almost trivial. The inclusion $COP(C^1) \subseteq C^1(\mathbb{R})$ follows from the fact that the identity $f(x) = x$ is C^1, while the inclusion $COP(C^1) \supseteq C^1(\mathbb{R})$ follows from the chain rule. To prove boundedness of C_h, we use the notation (5.15) and (5.16) for a function $h \in C^1(\mathbb{R})$. From the chain rule, it follows then that $\|f\|_{C^1} \le r$ implies

$$|(h \circ f)(a)| \le \tilde{k}(r), \quad |(h \circ f)'(t)| = |h'(f(t))|\,|f'(t)| \le r\tilde{k}_1(r)$$

which shows that C_h is bounded. To prove that C_h is continuous in the norm (5.52), let $(f_n)_n$ be a sequence of C^1 functions which converges in the norm (5.52) to some function $f \in C^1$. Then $(f_n)_n$ is bounded, say $\|f_n\|_{C^1} \le r$. Putting

$$g_n := C_h f_n - C_h f = (h \circ f_n) - (h \circ f), \tag{5.58}$$

we get

$$g_n' = (h' \circ f_n) f_n' - (h' \circ f) f',$$

and we have to show that $\|g_n'\|_C \to 0$ as $n \to \infty$. Let $\varepsilon > 0$. Since $h \in C^1(\mathbb{R})$ and $\|f_n - f\|_C \to 0$, by the mean value theorem, we may find $n_0 \in \mathbb{N}$ such that $|h'(f_n(t)) - h'(f(t))| \le \varepsilon$ for $n \ge n_0$ and $a \le t \le b$. However, this implies that

$$\begin{aligned} |g_n'(t)| &\le |h'(f_n(t))|\,|f_n'(t) - f'(t)| + |h'(f_n(t)) - h'(f(t))|\,|f'(t)| \\ &\le \tilde{k}_1(r)\|f_n' - f'\|_C + \varepsilon\|f'\|_C, \end{aligned}$$

which shows that $\|g_n'\|_C \to 0$ as $n \to \infty$. The relation $|g_n(a)| \to 0$ is obvious. □

Now, we study the set $COP(X^1)$ for $X \in \{AC, BV, WBV_p, RBV_p, Lip\}$. As an immediate consequence of the inclusions (2.93), we get the inclusions

$$Lip^1([a,b]) \subseteq RBV_p^1([a,b]) \subseteq AC^1([a,b]) \subseteq C^1([a,b]) \cap BV^1([a,b]) \tag{5.59}$$

for $p > 1$. In the next example, we show that all inclusions in (5.59) are strict.

Example 5.31. For $\tau \geq 1$, consider the function[11] $f_\tau : [0,1] \to \mathbb{R}$ defined by $f_\tau(x) := x^\tau$. A straightforward calculation shows that

$$f_\tau \in Lip^1([0,1]) \Leftrightarrow f_{\tau-1} \in Lip([0,1]) \Leftrightarrow \tau \geq 2,$$
$$f_\tau \in RBV^1_p([0,1]) \Leftrightarrow f_{\tau-1} \in RBV_p([0,1]) \Leftrightarrow \tau > 2 - \frac{1}{p},$$

and

$$f_\tau \in AC^1([0,1]) \Leftrightarrow f_{\tau-1} \in AC([0,1]) \Leftrightarrow \tau > 1.$$

So, for $2-\frac{1}{p} < \tau < 2$, we have $f_\tau \in RBV^1_p([0,1]) \setminus Lip^1([0,1])$, while for $1 < \tau \leq 2-\frac{1}{p}$, we have $f_\tau \in AC^1([0,1]) \setminus RBV^1_p([0,1])$.

To show that the last inclusion in (5.59) is strict, we cannot use the function f_τ because the statements $f_\tau \in AC^1$, $f_\tau \in C^1$, $f_\tau \in BV^1$, and $\tau > 1$ are all equivalent. However, we may use the Cantor function $\varphi : [0,1] \to \mathbb{R}$ from (3.6). Since φ is a non-negative and continuous map of $[0,1]$ onto itself, the function $f : [0,1] \to \mathbb{R}$ defined by

$$f(x) := \int_0^x \varphi(t)\,dt \quad (0 \leq x \leq 1) \tag{5.60}$$

is an increasing C^1 function. On the other hand, its derivative $f' = \varphi$ does not map nullsets into nullsets, and thus fails to be absolutely continuous, by the Vitali–Banach–Zaretskij theorem (Theorem 3.9). Consequently, $f \in [C^1([0,1]) \cap BV^1([0,1])] \setminus AC^1([0,1])$. ♥

Since we are going to deal with functions which have first derivatives everywhere and second derivatives almost everywhere, we will use the fact that under reasonable conditions on f and h, we not have only the chain rule (5.38) for the first derivative, but also the chain rule

$$(h \circ f)''(x) = (h''(f(x))f'^2(x) + (h(f(x))f''(x) \tag{5.61}$$

for the second derivative. Before studying the COP in spaces of differentiable functions different from C^1, we state a general lemma. Recall that $P_n([a,b])$ denotes the linear space of all polynomials of degree $\leq n$ on $[a,b]$.

Proposition 5.32. *Suppose that a function space of type $X^1 = X^1([a,b])$ contains the set $P_1([a,b])$ of all affine functions. Assume that $h : \mathbb{R} \to \mathbb{R}$ generates a composition operator (5.1) which maps X^1 into itself. Then $h \in X^1_{loc}(\mathbb{R})$, i.e. the derivative h' of h belongs to the space $X = X([c,d])$ for any compact interval $[c,d] \subset \mathbb{R}$.*

---

11 For some properties of this function in case $\tau \leq 1$, see Examples 2.78 and 3.35.

Proof. Given $[c,d] \subset \mathbb{R}$, consider the affine function $\ell^{-1} : [a,b] \to [c,d]$ defined by (5.5). By assumption, $\ell^{-1} \in P_1([a,b]) \subseteq X^1([a,b])$, and so $g := h \circ \ell^{-1} \in X^1([a,b])$ as well. Consequently, the function $h = g \circ \ell$ belongs to $X^1([c,d])$ as claimed. □

Now, we discuss the COP for the space $BV^1([a,b])$ of all primitives of BV-functions equipped with the norm (5.57). Here, the situation is much more difficult than in the space $C^1([a,b])$. It is not hard to see that the condition $h \in Lip^1_{\text{loc}}(\mathbb{R})$ is sufficient for C_h to map the space $BV^1([a,b])$ into itself, while the condition $h \in BV^1_{\text{loc}}(\mathbb{R})$ is necessary, by Proposition 5.32. We begin with an example which shows that the condition $h \in Lip^1_{\text{loc}}(\mathbb{R})$ is not necessary for the mapping condition $C_h(BV^1) \subseteq BV^1$, and hence too strong.

Example 5.33. Let $[a,b] = [-1,1]$ and $h(u) := \max\{u,0\}$; clearly, $h' = \chi_{(0,\infty)} \notin Lip_{\text{loc}}(\mathbb{R})$. On the other hand, for every $f \in BV^1([-1,1]) \subseteq AC([-1,1])$, we may apply (5.38) and get

$$(h \circ f)'(x) = f'(x)\chi_{(0,\infty)}(f(x)) = \begin{cases} f'(x) & \text{if } f(x) > 0\,, \\ 0 & \text{if } f(x) \le 0\,. \end{cases}$$

Consequently, $f' \in BV([-1,1])$ implies $(h \circ f)' \in BV([-1,1])$, and so C_h maps $BV^1([-1,1])$ into itself. ♥

We remark that the operator C_h in Example 5.33 is also bounded in the norm (5.57), and even continuous at zero since $\|C_h f\|_{BV^1} \le \|f\|_{BV^1}$. However, it is somewhat surprising that C_h is not continuous everywhere:

Example 5.34. Let h be defined as in Example 5.33, and let

$$f(x) := x\,, \quad f_n(x) := x + \frac{1}{n} \quad (-1 \le x \le 1)\,.$$

Clearly,

$$\|f_n - f\|_{BV^1} = |f_n(-1) - f(-1)| = \frac{1}{n} \to 0 \quad (n \to \infty)\,.$$

The function $g_n := h \circ f_n - h \circ f$ and its derivative have the form

$$g_n(x) = \begin{cases} 0 & \text{for } -1 \le x < -\frac{1}{n}\,, \\ x + \frac{1}{n} & \text{for } -\frac{1}{n} \le x < 0\,, \\ \frac{1}{n} & \text{for } 0 \le x \le 1\,, \end{cases}$$

and

$$g_n'(x) = \begin{cases} 0 & \text{for } -1 \le x < -\frac{1}{n}\,, \\ 1 & \text{for } -\frac{1}{n} \le x < 0\,, \\ 0 & \text{for } 0 \le x \le 1\,, \end{cases}$$

respectively. Consequently,

$$\|g_n\|_{BV^1} \ge \operatorname{Var}(g_n'; [-1,1]) = 2\,,$$

which shows that $h \circ f_n \not\to h \circ f$ in the norm (5.57) as $n \to \infty$. ♥

The following remarkable Theorems 5.35 and 5.37 which are due to Burenkov [74, 75] give a complete solution of the COP in the spaces BV^1 and AC^1.

Theorem 5.35. *The operator* (5.1) *maps the space* $BV^1([a,b])$ *into itself if and only if* $h \in BV^1_{loc}(\mathbb{R})$. *Moreover, in this case, the operator* (5.1) *is automatically bounded in the norm* (5.57).

Proof. The "only if" part is a consequence of Proposition 5.32, and so we must only prove the "if" part. Let $f \in BV^1([a,b])$, $h \in BV^1([c,d])$, where $f([a,b]) \subseteq [c,d]$, and $g := h \circ f$. First, we note that the function f' is then continuous: since f' has a primitive, the Darboux intermediate value theorem implies that f' cannot have removable discontinuities or discontinuities of first kind (jumps), and since f' is of bounded variation, f' cannot have discontinuities of second kind either.[12] So, we may apply Exercise 1.53 for $\phi := h' \circ f$ and $\psi := f'$ and obtain

$$\mathrm{Var}(g'; [a,b]) \le \|h' \circ f\|_\infty \mathrm{Var}(f'; [a,b]) + \sum_{k=1}^{\infty} \mathrm{Var}(g'; [a_k, b_k]), \tag{5.62}$$

where (a_k, b_k) denote the connected components of $(a,b) \setminus (f')^{-1}(0)$, i.e. the open intervals between two subsequent zeros of f'. Since $h' \circ f$ is bounded on $[a,b]$, being the composition of two bounded functions, and $\mathrm{Var}(f'; [a,b])$ is finite, by assumption, the only term we have to estimate is the series (or sum) in (5.62).

Given $P = \{t_0, t_1, \dots, t_m\} \in \mathcal{P}([a_k, b_k])$, we get

$$\begin{aligned}
\mathrm{Var}(g', P; [a_k, b_k]) &= \sum_{j=1}^{m} |g'(t_j) - g'(t_{j-1})| \\
&\le \sum_{j=1}^{m} \left|h'(f(t_j)) - h'(f(t_{j-1}))\right| |f'(t_j)| + \sum_{j=1}^{m} \left|h'(f(t_{j-1}))\right| |f'(t_j) - f'(t_{j-1})| \\
&\le \sup_{j=0,\dots,m} |f'(t_j)| \mathrm{Var}(h' \circ f, P; [a_k, b_k]) + \sup_{c \le u \le d} |h'(u)| \mathrm{Var}(f', P; [a_k, b_k]) \\
&\le \sup_{j=0,\dots,m} |f'(t_j)| \mathrm{Var}(h'; [c,d]) + \sup_{c \le u \le d} |h'(u)| \mathrm{Var}(f', P; [a_k, b_k]),
\end{aligned}$$

where in the last inequality, we have used the fact that $f'(x) \ne 0$ on (a_k, b_k), and hence f is monotone on $[a_k, b_k]$.

We distinguish two cases. If $f'(x) \ne 0$ on $[a,b]$, then the series in (5.62) consists of just one term, and the above estimate simplifies to

$$\mathrm{Var}(g'; [a,b]) \le \|f'\|_C \mathrm{Var}(h'; [c,d]) + \sup_{c \le u \le d} |h'(u)| \mathrm{Var}(f'; [a,b]), \tag{5.63}$$

12 This reasoning implies the somewhat surprising inclusion $BV^1 \subseteq C^1$, although the analogous inclusion $BV \subseteq C$ is of course not true.

which shows that $g \in BV^1([a,b])$. On the other hand, assume that f' vanishes somewhere in $[a,b]$, say $f'(a_k) = 0$. Then

$$\sup_{j=0,\dots,m} |f'(t_j)| = \sup_{j=0,\dots,m} |f'(t_j) - f'(a_k)| \le \mathrm{Var}(f';[a_k,b_k]) .$$

Consequently, passing to the supremum over $P \in \mathcal{P}([a_k,b_k])$, we get

$$\mathrm{Var}(g';[a_k,b_k]) \le \mathrm{Var}(f';[a_k,b_k]) \left[\mathrm{Var}(h';[c,d]) + \sup_{c\le u\le d} |h'(u)| \right],$$

and summing up over k on both sides yields

$$\mathrm{Var}(g';[a,b]) \le \mathrm{Var}(f';[a,b]) \left[\mathrm{Var}(h';[c,d]) + \sup_{c\le u\le d} |h'(u)| \right]. \tag{5.64}$$

It remains to show that the operator C_h is bounded under the assumptions of Theorem 5.35. To this end, we use the norm

$$|||f|||_{BV^1} := |f(a)| + |||f'|||_{BV} = |f(a)| + \|f'\|_\infty + \mathrm{Var}(f';[a,b])$$

which is equivalent to the norm (5.57). In the first case, from (5.63), we get

$$\mathrm{Var}(g';[a,b]) \le \left(\|f'\|_C + \mathrm{Var}(f';[a,b]) \right) \left(\sup_{c\le u\le d} |h'(u)| + \mathrm{Var}(h';[c,d]) \right),$$

and hence

$$|||g|||_{BV^1} \le 2|||f|||_{BV^1}|||h|||_{BV^1} .$$

The reasoning in the second case is similar and follows from (5.64). □

Before stating a parallel result for the space AC^1 of functions with absolutely continuous derivatives, we make some general remarks on absolutely continuous functions. If $f \in AC^1([a,b])$ and $g \in AC_{\mathrm{loc}}(\mathbb{R})$, we remark that $g \circ f$ need not be absolutely continuous. We give an example [75] involving the oscillation functions (0.86).

Example 5.36. Let $f_{\alpha,\beta} : [0,1] \to \mathbb{R}$ be defined by (0.86), and let $g : \mathbb{R} \to \mathbb{R}$ be defined by $g(u) := |u|^\gamma$. Here, we choose α, β and γ in such a way that[13]

$$0 < \gamma < \frac{1}{2}, \quad \beta < -\frac{2\gamma}{1-2\gamma}, \quad 2 - 2\beta < \alpha \le -\frac{\beta}{\gamma}. \tag{5.65}$$

From Exercise 0.54 (c), it follows that $f_{\alpha,\beta} \in C^2([0,1]) \subseteq AC^1([0,1])$ for this choice of α and β. Clearly, $g \in AC([-1,1])$, by the Vitali–Banach–Zaretskij theorem (Theorem 3.9). On the other hand, the function $g \circ f_{\alpha,\beta}$ is not absolutely continuous since its derivative on $(0,1]$ has the form

$$(g \circ f_{\alpha,\beta})'(x) = \alpha\gamma x^{\alpha\gamma-1} \sin^\gamma x^\beta + \beta\gamma x^{\alpha\gamma+\beta-1} \sin^{\gamma-1} x^\beta \cos x^\beta$$

and so does not belong to $L_1([0,1])$, by our assumption $\alpha\gamma \le -\beta$ and Exercise 0.7. ♥

13 Note that the first condition in (5.65) implies $\beta < 0$, while the second condition in (5.65) implies $\alpha > 0$.

One may even construct, by means of a refinement of Example 5.36, functions $f \in C^\infty([a,b])$ and $g \in AC_{\mathrm{loc}}(\mathbb{R})$ such that $g \circ f \notin AC([a,b])$. However, it is a remarkable fact that we *always* have

$$(g \circ f)f' \in AC([a,b]) \tag{5.66}$$

for $f \in AC^1([a,b])$ and $g \in AC_{\mathrm{loc}}(\mathbb{R})$, and so multiplying by f' has a "smoothing effect." Now, applying this to the derivative $g = h'$ of $h \in AC^1_{\mathrm{loc}}(\mathbb{R})$, we may conclude from (5.66) that $f \in AC^1([a,b])$ and $h \in AC^1_{\mathrm{loc}}(\mathbb{R})$ implies $(h \circ f)' \in AC([a,b])$; this is basically the contents of the following

Theorem 5.37. *The operator* (5.1) *maps the space* $AC^1([a,b])$ *into itself if and only if* $h \in AC^1_{loc}(\mathbb{R})$. *Moreover, in this case, the operator* (5.1) *is automatically bounded in the norm* (5.53).

Proof. Again, the "only if" part is a consequence of Proposition 5.32, so we must only prove the "if" part. As before, let $f \in AC^1([a,b])$, $h \in AC^1([c,d])$, where $f([a,b]) \subseteq [c,d]$, and $g := h \circ f$. We use the Vitali–Banach–Zaretskij theorem (Theorem 3.9) to prove that $g \in AC^1([a,b])$.

From $f' \in AC([a,b]) \subseteq C([a,b]) \cap BV([a,b])$ and $h' \in AC([c,d]) \subseteq C([c,d]) \cap BV([c,d])$, it follows immediately that g' is continuous and has bounded variation by the chain rule (5.38) and the fact that both spaces C and BV are algebras with respect to multiplication. However, from Exercise 3.21, we know that the set $Lu([a,b])$, see Definition 3.8, is also stable under multiplication, and so Theorem 3.9 shows that $g' \in AC([a,b])$ as claimed. The proof of the boundedness of the operator (5.1) in the norm (5.53) is the same as in Theorem 5.35. □

Again, we do not know whether or not the composition operator (5.1) is also continuous in the norm (5.53) whenever it maps the space $AC^1([a,b])$ into itself.

We pass now to the space $RBV^1_p([a,b])$ for $1 < p < \infty$. The restriction $1 < p < \infty$ is important insofar as the space $BV([a,b])$ plays an exceptional role: as we have seen in Theorem 5.10, the inclusion $C_h(RBV_p) \subseteq RBV_p$ holds if and only if h satisfies (5.13). The following theorem is in sharp contrast to this.

Theorem 5.38. *For* $1 < p < \infty$, *the operator* (5.1) *maps the space* $RBV^1_p([a,b])$ *into itself if and only if* $h \in RBV^1_{p,loc}(\mathbb{R})$. *Moreover, in this case, the operator* (5.1) *is automatically bounded in the norm* (5.55).

Proof. Without loss of generality, we assume again that $[a,b] = [0,1]$. The necessity of the condition $h \in RBV^1_{p,loc}(\mathbb{R})$ for the inclusion $C_h(RBV^1_p) \subseteq RBV^1_p$ follows from the fact that the identity $f(x) = x$ belongs to $RBV^1_p([a,b])$ for any interval $[a,b]$.

To prove sufficiency, suppose that $h \in RBV^1_{p,loc}(\mathbb{R})$, and let $f \in RBV^1_p([0,1])$, and so $h'' \in L_{p,loc}(\mathbb{R})$ and $f'' \in L_p([0,1])$, by the Riesz theorem (Theorem 3.34). Since $h' \in RBV_{p,loc}(\mathbb{R}) \subset AC(\mathbb{R})$ and $f \in C([0,1])$, we have $h' \circ f \in C([0,1])$ and so $(h' \circ f)f'' \in L_p([0,1])$. Similarly, from $h'' \in L_{p,loc}(\mathbb{R})$, $f \in RBV^1_p([0,1])$ and $f' \in RBV_p([0,1])$ it follows that $(h'' \circ f)f'^2 \in L_p([0,1])$. So we have proved that $(h \circ f)'' \in L_p([0,1])$, by (5.61). The fact that $(h \circ f)' \in RBV_p([0,1])$ is proved in the same way as in Theorem 5.35.

Finally, let us show that the operator C_h is bounded in the norm (5.55) whenever it maps RBV_p^1 into itself. Given $f \in RBV_p^1([0,1])$ with $\|f\|_{RBV_p^1} \le r$, and hence $|f(t)| \le r$ and $|f'(t)| \le r$ for $0 \le t \le 1$, we get the estimates

$$|(h \circ f)(0)| \le \tilde{k}(r), \quad |(h \circ f)'(0)| = |h'(f(0))|\,|f'(0)| \le r\tilde{k}_1(r)$$

and

$$\int_0^1 |h''(f(t))|^p |f'(t)|^{2p}\,dt \le r^{2p} \int_0^1 |h''(f(t))|^p\,dt \le r^{2p} c_p(r),$$

where we use the notation (5.15) and (5.16), and the constant $c_p(r)$ is finite since $h'' \circ f \in L_{p,loc}(\mathbb{R})$. Moreover,

$$\int_0^1 |h'(f(t))|^p |f''(t)|^p\,dt \le \tilde{k}_1(r)^p \int_0^1 |f''(t)|^p\,dt = \tilde{k}_1(r)^p \|f''\|_{L_p}^p \le r^p \tilde{k}_1(r)^p .$$

So, from (5.38) and (5.61), we conclude that C_h is bounded in the norm (5.55). □

Again, the problem arises whether or not the operator (5.1) is also continuous in the norm (5.55) whenever it maps the Riesz space $RBV_p^1([a,b])$ into itself. This seems to be an open problem. On the other hand, the continuity problem for the operator (5.1) in the Wiener space $WBV_p^1([a,b])$ for $p > 1$ was completely solved in [56], see Theorem 5.56 in Section 5.6 below.

Now, we solve the COP for the space Lip_α^1, where for simplicity, we restrict ourselves to the case $\alpha = 1$. Again, the following theorem is in sharp contrast to Theorem 5.10.

Theorem 5.39. *The operator* (5.1) *maps the space* $Lip^1([a,b])$ *into itself if and only if* $h \in Lip_{\text{loc}}^1(\mathbb{R})$. *Moreover, in this case, the operator* (5.1) *is automatically bounded in the norm* (5.56).

Proof. Again, without loss of generality, we may assume that $[a,b] = [0,1]$. The necessity of the condition $h \in Lip_{\text{loc}}^1(\mathbb{R})$ for the inclusion $C_h(Lip^1) \subseteq Lip^1$ follows as before. Suppose that $h \in Lip_{\text{loc}}^1(\mathbb{R})$, and let $f \in Lip^1([0,1])$ and $g := C_h f$, and thus $g' = (h' \circ f)f'$. From $h \in Lip_{\text{loc}}^1(\mathbb{R})$, we get $C_{h'}(Lip) \subseteq Lip$, by Theorem 5.10. Combining this with $f' \in Lip([0,1])$, we conclude that $g' \in Lip([0,1])$ since $Lip([0,1])$ is an algebra. So, we have shown that $g \in Lip^1([0,1])$, i.e. $C_h(Lip^1) \subseteq Lip^1$.

Now, we show that the operator C_h is bounded in the norm (5.56) under the assumption $h \in Lip_{\text{loc}}^1(\mathbb{R})$. Given $f \in Lip^1([0,1])$ with $\|f\|_{Lip^1} \le r$ and using again (5.38), we have

$$\|f'\|_C \le |f'(0)| + lip(f') = \|f'\|_{Lip} \le r, \quad \|h' \circ f\|_C \le \tilde{k}_1(r).$$

Moreover, since the operator $C_{h'}$ is bounded in the space $Lip([0,1])$, by Theorem 5.24, we have $lip(h' \circ f) \leq \|C_{h'} f\|_{Lip}$, which implies that C_h is bounded in the norm (5.56). □

Observe that again we did not claim automatic continuity of C_h in Theorem 5.39. However, we are able to present a counterexample which shows that the operator (5.1) may map the space Lip^1 into itself without being continuous in the norm (5.56). This counterexample imitates Example 5.25.

Example 5.40. Define $h : \mathbb{R} \to \mathbb{R}$ by

$$h(u) := \begin{cases} 0 & \text{for } u \leq 0\,, \\ \frac{1}{2}u^2 & \text{for } 0 < u < 1\,, \\ u - \frac{1}{2} & \text{for } u \geq 1\,. \end{cases}$$

Since $h'(u) \equiv 0$ for $u \leq 0$ and $h'(u) = \min\{u, 1\}$ for $u > 0$, we certainly have $h \in Lip^1_{loc}(\mathbb{R})$ (even $h \in Lip^1(\mathbb{R})$), and so C_h maps $Lip^1([0,1])$ into itself and is bounded, by Theorem 5.39. However, C_h is not continuous in the norm (5.56). To see this, consider the functions $f, f_n : [0,1] \to \mathbb{R}$ defined by

$$f(x) := x, \quad f_n(x) := \frac{n+1}{n}x \quad (0 \leq x \leq 1)\,.$$

Clearly, $\|f_n - f\|_{Lip^1} = \frac{1}{n} \to 0$ as $n \to \infty$. On the other hand, the function g_n defined by (5.58) and its derivative satisfy

$$g_n(x) = \begin{cases} \frac{2n+1}{2n^2}x^2 & \text{for } 0 \leq x \leq \tau_n\,, \\ \frac{n+1}{n}x - \frac{1}{2}(1 + x^2) & \text{for } \tau_n < x \leq 1\,, \end{cases}$$

and

$$g_n'(x) = \begin{cases} \frac{2n+1}{n^2}x & \text{for } 0 \leq x \leq \tau_n\,, \\ \frac{n+1}{n} - x & \text{for } \tau_n < x \leq 1\,, \end{cases}$$

where $\tau_n := \frac{n}{n+1}$. Consequently,

$$\|C_h f_n - C_h f\|_{Lip^1} \geq lip(g_n') \geq \frac{|g_n'(1) - g_n'(\tau_n)|}{1 - \tau_n} = \frac{n+1}{n+1} \equiv 1\,,$$

which shows that $\|C_h f_n - C_h f\|_{Lip^1} \not\to 0$ as $n \to \infty$. ♥

Observe that the function h in Example 5.40 belongs to $Lip^1_{loc}(\mathbb{R})$, and so also to $C^1(\mathbb{R})$, but has no second derivative at 0 and 1. This is not accidental, as the following theorem shows which provides a necessary and sufficient continuity condition and is parallel to Theorem 5.26.

Theorem 5.41. *The operator* (5.1) *maps the space* $Lip^1([a,b])$ *into itself and is continuous in the norm* (5.56) *if and only if* $h \in C^2(\mathbb{R})$.

Since the proof of this theorem is quite similar to that of Theorem 5.26, we leave it as an exercise to the reader (Exercise 5.24).

Our previous discussion shows that in all spaces under consideration, we get boundedness of the operator C_h for free, while continuity is a delicate problem: in the largest space C^1 in (5.59), continuity holds, in the smallest space Lip^1 in (5.59), continuity fails, and in the intermediate spaces RBV_p^1 and AC^1 in (5.59), we do not know the answer. For the reader's ease, we summarize our results in the following synoptic tables which complement the Tables 5.2–5.8 above.

Table 5.9. The operator C_h in $C^1([a,b])$.

C_h bounded in C^1	$\Leftrightarrow$	$C_h(C^1) \subseteq C^1$	$\Leftrightarrow$	C_h continuous in C^1
		$\Updownarrow$		
		$h \in C^1(\mathbb{R})$		

Table 5.10. The operator C_h in $Lip_\alpha^1([a,b])$ $(0 < \alpha \leq 1)$.

C_h bounded in Lip_α^1	$\Leftrightarrow$	$C_h(Lip_\alpha^1) \subseteq Lip_\alpha^1$	$\Leftarrow$	C_h continuous in Lip_α^1
		$\Updownarrow$		$\Updownarrow$
		$h \in Lip_{\text{loc}}^1(\mathbb{R})$	$\Leftarrow$	$h \in C^2(\mathbb{R})$

Table 5.11. The operator C_h in $BV^1([a,b])$.

C_h bounded in BV^1	$\Leftrightarrow$	$C_h(BV^1) \subseteq BV^1$	$\Leftarrow$	C_h continuous in BV^1
		$\Updownarrow$		
		$h \in BV_{\text{loc}}^1(\mathbb{R})$		

Table 5.12. The operator C_h in $AC^1([a,b])$.

C_h bounded in AC^1	$\Leftrightarrow$	$C_h(AC^1) \subseteq AC^1$	$\Leftarrow$	C_h continuous in AC^1
		$\Updownarrow$		
		$h \in AC_{\text{loc}}^1(\mathbb{R})$		

Table 5.13. The operator C_h in $RBV_p^1([a,b])$ $(p > 1)$.

C_h bounded in RBV_p^1	$\Leftrightarrow$	$C_h(RBV_p^1) \subseteq RBV_p^1$	$\Leftarrow$	C_h continuous in RBV_p^1
		$\Updownarrow$		
		$h \in RBV_{p,loc}^1(\mathbb{R})$		

Observe that the Tables 5.11–5.13 do not contain conditions on h, both necessary and sufficient, under which the operator C_h is continuous in the norm of the corresponding space. As far as we know, such continuity criteria are not known. Of course, one may again try to find conditions which are just sufficient (Exercises 5.21–5.23).

5.4 Global Lipschitz continuity

Suppose that the composition operator C_h given by (5.1) maps a normed space X into a normed space Y. If we want to apply fixed point theorems to nonlinear problems in infinite dimensional Banach spaces, just continuity of the operators involved is too weak to guarantee the existence of fixed points. In case of the Schauder fixed point principle, we need some compactness requirement which in infinite dimensional spaces is often quite restrictive. On the other hand, in case of the Banach–Caccioppoli fixed point principle, one usually imposes a (global) Lipschitz condition of the type

$$\|C_h f - C_h g\|_X \leq K\|f - g\|_X \quad (f, g \in X) \tag{5.67}$$

with $K < 1$ to guarantee the existence (and uniqueness) of a fixed point of C_h in X. Unfortunately, if C_h is the nonlinear composition operator (5.1) (or, more generally, the nonlinear superposition operator (5.2), the global Lipschitz condition (5.67) leads to a strong degeneracy for the generating function h in many function spaces X.

In fact, from (5.67), it often follows that the function h must be *affine*, which means that

$$h(u) = \alpha + \beta u \quad (\alpha, \beta \geq 0) \tag{5.68}$$

in case of the composition operator (5.1), or

$$h(t, u) = \alpha(t) + \beta(t)u \quad (\alpha, \beta \in Y) \tag{5.69}$$

in case of the superposition operator (5.2). This was shown, even in the nonautonomous case of the operator (5.2), for

- the space $Lip_\alpha([a, b])$ of Hölder continuous functions of order $\alpha < 1$ with norm (0.71) in [193];
- the space $C^n([a, b])$ of n-times continuously differentiable functions with norm (0.63) in [194];
- the space $WBV_{2,p}([a, b])$ of functions of bounded $(2, p)$-variation with norm (2.143) in [204];
- the space $Lip^n([a, b])$ of functions with Lipschitz continuous n-th derivative with norm (0.78) in [162];
- the space $Lip_\alpha^n([a, b])$ of functions with Hölder continuous n-th derivative in [187];
- the space $AC^n([a, b])$ of functions with absolutely continuous n-th derivative in [289];
- the space $RBV_p([a, b])$ of functions of bounded Riesz variation for $1 < p < \infty$ with norm (2.90) in [205][14] and [224]; and
- the space $RBV_\phi([a, b])$ of functions of bounded Riesz–Medvedev variation with norm (2.99) by [217].

14 To be precise, in [205] the authors consider the Sobolev space W_p^1 of functions with distributional first derivative in L_p; as we have seen in Theorem 3.34, this is the same as RBV_p.

Since Janusz Matkowski was the first, to the best of our knowledge, to discover this degeneracy phenomenon, we will use the following terminology.

Definition 5.42. We say that a pair (X, Y) of two normed spaces $(X, \|\cdot\|)$ and $(Y, \|\cdot\|)$ has the *Matkowski property* if whenever the operator (5.1) [the operator (5.2), respectively] maps the space X into the space Y and satisfies (5.67) [(5.67) with C_h replaced by S_h, respectively], the corresponding function h must have the form (5.68) [the form (5.69), respectively]. In case $X = Y$, we simply say that X has the Matkowski property. ∎

Thus, our collection above shows that many spaces occurring frequently in applications have the Matkowski property. We point out, however, that there are important function spaces which do *not* have the Matkowski property. Two examples of such spaces are contained in the following Theorems 5.43 and 5.44; we state them for the general operator (5.2).

Theorem 5.43. *Suppose that the operator S_h given by* (5.2) *maps the space $C([a, b])$ into itself. Then S_h satisfies* (5.67) *if and only if the corresponding function h satisfies*

$$|h(t, u) - h(t, v)| \leq g(t)|u - v| \quad (a \leq t \leq b,\ u, v \in \mathbb{R}) \tag{5.70}$$

for some continuous function $g : [a, b] \to \mathbb{R}$.

Theorem 5.44. *Suppose that the operator S_h given by* (5.2) *maps the space $L_p([a, b])$ into the space $L_q([a, b])$, where $1 \leq q \leq p < \infty$. Then S_h satisfies* (5.67) *if and only if the corresponding function h satisfies* (5.70), *where $g \in L_{pq/(p-q)}([a, b])$.*

The proof of Theorems 5.43 and 5.44 may be found in [20]. We remark that necessary and sufficient conditions under which the superposition operator (5.2) fulfills the hypotheses of these theorems are well known.

Clearly, condition (5.70) may be replaced by the simpler condition

$$|h(t, u) - h(t, v)| \leq k|u - v| \quad (a \leq t \leq b,\ u, v \in \mathbb{R}), \tag{5.71}$$

where $k := \|g\|_C$, i.e. by a *global* Lipschitz condition for the function $f(t, \cdot)$ on $\mathbb{R}$. In case of the autonomous composition operator (5.1), condition (5.71) becomes

$$|h(u) - h(v)| \leq k|u - v| \quad (u, v \in \mathbb{R}), \tag{5.72}$$

which is a global Lipschitz condition for f on the whole real axis. Of course, (5.72) is much more restrictive than the local condition (5.13).

We now prove a general theorem which provides a unified approach to spaces with the Matkowski property for the autonomous operator (5.1). Without loss of generality, we take $[a, b] = [0, 1]$. We consider the space $P_n([0, 1])$ of all polynomials of degree $\leq n$

on $[0, 1]$ equipped with the C^n-norm (0.63). In particular, on the space $P_1([0, 1])$ of all affine functions $f(x) = cx+d$, we consider the C^1-norm $\|f\|_{C^1} = |c|+|d|$. It is interesting to note that the global Lipschitz condition (5.67) for C_h may even be weakened: just uniform continuity[15] of C_h suffices to imply (5.68).

Theorem 5.45. *Suppose that the operator* (5.1) *maps a normed space X into a normed space Y and is uniformly continuous on X. Assume that the space* $P_1([0, 1])$ *of affine functions equipped with the* C^1 *norm* (0.65) *is imbedded into X, and Y is imbedded into the Hölder space* $Lip_\gamma([0, 1])$ *for some* $\gamma \in (0, 1]$ *with norm* (0.71). *Then there exist constants* $\alpha, \beta \in \mathbb{R}$ *such that* (5.68) *holds true.*

Proof. From our assumptions, it follows that we can find a $\delta > 0$ such that $\|C_h f - C_h g\|_{Lip_\gamma} \leq 1$ for all $f, g \in P_1([0, 1])$ satisfying $\|f - g\|_{C^1} \leq \delta$.

Fix $\omega > 0$ and $v \in [-\delta, \delta]$, and define $f_\omega, g_\omega \in P_1([0, 1])$ by $f_\omega(t) := \omega t + v$ and $g_\omega(t) := \omega t$. Since $\|f_\omega - g_\omega\|_{C^1} = |v| \leq \delta$, we know that $lip_\gamma(C_h f_\omega - C_h g_\omega) \leq 1$, and hence

$$|h(\omega s + v) - h(\omega s) - h(\omega t + v) + h(\omega t)| \leq |s - t|^\gamma .$$

Putting, in particular, $s = u/\omega$ and $t = 0$, we conclude that

$$|h(u + v) - h(u) - h(v) + h(0)| \leq \left|\frac{u}{\omega}\right|^\gamma \to 0 \quad (\omega \to \infty) .$$

Suppose first that $h(0) = 0$. Then the last equality shows that

$$h(u + v) = h(u) + h(v) \quad (u, v \in \mathbb{R},\ |v| \leq \delta)$$

which by standard arguments, implies that $h(u) = \beta u$ with $\beta = h(1)$. Replacing h in case $h(0) \neq 0$ by the function $u \mapsto h(u) - h(0)$ the statement follows with $\alpha = h(1) - h(0)$. □

The range of applicability of Theorem 5.45 seems to be rather limited because the hypothesis $Y \hookrightarrow Lip_\gamma([0, 1])$ is quite strong. However, some information may be deduced from this theorem which we state as

Corollary 5.46. *The operator* (5.1) *is uniformly continuous on either of the spaces Lip,* Lip_α, C^1, AC^1, BV^1, Lip^1, *or* RBV_p^1 *if and only if the corresponding function h has the form* (5.68).

To cover a larger family of spaces with the Matkowski property, we use the space WBV_p of bounded p-variation in Wiener's sense (in particular, the classical space $WBV_1 = BV$).

15 Here, we mean uniform continuity *on the whole space*; this has to be carefully distinguished from uniform continuity on bounded subsets which we considered in Theorem 5.26. Also, note that (5.68) trivially implies that the corresponding operator C_h is uniformly continuous on X.

Theorem 5.47. *Assume that the space X contains all piecewise linear functions, and the space Y is imbedded into the space $WBV_p([a,b])$ for some $p \geq 1$ with norm (1.65). Then the pair (X,Y) has the Matkowski property.*

Proof. Suppose that the composition operator (5.1) maps X into Y and satisfies the Lipschitz condition (5.67). From our hypothesis $Y \hookrightarrow WBV_p([a,b])$, it then follows that

$$\|C_h f - C_h g\|_{WBV_p} \leq L\|f-g\|_X \quad (f,g \in X) \tag{5.73}$$

for some $L > 0$. Let $a \leq s < t \leq b$, and let $P_m := \{t_0, t_1, \ldots, t_{2m}\} \in \mathcal{P}([s,t])$ be the equidistant partition defined by

$$t_0 = s, \quad t_j - t_{j-1} = \frac{t-s}{2m} \quad (j = 1, 2, \ldots, 2m).$$

Given $u, v \in \mathbb{R}$ with $u \neq v$, define $f, g : [a,b] \to \mathbb{R}$ by

$$f(x) := \begin{cases} v & \text{if } x = t_j \text{ for some even } j, \\ \frac{u+v}{2} & \text{if } x = t_j \text{ for some odd } j, \\ \text{linear} & \text{otherwise}, \end{cases}$$

and

$$g(x) := \begin{cases} \frac{u+v}{2} & \text{if } x = t_j \text{ for some even } j, \\ u & \text{if } x = t_j \text{ for some odd } j, \\ \text{linear} & \text{otherwise}, \end{cases}$$

respectively. Then the difference $f - g$ trivially satisfies

$$|f(x) - g(x)| \equiv \frac{|u-v|}{2} \quad (a \leq x \leq b).$$

Consequently, substituting these functions f and g into (5.73) yields

$$\|C_h f - C_h g\|_{WBV_p} \leq \frac{K|u-v|}{2},$$

where the constant K is given by $K := L\|f_1\|_X$ and f_1 denotes the constant function $f(x) \equiv 1$. In particular, for the partition P_m, as above, we have

$$\begin{aligned} &\sum_{j=1}^{m} \left|(h \circ f)(t_{2j}) - (h \circ g)(t_{2j}) - (h \circ f)(t_{2j-1}) + (h \circ g)(t_{2j-1})\right|^p \\ &\quad = \mathrm{Var}_p^W(C_h f - C_h g, P_m; [s,t]) \leq \frac{K^p|u-v|^p}{2^p}. \end{aligned} \tag{5.74}$$

However, by definition of the functions f and g, we have

$$\begin{aligned} &(h \circ f)(t_{2j}) - (h \circ g)(t_{2j}) - (h \circ f)(t_{2j-1}) + (h \circ g)(t_{2j-1}) \\ &\quad = h(v) - h(\tfrac{u+v}{2}) - h(\tfrac{u+v}{2}) + h(u) \end{aligned}$$

which is actually independent of m. So, combining this with (5.74), we obtain

$$\sum_{j=1}^{m} |h(v) - h(\tfrac{u+v}{2}) - h(\tfrac{u+v}{2}) + h(u)|^p \leq \tfrac{K^p|u-v|^p}{2^p}. \tag{5.75}$$

Letting now $m \to \infty$ in (5.75), we conclude that the function h satisfies the Cauchy functional equation

$$h(u) + h(v) = 2h(\tfrac{u+v}{2}).$$

Since h is continuous, it follows that $h(u) = \alpha + \beta u$ with $\beta = h(0)$ and $\alpha = h(1) - h(0)$ as desired. □

In Chapter 6, we shall obtain a far reaching generalization of Theorem 5.47 for the case of the superposition operator (5.2). Theorem 5.47 applies to the composition operator (5.1) in any space X which is continuously imbedded into WBV_p for some $p \geq 1$, such as WBV_p itself, BV, RBV_p, or AC. Moreover, Proposition 2.33 shows that Theorem 5.47 applies as well to the Waterman space $\Lambda_q BV$ for any $q \in (0, 1)$ since $\Lambda_q BV \hookrightarrow WBV_p$ for $p \geq 1/(1-q)$.

5.5 Local Lipschitz continuity

As we have seen, imposing the global Lipschitz condition (5.67) to the composition operator C_h in BV leads to a very restrictive condition for the generating function h. So, the question arises regarding how to replace (5.67) by some milder condition for C_h which does not lead to such a narrow class of generating functions h.

A good idea is to replace the *global* Lipschitz condition (5.67) by a *local* Lipschitz condition of type

$$\|C_h f - C_h g\|_X \leq K(r)\|f - g\|_X \quad (f, g \in X;\ \|f\|, \|g\| \leq r), \tag{5.76}$$

i.e. imposing Lipschitz conditions only on closed balls of radius $r > 0$ and allowing the corresponding Lipschitz constant $K(r)$ to depend on r (and, as a matter of fact, tending to infinity as $r \to \infty$). It turns out that the local condition (5.76) is much more reasonable than the global condition (5.67) for C_h, insofar as it does not lead to a strong degeneracy for the generating function h. We will show this now for the classical space $BV([a, b])$ in case of the composition operator (5.1). First, we need a technical lemma which seems to be of some interest on its own. Apart from condition (5.76) for the operator C_h, at many places, we will need the local Lipschitz condition (5.13) for the function h as well as the local Lipschitz condition (5.14) for its derivative. Moreover, we will use the characteristics $\tilde{k}(r)$ and $\tilde{k}_1(r)$ defined in (5.15) and (5.16), respectively.

Lemma 5.48. *Suppose that the derivative of a function $h \in C^1(\mathbb{R})$ satisfies the local Lipschitz condition (5.14). Then for $|x_1|, |x_2|, |y_1|, |y_2| \le r$, we have the estimate*

$$|h(x_1) - h(y_1) - h(x_2) + h(y_2)| \\ \le k_1(r)\left(|x_1 - x_2| + |y_1 - y_2|\right)\left(|x_1 - y_1| + |x_2 - y_2|\right) + \tilde{k}_1(r)|x_1 - y_1 - x_2 + y_2| .$$

Proof. We distinguish the cases

$$|x_1 - y_1| + |x_2 - y_2| \le |x_1 - x_2| + |y_1 - y_2| \tag{5.77}$$

and

$$|x_1 - y_1| + |x_2 - y_2| > |x_1 - x_2| + |y_1 - y_2| . \tag{5.78}$$

In the first case, we choose, by the mean value theorem, some ξ_i between x_i and y_i (hence, $|\xi_i| \le r$) with

$$h(x_i) - h(y_i) = h'(\xi_i)(x_i - y_i) \quad (i = 1, 2) .$$

Using (5.77), a straightforward but cumbersome case distinction shows that

$$|\xi_1 - \xi_2| \le |x_1 - x_2| + |y_1 - y_2| .$$

Consequently,

$$\begin{aligned} &|h(x_1) - h(y_1) - h(x_2) + h(y_2)| \\ &\quad = |h'(\xi_1)(x_1 - y_1) - h'(\xi_2)(x_2 - y_2)| \\ &\quad = |\left[h'(\xi_1) - h'(\xi_2)\right](x_1 - y_1) + h'(\xi_2)(x_1 - y_1 - x_2 + y_2)| \\ &\quad \le k_1(r)\left(|x_1 - x_2| + |y_1 - y_2|\right)|x_1 - y_1| + \tilde{k}_1(r)|x_1 - y_2 - x_2 + y_2| . \end{aligned} \tag{5.79}$$

In the second case, we choose, again by the mean value theorem, some η_x between x_1 and x_2 and some η_y between y_1 and y_2 (hence, $|\eta_x|, |\eta_y| \le r$) satisfying

$$h(x_1) - h(x_2) = h'(\eta_x)(x_1 - x_2), \quad h(y_1) - h(y_2) = h'(\eta_y)(y_1 - y_2) .$$

As before, a straightforward but cumbersome case distinction, now building on (5.78), shows that

$$|\eta_x - \eta_y| \le |x_1 - y_1| + |x_2 - y_2| .$$

So, in this case, we obtain

$$\begin{aligned} &|h(x_1) - h(y_1) - h(x_2) + h(y_2)| \\ &\quad = |h'(\eta_x)(x_1 - x_2) - h'(\eta_y)(y_1 - y_2)| \\ &\quad = |\left[h'(\eta_x) - h'(\eta_y)\right](x_1 - x_2) + h'(\eta_y)(x_1 - y_1 - x_2 + y_2)| \\ &\quad \le k_1(r)\left(|x_1 - y_1| + |x_2 - y_2|\right)|x_1 - x_2| + \tilde{k}_1(r)|x_1 - y_1 - x_2 + y_2| . \end{aligned} \tag{5.80}$$

Estimating the terms after $k_1(r)$ in (5.79) and (5.80) in a unified manner, we obtain the statement. □

Theorem 5.49. *Suppose that the operator* (5.1) *maps the space* $BV([a,b])$ *into itself and is continuous and bounded with respect to the norm* (1.16). *Then the operator* (5.1) *satisfies* (5.76) *if and only if* h' *exists on* $\mathbb{R}$ *and satisfies* (5.14).

Proof. By the COP-invariance of the space BV (Example 5.3), there is no loss of generality to assume that $[a,b] = [0,1]$. Suppose first that the derivative h' of h satisfies the local Lipschitz condition (5.14), and define $\tilde{k}_1(r)$ as in (5.16). Fix $f, g \in BV([0,1])$ with $\|f\|_{BV} \le r$ and $\|g\|_{BV} \le r$. Given a partition $P = \{t_0, t_1, \dots, t_m\} \in \mathcal{P}([0,1])$, we apply Lemma 5.48 to the choice

$$f(t_j) =: x_1, \quad f(t_{j-1}) =: x_2, \quad g(t_j) =: y_1, \quad g(t_{j-1}) =: y_2 \quad (j = 1, 2, \dots, m).$$

As a result, we get the estimate

$$\begin{aligned}
&|h(f(t_j)) - h(g(t_j)) - h(f(t_{j-1})) + h(g(t_{j-1}))| \\
&\quad \le k_1(r)\big(|f(t_j) - f(t_{j-1})| + |g(t_j) - g(t_{j-1})|\big)\big(|f(t_j) - g(t_j)| + |f(t_{j-1}) - g(t_{j-1})|\big) \\
&\qquad + \tilde{k}_1(r)|f(t_j) - g(t_j) - f(t_{j-1}) + g(t_{j-1})|.
\end{aligned}$$

Taking the sum over $j = 1, 2, \dots, m$ and using the norm $\|\cdot\|_\infty$ from (0.39), we obtain

$$\begin{aligned}
&\sum_{j=1}^{m} |(h \circ f)(t_j) - (h \circ g)(t_j) - (h \circ f)(t_{j-1}) + (h \circ g)(t_{j-1})| \\
&\quad \le 2k_1(r)\|f - g\|_\infty \sum_{j=1}^{m} \big(|f(t_j) - f(t_{j-1})| + |g(t_j) - g(t_{j-1})|\big) \\
&\qquad + \tilde{k}_1(r) \sum_{j=1}^{m} |f(t_j) - g(t_j) - f(t_{j-1}) + g(t_{j-1})| \\
&\quad \le 2k_1(r)\|f - g\|_\infty (\operatorname{Var}(f) + \operatorname{Var}(g)) + \tilde{k}_1(r)\operatorname{Var}(f - g) \le \hat{k}_1(r)\|f - g\|_{BV}
\end{aligned}$$

with $\hat{k}_1(r) := \max\{4rk_1(r), \tilde{k}_1(r)\}$. This proves the first part of the theorem.

Now, we suppose that C_h maps the space $BV([0,1])$ into itself and satisfies a local Lipschitz condition (5.76) in the norm (1.16). By Theorem 5.9, h then satisfies the local Lipschitz condition (5.13). We will only use the fact that h is absolutely continuous on $[-r,r]$. There is a nullset $N \subset \mathbb{R}$ such that h' exists on $[-r,r] \setminus N$. By Exercise 3.32, it suffices to show that the restriction function $f'|_{[-r,r]\setminus N}$ to $[-r,r] \setminus N$ satisfies a Lipschitz condition.

We show this for the Lipschitz constant $L := K(3r+1)$, where $K = K(r)$ is the constant from (5.76). Thus, assume by contradiction that there are $u_0, v_0 \in [-r,r] \setminus N$ with

$$|h'(u_0) - h'(v_0)| > L|u_0 - v_0|.$$

Let $P = \{t_0, t_1, \dots, t_m\} \in \mathcal{P}([0,1])$ be any partition where

$$m := \operatorname{ent} \frac{1}{|u_0 - v_0|} \in \mathbb{N}$$

denotes the integer part of $1/|u_0 - v_0|$, and let $f : [0, 1] \to \mathbb{R}$ be defined by

$$f(x) := \begin{cases} u_0 & \text{if } x = t_j \text{ for some even } j \\ v_0 & \text{if } x = t_j \text{ for some odd } j \\ \text{linear} & \text{otherwise}. \end{cases} \tag{5.81}$$

Then f has bounded variation, being even Lipschitz continuous on $[0, 1]$, with

$$\mathrm{Var}(f; [0, 1]) = \mathrm{Var}(f, P; [0, 1]) = m|u_0 - v_0| < 1 + 2r\,,$$

and so $\|f\|_{BV} < 1 + 3r$. Hence, if n is sufficiently large, the function $f_n \in Lip([0, 1])$ defined by $f_n(t) := f(t) + 1/n$ also satisfies $\|f_n\|_{BV} < 1 + 3r$. By hypothesis, the function $g_n : [0, 1] \to \mathbb{R}$ defined by $g_n := n\,(C_h f_n - C_h f)$ thus satisfies

$$\|g_n\|_{BV} = n\|C_h f_n - C_h f\|_{BV} \le Ln\|f_n - f\|_{BV} = L$$

for all sufficiently large n. In particular,

$$L \ge \mathrm{Var}(g_n; [0, 1])) \ge \sum_{j=1}^{m} |g_n(t_j) - g_n(t_{j-1})|$$

for these n. However, the definition of the derivative implies that

$$\lim_{n\to\infty} g_n(t_j) = \begin{cases} h'(u_0) & \text{if } j \text{ is even}, \\ h'(v_0) & \text{if } j \text{ is odd}; \end{cases}$$

thus, we finally obtain

$$L \ge \sum_{j=1}^{m} |h'(u_0) - h'(v_0)| > Lm|u_0 - v_0| \ge L\,,$$

that is, a contradiction. The proof is complete. □

Observe that the proof of Theorem 5.49 provides an interesting interconnection between the Lipschitz constant $k_1(r)$ of h' in (5.14) and the Lipschitz constant $K(r)$ of C_h in (5.76): from (5.14), it follows that (5.76) holds with

$$K(r) \le \max\,\{4rk_1(r), \tilde{k}_1(r)\}\,, \tag{5.82}$$

where $\tilde{k}_1(r)$ is given by (5.16). Conversely, from (5.76), it follows that (5.14) holds with

$$k_1(r) \le \frac{2K(2r) + 1}{r}\,. \tag{5.83}$$

Such estimates are useful for applying fixed point theorems to problems involving nonlinear composition operators. Moreover, they sometimes make it possible to prove the degeneracy results from the previous section quite easily (see, e.g. Exercise 5.28).

In the same way as we have obtained an essential extension of Theorem 5.9 in Theorem 5.10, we may give the following important generalization of Theorem 5.49:

Theorem 5.50. *The following six statements about the operator* (5.1) *and the generating function $h : \mathbb{R} \to \mathbb{R}$ are equivalent.*

(a) *The operator* (5.1) *satisfies the local Lipschitz condition* (5.76) *between the spaces $Lip([a, b])$ and $BV([a, b])$.*
(b) *The operator* (5.1) *satisfies the local Lipschitz condition* (5.76) *in the space $Lip([a, b])$.*
(c) *The operator* (5.1) *satisfies the local Lipschitz condition* (5.76) *in the space $RBV_p([a, b])$.*
(d) *The operator* (5.1) *satisfies the local Lipschitz condition* (5.76) *in the space $AC([a, b])$.*
(e) *The operator* (5.1) *satisfies the local Lipschitz condition* (5.76) *in the space $BV([a, b])$.*
(f) *The derivative of the function h exists and satisfies* (5.14).

Proof. We take again $[a, b] = [0, 1]$. The inclusions (5.23) show that any of the statements (b), (c), (d) or (e) implies (a). So, we first prove, as in the proof of Theorem 5.10, that (a) implies (f). To this end, we analyze again the proof of Theorem 5.49. If we take the partition $P = \{t_0, t_1, \dots, t_m\} \in \mathcal{P}([0, 1])$ equidistant, i.e. such that $m(t_j - t_{j-1}) = 1$, then the function f defined by (5.81) is Lipschitz continuous with

$$lip(f) = m|u_0 - v_0| < 1 + 2r,$$

and so $\|f\|_{Lip} < 1 + 3r$. Hence, if n is sufficiently large, the function $f_n \in Lip([0, 1])$ defined by $f_n(t) := f(t) + 1/n$ also satisfies $\|f_n\|_{Lip} < 1 + 3r$. By hypothesis, the function $g_n : [0, 1] \to \mathbb{R}$ defined by $g_n := n(C_h f_n - C_h f)$ then satisfies

$$\|g_n\|_{BV} = n\|C_h f_n - C_h f\|_{BV} \le nL\|f_n - f\|_{Lip} = L,$$

where $L = K(3r + 1)$ as before.[16] The remaining part of the proof goes like that of Theorem 5.49. So, we conclude that (a) implies (f).

Conversely, if the derivative of h exists on $\mathbb{R}$ and satisfies (5.14), then C_h satisfies (5.76) in the space BV, by Theorem 5.49. However, all inclusions in (5.23) are indeed continuous imbeddings, and so C_h satisfies (5.76) also in the space Lip, RBV_p, and AC. This shows that (f) implies any of the other statements (a)–(e), and we are done. □

For the sake of completeness, we now prove that the same result holds in the Hölder space Lip_α.

Theorem 5.51. *Suppose that the operator* (5.1) *maps the space $Lip_\alpha([a, b])$ into itself and is continuous and bounded with respect to the norm* (0.71). *Then the operator* (5.1) *satisfies* (5.76) *if and only if h' exists on $\mathbb{R}$ and satisfies* (5.14).

Proof. Suppose first that the derivative h' of h satisfies the local Lipschitz condition (5.14), and define $\tilde{k}(r)$ as in (5.15). Fix $s, t \in [a, b]$ and $f, g \in Lip_\alpha([a, b])$ with $\|f\|_{Lip_\alpha} \le$

16 Here, we consider the Lipschitz condition (5.76) between different spaces: the norm on the left-hand side of (5.76) is taken in $BV([0, 1])$, while the norm on the right-hand side of (5.76) is taken in $Lip([0, 1])$.

r and $\|g\|_{Lip_\alpha} \leq r$. Then applying Lemma 5.48 to the choice $f(s) =: x_1$, $f(t) =: x_2$, $g(s) =: y_1$, and $g(t) =: y_2$ (with $s \neq t$) yields

$$\begin{aligned}&|(h \circ f)(s) - (h \circ g)(s) - (h \circ f)(t) + (h \circ g)(t)| \\ &\qquad \leq 2k(r)\,(|f(s) - f(t)| + |g(s) - g(t)|)\,\|f - g\|_C + \tilde{k}(r)|f(s) - g(s) - f(t) + g(t)|\,.\end{aligned}$$

Dividing by $|s - t|^\alpha$ and passing to the supremum over $s \neq t$, we arrive at

$$\begin{aligned}h_\alpha(C_h f - C_h g) &\leq 2k(r)\,(h_\alpha(f) + h_\alpha(g))\,\|f - g\|_C + \tilde{k}(r)h_\alpha(f - g) \\ &\leq \hat{k}(r)\|f - g\|_{Lip_\alpha}\,,\end{aligned}$$

with the same $\hat{k}(r)$ as in the proof of Theorem 5.49, which proves the "if" part of Theorem 5.51.

To prove the "only if" part, let us now suppose that C_h satisfies a local Lipschitz condition (5.76) in the norm (0.71) of the space $Lip_\alpha([a, b])$. Putting again $f(t) \equiv u$ and $g(t) \equiv v$ in (5.76), we see that h satisfies the local Lipschitz condition (5.13); we claim that h' satisfies (5.14) for each $r > 0$.

Fix $r > 0$ and choose $f \in Lip_\alpha([a, b])$ and $\delta > 0$ such that $\|f\|_{Lip_\alpha} < r$ and $\|f_\delta\|_{Lip_\alpha} \leq r$, where $f_\delta(t) := f(t) + \delta$. By assumption, we then have

$$\|C_h f_\delta - C_h f\|_{Lip_\alpha} \leq K(r)\|f_\delta - f\|_{Lip_\alpha} = K(r)\delta\,,$$

and hence

$$\frac{|h(f(t) + \delta) - h(f(t))|}{\delta} \leq \frac{\|C_h f_\delta - C_h f\|_C}{\delta} \leq K(r) \tag{5.84}$$

for all $t \in [a, b]$ as well as

$$\frac{|h(f(s) + \delta) - h(f(s)) - h(f(t) + \delta) + h(f(t))|}{\delta|s - t|^\alpha} \leq \frac{h_\alpha(C_h f_\delta - C_h f)}{\delta} \leq K(r) \tag{5.85}$$

for all $s, t \in [a, b]$. Letting $\delta \to 0$ in (5.84) and (5.85), we conclude that

$$|h'(f(t))| \leq K(r), \qquad \frac{|h'(f(s)) - h'(f(t))|}{|s - t|^\alpha} \leq K(r)\,.$$

This shows that the composition operator $C_{h'}$ generated by h' maps the space $Lip_\alpha([a, b])$ into itself, by Theorem 5.24, and so the assertion follows. □

In the following Tables 5.14–5.16 we summarize our results on the "interaction" between properties of the function $h : \mathbb{R} \to \mathbb{R}$ and those of the corresponding composition operator $C_h : X \to X$. We do this for the three spaces $X = C$, $X = Lip$, and $X = BV$ which have been the main focus in this chapter. Similar results hold for the more general spaces $X = Lip_\alpha$ $(0 < \alpha < 1)$ and $X = WBV_p$ $(1 < p < \infty)$. The hypotheses on h and C_h get more and more restrictive in each row, i.e. the first row contains the mildest requirement, and the last row the strongest, at least formally.

Table 5.14. The operator C_h in $C([a,b])$.

$h \in C(\mathbb{R})$	$\Leftrightarrow$	$C_h : C \to C$ bounded
$h \in C(\mathbb{R})$	$\Leftrightarrow$	$C_h : C \to C$ continuous
$h \in C(\mathbb{R})$	$\Leftrightarrow$	$C_h : C \to C$ uniformly continuous on bounded sets
$h \in C(\mathbb{R})$	$\Leftrightarrow$	$C_h : C \to C$ uniformly continuous on C
$h \in Lip_{\mathrm{loc}}(\mathbb{R})$	$\Leftrightarrow$	$C_h : C \to C$ locally Lipschitz continuous
$h \in Lip(\mathbb{R})$	$\Leftrightarrow$	$C_h : C \to C$ globally Lipschitz continuous

Table 5.15. The operator C_h in $Lip([a,b])$.

$h \in Lip_{\mathrm{loc}}(\mathbb{R})$	$\Leftrightarrow$	$C_h : Lip \to Lip$ bounded
$h \in C^1(\mathbb{R})$	$\Leftrightarrow$	$C_h : Lip \to Lip$ continuous
$h \in C^1(\mathbb{R})$	$\Leftrightarrow$	$C_h : Lip \to Lip$ uniformly continuous on bounded sets
h affine	$\Leftrightarrow$	$C_h : Lip \to Lip$ uniformly continuous on Lip
$h \in Lip^1_{\mathrm{loc}}(\mathbb{R})$	$\Leftrightarrow$	$C_h : Lip \to Lip$ locally Lipschitz continuous
h affine	$\Leftrightarrow$	$C_h : Lip \to Lip$ globally Lipschitz continuous

Table 5.16. The operator C_h in $BV([a,b])$.

$h \in Lip_{\mathrm{loc}}(\mathbb{R})$	$\Leftrightarrow$	$C_h : BV \to BV$ bounded
$h \in C^1(\mathbb{R})$	$\Rightarrow$	$C_h : BV \to BV$ continuous
$h \in C^1(\mathbb{R})$	$\Rightarrow$	$C_h : BV \to BV$ uniformly continuous on bounded sets
h affine	$\Rightarrow$	$C_h : BV \to BV$ uniformly continuous on BV
$h \in Lip^1_{\mathrm{loc}}(\mathbb{R})$	$\Leftrightarrow$	$C_h : BV \to BV$ locally Lipschitz continuous
h affine	$\Leftrightarrow$	$C_h : BV \to BV$ globally Lipschitz continuous

We make some comments on these tables. Although the first four conditions on the right-hand side of Table 5.14 are formally independent of each other, the left-hand side show that they are actually all equivalent, and they all simply follow from the inclusion $C_h(C) \subseteq C$. Moreover, global respectively local Lipschitz continuity of C_h in $C([a,b])$ is reflected by exactly the same property of the underlying function h, so no degeneracy for h occurs.

In Table 5.15, the situation is completely different. Only continuity and uniform continuity of C_h on bounded subsets of $Lip([a,b])$ are equivalent, as we have proved in Theorem 5.26. Boundedness of C_h simply follows from the inclusion $C_h(Lip) \subseteq Lip$. Moreover, local Lipschitz continuity of C_h in $Lip([a,b])$ is equivalent to local Lipschitz continuity of h' on the real line, and therefore holds for a reasonably large variety of nonlinear functions h. On the other hand, imposing a global Lipschitz condition for C_h leads to a strong degeneracy for h.

Finally, less is known in the situation described in Table 5.16. As in Table 5.15, boundedness of C_h simply follows from the inclusion $C_h(BV) \subseteq BV$, local Lipschitz continuity of C_h in $BV([a,b])$ is equivalent to local Lipschitz continuity of h' on the real line, and imposing a global Lipschitz condition for C_h leads to a strong degeneracy for

h. Unfortunately, only sufficient conditions on h are known which imply the continuity of C_h, in one or the other sense, in $BV([a,b])$.

At this moment, it is time to take a deep breath and to summarize, even at the risk of being redundant, what we have learned in this chapter about the autonomous composition operator (5.1) in many function spaces.

- The operator C_h maps the space $C([a,b])$ into itself if and only if the function h is continuous on $\mathbb{R}$. In this case, C_h is automatically bounded and continuous. Moreover, C_h is globally Lipschitz continuous if and only if h is globally Lipschitz on $\mathbb{R}$, and locally Lipschitz continuous if and only if h is locally Lipschitz on $\mathbb{R}$.
- The operator C_h maps the space $C^1([a,b])$ into itself if and only if the function h is continuously differentiable on $\mathbb{R}$. In this case, C_h is automatically bounded and continuous. Moreover, C_h is globally Lipschitz continuous if and only if h is affine, and locally Lipschitz continuous if and only if h' is locally Lipschitz on $\mathbb{R}$.
- The operator C_h maps the space $Lip([a,b])$ into itself if and only if the function h is locally Lipschitz on $\mathbb{R}$. In this case, C_h is automatically bounded, but not necessarily continuous (Example 5.25). Moreover, C_h is globally Lipschitz continuous if and only if h is affine, and locally Lipschitz continuous if and only if h' is locally Lipschitz on $\mathbb{R}$.
- The operator C_h maps the space $Lip_\alpha([a,b])$ $(0 < \alpha < 1)$ into itself if and only if the function h is locally Lipschitz on $\mathbb{R}$. In this case, C_h is automatically bounded, but not necessarily continuous. Moreover, C_h is globally Lipschitz continuous if and only if h is affine, and locally Lipschitz continuous if and only if h' is locally Lipschitz on $\mathbb{R}$.
- The operator C_h maps the space $Lip^1([a,b])$ into itself if and only if the function h' is locally Lipschitz on $\mathbb{R}$. In this case, C_h is automatically bounded, but not necessarily continuous (Example 5.40). Moreover, C_h is globally Lipschitz continuous if and only if h is affine, and locally Lipschitz continuous if and only if h'' is locally Lipschitz on $\mathbb{R}$.
- The operator C_h maps the space $AC([a,b])$ into itself if and only if the function h is locally Lipschitz on $\mathbb{R}$. In this case, C_h is automatically bounded. Moreover, C_h is globally Lipschitz continuous if and only if h is affine, and locally Lipschitz continuous if and only if h' is locally Lipschitz on $\mathbb{R}$. The problem of characterizing continuity of C_h is open.
- The operator C_h maps the space $AC^1([a,b])$ into itself if and only if the function h' is absolutely continuous on $\mathbb{R}$. In this case, C_h is automatically bounded. Moreover, C_h is globally Lipschitz continuous if and only if h is affine, and locally Lipschitz continuous if and only if h'' is locally Lipschitz on $\mathbb{R}$. The problem of characterizing continuity of C_h is open.
- The operator C_h maps the space $BV([a,b])$ into itself if and only if the function h is locally Lipschitz on $\mathbb{R}$. In this case, C_h is automatically bounded. Moreover, C_h is globally Lipschitz continuous if and only if h is affine, and locally Lipschitz

continuous if and only if h' is locally Lipschitz on $\mathbb{R}$. The problem of characterizing continuity of C_h is open.
- The operator C_h maps the space $BV^1([a,b])$ into itself if and only if the function h' has locally bounded variation on $\mathbb{R}$. In this case, C_h is automatically bounded. Moreover, C_h is globally Lipschitz continuous if and only if h is affine, and locally Lipschitz continuous if and only if h'' is locally Lipschitz on $\mathbb{R}$. The problem of characterizing continuity of C_h is open.
- The operator C_h maps the space $WBV_p([a,b])$ $(1 < p < \infty)$ into itself if and only if the function h is locally Lipschitz on $\mathbb{R}$. In this case, C_h is automatically bounded. Moreover, C_h is globally Lipschitz continuous if and only if h is affine, and locally Lipschitz continuous if and only if h' is locally Lipschitz on $\mathbb{R}$. The problem of characterizing continuity of C_h is open.
- The operator C_h maps the space $WBV_p^1([a,b])$ $(1 < p < \infty)$ into itself if and only if the function h' has locally bounded p-variation in Wiener's sense on $\mathbb{R}$. In this case, C_h is automatically bounded. Moreover, C_h is globally Lipschitz continuous if and only if h is affine, and locally Lipschitz continuous if and only if h'' is locally Lipschitz on $\mathbb{R}$. The problem of characterizing continuity of C_h is solved in Theorem 5.56 below.
- The operator C_h maps the space $RBV_p([a,b])$ $(1 < p < \infty)$ into itself if and only if the function h is locally Lipschitz on $\mathbb{R}$. In this case, C_h is automatically bounded. Moreover, C_h is globally Lipschitz continuous if and only if h is affine, and locally Lipschitz continuous if and only if h' is locally Lipschitz on $\mathbb{R}$. The problem of characterizing continuity of C_h is open.
- The operator C_h maps the space $RBV_p^1([a,b])$ $(1 < p < \infty)$ into itself if and only if the function h' has locally bounded p-variation in Riesz's sense on $\mathbb{R}$. In this case, C_h is automatically bounded. Moreover, C_h is globally Lipschitz continuous if and only if h is affine, and locally Lipschitz continuous if and only if h'' is locally Lipschitz on $\mathbb{R}$. The problem of characterizing continuity of C_h is open.
- The operator C_h maps the space $\Lambda BV([a,b])$ into itself if and only if the function h is locally Lipschitz on $\mathbb{R}$. In this case, C_h is automatically bounded. Moreover, C_h is globally Lipschitz continuous if and only if h is affine, and locally Lipschitz continuous if and only if h' is locally Lipschitz on $\mathbb{R}$. The problem of characterizing continuity of C_h is open.

We point out that this summary mostly contains conditions which are *both necessary and sufficient*. For the nonautonomous superposition operator (5.2), we are going to do the same in the next chapter; as we will see, the result will be quite different.

5.6 Comments on Chapter 5

Although some of the results proved here for the composition operator (5.1) will be repeated in a more general setting in the next chapter for the superposition operator (5.2), we will give references for the most important theorems.

The composition operator problem (COP) was formulated in a slightly different notation in [55, 56]. As we have seen, the local Lipschitz condition (5.13) for h is necessary and sufficient for the inclusion $C_h(X) \subseteq X$ in many function spaces X arising in applications. However, there are some exceptions. On the one hand, for some spaces X, condition (5.13) is *too strong*; for example, $C_h(C) \subseteq C$ if and only if h is continuous, see Theorem 5.20.[17] On the other hand, for some other spaces X, it may also happen that condition (5.13) is *too weak*; this refers to all spaces of type X^1 which we considered in Section 5.3. Apart from these spaces, there are other "exotic" spaces with the same property. To illustrate this, we recall a surprising result on functions having a primitive; the proof can be found in the paper [21].

Theorem 5.52. *Let $D^{-1}([a,b])$ denote the space of all functions $f : [a,b] \to \mathbb{R}$ with a primitive, i.e. $f = F'$ for some differentiable function F. Then the operator* (5.1) *maps the space $D^{-1}([a,b])$ into itself if and only if the corresponding function h is affine, i.e. has the form* (5.68).

As mentioned before, Theorem 5.9 due to Josephy [155] is the first example of a function space X satisfying $COP(X) = Lip_{\mathrm{loc}}(\mathbb{R})$, namely, $X = BV$. The analogous result for $X = AC$ has been given in [219], for $X = RBV_p$ in [221], and for $X = Lip$ and $X = Lip_\alpha$ in [293–295].[18] Of course, Theorem 5.10 which has been proved in [15] provides a unified approach to all of these results. The similar Theorem 5.12 is also taken from [14]. In [227], it is shown that $C_h(RBV_p) \subseteq BV$ if and only if $h \in Lip_{\mathrm{loc}}(\mathbb{R})$; of course, this result follows from our Theorem 5.10. On the other hand, the requirement $C_h(BV) \subseteq C$ is so restrictive that it is fulfilled only for constant functions h. To see this, suppose that $h(u) \neq h(v)$ for some $u, v \in \mathbb{R}$. For fixed $c \in (a,b)$, the function

$$f(x) := u\chi_{[a,c]}(x) + v\chi_{(c,b]}(x) = \begin{cases} u & \text{for } a \le x \le c\,, \\ v & \text{for } c < x \le b \end{cases}$$

then belongs to $BV([a,b])$, but the function

$$C_h f(x) := \begin{cases} h(u) & \text{for } a \le x \le c\,, \\ h(v) & \text{for } c < x \le b \end{cases}$$

17 Another example is given by Proposition 4.16 (a) which shows that $C \subseteq COP(RS_\alpha)$.

18 Unfortunately, the proof in these papers is false; a correct proof has been given in the recent survey paper [14].

is discontinuous at c. This implies, in particular, the result from [227] that $C_h(BV) \subseteq RBV_p$ (for some $p > 1$) if and only if h is constant.

Theorem 5.13 may be found in [221], Theorems 5.14 and 5.15 in [251]. Table 5.1 is contained together with a detailed discussion of related results in the "folkloristic" paper [11]. There are many other function spaces X with the characteristic property $COP(X) = Lip_{\text{loc}}(\mathbb{R})$. For instance, Ul'yanov [306–308] has shown that this holds for the space $X = A_\omega([0, 2\pi)$ of all integrable functions whose Fourier–Haar coefficients satisfy a summability condition with weight ω.

The validity of the chain rule (5.38) and the change of variable formula (5.42) for absolutely continuous functions are dealt with of course in almost every textbook on real analysis and Lebesgue integration.

Our discussion in Section 5.2 may be summarized as follows: in many function spaces X, the mere inclusion $C_h(X) \subseteq X$ implies the boundedness of C_h in the underlying norm, while continuity of C_h in the underlying norm is a very delicate (and often open) question. As we pointed out before, three cases typically arise here: either we get continuity of C_h, in the same manner as boundedness, "for free," or one knows precisely what additional condition on h has to be imposed to get continuity of C_h, or the answer is simply unknown. The spaces C and C^1 fall into the first category, the spaces Lip and Lip_α into the second, and the spaces BV and AC (and many more) into the third.

A simple sufficient condition for the continuity of C_h in the space BV which covers sufficiently many examples can be given; we state this in the following

Proposition 5.53. *Suppose that $h : \mathbb{R} \to \mathbb{R}$ is real analytic. Then the corresponding operator* (5.1) *maps the space $BV([a, b])$ into itself and is both bounded and continuous in the norm* (1.16).

Proof. The assumption means that h admits an expansion into a power series

$$h(u) = \sum_{k=0}^{\infty} a_k u^k$$

with infinite radius of convergence. Since h is then locally Lipschitz on $\mathbb{R}$, the operator C_h maps $BV([a, b])$ into itself and is bounded, by Theorem 5.9. It remains to show that C_h is continuous in the norm (1.16).

Clearly, the composition operator C_{h_n} generated by the n-th truncation

$$h_n(u) = \sum_{k=0}^{n} a_k u^k$$

of h maps $BV([a, b])$ into itself since $(BV([a, b]), \|\cdot\|_{BV})$ is an algebra. Moreover, the difference $h - h_n$ is locally Lipschitz because

$$\sup_{|u| \le r} |h(u) - h_n(u) - h(v) + h_n(v)| \le k_1(r)|u - v| \quad (r > 0)$$

with

$$k_1(r) := \sup_{|w|\le r} |h'(w) - h_n'(w)| \quad (r > 0).$$

This implies that

$$\begin{aligned}\|C_h f - C_{h_n} f\|_{BV} &= |h(f(a)) - h_n(f(a))| + \mathrm{Var}(C_h f - C_{h_n} f; [a,b]) \\ &\le |h(f(a)) - h_n(f(a))| + k_1(\|f\|_\infty)\, \mathrm{Var}(f; [a,b]).\end{aligned}$$

However, this shows that the sequence $(C_{h_n} f)_n$ converges to $C_h f$ in the norm (1.16) uniformly on each bounded subset of $BV([a,b])$. Consequently, the operator C_h is continuous in the norm (1.16) as well. □

In Section 5.3, we have seen that the composition operator problem has a completely different solution in spaces of differentiable functions like C^1, Lip^1, BV^1, or AC^1. Here, several new phenomena occur. First, a necessary and sufficient condition for the inclusion $C_h(X^1) \subseteq X^1$ is often that h belongs locally to the same space X^1. Second, working in a space of type X^1, we must make extensive use of the chain rule (5.38), and here the appearance of the factor $f'(x)$ on the right-hand side of (5.38) has a certain "regularizing effect." For instance, before Theorem 5.37, we have discussed the phenomenon that $f \in C^\infty$ and $g \in AC_{\mathrm{loc}}(\mathbb{R})$ does not imply that $g \circ f \in AC$, but multiplying by f' has the effect that $(g \circ f)f' \in AC$. This effect also occurs in other function spaces. For example, for $f(x) := x^2$ and $g(y) := 1/\sqrt{|y|}$ $(0 < y \le 1)$, we have

$$f \in C^\infty([0,1]),\ g \in L_1([0,1]),\ g \circ f \notin L_1([0,1]),\ (g \circ f)f' \in L_1([0,1]).$$

For a slight generalization of this, see Exercise 5.15.

Let us go back for a moment to the COP which consists of finding a necessary and sufficient condition for a fixed *outer function* h under which $h \circ f$ belongs to a given class of functions whenever f belongs to the same class. Sometimes, the following dual problem is also of interest: determine a necessary and sufficient condition for a fixed *inner function* f under which $h \circ f$ belongs to a given class of functions whenever h belongs to the same class. For example, for the class AC, the solution of this problem goes back to Fichtengol'ts [114] who proved the following

Theorem 5.54. *Given $f : [a,b] \to [c,d]$, the following two conditions are equivalent.*
(a) *For every $h \in AC([c,d])$, we have $h \circ f \in AC([a,b])$.*
(b) *The function f is absolutely continuous, and the set $f^{-1}(y) \cap [a,b]$ only consists of isolated points for each $y \in [c,d]$.*

As Example 5.8 shows, the last condition cannot be dropped: both functions f and h in Example 5.8 are absolutely continuous, but $f^{-1}(0)$ has the accumulation point 0.

Since Janusz Matkowski seems to have been the first author who discovered the degeneracy phenomenon described in Section 5.4, the term *spaces with the Matkowski property* (our Definition 5.42) was introduced in [226]. At the beginning of Section 5.4, we have given a short list of such spaces. The general Theorems 5.45 and 5.47 which are taken from [15] (see also [56] and [226]) show that we may add the spaces RBV^n_p, BV^n, WBV_p, BV, AC, and $\Lambda_q BV$ to this list. We point out, however, that the situation for the spaces BV and WBV_p becomes more complicated in case of the nonautonomous superposition operator (5.2), see Definition 6.20, Theorem 6.19, and Example 6.21 in the next chapter.

The fact that the global Lipschitz condition (5.67) for the operator C_h leads to the very special form (5.68) of h has unpleasant consequences for applications: it means that whenever we want to apply Banach's contraction mapping principle to a problem involving composition operators in a space with the Matkowski property, we can do this only if the problem is actually *linear*, and therefore not very interesting. This is the reason why one has to replace the global condition (5.67) by the local condition (5.76), as we have done in Section 5.5.

Given a function $f : [a,b] \to \mathbb{R}$ and a partition $P = \{t_0, t_1, \ldots, t_m\}$ of $[a,b]$, recall that the second variation $\mathrm{Var}^W_{2,1}(f, P; [a,b])$ of f on $[a,b]$ with respect to P is defined by (2.141). In (2.150), we have established the important equality $\mathrm{Var}^W_{2,1}(f; [a,b]) = \|f''\|_{L_1}$ for $f \in AC^1([a,b])$ which has been proven by Russell and may be viewed as a "second order analogue" to the classical formula (3.33). One could ask whether or not the requirement that $f \in AC^1([a,b])$ is essential in Russell's result. The following example shows that, in fact, functions exist with bounded second variation whose derivatives are not absolutely continuous.

Example 5.55. Let $\varphi : [0,1] \to \mathbb{R}$ be the Cantor function (3.6), and let $f : [0,1] \to \mathbb{R}$ be defined as in (5.60). We have already seen in Example 5.31 that $f \notin AC^1([0,1])$. On the other hand, since φ is monotonically increasing, f is convex, and thus belongs to $WBV_{2,1}([0,1])$. ♥

Example 5.55 shows that we cannot use Theorem 5.37 to solve the COP for the space $WBV_{2,1}$. As far as we know, the problem of describing $COP(WBV_{2,1})$ is still open.

As we have seen, the problem of characterizing those functions h for which the corresponding operator C_h is continuous in norm is open for many spaces. We state this for absolutely continuous functions as

Problem 5.1. *Suppose that $h : \mathbb{R} \to \mathbb{R}$ satisfies condition* (5.13). *Does this imply that the corresponding operator C_h is continuous in the norm* (3.42) *or the norm* (3.43) *of the space $AC([a,b])$?*

Problem 5.2. *Same question as in Problem* 5.1 *for the spaces $BV([a,b])$ and $RBV_p([a,b])$ $(p > 1)$.*

Surprisingly enough, in [56], this problem was solved for the space WBV^1_p in case $p > 1$, i.e. for primitives of WBV_p-functions. More precisely, the authors of [56] prove the following

Theorem 5.56. *For $1 < p < \infty$, the operator* (5.1) *maps the space $WBV_p^1([a,b])$ into itself if and only if $h \in WBV_{p,loc}^1(\mathbb{R})$. In this case, the operator* (5.1) *is automatically bounded in the norm* (5.54). *Moreover, the operator* (5.1) *is continuous in the norm* (5.54) *if and only if h belongs to the closure of $WBV_{p,loc}^1(\mathbb{R}) \cap C^\infty(\mathbb{R})$ in the norm* (5.54).

Of course, the closure of $WBV_{p,loc}^1(\mathbb{R}) \cap C^\infty(\mathbb{R})$ in the norm (5.54) is a proper subspace of $WBV_{p,loc}^1(\mathbb{R})$. We believe that a similar restriction is needed in the Riesz space $RBV_p^1([a,b])$, but we were unable to prove it.

Problem 5.3. *Find a condition on h, both necessary and sufficient, under which the corresponding operator* (5.1) *is continuous in the space $RBV_p^1([a,b])$ with norm* (5.55).

Problem 5.4. *Find a condition on h, both necessary and sufficient, under which the corresponding operator* (5.1) *maps the space $WBV_{2,1}([a,b])$ into itself.*

The COP for the space κBV of functions of bounded Korenblum variation was solved recently in the paper [17], where it is shown that $C_h(\kappa BV) \subseteq \kappa BV$ if and only if $h \in Lip_{\text{loc}}(\mathbb{R})$. Moreover, in this case, the operator (5.1) is automatically bounded in the norm (2.136).

Most of the material covered in Section 5.5 has been taken from the paper [18]. Locally Lipschitz composition operators in the Schramm space ΦBV (which is missing in the summary at the end of Section 5.5) are discussed in [230].

In this chapter, we have only considered functions over some compact interval. Superpositions of functions of several variables have been studied by Chistyakov in a series of papers [85–95]. Since these results refer to the nonautonomous case of the operator (5.2), we postpone them to the next chapter.

However, following [191], we briefly consider a special result on composition operators which are generated by an *outer function h* of two variables. In the legendary *Scottish Book*, the mathematician Max Eidelheit posed the following problem for functions which are absolutely continuous "on traces:" given a function $h : [0,1] \times [0,1] \to \mathbb{R}$, suppose that $h(u,\cdot) \in AC([0,1])$ for every $u \in [0,1]$, and $h(\cdot,v) \in AC([0,1])$ for every $v \in [0,1]$, and let $f : [0,1] \to [0,1]$ and $g : [0,1] \to [0,1]$ also be absolutely continuous functions. Does it follow that the function $x \mapsto h(f(x), g(x))$ is absolutely continuous on $[0,1]$? If not, then perhaps this is true under the additional assumption that

$$\int_0^1 \int_0^1 |h_u(u,v)|^p \, dv \, du < \infty, \qquad \int_0^1 \int_0^1 |h_v(u,v)|^p \, dv \, du < \infty \tag{5.86}$$

for some $p > 1$?

The next example shows that the answer to Eidelheit's first question is negative.

Example 5.57. Define $h : [0,1] \times [0,1] \to \mathbb{R}$ by

$$h(u,v) := \begin{cases} \frac{2uv}{u^2+v^2} & \text{for } (u,v) \neq (0,0)\,, \\ 0 & \text{for } (u,v) = (0,0)\,, \end{cases} \tag{5.87}$$

and let $f(x) = g(x) := x$. Then $|h(u,v) - h(u',v)| \leq \frac{2}{v}|u - u'|$ for $0 < v \leq 1$, and similarly $|h(u,v) - h(u,v')| \leq \frac{2}{u}|v - v'|$ for $0 < u \leq 1$; moreover, $h(u,0) = h(0,v) \equiv 0$. Consequently, all hypotheses of Eidelheit's problem are satisfied, but the function $h(f(x), g(x)) = h(x,x) = \chi_{(0,1]}$ is not continuous, let alone absolutely continuous. ♥

Note that the integrability condition (5.86) *fails* for the function (5.87) for every $p > 1$ (Exercise 5.27), and so this function does not provide a counterexample to Eidelheit's second question. However, one may show that Eidelheit's second question also has a negative answer. In fact, consider the function $h : [0,1] \times [0,1] \to \mathbb{R}$ defined by

$$h(u,v) := \int_0^u \log t \, dt \quad (0 \leq u \leq 1) \tag{5.88}$$

which depends on v only formally. Then

$$\int_0^1 \int_0^1 |h_u(u,v)|^p \, dv \, du = \int_0^1 |\log t|^p \, dt < \infty, \quad \int_0^1 \int_0^1 |h_v(u,v)|^p \, dv \, du = 0$$

for all $p > 1$, but[19] $h_u(\cdot, v) \notin L_\infty([0,1])$.

Now, the trick consists of choosing $f = g \in AC([0,1])$ in such a way[20] that the map $x \mapsto h(f(x), g(x))$ is not absolutely continuous on $[0,1]$. From Theorem 5.10, we know that the outer function h must not be locally Lipschitz in this case; but this is true by Theorem 3.20.

5.7 Exercises to Chapter 5

We state some exercises on the topics covered in this chapter; exercises marked with an asterisk * are more difficult.

Exercise 5.1. Let $f \in AC([a,b])$ and let $\ell : [c,d] \to [a,b]$ be increasing and absolutely continuous with $\ell(c) = a$ and $\ell(d) = b$. Show that $f \circ \ell \in AC([c,d])$. Is the same true if ℓ is only increasing and continuous?

Exercise 5.2. Let $f \in AC([a,b])$ and $h \in AC_{\mathrm{loc}}(\mathbb{R})$. Prove that $h \circ f \in AC([a,b])$ if and only if $h \circ f \in BV([a,b])$. Compare with Example 5.8.

Exercise 5.3. Let $f : [a,b] \to [c,d]$ be monotone and $h \in BV([c,d])$. Show that $h \circ f \in BV([a,b])$ and compare with Example 5.8.

---

19 We have already seen in the similar Example 0.12 that $\log \in L_p([0,1])$ for every $p \geq 1$, but $\log \notin L_\infty([0,1])$.

20 In case of the function $h(u) = \sqrt{u}$, we have achieved this by choosing f as in (5.11); for the function (5.88), this has to be modified accordingly.

Exercise 5.4. Prove or disprove the following converse of Lemma 5.2: every COP-invariant function space X is shift-invariant.

Exercise 5.5. Is the space $\Phi BV([a,b])$ COP-invariant for all Schramm sequences Φ? If not, find additional conditions on Φ under which this is true.

Exercise 5.6*. Formulate and prove an analogous result to Theorem 5.14 for the space $\Phi BV([a,b])$, where Φ is an arbitrary Schramm sequence.

Exercise 5.7. Solve again Exercise 3.5 by means of Theorem 5.10 and Example 5.8.

Exercise 5.8. Let ϕ be a Young function which satisfies the δ_2-condition (2.4), but not the 0_1 condition (2.17). Suppose that the operator (5.1) maps the space $\phi BV([a,b])$ into itself. Imitating the proof of Proposition 5.7, show that the generating function h is then continuous.

Exercise 5.9. Is the hypothesis $h \in C^1(\mathbb{R})$ sufficient for the corresponding operator C_h to be continuous in the norm (3.42) of the space $AC([a,b])$?

Exercise 5.10. Is the hypothesis $h \in C^1(\mathbb{R})$ sufficient for the corresponding operator C_h to be continuous in the norm (2.90) of the space $RBV_p([a,b])$?

Exercise 5.11. Suppose that $h : \mathbb{R} \to \mathbb{R}$ has the property that the corresponding composition operator (5.1) maps a function space X into itself and is bounded in the norm of X. For $r > 0$, we put

$$\mu_r(h, X) := \sup\{\|C_h f\|_X : \|f\|_X \le r\} = \sup\{\|h \circ f\|_X : \|f\|_X \le r\}$$

and call this characteristic the *growth function* of h in X. Calculate the functions $\mu_r(h, C)$ and $\mu_r(h, C^1)$ under the hypotheses of Theorem 5.20 and Theorem 5.30, respectively.

Exercise 5.12. Using the notation of Theorem 5.24, give an upper estimate for the growth function $\mu_r(h, Lip_\alpha)$ from Exercise 5.11 in the space $Lip_\alpha([0,1])$.

Exercise 5.13. Let $f \in AC([a,b])$ with $f([a,b]) \subseteq [c,d]$, and let $g \in L_1([c,d])$. If $(g \circ f)f' \in L_1([a,b])$, then

$$\int_{f(\alpha)}^{f(\beta)} g(y)\,dy = \int_{\alpha}^{\beta} g(f(t))f'(t)\,dt$$

for every interval $[\alpha,\beta] \subset [a,b]$. Compare with Theorem 5.18.

Exercise 5.14. In the notation of Exercise 5.13, prove that $(g \circ f)f' \in L_1([a,b])$ if $f \in AC([a,b])$ is monotone with $f([a,b]) \subseteq [c,d]$ and $g \in L_1([c,d])$.

Exercise 5.15. For $\alpha \in \mathbb{R}^+$ and $\beta \in \mathbb{R}$, define $f : [0,1] \to [0,1]$ by $f(x) := x^\alpha$ and $g : (0,1] \to \mathbb{R}$ by $g(y) := y^\beta$. Determine all pairs (α,β) for which $g \circ f \in L_1([0,1])$ and all pairs (α,β) for which $(g \circ f)f' \in L_1([0,1])$.

Exercise 5.16. Suppose that $f \in C^\infty([a,b])$ and $g \in BV_{loc}(\mathbb{R})$. Does it follow that $(g \circ f)f' \in BV([a,b])$?

Exercise 5.17. Using the notation of Theorem 5.35, give an upper estimate for the growth function $\mu_r(h, BV^1)$ from Exercise 5.11 in the space $BV^1([0,1])$.

Exercise 5.18. Using the notation of Theorem 5.37, give an upper estimate for the growth function $\mu_r(h, AC^1)$ from Exercise 5.11 in the space $AC^1([0,1])$.

Exercise 5.19. Using the notation of Theorem 5.38, give an upper estimate for the growth function $\mu_r(h, RBV_p^1)$ from Exercise 5.11 in the space $RBV_p^1([0,1])$.

Exercise 5.20. Using the notation of Theorem 5.39, give an upper estimate for the growth function $\mu_r(h, Lip^1)$ from Exercise 5.11 in the space $Lip^1([0,1])$.

Exercise 5.21. Is the hypothesis $h \in C^2(\mathbb{R})$ sufficient for the corresponding operator C_h to be continuous in the norm (5.53) of the space $AC^1([a,b])$?

Exercise 5.22. Is the hypothesis $h \in C^2(\mathbb{R})$ sufficient for the corresponding operator C_h to be continuous in the norm (5.54) of the space $WBV_p^1([a,b])$?

Exercise 5.23. Is the hypothesis $h \in C^2(\mathbb{R})$ sufficient for the corresponding operator C_h to be continuous in the norm (5.55) of the space $RBV_p^1([a,b])$?

Exercise 5.24. Imitate the proof of Theorem 5.26 to prove Theorem 5.41.

Exercise 5.25. Suppose that the composition operator C_h generated by some function $h : \mathbb{R} \to \mathbb{R}$ maps the space $C^1([a,b])$ into itself and is continuous and bounded with respect to one of the norms in (0.65). Moreover, assume that h' exists on $\mathbb{R}$ and satisfies (5.14). Show that the operator C_h generated by h satisfies then (5.76).

Exercise 5.26. Prove the following converse of Exercise 5.6. Suppose that the composition operator C_h generated by h maps $C^1([a,b])$ into itself and satisfies (5.76). Show that then h' exists on $\mathbb{R}$ and satisfies (5.14).

Exercise 5.27. Let $h : [0,1] \times [0,1] \to \mathbb{R}$ be defined by (5.87). Show that

$$\int_0^1 \int_0^1 |h_u(u,v)|^p \, dv \, du = \int_0^1 \int_0^1 |h_v(u,v)|^p \, dv \, du = \infty$$

for every $p > 1$.

Exercise 5.28. Deduce from the estimate (5.83) the degeneracy result contained in Theorem 5.47 for the space $BV = WBV_1$.

6 Nonlinear superposition operators

When we replace the autonomous operator $C_h f(x) = h(f(x))$ by the nonautonomous operator $S_h f(x) = h(x, f(x))$, we encounter many new phenomena. For example, even necessary and sufficient criteria on h under which the corresponding operator S_h maps a function space into itself are only known in quite exceptional cases. Moreover, while the operator C_h is often both automatically bounded and continuous in standard function spaces, it may happen that the operator S_h is bounded but discontinuous, or continuous but unbounded. Imposing a global Lipschitz condition on S_h in norm leads to the same degeneracy phenomena as for the operator C_h, which is not true for local Lipschitz conditions in norm. In this chapter, we will also study the uniform boundedness and uniform continuity of the operator S_h. As a rule, for all of these properties of S_h, one may only formulate sufficient conditions, while their necessity is often an open problem.

6.1 Boundedness and continuity

Given a function $h : [a, b] \times \mathbb{R} \to \mathbb{R}$, we are now going to study the (nonautonomous) *superposition operator* S_h defined by

$$S_h f(x) = h(x, f(x)) \quad (a \leq x \leq b), \tag{6.1}$$

where f is taken from some space of real functions on $[a, b]$. It turns out that the operator (6.1) is far more complicated than its autonomous counterpart (5.1) which we studied in great detail in Chapter 5. The main reason for this is the "interaction" between the two variables of the function h which, in some cases, may lead to a quite pathological behavior of the corresponding operator S_h.

Fortunately, there are some exceptions from this rule: in a few simple function spaces, the operator (6.1) behaves in exactly the same way as the operator (5.1). For example, the following two results hold which are precise analogues to Theorems 5.20 and 5.21:

Theorem 6.1. *The operator* (6.1) *maps the space* $C([a, b])$ *into itself if and only if the function* h *is continuous on* $[a, b] \times \mathbb{R}$. *In this case, the operator* (6.1) *is automatically bounded and continuous in the norm* (0.45).

Theorem 6.2. *For* $1 \leq p, q < \infty$, *the superposition operator* (6.1) *maps the space* $L_p([a, b])$ *into the space* $L_q([a, b])$ *if and only if the function* h *satisfies the growth condition*

$$|h(x, u)| \leq \alpha(x) + \beta |u|^{p/q} \quad (a \leq x \leq b, u \in \mathbb{R}) \tag{6.2}$$

for some function $\alpha \in L_q([a, b])$ *and some constant* $\beta > 0$. *In this case, the operator* (6.1) *is automatically bounded and continuous in the norm* (0.11).

The proof of Theorem 6.1 is a simple consequence of the Tietze–Urysohn extension lemma, while the proof of Theorem 6.2 relies on a classical result of Krasnosel'skij ([165–168] or [170]). Note that in case $p = q$, condition (6.2) reads

$$|h(x,u)| \leq \alpha(x) + \beta|u| \quad (a \leq x \leq b, u \in \mathbb{R}), \tag{6.3}$$

involving a function $\alpha \in L_p([a,b])$ and a constant $\beta > 0$; this means that $S_h(L_p) \subseteq L_p$ if and only if the map $u \mapsto h(x,u)$ has *sublinear growth* for large values of $|u|$.

The following results and counterexamples are in sharp contrast to Theorems 6.1 and 6.2. In fact, we take this opportunity to recall some pathological phenomena of the nonautonomous superposition operator (6.1) in other function spaces which are somewhat beyond the scope of this monograph. All of these phenomena are possible only due to a complicated "interaction" between the variables x and u in the function $(x,u) \mapsto h(x,u)$ which is "hidden" in the autonomous case of the composition operator (5.1).

To begin with, consider for $0 < \alpha \leq 1$ the Banach space $Lip_\alpha([a,b])$ of all Hölder continuous (in particular, Lipschitz continuous for $\alpha = 1$) functions on $[a,b]$, equipped with one of the equivalent norms (0.71) or (0.77). Interestingly, in this space, the superposition operator (6.1) exhibits a rather pathological behavior:

Example 6.3. Let $h : [0,1] \times \mathbb{R} \to \mathbb{R}$ be defined by

$$h(x,u) := \begin{cases} 0 & \text{if } u \leq x^{\alpha/2}, \\ \frac{1}{u^{2/\alpha}} - \frac{x}{u^{4/\alpha}} & \text{if } u > x^{\alpha/2}. \end{cases} \tag{6.4}$$

A somewhat cumbersome calculation then shows that the operator (6.1) generated by this function maps $Lip_\alpha([0,1])$ into itself, but h is *discontinuous* at $(0,0)$, and so S_h does not map $C([0,1])$ into itself! ♥

We point out that the reason for the pathological behavior of the function h in Example 6.3 is the *lack of boundedness* of the corresponding superposition operator S_h in the norm (0.71). In fact, the sequence $(f_n)_n$ of constant functions $f_n(x) \equiv 1/n$ satisfies

$$\|f_n\|_{Lip_\alpha} = \frac{1}{n}, \quad \|S_h f_n\|_{Lip_\alpha} \geq |S_h f_n(0)| = n^{2/\alpha} \to \infty \quad (n \to \infty).$$

If we *add* boundedness to the acting condition $S_h(Lip_\alpha) \subseteq Lip_\alpha$, however, we get the result that we expect (see [20, Theorem 7.3] for the proof):

Theorem 6.4. *The operator* (6.1) *maps the space* $Lip_\alpha([a,b])$ *into itself and is bounded with respect to the norm* (0.71) *if and only if the function h satisfies the mixed local Hölder–Lipschitz condition*

$$|h(s,u) - h(t,v)| \leq k(r)\left(|s-t|^\alpha + |u-v|\right) \quad (a \leq s,t \leq b;\ |u|,|v| \leq r). \tag{6.5}$$

In particular, the function h is then necessarily continuous on $[a,b] \times \mathbb{R}$.

Of course, in the special case of the autonomous composition operator (5.1), condition (6.5) takes the simpler form

$$|h(u) - h(v)| \leq k(r)|u - v| \quad (|u|, |v| \leq r) \tag{6.6}$$

which is precisely condition (5.13) and so reproduces Theorem 5.24. One could ask if it is possible to give a necessary and sufficient condition for the mere inclusion $S_h(Lip_\alpha) \subseteq Lip_\alpha$, i.e. without requiring boundedness of the operator S_h. Such conditions are in fact known, but there are very clumsy and hard to verify in practice. For the sake of completeness, we cite a result whose (quite technical) proof can be found in [20, Theorem 7.1]. To this end, we denote for $(s_0, u_0) \in [a, b] \times \mathbb{R}$, $r > 0$, $\delta > 0$, and $\alpha \in (0, 1]$, the "bow-tie" region

$$W_\alpha(s_0, u_0, r, \delta) := \{(s, u) \in [a, b] \times \mathbb{R} : |s - s_0| \leq \delta, |u - u_0| \leq r|s - s_0|^\alpha\}.$$

Proposition 6.5. *The operator* (6.1) *maps the space* $Lip_\alpha([a, b])$ *into itself if and only if for all* $(s_0, u_0) \in [a, b] \times \mathbb{R}$ *and all* $r > 0$, *one may find* $\delta > 0$ *and* $k(r) > 0$ *such that*

$$|h(s, u) - h(t, v)| \leq k(r)\left(|s - t|^\alpha + \frac{|u - v|}{r}\right)$$

for all $(s, u), (t, v) \in W_\alpha(s_0, u_0, r, \delta)$.

Now, we study the superposition operator (6.1) in the space $C^1([a, b])$ of continuously differentiable functions $f : [a, b] \to \mathbb{R}$, equipped with the first norm in (0.65). Since we are going to deal with differentiable functions, we have to consider the derivative

$$g'(x) = D_1 h(x, f(x)) + D_2 h(x, f(x)) f'(x) \tag{6.7}$$

of $g = S_h f$, where $D_1 h$ denotes the partial derivative of h with respect to the first argument and $D_2 h$ the partial derivative of h with respect to the second argument.

Surprisingly enough, in contrast to the space $C([a, b])$, the operator S_h has a quite unexpected behavior in the space $C^1([a, b])$. For example, it may happen that S_h maps C^1 into itself, although the generating function h is *discontinuous* (and so S_h *does not* map C into itself)! Since such pathologies seem to be interesting, we briefly recall an example due to Matkowski (see Section 8.2 in [20]).

Example 6.6. Let $h : [0, 1] \times \mathbb{R} \to \mathbb{R}$ be defined by

$$f(x, u) := \begin{cases} 0 & \text{if } u \leq 0, \\ 3\frac{u^2}{x} - 2\frac{u^3}{x\sqrt{x}} & \text{if } 0 < u < \sqrt{x} \\ 1 & \text{if } u \geq \sqrt{x}. \end{cases} \tag{6.8}$$

As in Example 6.3, a rather cumbersome but straightforward calculation then shows that the operator (6.1) generated by this function maps $C^1([0, 1])$ into itself, but h is obviously *discontinuous* at $(0, 0)$, and so S_h does not map $C([0, 1])$ into itself! ♥

Theorem 6.4 shows that the reason of the pathological behavior of the function h in Example 6.3 is the *lack of boundedness* of the corresponding superposition operator S_h in the norm (0.71). On the other hand, the reason for the pathological behavior of the function h in Example 6.6 is the *lack of continuity* of the corresponding superposition operator S_h in the norm (0.65). In fact, considering again the sequence $(f_n)_n$ of constant functions $f_n(x) \equiv 1/n$, we obtain

$$\|f_n\|_{C^1} = \frac{1}{n} \to 0, \quad \|S_h f_n\|_{C^1} \geq |S_h f_n(0)| = 1 \not\to 0 \quad (n \to \infty).$$

If we *add* continuity to the acting condition $S_h(C^1) \subseteq C^1$, however, we get the result that we expect; this time, we give a complete proof.

Theorem 6.7. *The operator* (6.1) *maps the space* $C^1([a,b])$ *into itself and is continuous with respect to the norm* (0.65) *if and only if the function* h *is continuously differentiable on* $[a,b] \times \mathbb{R}$.

Proof. The "if" part is an immediate consequence of the chain rule (6.7) and the definition of the norm (0.65) in the space $C^1([a,b])$; thus, we only have to prove the "only if" part.

Suppose that S_h maps the space $C^1([a,b])$ into itself and is continuous with respect to the norm (0.65). We have to show that the partial derivatives D_1h and D_2h of h are continuous functions.

Writing f_u for the constant function $f_u(x) \equiv u$ and $g_u = S_h f_u$, we have $g_u(x) = h(x,u)$, and thus $g_u'(x) = D_1h(x,u)$. Given $\varepsilon > 0$, choose $\delta > 0$ such that $|s-t| \leq \delta$ and $|u-v| = \|f_u - f_v\|_{C^1} \leq \delta$ implies

$$|g_u'(s) - g_u'(t)| \leq \varepsilon, \quad \|g_u - g_v\|_{C^1} \leq \varepsilon,$$

which is possible by the uniform continuity of g_u' on $[a,b]$ and our continuity assumption on the operator S_h. Then

$$\begin{aligned} &|D_1h(s,u) - D_1h(t,v)| \\ &\quad = |g_u'(s) - g_v'(t)| \leq |g_u'(s) - g_u'(t)| + |g_u'(t) - g_v'(t)| \\ &\quad \leq \varepsilon + \|g_u' - g_v'\|_C \leq \varepsilon + \|g_u - g_v\|_{C^1} \leq 2\varepsilon \end{aligned}$$

for $|s-t| \leq \delta$ and $|u-v| \leq \delta$, which shows that D_1h is indeed continuous on $[a,b] \times \mathbb{R}$.

The proof of the continuity of D_2h on $[a,b] \times \mathbb{R}$ is somewhat harder. Writing $f_{t,v}$ for the linear function $f_{t,v}(x) = v - t + x$ and $g_{t,v} = S_h f_{t,v}$, we have $g_{t,v}(x) = h(x, v-t+x)$, and hence

$$g_{t,v}'(x) = D_1h(x, v-t+x) + D_2h(x, v-t+x)$$

by (6.7). For fixed $s_0 \in [a,b]$, $u_0 \in \mathbb{R}$, and $\tau > 0$, we have

$$\begin{aligned} h(s_0,u_0+\tau)-h(s_0,u_0) &= h(s_0+\tau,u_0+\tau)-h(s_0,u_0)-h(s_0+\tau,u_0+\tau)+h(s_0,u_0+\tau) \\ &= g_{s_0,u_0}(s_0+\tau)-g_{s_0,u_0}(s_0)-\int_{s_0}^{s_0+\tau} D_1h(\sigma,u_0+\tau)\,d\sigma \\ &= g_{s_0,u_0}(s_0+\tau)-g_{s_0,u_0}(s_0)-D_1h(s_0+\theta,u_0)\tau \\ &\quad -\int_{s_0}^{s_0+\tau}[D_1h(\sigma,u_0+\tau)-D_1h(\sigma,u_0)]\,d\sigma \end{aligned}$$

with $0 \le \theta \le \tau$. However, the partial derivative $D_1 h$ is continuous by what we have just shown, and so

$$D_2h(s_0,u_0)=\lim_{\tau\to 0}\frac{h(s_0,u_0+\tau)-h(s_0,u_0)}{\tau}=g'_{s_0,u_0}(s_0)-D_1h(s_0,u_0)\,.$$

Given $\varepsilon > 0$, we may choose $\delta > 0$ such that $|s-t| \le \delta$ and $|u-v| \le \delta$ implies

$$|g'_{s,u}(s)-g'_{t,v}(t)|\le\varepsilon,\quad |D_1h(s,u)-D_1h(t,v)|\le\varepsilon,\quad \|g_{s,u}-g_{t,v}\|_{C^1}\le\varepsilon$$

because $D_1 h$ is continuous on $[a,b]\times\mathbb{R}$, the operator S_h is continuous in $C^1([a,b])$, and

$$\|f_{s,u}-f_{t,v}\|_{C^1}=|u-v-s+t|\le|u-v|+|s-t|\le 2\delta\,.$$

Combining these estimates, we obtain

$$\begin{aligned} |D_2h(s,u)-D_2h(t,v)| &= |g'_{s,u}(s)-D_1h(s,u)-g'_{t,v}(t)+D_1h(t,v)| \\ &\le |g'_{s,u}(s)-g'_{s,u}(t)|+|g'_{s,u}(t)-g'_{t,v}(t)|+|D_1h(s,u)-D_1h(t,v)| \\ &\le \varepsilon+\|g_{s,u}-g_{t,v}\|_{C^1}+\varepsilon\le 3\varepsilon\,. \end{aligned}$$

Therefore, we have shown that the partial derivative $D_2 h$ exists and is continuous on $[a,b]\times\mathbb{R}$ as well, and the proof is complete. □

We may summarize the contents of Theorems 6.1, 6.2, 6.4, and 6.7 in the following form: in addition to the mapping condition $S_h(X) \subseteq X$, in case $X = C^1$, we have to *require* the continuity of S_h, while in case $X = Lip_\alpha$, we have to *require* the boundedness of S_h to obtain a "natural" condition for h. On the other hand, in case $X = C$ or $X = L_p$, we get both the continuity and boundedness of the operator S_h "for free" as a consequence of the mere mapping condition $S_h(X) \subseteq X$, which means that in this case, the situation is the same as for the autonomous composition operator (5.1).

Our discussion shows that, in contrast to the case of the autonomous composition operator (5.1), almost nothing is known for the nonautonomous superposition operator (6.1). Therefore, we cannot repeat the complete list of Tables 5.2–5.8 here; the only exception concerns the spaces L_p, C, Lip_α, and C^1.

Table 6.1. The operator S_h in $L_p([a,b])$.

S_h bounded in L_p	$\Leftrightarrow$	$S_h(L_p) \subseteq L_p$	$\Leftrightarrow$	S_h continuous in L_p
		$\Updownarrow$		
		$\vert h(x,u)\vert \le \alpha(x) + \beta \vert u\vert$		

Table 6.2. The operator S_h in $C([a,b])$.

S_h bounded in C	$\Leftrightarrow$	$S_h(C) \subseteq C$	$\Leftrightarrow$	S_h continuous in C
		$\Updownarrow$		
		$h \in C([a,b] \times \mathbb{R})$		

Table 6.3. The operator S_h in $Lip_\alpha([a,b])$ $(0 < \alpha \le 1)$.

S_h bounded in Lip_α	$\Rightarrow$	$S_h(Lip_\alpha) \subseteq Lip_\alpha$	$\Leftarrow$	S_h continuous in Lip_α
$\Updownarrow$		$\Updownarrow$		
$h(\cdot,u) \in Lip_\alpha$		see Proposition 6.5		
$h(x,\cdot) \in Lip_{\mathrm{loc}}(\mathbb{R})$				

Table 6.4. The operator S_h in $C^1([a,b])$.

S_h bounded in C^1	$\Rightarrow$	$S_h(C^1) \subseteq C^1$	$\Leftarrow$	S_h continuous in C^1
				$\Updownarrow$
				$h \in C^1([a,b] \times \mathbb{R})$

We point out again the differences in Table 6.3 and Table 6.4: in the space Lip_α, we have a precise characterization of bounded superposition operators, but not for continuity. In the space C^1, we have a precise characterization of continuous superposition operators, but not for boundedness.

Now, we study the operator (6.1) in the space BV which is our main focus. The mixed condition (6.5) for the inclusion $S_h(Lip_\alpha) \subseteq Lip_\alpha$ may be rephrased in the following form:

– The function $h(\cdot,u) : [a,b] \to \mathbb{R}$ is Hölder continuous on $[a,b]$, uniformly with respect to u belonging to compact subsets of $\mathbb{R}$, i.e.

$$\sup_{|u|\le r} lip_\alpha(h(\cdot,u);[a,b]) \le \upsilon(r) < \infty$$

for each $r > 0$.

– The function $h(x,\cdot) : \mathbb{R} \to \mathbb{R}$ is locally Lipschitz continuous on $\mathbb{R}$, uniformly with respect to $x \in [a,b]$, i.e.

$$\sup_{a\le x\le b} |h(x,u) - h(x,v)| \le k(r)|u-v| \quad (|u|,|v| \le r)$$

for each $r > 0$.

This suggests that the following two conditions are also sufficient for the inclusion $S_h(BV) \subseteq BV$.

- The function $h(\cdot, u) : [a, b] \to \mathbb{R}$ has bounded variation on $[a, b]$, uniformly with respect to u belonging to compact subsets of $\mathbb{R}$, i.e.

$$\sup_{|u|\leq r} \operatorname{Var}(h(\cdot, u); [a, b]) \leq v(r) < \infty \tag{6.9}$$

  for each $r > 0$.
- The function $h(x, \cdot) : \mathbb{R} \to \mathbb{R}$ is locally Lipschitz continuous on $\mathbb{R}$, uniformly with respect to $x \in [a, b]$, i.e.

$$\sup_{a\leq x\leq b} |h(x, u) - h(x, v)| \leq k(r)|u - v| \quad (|u|, |v| \leq r) \tag{6.10}$$

  for each $r > 0$.

As a matter of fact, these condition have been formulated, without proof, by Lyamin in [189], and afterwards cited and used by many authors (e.g. in the book [20]). However, it came out as a surprise only quite recently that this is *false*. Indeed, Maćkowiak [190] has shown this by means of a sophisticated counterexample which looks as follows.

Example 6.8. We construct a function $h : [0, 1] \times \mathbb{R} \to \mathbb{R}$ which fulfills the two conditions (6.9) and (6.10) as well as a function $f \in BV([0, 1])$ such that $S_h f \notin BV([0, 1])$. For $n = 2, 3, \ldots$, let

$$c_n := 1 - \frac{1}{n}, \quad d_n := \frac{1}{2n} \quad I_n := (c_n - d_n, c_n + d_n) = \left(\frac{2n-3}{2n}, \frac{2n-1}{2n}\right).$$

Clearly, I_n is a subinterval of $(0, 1)$ of length $1/n$. Now, we define $h : [0, 1] \times \mathbb{R} \to \mathbb{R}$ by

$$h(x, u) := \begin{cases} \frac{1}{n}\left(1 - \frac{|u-c_n|}{d_n}\right) & \text{if } x = c_n \text{ and } u \in I_n, \\ 0 & \text{otherwise}. \end{cases} \tag{6.11}$$

Geometrically, the sets $\{c_n\} \times I_n$ on which h does not vanish are vertical segments without endpoints in the plane which get smaller as $1/n$ and are shifted to the right and above as n increases. Since

$$\lim_{u\to c_n\pm d_n} h(c_n, u) = \frac{1}{n}\left(1 - \frac{d_n}{d_n}\right) = 0,$$

$h(x, \cdot)$ is continuous on $\mathbb{R}$. Even more, the equality

$$\sup \{|D_2 h(c_n, u)| : u \in I_n \setminus \{c_n\}\} = \frac{1}{n}\frac{1}{d_n} = 2$$

shows that the function $h(x, \cdot)$ is (globally) Lipschitz continuous on $\mathbb{R}$, uniformly with respect to $x \in [0, 1]$, with Lipschitz constant 2, and so (6.10) holds true.

Now, we prove that $h(\cdot, u) \in BV([0,1])$, uniformly with respect to[1] $u \in [0,1]$, and so (6.9) holds true as well. To this end, we fix $m \in \mathbb{N}$ and show that

$$I_m \cap I_n = \emptyset \quad (n \geq 4m). \tag{6.12}$$

In fact, for $n = 4m$, this follows from the inequalities

$$c_n - d_n = 1 - \frac{3}{2n} = 1 - \frac{3}{8m} > 1 - \frac{1}{2m} = c_m + d_m.$$

The function $g : [1, \infty) \to \mathbb{R}$ defined by $g(s) := 1 - \frac{1}{s} - \frac{1}{2s}$ is continuous and strictly increasing, so for $n > 4m$, we get

$$c_n - d_n = g(n) > g(4m) = 1 - \frac{1}{4m} - \frac{1}{8m} = 1 - \frac{3}{8m} > 1 - \frac{1}{2m} = c_m + d_m,$$

which proves (6.12). Since $m \geq 2$, we further obtain

$$\sum_{n=m}^{4m} \frac{1}{n} = \frac{1}{m} + \frac{1}{m+1} + \ldots + \frac{1}{4m} \leq \underbrace{\frac{1}{m} + \frac{1}{m} + \ldots + \frac{1}{m}}_{3m+1 \ terms} = \frac{3m+1}{m} < 4. \tag{6.13}$$

Now, let $u \in I_m$ for some $m \geq 2$. Then $u \notin I_n$ for $n \geq 4m$, by (6.12). Moreover, (6.13) implies

$$\mathrm{Var}(h(\cdot, u); [0,1]) \leq 2 \sum_{n=q(m)}^{4m} \frac{1}{n} = 2\left(\sum_{n=q(m)}^{4q(m)-1} \frac{1}{n} + \sum_{n=4q(m)}^{m-1} \frac{1}{n} + \sum_{n=m}^{4m} \frac{1}{n} \right) \leq 22,$$

where $q(m) := \max\{\mathrm{ent}(m/4), 2\}$. Since the upper bound does not depend on u, the assertion is proved.

It remains to show that the operator (6.1) generated by h does not map the space $BV([0,1])$ into itself. Consider the function $f(x) := x$ which trivially belongs to $BV([0,1])$. By definition of h, we then have

$$S_h f(x) = h(x, x) = \begin{cases} \frac{1}{n} & \text{if } x = c_n, \\ 0 & \text{otherwise}. \end{cases}$$

Consider a partition of the form $P_n := \{0, t_1, c_2, t_2, c_3, \ldots, c_n, t_n, 1\} \in \mathcal{P}([0,1])$ with $c_j < t_j < c_{j+1}$ for $j = 1, 2, \ldots, n$. Then $h(t_j, t_j) = 0$ implies

$$\mathrm{Var}(S_h f, P_n; [0,1]) \geq \sum_{k=2}^{n} \frac{1}{k} \to \infty \quad (n \to \infty),$$

and so $S_h f \notin BV([0,1])$. ♥

---

1 Obviously, this suffices since $h(x, u) \equiv 0$ for $u \notin [0,1]$.

Observe that Example 6.8 shows even more: *for h given by* (6.11), *the corresponding superposition operator S_h does not map any function space $X \subseteq BV$ which contains the identity $f(x) = x$ into itself.*[2] This is of course in sharp contrast to Theorem 5.9 which asserts that in the autonomous case $h : \mathbb{R} \to \mathbb{R}$, local Lipschitz continuity of h on the real axis is both necessary and sufficient for the operator (5.1) to map any space X with $Lip \subseteq X \subseteq BV$ into itself.

Now, we are interested in the problem to what extent the above conditions (6.9) and (6.10) are necessary for the inclusion $S_h(BV) \subseteq BV$. Of course, since constant functions belong to $BV([a, b])$, from this inclusion, it follows that $h(\cdot, u) \in BV([a, b])$ for every $u \in \mathbb{R}$. The following simple example shows, however, that one cannot expect local Lipschitz continuity of $h(x, \cdot)$ for every $x \in [a, b]$:

Example 6.9. Let $h_0(u) := \min\left\{\sqrt{|u|}, 1\right\}$ be the seagull function from Example 5.8, and define $h : [0, 1] \times \mathbb{R} \to \mathbb{R}$ by

$$h(x, u) := \begin{cases} h_0(u) & \text{if } x = 0\,, \\ 0 & \text{if } 0 < x \le 1\,. \end{cases} \tag{6.14}$$

Then the corresponding operator (6.1) maps the space $BV([0, 1])$ into itself since

$$\mathrm{Var}(S_h f, P; [0, 1]) = |h(0, f(0))| = |f(0)|$$

for $f \in BV([0, 1])$ and every $P \in \mathcal{P}([0, 1])$. The same calculation shows that $\|f\|_{BV} \le r$ implies $\|S_h f\|_{BV} \le 2\sqrt{r}$, and so the operator S_h is bounded in the norm (1.16) of $BV([0, 1])$. On the other hand, the function $h(0, \cdot) = h_0$ is certainly not locally Lipschitz at zero, and so (6.10) is not true. ♥

Observe that the Lipschitz continuity of the function $h(x, \cdot)$ fails in Example 6.9 only at one point, namely, $x = 0$. Clearly, one could generalize Example 6.9 by fixing points $t_1, t_2, \ldots, t_n \in [0, 1]$ and considering the function

$$h(x, u) := \begin{cases} h_i(u) & \text{if } x = t_i \ (i = 1, 2, \ldots, n)\,, \\ 0 & \text{otherwise}\,, \end{cases} \tag{6.15}$$

where $h_1, h_2, \ldots, h_n$ are *any* functions (even discontinuous). In spite of the finitely many "singularities" $t_1, t_2, \ldots, t_n \in [0, 1]$, the superposition operator S_h defined by (6.15) then maps $BV([0, 1])$ into itself and is bounded.

The following theorem shows that excluding countably many points $x \in [a, b]$ of this type for $h(x, \cdot)$, one gets a necessary condition for S_h to be bounded in $BV([a, b])$. This means, roughly speaking, that the function (6.14) and the more general function (6.15) are in a certain sense "optimal."

---

2 This makes Example 6.6 obsolete, but not Example 6.3 because $Lip_\alpha \not\subseteq BV$ for $\alpha < 1$, see Example 1.24.

Theorem 6.10. *Suppose that the operator* (6.1) *generated by* $h : [a, b] \times \mathbb{R} \to \mathbb{R}$ *is bounded in the space* $BV([a, b])$. *Then the function* h *may be represented as a sum*

$$h(x, u) = \tilde{h}(x, u) + \hat{h}(x, u), \tag{6.16}$$

where $\tilde{h}$ *and* $\hat{h}$ *have the following properties.*
(a) *The function* $\tilde{h}$ *satisfies* (6.9) and (6.10).
(b) *The function* $\hat{h}(x, u)$ *is zero on* $([a, b] \setminus C) \times \mathbb{R}$, *where* $C \subset [a, b]$ *is some countable subset.*
(c) *The superposition operators* $S_{\tilde{h}}$ *and* $S_{\hat{h}}$ *generated by* $\tilde{h}$ *and* $\hat{h}$, *respectively, are both bounded in* $BV([a, b])$.

Proof. For the proof, we take $[a, b] = [0, 1]$ and consider the space $BV([0, 1])$ equipped with the norm

$$|||f|||_{BV} := \max\{\|f\|_\infty, \mathrm{Var}(f; [0, 1])\}, \tag{6.17}$$

with $\|\cdot\|_\infty$ given by (0.39). The norm (6.17) is equivalent to the usual norm (1.16) because $\|f\|_{BV} \leq |||f|||_{BV} \leq 2\|f\|_{BV}$ by (1.9).

Suppose that, for any countable set $C \subset [0, 1]$, the function $h(t, \cdot)$ does *not* belong to $Lip_{\mathrm{loc}}(\mathbb{R})$ for $t \in [0, 1] \setminus C$. In particular, we then find $k \in \mathbb{N}$ such that $h(t, \cdot) \notin Lip([-k, k])$ for $t \in [0, 1] \setminus C$. Keeping this k fixed for the moment, we conclude that the set

$$E_{m,k} := \{t \in [0, 1] : |h(t, u) - h(t, v)| > m|u - v| \quad \text{for some } u, v \in [-k, k]\} \tag{6.18}$$

is *uncountable* for all $m \in \mathbb{N}$. For any $u, v \in [-k, k]$, we may choose points $u_j, v_j \in [-k, k]$ $(j = 1, 2, \ldots, 10k)$ such that $|u_j - v_j| \leq 1/5$ and

$$|h(t, u) - h(t, v)| \leq 10k \max\{|h(t, u_j) - h(t, v_j)| : j = 1, 2, \ldots, 10k\}$$

for all $t \in [0, 1]$. Therefore, for each $m \in \mathbb{N}$ and all $t \in E_{m,k}$, there are $u, v \in [-k, k]$ such that

$$|u - v| \leq \frac{1}{5}, \quad |h(t, u) - h(t, v)| > \frac{m}{10k}|u - v|.$$

Now, we use our assumption that the operator S_h maps $BV([0, 1])$ into itself and is bounded. In particular, S_h maps the ball $\tilde{B}_k := \{f \in BV([0, 1]) : |||f|||_{BV} \leq k\}$ into some ball $B_R := \{f \in BV([0, 1]) : \|f\|_{BV} \leq R\}$. Fix $m \geq 10kR\rho$, where $\rho = 16$, and let $\mathcal{E} := E_{m,k}$ for this m, with $E_{m,k}$ given by (6.18). For $r = 5, 6, \ldots$, denote by $\mathcal{E}_r$ the set of all $t \in \mathcal{E}$ such that u and v can be chosen with

$$\frac{1}{2r} < |u - v| \leq \frac{1}{r}.$$

Since $\mathcal{E}$ is the union of all sets $\mathcal{E}_r$, $r = 5, 6, \ldots$, some $\mathcal{E}_r$ must be uncountable, and we can assume $\mathcal{E} = \mathcal{E}_r$. Choose $\delta \in (0, 1/9)$ such that $2r\delta = 1$, and fix $u = x_t$ and $v = y_t$ such that $\delta < y_t - x_t < 2\delta$ for all t. There is a finite set $F \subset [-k, k]$ such that each point in $[-k, k]$ has distance $< \delta/3$ from some point in F. This means that for some $u \in F$, there is an uncountable set $A_u \subseteq [0,1]$ with $x_t \le u \le y_t$ for all $t \in A_u$. Consider a partition $P = \{s_0, s_1, \dots, s_n, s_{n+1}\} \in \mathcal{P}([0,1])$, where $s_0 = 0$, $s_{n+1} = 1$, $s_1, \dots, s_n \in A_u$, and n is so large that $n\rho\delta > 2$.

Now, we define a function $g : [0,1] \to \mathbb{R}$ recursively as follows. First, let $g(s_0) = g(0) := u$. Suppose we have defined $g(s_0), \dots, g(s_{j-1})$ for $j \le n$, where $g(s_i) = x_{s_i}$ or $g(s_i) = y_{s_i}$, for each $i \in \{1, 2, \dots, j-1\}$, so that $|g(s_i) - u| \le 2\delta$. Since $|h(s_j, x_{s_j}) - h(s_j, y_{s_j})| > R\rho\delta$, we can now choose $g(s_j) = x_{s_j}$ or $g(s_j) = y_{s_j}$ in such a way that

$$|h(s_j, g(s_j)) - h(s_{j-1}, g(s_{j-1}))| > \frac{1}{2} R\rho\delta .$$

In either case, $|g(s_j) - g(s_{j-1})| \le 4\delta$. Thus, we have defined $g(s_j)$ for $j = 0, 1, \dots, n$. Now, for $\nu = 1, 2, \dots, n$, we have

$$\sum_{j=1}^{\nu} |h(s_j, g(s_j)) - h(s_{j-1}, g(s_{j-1}))| \ge \frac{1}{2} \nu R\rho\delta .$$

There is a smallest $\nu \in \{2, 3, \dots, n\}$ such that $\nu\rho\delta > 2$, hence $8\nu\delta > 1$, and for this ν, we have

$$\sum_{j=1}^{\nu} |S_h g(s_j) - S_h g(s_{j-1})| = \sum_{j=1}^{\nu} |h(s_j, g(s_j)) - h(s_{j-1}, g(s_{j-1}))| > R . \tag{6.19}$$

Since $k \ge 1$ and $\nu\rho\delta \le 4$, we further obtain

$$\sum_{j=1}^{\nu} |g(s_j) - g(s_{j-1})| \le 4\nu\delta \le 1 \le k . \tag{6.20}$$

Now, we are almost done. Using the function g, we define $f : [0,1] \to \mathbb{R}$ by

$$f(t) := \begin{cases} g(s_j) & \text{for } t = s_j,\ j = 0, 2, \dots, \nu, \\ g(s_\nu) & \text{for } s_\nu \le t \le 1, \\ \text{linear} & \text{otherwise}. \end{cases}$$

By monotonicity, the variation $\mathrm{Var}(f, P; [0,1])$ with respect to any partition $P \in \mathcal{P}([0,1])$ is dominated by one in which adjoining increments of the same sign are combined, so we can assume that adjoining increments are of opposite signs. Moreover, we may suppose that the partition points are local maxima or minima of f, and that only the points $s_0, \dots, s_\nu$ occur in the partition.

Each value $g(s_j)$ has distance $\le 2\delta$ from u, so

$$\mathrm{Var}(f; [0,1]) \le 4\nu\delta \le 1 ,$$

by (6.20), and $\|f\|_{BV} \le 1 \le k$. Since also $\|f\|_\infty \le k$, by construction, our definition of the norm (6.17) shows that $f \in \tilde{B}_k$. Thus, our choice of R implies that $S_h f \in B_R$, i.e.

$\mathrm{Var}(S_h f; [0,1]) \le \|S_h f\|_{BV} \le R$, contradicting (6.19). This contradiction shows that our assumption was false, and so the assertions (a) and (b) are proved.

It remains to prove (c). Since the operator S_h is bounded in the space $BV([0,1])$, by assumption, the ball $\{f \in BV([0,1]) : \|f\|_\infty \le |||f|||_{BV} \le r\}$ is mapped by the operator $S_{\tilde{h}}$ into a ball $\{g \in B([0,1]) : \|g\|_\infty \le R\}$, and so $S_{\tilde{h}}$ is bounded in the space $B([0,1])$. Any variation sum

$$\mathrm{Var}(S_{\tilde{h}} f, P; [0,1]) = \sum_{j=1}^{m} |\tilde{h}(s_j, f(s_j)) - \tilde{h}(s_{j-1}, f(s_{j-1}))|$$

with respect to some partition $P = \{s_0, s_1, \ldots, s_m\} \in \mathcal{P}([0,1])$ can be approximated arbitrarily closely by another variation sum

$$\mathrm{Var}(S_{\tilde{h}} f, Q; [0,1]) = \sum_{j=1}^{m} |\tilde{h}(t_j, f(t_j)) - \tilde{h}(t_{j-1}, f(t_{j-1}))|$$

with respect to some partition $Q = \{t_0, t_1, \ldots, t_m\} \in \mathcal{P}([0,1])$, where t_j does not belong to the exceptional set C from (b), t_j is close enough to s_j, and $t_j > s_j$ except if $s_m = 1$, when $t_m < s_m$. For any such partition Q, there is a function $g \in BV([0,1])$ with $|||g|||_{BV} \le |||f|||_{BV}$ and $g(t_j) = f(s_j)$ for $j = 1, 2, \ldots, m$. It follows that

$$\sup\{\|S_{\tilde{h}} f\|_{BV} : |||f|||_{BV} \le r\} \le \sup\{\|S_h g\|_{BV} : |||g|||_{BV} \le r\} < \infty,$$

which shows that the superposition operator $S_{\tilde{h}}$ is in fact bounded in $BV([0,1])$. The same then holds for the operator $S_{\hat{h}} = S_h - S_{\tilde{h}}$, and the proof is complete. □

In view of Example 6.8, the question arises how to strengthen[3] (6.9) or (6.10) to get a sufficient condition for the inclusion $S_h(BV) \subseteq BV$. An interesting condition was proposed by Bugajewska [66]; to state this condition, we need some notation.

Given $r > 0$, by $\mathfrak{U}_r$, we denote the family of all finite collections $U_r = \{u_1, u_2, \ldots, u_m\} \subset [-r, r]$; thus, the family $\mathfrak{U}_r$ may be regarded as some kind of "vertical" analogue of the partition family $\mathcal{P}([a,b])$ in the "horizontal" direction. For $h : [a,b] \times \mathbb{R} \to \mathbb{R}$, $P = \{t_0, t_1, \ldots, t_m\} \in \mathcal{P}([a,b])$, and $U_r = \{u_1, u_2, \ldots, u_m\} \in \mathfrak{U}_r$, we put

$$\mathrm{Var}(h, P, U_r; [a,b] \times \mathbb{R}) := \sum_{j=1}^{m} |h(t_j, u_j) - h(t_{j-1}, u_j)| \tag{6.21}$$

and

$$\mathrm{Var}(h, U_r; [a,b] \times \mathbb{R}) := \sup\{\mathrm{Var}(h, P, U_r; [a,b] \times \mathbb{R}) : P \in \mathcal{P}([a,b])\} \tag{6.22}$$

---

3 Since (6.10) is necessary and sufficient in the autonomous case, it seems reasonable to only strengthen condition (6.9).

where the supremum in (6.22) is taken over all partitions $P \in \mathcal{P}([a,b])$. Using this notation, we now replace (6.9) by the stronger condition[4]

$$\sup_{U_r \in \mathfrak{U}_r} \mathrm{Var}(h, U_r; [a,b] \times \mathbb{R}) \le v(r) < \infty \tag{6.23}$$

for each $r > 0$. The next theorem [66] shows that this condition, together with (6.10), in fact, suffices to guarantee what we need.

Theorem 6.11. *Suppose that h satisfies the conditions* (6.10) *and* (6.23). *Then the corresponding operator* (6.1) *maps the space* $BV([a,b])$ *into itself and is bounded.*

Proof. Without loss of generality, we take $[a,b] = [0,1]$. Fix $f \in BV([0,1])$ with $\|f\|_{BV} \le r$, $P = \{t_0, t_1, \dots, t_m\} \in \mathcal{P}([0,1])$, and $U_r = \{u_1, u_2, \dots, u_m\} \in \mathfrak{U}_r$, where $u_j = f(t_j)$ $(j = 1, 2, \dots, m)$. Then

$$\begin{aligned}
\mathrm{Var}(S_h f, P; [0,1]) &= \sum_{j=1}^{m} |h(t_j, f(t_j)) - h(t_{j-1}, f(t_{j-1}))| \\
&\le \sum_{j=1}^{m} |h(t_j, f(t_j)) - h(t_{j-1}, f(t_j))| + \sum_{j=1}^{m} |h(t_{j-1}, f(t_j)) - h(t_{j-1}, f(t_{j-1}))| \\
&\le v(r) + k(r) \sum_{j=1}^{m} |f(t_j) - f(t_{j-1})| \le v(r) + k(r) r\,,
\end{aligned}$$

where $v(r)$ is taken from (6.23) and $k(r)$ from (6.10). This not only shows that $S_h f$ belongs to $BV([0,1])$, but also the boundedness of S_h in the norm (1.16). □

Observe that for the function (6.14) in Example 6.9, we have

$$\sum_{j=1}^{m} |h(t_j, u_j) - h(t_{j-1}, u_j)| = |h(0, u_1)| = \min\{\sqrt{|u_1|}, 1\}\,,$$

and so (6.23) holds in this example with $v(r) := \min\{\sqrt{r}, 1\}$. Here is another example which illustrates Theorem 6.11.

Example 6.12. Suppose that h factorizes into two functions of the form $h(x,u) = h_1(x)h_2(u)$, where $h_1 \in BV([a,b])$ and $h_2 \in Lip_{\mathrm{loc}}(\mathbb{R})$. In this case, (6.10) is trivially true since

$$|h(x,u) - h(x,v)| = |h_1(x)|\,|h_2(u) - h_2(v)| \le \|h_1\|_\infty lip(h_2; [-r,r])|u - v|\,.$$

4 Indeed, if we restrict ourselves to singletons $U_r = \{u\} \in \mathfrak{U}_r$ with $|u| \le r$ in (6.23), we exactly get condition (6.9).

However, (6.23) is also true because for $P = \{t_0, t_1, \dots, t_m\} \in \mathcal{P}([0,1])$ and $U_r = \{u_1, u_2, \dots, u_m\} \in \mathfrak{U}_r$, we have

$$\sum_{j=1}^{m} |h(t_j, u_j) - h(t_{j-1}, u_j)| = \sum_{j=1}^{m} |h_1(t_j) - h_1(t_{j-1})|\,|h_2(u_j)|$$
$$\leq \sup_{|u|\leq r} |h_2(u)| \operatorname{Var}(h_1; [a,b]) .$$

This case occurs frequently in applications. We point out again that the corresponding condition (6.9) in this case is not sufficient, as Example 6.9 (with $h_1 := \chi_{\{0\}}$ and $h_2 := h_0$) shows. ♥

We may summarize our discussion on necessary or sufficient conditions for the inclusion $S_h(BV) \subseteq BV$ as follows:
- Conditions (6.9) and (6.10) are not sufficient for $S_h(BV) \subseteq BV$, but conditions (6.10) and (6.23) are.
- Condition (6.9) is necessary for $S_h(BV) \subseteq BV$, but condition (6.10) is not.

Now, we state a generalization of Theorem 6.11 to the space WBV_ϕ, see Section 2.1, which is also due to Bugajewska [66]. To this end, we suppose that $\phi : [0,\infty) \to [0,\infty)$ is a Young function which satisfies the δ_2-condition introduced in Definition 2.4. As a consequence, the function $M : (0,\infty) \times [1,\infty) \to [1,\infty)$ defined by

$$M(T,\lambda) = \sup \left\{ \frac{\phi(\lambda t)}{\phi(t)} : 0 < t \leq T \right\} \tag{6.24}$$

is well-defined and increasing[5] in both T and λ. Observe that for $f \in WBV_\phi([a,b])$ and $\lambda \geq 1$, we then have

$$\operatorname{Var}_\phi^W(\lambda f; [a,b]) \leq M(2\|f\|_\infty, \lambda) \operatorname{Var}_\phi^W(f; [a,b]) , \tag{6.25}$$

by the Definition (2.2) of the Wiener–Young variation. We also have to replace (6.21) and (6.22) by

$$\operatorname{Var}_\phi^W(h, P, U_r; [a,b] \times \mathbb{R}) := \sum_{j=1}^{m} \phi\left(|h(t_j, u_j) - h(t_{j-1}, u_j)|\right) \tag{6.26}$$

and

$$\operatorname{Var}_\phi^W(h, U_r; [a,b] \times \mathbb{R}) := \sup \{\operatorname{Var}_\phi^W(h, P, U_r; [a,b] \times \mathbb{R}) : P \in \mathcal{P}([a,b])\} \tag{6.27}$$

respectively, where the supremum in (6.27) is again taken over all partitions $P \in \mathcal{P}([a,b])$.

5 Note that $M(T,2)$ coincides with the characteristic $M(T)$ which we introduced in (2.6). In the Wiener space WBV_p, we have the explicit formula $M(T,\lambda) = \lambda^p$ for every $T > 0$.

Theorem 6.13. *Suppose that h satisfies* (6.10) *and*

$$\sup_{U_r \in \mathfrak{U}_r} \operatorname{Var}_\phi^W(h, U_r; [a,b] \times \mathbb{R}) \le v_\phi(r) < \infty \tag{6.28}$$

for each $r > 0$, *where* $\phi \in \delta_2$ *is a Young function, and* $\operatorname{Var}_\phi^W(h, U_r; [a,b] \times \mathbb{R})$ *is defined by* (6.27). *Then the corresponding operator* (6.1) *maps the space* $WBV_\phi([a,b])$ *into itself and is bounded in the norm* (2.11).

Proof. We take again $[a,b] = [0,1]$. Fix $f \in WBV_\phi([0,1])$ with $\|f\|_{WBV_\phi} \le r$, $P = \{t_0, t_1, \dots, t_m\} \in \mathcal{P}([0,1])$, and $U_r = \{u_1, u_2, \dots, u_m\} \in \mathfrak{U}_r$. For

$$\hat{k}(r) := \max\{k(r), 1\}, \quad T := \max\{v_\phi(r), 2r\hat{k}(r)\},$$

we then obtain

$$\begin{aligned}\phi(|h(t_j,u_j) - h(t_{j-1},u_{j-1})|) &\le \phi(|h(t_j,u_j) - h(t_{j-1},u_j)| + |h(t_{j-1},u_j) - h(t_{j-1},u_{j-1})|)\\ &\le \frac{1}{2}\phi\left(2|h(t_j,u_j) - h(t_{j-1},u_j)| + 2k(r)|u_j - u_{j-1}|\right)\\ &\le \frac{M(T,4\hat{k}(r))}{2}\left[\phi\left(|h(t_j,u_j) - h(t_{j-1},u_j)|\right) + \phi\left(|u_j - u_{j-1}|\right)\right],\end{aligned} \tag{6.29}$$

where we have used the monotonicity and convexity of ϕ as well as (6.10) and (6.25). Now, putting $u_j = f(t_j)$ and $u_{j-1} = f(t_{j-1})$ in (6.29) and summing up over $j = 1, 2, \dots, m$, we further get

$$\begin{aligned}&\operatorname{Var}_\phi^W(S_h f, P; [0,1])\\ &= \sum_{j=1}^m \phi(|h(t_j, f(t_j)) - h(t_{j-1}, f(t_{j-1}))|)\\ &\le \frac{M(T,4\hat{k}(r))}{2}\sum_{j=1}^m \left[\phi\left(|h(t_j,f(t_j)) - h(t_{j-1},f(t_j))|\right) + \phi\left(|f(t_j) - f(t_{j-1})|\right)\right]\\ &\le \frac{M(T,4\hat{k}(r))}{2}\left[v_\phi(r) + \operatorname{Var}_\phi^W(f;[0,1])\right] \le \frac{M(T,4\hat{k}(r))}{2}(v_\phi(r) + r).\end{aligned}$$

This not only shows that $S_h f \in WBV_\phi([0,1])$, but also proves the boundedness of S_h on balls of radius $r \le 1$.

To prove the boundedness of S_h on larger balls, we use a rescaling argument and the δ_2-property of ϕ. Let $f \in WBV_\phi([0,1])$ with $\|f\|_{WBV_\phi} \le r$, and hence $\|f\|_\infty \le r$ for some $r > 1$. Then (2.11) and (6.25) imply

$$\begin{aligned}\operatorname{Var}_\phi^W(f;[0,1]) = \operatorname{Var}_\phi^W\left(\frac{rf}{r};[0,1]\right) &\le M(2r,r)\operatorname{Var}_\phi^W\left(\frac{f}{r};[0,1]\right)\\ &\le M(2r,r)\left\|\frac{f}{r}\right\|_{WBV_\phi} \le M(2r,r)\frac{r}{r} = M(2r,r)\end{aligned}$$

since $\|f/r\|_{WBV_\phi} \le 1$. In this way, we have reduced the case $r > 1$ to the previous case, and the proof is complete. □

We do not know whether or not Theorem 6.13 is also true without the requirement $\phi \in \delta_2$.

6.2 Lipschitz continuity

Suppose that the superposition operator S_h given by (6.1) maps a normed space X into a normed space Y. In this section, we are going to study the (global) Lipschitz condition

$$\|S_h f - S_h g\|_Y \leq K\|f - g\|_X \quad (f, g \in X), \tag{6.30}$$

compare it with the (local) Lipschitz condition

$$\|S_h f - S_h g\|_Y \leq K(r)\|f - g\|_X \quad (f, g \in X,\ \|f\|_X, \|g\|_X \leq r), \tag{6.31}$$

and try to develop a parallel theory to that given in Sections 5.4 and 5.5 for the autonomous composition operator (5.1).

Clearly, if Y is an algebra, every affine function

$$h(x, u) = \alpha(x) + \beta(x)u \quad (\alpha, \beta \in Y) \tag{6.32}$$

generates a superposition operator S_h which satisfies (6.30) with $K := \|\beta\|_Y$. Conversely, as mentioned before, from (6.30), it follows in many function spaces Y that the function h must be of the form (6.32). We have encountered this degeneracy phenomenon in Section 5.4 and describe such spaces by the Matkowski property.

Fortunately, there are important function spaces which do *not* have the Matkowski property. Two examples of such spaces are contained in the following Theorems 6.14 and 6.15; the proofs of these theorems may be found in the paper [9].

Theorem 6.14. *Suppose that the operator* (6.1) *maps the space* $C([a, b])$ *into itself. Then this operator satisfies* (6.30) *if and only if the corresponding function h satisfies*

$$|h(x, u) - h(x, v)| \leq g(x)|u - v| \quad (a \leq x \leq b,\ u, v \in \mathbb{R}) \tag{6.33}$$

for some continuous nonnegative function $g : [a, b] \to \mathbb{R}$.

Theorem 6.15. *Suppose that the operator* (6.1) *maps the space* $L_p([a, b])$ *into the space* $L_q([a, b])$*, where* $1 \leq q \leq p < \infty$*. Then this operator satisfies* (6.30) *if and only if the corresponding function h satisfies*

$$|h(x, u) - h(x, v)| \leq g(x, w)|u - v| \quad (a \leq x \leq b,\ |u|, |v| \leq w), \tag{6.34}$$

where the superposition operator S_g *defined by the nonnegative function* g *maps* $L_p([a, b])$ *into* $L_{pq/(p-q)}([a, b])$*. In particular, in case* $p = q$*, we have* $S_g(L_p) \subseteq L_\infty$*, and so condition* (6.34) *is equivalent to* (6.33).

We remark that necessary and sufficient conditions under which the operator (6.1) fulfills the hypotheses of these theorems are well known, see Theorems 6.1 and 6.2 in the first section.

Clearly, the Lipschitz condition (6.33) may be replaced by the simpler condition

$$|h(x,u) - h(x,v)| \le k|u - v| \quad (a \le x \le b,\ u, v \in \mathbb{R}), \tag{6.35}$$

where $k := \|g\|_C$ in Theorem 6.14, and $k := \|g\|_{L_\infty}$ in Theorem 6.15. In the next proposition, we give a very mild condition on two spaces X and Y which implies that the function h satisfies (6.35) whenever the corresponding operator S_h satisfies the Lipschitz condition (6.30).

Proposition 6.16. *Suppose that the operator* (6.1) *maps a normed space X into a normed space Y and satisfies the Lipschitz condition* (6.30). *Assume that the space X contains the constant functions, and the space Y is imbedded into the space of bounded functions. Then the function h satisfies the Lipschitz condition* (6.35).

Proof. The proof is very simple. From (6.30) and our hypothesis $Y \hookrightarrow B([a,b])$, it follows that

$$\|S_h f - S_h g\|_\infty \le Kc(Y,B)\|f - g\|_X \quad (f, g \in X), \tag{6.36}$$

where $\|\cdot\|_\infty$ denotes the norm (0.39) and $c(Y,B)$ denotes the imbedding constant (0.36). Choosing $f(x) \equiv u$ and $g(x) \equiv v$ in (6.36) yields

$$|h(x,u) - h(x,v)| \le \|S_h f - S_h g\|_\infty \le Kc(Y,B)\|f_1\|_X |u - v|,$$

where f_1 denotes the constant function $f_1(x) \equiv 1$. So, (6.35) holds with $k := Kc(Y,B)\|f_1\|_X$. □

A typical example for applying Proposition 6.16 is $X = Y = C([a,b])$; in this case, we have $c(C,B) = 1$ and $\|f_1\|_C = 1$, and thus $k = K$. On the other hand, in case $X = Y = C^1([a,b])$, say, Proposition 6.16 is too “rough;” in fact, one may prove that (6.30) in this case leads to affine functions h, not just Lipschitz continuous functions h.

To show this, we need a refinement of Proposition 6.16 which is based on the use of special polynomials rather than constant functions [226]. This refinement covers several spaces with the Matkowski property, as we shall see. Without loss of generality, we take $[a,b] = [0,1]$. By $P_n([0,1])$, we denote the linear space of all polynomials of degree $\le n$; in particular, $P_1([0,1])$ is the space of all affine functions. We consider $P_n([0,1])$ equipped with the C^n-norm (0.63); in particular, the polynomial $f(x) = Ax+B$ has the norm $\|f\|_{C^1} = |f(0)| + \|f'\| = |A| + |B|$.

Fix $x_1, x_2 \in [0,1]$ and $u_1, u_2 \in \mathbb{R}$, where $x_1 \ne x_2$, and define $f \in P_1([0,1])$ by

$$f(x) := \frac{u_1 - u_2}{x_1 - x_2}x + \frac{x_1u_2 - x_2u_1}{x_1 - x_2} = \frac{u_1(x - x_2) + u_2(x_1 - x)}{x_1 - x_2}. \tag{6.37}$$

It is not hard to see that this polynomial satisfies the conditions

$$f(x_1) = u_1, \quad f(x_2) = u_2, \quad \|f\|_{C^1} = \frac{|u_1 - u_2| + |x_1 u_2 - x_2 u_1|}{|x_1 - x_2|}.$$

Denoting by g the analogous polynomial with u_1 replaced by v_1 and u_2 replaced by v_2 in (6.37), i.e.

$$g(x) := \frac{v_1 - v_2}{x_1 - x_2} x + \frac{x_1 v_2 - x_2 v_1}{x_1 - x_2} = \frac{v_1(x - x_2) + v_2(x_1 - x)}{x_1 - x_2}, \tag{6.38}$$

we have

$$g(x_1) = v_1, \quad g(x_2) = v_2, \quad \|g\|_{C^1} = \frac{|v_1 - v_2| + |x_1 v_2 - x_2 v_1|}{|x_1 - x_2|}.$$

Consequently,

$$\|f - g\|_{C^1} = \frac{|u_1 - u_2 - v_1 + v_2| + |x_1(u_2 - v_2) - x_2(u_1 - v_1)|}{|x_1 - x_2|}.$$

Moreover, fixing $x \in [0, 1]$, multiplying by $|x_1 - x_2|$, and letting $x_1 \to x$ and $x_2 \to x$, we obtain

$$\lim_{x_1, x_2 \to x} |x_1 - x_2| \|f - g\|_{C^1} = (1 + |x|)|u_1 - u_2 - v_1 + v_2|. \tag{6.39}$$

Using this construction, we are now in a position to characterize many spaces with the Matkowski property.

Theorem 6.17. *Let X and Y be two function spaces over $[a, b]$. Assume that the space $P_1([a, b])$ of affine functions with the C^1-norm (0.65) is imbedded into X, and Y is imbedded into the space $Lip([a, b])$ with norm (0.70). Then (X, Y) has the Matkowski property.*

Proof. Without loss of generality, we take $[a, b] = [0, 1]$. Suppose that the superposition operator (6.1) maps X into Y and satisfies the global Lipschitz condition (6.30). From our hypotheses $P_1([0, 1]) \hookrightarrow X$ and $Y \hookrightarrow Lip([0, 1])$, it follows that

$$\|S_h f - S_h g\|_{Lip} \le L\|f - g\|_{C^1} \quad (f, g \in P_1([0, 1])) \tag{6.40}$$

for $L := Kc(P_1, X)c(Y, Lip)$. In particular, since constant functions belong to $P_1([0, 1])$, we see that $h(\cdot, u) \in Lip([0, 1])$ for each $u \in \mathbb{R}$, and so the function $h(\cdot, u)$ is continuous.

Fix $x_1, x_2 \in [0, 1]$ and $u_1, u_2, v_1, v_2 \in \mathbb{R}$, where $x_1 \neq x_2$, and define $f, g \in P_1([0, 1])$ as in (6.37) and (6.38), respectively. By definition (0.70) of the norm in $Lip([0, 1])$, we get the estimates

$$\begin{aligned} &\left| \frac{h(x_1, u_1) - h(x_2, u_2) - h(x_1, v_1) + h(x_2, v_2)}{x_1 - x_2} \right| \\ &\quad = \left| \frac{h(x_1, f(x_1)) - h(x_2, f(x_2)) - h(x_1, g(x_1)) + h(x_2, g(x_2))}{x_1 - x_2} \right| \\ &\quad \le lip(S_h f - S_h g) \le \|S_h f - S_h g\|_{Lip} \le L\|f - g\|_{C^1}, \end{aligned}$$

by (6.40). Consequently,

$$|h(x_1, u_1) - h(x_2, u_2) - h(x_1, v_1) + h(x_2, v_2)| \leq L|x_1 - x_2| \|f - g\|_{C^1} .$$

Fixing now $x \in [0, 1]$ and letting $x_1 \to x$ and $x_2 \to x$, we obtain, by (6.39),

$$|h(x, u_1) - h(x, u_2) - h(x, v_1) + h(x, v_2)| \leq L(1 + |x|)|u_1 - u_2 - v_1 + v_2| . \tag{6.41}$$

Substituting $u_1 := y + z$, $u_2 := y$, $v_1 := z$ and $v_2 := 0$, and observing that the right-hand side of (6.41) then becomes zero, we arrive at the equality

$$h(x, y + z) - h(x, y) - h(x, z) = -\alpha(x) , \tag{6.42}$$

where we have used the shortcut $\alpha(x) := h(x, 0)$. Assume first that $\alpha(x) \equiv 0$. Then (6.42) shows that for each $x \in [0, 1]$, the function $h(x, \cdot)$ satisfies the Cauchy functional equation. Moreover, putting $v_1 = v_2 = 0$ in (6.41), we see that $h(x, \cdot)$ is (Lipschitz) continuous. We conclude that $h(x, u) = \beta(x)u$ for some function $\beta : [0, 1] \to \mathbb{R}$. In the general case when $\alpha(x) \not\equiv 0$, we pass from h to the function $\tilde{h}$ defined by $\tilde{h}(x, u) := h(x, u) - \alpha(x)$, and the statement follows. The assertion $\alpha, \beta \in Y$ follows from the fact that $\alpha(x) = h(x, 0)$ and $\beta(x) = h(x, 1) - h(x, 0)$. □

Theorem 6.17 applies to several spaces we have considered so far. In fact, if all functions in a space X have bounded derivatives, they are also Lipschitz continuous. Combining this argument with Proposition 0.40 and Exercises 0.46, 2.31, and 3.53 leads to the following

Corollary 6.18. *The spaces Lip, C^1, Lip^1 Lip^1_α, BV^1, RBV^1_p, WBV^1_p and AC^1 have the Matkowski property.*

Corollary 6.18 contains Matkowski's degeneracy results for the space *Lip* proved in [195] and for the space C^1 proved in [196]. However, there are some spaces with the Matkowski property which are not covered by Theorem 6.17 since the requirement $Y \hookrightarrow Lip$ is too restrictive. For instance, the spaces Lip_α $(\alpha < 1)$ and AC are excluded by this condition, but do have the Matkowski property, as was shown in [193] and [124], respectively.

Theorem 6.17 has another interesting consequence. Suppose that the operator S_h maps the space $C^n([a, b])$ into the space $C^m([a, b])$, where $m > n$, and satisfies (6.40) with $\| \cdot \|_{Lip}$ replaced by $\| \cdot \|_{C^m}$. Then our proof shows that the function h does not depend on the second variable, which means that the operator S_h is *constant*. This strong degeneracy phenomenon has been proved directly by Matkowski in [196].

Now, we study Lipschitz continuous superposition operators in the space $BV([a, b])$ of functions of bounded variation. Here, the degeneracy one encounters is somewhat different. Recall that given a function $f : [a, b] \to \mathbb{R}$, the *right regularization* $f^\# : [a, b] \to \mathbb{R}$ of f is defined by

$$f^\#(x) := \begin{cases} \lim_{s \to x+} f(s) & \text{for } a \leq x < b , \\ f(b) & \text{for } x = b , \end{cases} \tag{6.43}$$

while the *left regularization* $f^\flat : [a,b] \to \mathbb{R}$ of f is defined by

$$f^\flat(x) := \begin{cases} \lim_{s\to x-} f(s) & \text{for } a < x \le b\,, \\ f(a) & \text{for } x = a\,. \end{cases} \tag{6.44}$$

Of course, these regularizations are different from f at a point x only if f is not right- or left-continuous at x, respectively. In what follows, we will restrict ourselves to the right regularization (6.43). As we have seen[6] in Proposition 4.28, the right regularization (6.43) of a BV-function is also a BV-function and satisfies

$$\|f^\#\|_{BV} \le \|f\|_{BV}\,, \tag{6.45}$$

where the inequality may be strict. The following remarkable result was proved by Matkowski and Miś in [207], see also [226].

Theorem 6.19. *If h has the form* (6.32) *with $\alpha, \beta \in BV([a,b])$, the corresponding operator* (6.1) *satisfies a Lipschitz condition of type* (6.30) *in the space $BV([a,b])$ with norm* (1.16). *Conversely, suppose that the operator* (6.1) *maps the space $BV([a,b])$ with norm* (1.16) *into itself and satisfies a Lipschitz condition of type* (6.30). *Then the following is true.*
(a) *The function $h(x,\cdot)$ satisfies the Lipschitz condition* (6.35) *with $k = K$.*
(b) *The right regularization* (6.43) *of $h(\cdot,u)$ has the form*

$$h^\#(x,u) = \alpha(x) + \beta(x)u \quad (a \le x \le b,\ u \in \mathbb{R}) \tag{6.46}$$

for some functions $\alpha, \beta \in BV([a,b])$. An analogous statement is true for the left regularization (6.44).

Proof. The first statement follows from the fact that $BV([a,b])$ is an algebra, while (a) follows from Proposition 6.16 and the equality $c(BV,B) = \|f_1\|_{BV} = 1$. The nontrivial statement is of course (b) for the converse. However, assertion (b) follows from a more general result (Theorem 6.29) which we are going to prove in the next section. □

In view of Theorem 6.19, it seems reasonable to introduce a weaker form of Definition 5.42.

Definition 6.20. We say that a pair (X,Y) of two normed spaces $(X, \|\cdot\|_X)$ and $(Y, \|\cdot\|_Y)$ has the *weak Matkowski property* if whenever the operator (6.1) maps the space X into the space Y and satisfies (6.30), the corresponding right regularization (6.43) of $h(\cdot,u)$ must have the form (6.46). In case $X = Y$, we simply say that X has the weak Matkowski property. ■

6 To be precise, the regularization considered in Proposition 4.28 is different since we subtract $f(a)$ everywhere in (4.63), and thus "nail down" $f^\#$ to be zero in a. However, the estimate (6.45) is of course also true here since functions which differ by a constant have the same total variation.

Of course, in the autonomous case of the composition operator (5.1), there is no difference between the Matkowski property and the weak Matkowski property. Theorem 6.19 (b) states that the Banach space $(BV([a, b]), \|\cdot\|_{BV})$ has the weak Matkowski property. The following example which is a slight modification of an example given in [207] shows that this space does not have the Matkowski property in the sense of Definition 5.42. Therefore, in the nonautonomous case of the superposition operator (6.1), these notions *are* different.

Example 6.21. Let $\{r_0, r_1, r_2, \ldots\}$ be an enumeration of all rational numbers in $[0, 1]$ ($r_0 := 0$), and let $h_0 : \mathbb{R} \to \mathbb{R}$ be any function satisfying $h_0(0) = 0$ and $|h_0(u) - h_0(v)| \leq L|u - v|$. We define $h : [0, 1] \times \mathbb{R} \to \mathbb{R}$ by

$$h(x, u) := \begin{cases} \frac{h_0(u)}{2^k} & \text{if } x = r_k\,, \\ 0 & \text{otherwise}\,. \end{cases} \tag{6.47}$$

For any partition $P = \{t_0, t_1, \ldots, t_m\} \in \mathcal{P}([0, 1])$ and $f \in BV([0, 1])$, we have

$$\begin{aligned} \sum_{j=1}^{m} |S_h f(t_j) - S_h f(t_{j-1})| &\leq 2 \sum_{k=0}^{\infty} |h(r_k, f(r_k))| \\ &= 2 \sum_{k=0}^{\infty} \frac{|h_0(f(r_k))|}{2^k} \leq 4L\|f\|_{\infty}\,, \end{aligned} \tag{6.48}$$

which shows that S_h maps the space $BV([0, 1])$ into itself and is bounded. Furthermore, for $f, g \in BV([0, 1])$ and $P \in \mathcal{P}([0, 1])$, as above, we obtain the estimate

$$\begin{aligned} &\mathrm{Var}(S_h f - S_h g, P; [0, 1]) \\ &= \sum_{j=1}^{m} |S_h f(t_j) - S_h g(t_j) - S_h f(t_{j-1}) + S_h g(t_{j-1})| \\ &\leq 2 \sum_{j=1}^{m} |h(t_j, f(t_j)) - h(t_j, g(t_j))| \leq 2 \sum_{k=0}^{\infty} |h(r_k, f(r_k)) - h(r_k, g(r_k))| \\ &\leq 2 \sum_{k=0}^{\infty} \frac{|h_0(f(r_k)) - h_0(g(r_k))|}{2^k} \leq 2L \sum_{k=0}^{\infty} \frac{|f(r_k) - g(r_k))|}{2^k} \\ &= 2L|f(0) - g(0)| + 2L \sum_{k=1}^{\infty} \frac{|f(r_k) - g(r_k))|}{2^k} \leq 2L\|f - g\|_{BV}\,. \end{aligned}$$

This together with the trivial estimate $|S_h f(0) - S_h g(0)| \leq L|f(0) - g(0)|$ shows that S_h satisfies the global Lipschitz condition (6.30) with $K = 2L$, although h is not of the form (6.32). ♥

It is not hard to see that $h^{\#}(x, u) = h^{\flat}(x, u) \equiv 0$ for the function h in Example 6.21, in accordance with Theorem 6.19 (b).

In the next section (Theorem 6.29), we will formulate a very general condition for a pair of Banach spaces under which this pair even has a stronger property than the weak Matkowski property.

We do not reproduce the large summary at the end of Section 5.5 since conditions which are both necessary and sufficient for boundedness, continuity etc. are much more difficult to find (and, in fact, mostly unknown) in the nonautonomous case of the operator S_h. Therefore, the regions of *terra incognita* in such a summary would be vast: only in quite exceptional cases have we found explicit counterexamples for the operator (6.1), like the striking Examples 6.3 and 6.6. A small part of this "terra incognita" is covered by open problems stated in Section 6.5 below.

6.3 Uniform boundedness and continuity

In the last section, we have encountered several spaces with the Matkowski property (respectively, the weak Matkowski property), meaning that a global Lipschitz condition on S_h implies that the generating function h (respectively, its regularization $h^{\#}$) is affine.

We point out, however, that in some spaces, the degeneracy described by the Matkowski property also happens under the weaker assumption that S_h is uniformly bounded or uniformly continuous.[7] Later in this section, we will take a closer look at this phenomenon, and we will also consider several concepts of uniform boundedness of C_h and S_h. One such concept reads as follows [203]:

Definition 6.22. Let X and Y be normed spaces and $A : X \to Y$ a (usually, nonlinear) operator. If there exists an increasing function $\gamma : [0, \infty) \to [0, \infty)$ such that $\gamma(0) = 0$ and

$$\|Af - Ag\| \leq \gamma(\|f - g\|) \quad (f, g \in X)\,, \tag{6.49}$$

the operator A is called *uniformly bounded.* ■

The following proposition establishes a link between uniform boundedness and uniform continuity.

Proposition 6.23. *Every operator which is uniformly continuous (for example, Lipschitz continuous) is also uniformly bounded. Conversely, if the function γ in (6.49) is continuous at 0, then the uniform boundedness of A implies its uniform continuity.*

Proof. Suppose that $A : X \to Y$ is uniformly continuous. This means that, given $\varepsilon > 0$, we find $\delta > 0$ such that $\|f - g\| \leq \delta$ implies $\|Af - Ag\| \leq \varepsilon$ for $f, g \in X$. Then putting

$$\gamma(t) := \sup\{\|Af - Ag\| : \|f - g\| \leq t\} \quad (t \geq 0)\,, \tag{6.50}$$

7 In the autonomous case of the operator C_h, a typical result of this type is Corollary 5.46. In the nonautonomous case, this was first observed and studied in [200].

the function $\gamma : [0, \infty) \to [0, \infty)$ is well-defined and finite. Moreover, γ is obviously monotonically increasing. So, putting $t := \|f - g\|$ in (6.50), we obtain $\gamma(\|f - g\|) \geq \|Af - Ag\|$ as claimed.

Conversely, suppose that (6.49) holds for some increasing function γ which is continuous at 0. Given $\varepsilon > 0$, choose $\delta > 0$ such that $0 \leq t \leq \delta$ implies $0 \leq \gamma(t) \leq \varepsilon$. Fix any $f, g \in X$ satisfying $t := \|f - g\| \leq \delta$. Then

$$\|Af - Ag\| \leq \gamma(\|f - g\|) \leq \varepsilon\,. \tag{6.51}$$

However, this precisely means that A is uniformly continuous on X. □

Here is a simple example of an operator which is uniformly bounded, but not uniformly continuous.

Example 6.24. Consider the autonomous composition operator (5.1) generated by the function $h(u) = \sin u$ between $X = L_1([0,1])$ and $Y = L_\infty([0,1])$. Taking $\gamma(0) = 0$ and $\gamma(t) \equiv 2$ for $t > 0$, we see that (6.49) is true for $A = C_h$, and so C_h is uniformly bounded between X and Y in the sense of Definition 6.22. However, for the sequence of functions $f_n = \frac{\pi}{2}\chi_{[0,1/n]}$, we have

$$\|f_n\|_{L_1} \to 0 \quad (n \to \infty)\,,$$

but

$$\|C_h f_n\|_{L_\infty} = \|\chi_{[0,1/n]}\|_{L_\infty} \equiv 1\,. \tag{6.52}$$

This shows that C_h is not continuous at $f(x) \equiv 0$. ♥

From Proposition 6.23, it follows that (6.49) cannot hold for the operator $A = C_h$ in Example 6.24 with any function γ which is continuous at 0.

Now, we will study the case when A is the composition operator (5.1) or the superposition operator (6.1). We start with the following analogue of Proposition 6.16.

Proposition 6.25. *Suppose that the operator* (6.1) *maps a normed space X into a normed space Y and satisfies the uniform boundedness condition* (6.49). *Assume that the space X contains the constant functions, and the space Y is imbedded into the space of bounded functions. Then the function h is uniformly bounded in the sense that*

$$|h(x,u) - h(x,v)| \leq \tilde{\gamma}(|u - v|) \quad (a \leq x \leq b,\ u, v \in \mathbb{R})\,, \tag{6.53}$$

where $\tilde{\gamma} : [0, \infty) \to [0, \infty)$ is increasing with $\tilde{\gamma}(0) = 0$.

Proof. From (6.49) and our hypothesis $Y \hookrightarrow B([a,b])$, it follows that

$$\|S_h f - S_h g\|_\infty \leq Kc(Y,B)\gamma(\|f - g\|_X) \quad (f, g \in X)\,,$$

where we use the same notation as in Proposition 6.16. Choosing $f(x) \equiv u$ and $g(x) \equiv v$ in this estimate yields

$$|h(x,u) - h(x,v)| \leq \|S_h f - S_h g\|_\infty \leq Kc(Y,B)\gamma(\|f_1\|_X |u - v|)\,,$$

and so (6.53) holds with $\tilde{\gamma}(t) := Kc(Y,B)\gamma(\|f_1\|_X t)$. □

Table 6.5. The Matkowski property.

uniform Matkowski property	$\Rightarrow$	Matkowski property
$\Downarrow$		$\Downarrow$
uniform weak Matkowski property	$\Rightarrow$	weak Matkowski property

In connection with uniform boundedness, we introduce a new notion which is parallel to both Definition 5.42 and Definition 6.20.

Definition 6.26. We say that a pair (X, Y) of two normed spaces $(X, \|\cdot\|)$ and $(Y, \|\cdot\|)$ has the *uniform Matkowski property* if whenever the operator (5.1) [the operator (6.1), respectively] maps the space X into the space Y and is uniformly bounded, the corresponding function h must have the form (5.68) [the form (5.69), respectively]. In case $X = Y$, we simply say that X has the uniform Matkowski property.

Similarly, we say that a pair (X, Y) of two normed spaces $(X, \|\cdot\|)$ and $(Y, \|\cdot\|)$ has the *uniform weak Matkowski property* if whenever the operator (6.1) maps the space X into the space Y and is uniformly bounded, the corresponding right regularization (6.43) of h must have the form (6.46). In case $X = Y$, we simply say that X has the uniform weak Matkowski property. ■

Putting $\gamma(t) := Kt$ in (6.49), we see that the uniform (weak) Matkowski property implies the (weak) Matkowski property. So, we have the following hierarchy between these properties.

Now, we state two sufficient conditions on a pair of spaces (X, Y) to have the respectively uniform weak Matkowski property. By $PL([a, b])$, we denote the set of all continuous piecewise linear functions, see Exercise 3.66. The following result is similar to Theorem 6.17.

Theorem 6.27. *Let X and Y be two function spaces over $[a, b]$. Assume that X contains the space $PL([a, b])$, and Y is imbedded into the space $AC([a, b])$ with norm* (3.42). *Then (X, Y) has the uniform Matkowski property.*

Proof. We take again $[a, b] = [0, 1]$. Recall that the norm (3.42) on $AC([0, 1])$ is given by

$$\|f\|_{AC} = \|f\|_{BV} = |f(0)| + \mathrm{Var}(f; [0, 1]) = |f(0)| + \int_0^1 |f'(t)|\, dt\,.$$

Suppose that the superposition operator (6.1) maps X into Y and satisfies (6.49) for some function $\gamma : [0, \infty) \to [0, \infty)$. From our hypothesis $Y \hookrightarrow AC([0, 1])$ and the monotonicity of γ, it follows that

$$\int_0^1 |(S_h f)'(t) - (S_h g)'(t)|\, dt \le \|S_h f - S_h g\|_{AC} \le c(Y, AC)\gamma(\|f - g\|_{AC}) \tag{6.54}$$

for all $f, g \in X$, where $c(Y, AC)$ denotes the imbedding constant (0.36). Fix numbers $y_1, y_2, z_1, z_2 \in \mathbb{R}$ and $x_1, x_2, \ldots, x_{2n} \in (0, 1)$ such that

$$0 < x_1 < x_2 < \ldots < x_{2n} < 1 .$$

Consider the piecewise linear functions $f, g : [0, 1] \to \mathbb{R}$ defined by

$$f(x) := \begin{cases} y_1 & \text{for } x = 0 \text{ or } x \in \{x_1, x_3, \ldots, x_{2n-1}\}, \\ y_2 & \text{for } x = 1 \text{ or } x \in \{x_2, x_4, \ldots, x_{2n}\}, \\ \text{linear} & \text{otherwise}, \end{cases}$$

and similarly

$$g(x) := \begin{cases} z_1 & \text{for } x = 0 \text{ or } x \in \{x_1, x_3, \ldots, x_{2n-1}\}, \\ z_2 & \text{for } x = 1 \text{ or } x \in \{x_2, x_4, \ldots, x_{2n}\}, \\ \text{linear} & \text{otherwise}. \end{cases}$$

So, f is the unique piecewise linear functions whose graph is determined by the vertices

$$(0, y_1), \quad (x_1, y_1), \quad (x_2, y_2), \quad \ldots, \quad (x_{2k-1}, y_1), \quad (x_{2k}, y_2), \quad \ldots, \quad (x_{2n}, y_2), \quad (1, y_2),$$

and analogously for g. Thus, $f(x_k) = y_1$ if and only if $f(x_{k+1}) = y_2$ for $k = 1, 2, \ldots, 2n-1$, and similarly for g with y_1 replaced by z_1 and y_2 replaced by z_2. Since f and g are constant on the intervals $[0, x_1]$ and $[x_{2n}, 1]$, and affine on each of the other intervals $[x_k, x_{k+1}]$, by definition of the norm (3.42), we have

$$\begin{aligned} \|f - g\|_{AC} &= |y_1 - z_1| + \sum_{k=1}^{2n-1} \int_{x_k}^{x_{k+1}} |f'(t) - g'(t)|\, dt \\ &= |y_1 - z_1| + \sum_{k=1}^{2n-1} |y_1 - z_1 - y_2 + z_2| = |y_1 - z_1| + (2n-1)\,|y_1 - z_1 - y_2 + z_2| . \end{aligned}$$

On the other hand,

$$\begin{aligned} &\int_0^1 |(S_h f)'(t) - (S_h g)'(t)|\, dt \\ &= \sum_{k=1}^{2n-1} \int_{x_k}^{x_{k+1}} \left| \frac{d}{dt} [h(t, f(t)) - h(t, g(t))] \right| dt \\ &\geq \sum_{k=1}^{2n-1} \left| \int_{x_k}^{x_{k+1}} \frac{d}{dt} [h(t, f(t)) - h(t, g(t))]\, dt \right| \\ &= \sum_{k=1}^{2n-1} |h(x_{k+1}, f(x_{k+1})) - h(x_{k+1}, g(x_{k+1})) - h(x_k, f(x_k)) + h(x_k, g(x_k))| \\ &= \sum_{k=1}^{2n-1} |h(x_{k+1}, y_1) - h(x_{k+1}, z_1) - h(x_k, y_2) + h(x_k, z_2)| . \end{aligned}$$

Inserting this into (6.54) yields

$$\sum_{k=1}^{2n-1} \left|h(x_{k+1}, y_1) - h(x_{k+1}, z_1) - h(x_k, y_2) + h(x_k, z_2)\right|$$
$$\le c(Y, AC)\gamma(|y_1 - z_1| + (2n-1)\left|y_1 - z_1 - y_2 + z_2\right|).$$

Fixing $x \in [0, 1]$ and letting x_k tend to x for all $k \in \{1, 2, \dots, 2n\}$ gives

$$\begin{aligned}(2n-1)\left|h(x, y_1) - h(x, z_1) - h(x, y_2) + h(x, z_2)\right| \\ \le c(Y, AC)\gamma(|y_1 - z_1| + (2n-1)\left|y_1 - z_1 - y_2 + z_2\right|).\end{aligned} \tag{6.55}$$

Taking now arbitrary different points $u, v \in \mathbb{R}$ and making in (6.55) the special choice

$$y_1 := \frac{u+v}{2}, \quad y_2 := u, \quad z_1 := v, \quad z_2 := \frac{u+v}{2},$$

we obtain

$$(2n-1)\left|h\left(x, \frac{u+v}{2}\right) - h(x, u) - h(x, v) + h\left(x, \frac{u+v}{2}\right)\right| \le c(Y, \mathrm{AC})\gamma\left(\frac{|u+v|}{2}\right).$$

Since $n \in \mathbb{N}$ is arbitrary, we conclude that necessarily

$$2h\left(x, \frac{u+v}{2}\right) = h(x, v) + h(x, u).$$

This means that the function $h(x, \cdot)$ satisfies the Jensen functional equation, and therefore

$$h(x, u) = \alpha(x) + \beta(x)u \quad (a \le x \le b,\ u \in \mathbb{R})$$

for some functions $\alpha, \beta : [a, b] \to \mathbb{R}$ as claimed. The assertion $\alpha, \beta \in Y$ follows from the fact that $\alpha(x) = h(x, 0)$ and $\beta(x) = h(x, 1) - h(x, 0)$. □

By analyzing the hypothesis $PL \hookrightarrow X = Y \hookrightarrow AC$ in Theorem 6.27, we obtain the following

Corollary 6.28. *The spaces Lip, BV^1, WBV_p^1, and AC have the uniform Matkowski property.*

Unfortunately, the condition $PL \subseteq X$ in Theorem 6.27 excludes all spaces $X \subseteq C^1$ like RBV_p^1 for $p > 1$ or Lip_α^1 for $\alpha \le 1$. On the other hand, the requirement $Y \subseteq AC$ in Theorem 6.27 is less restrictive than the requirement $Y \subseteq Lip$ in Theorem 6.17. In particular, Theorem 6.27 covers the main result from [124] which states that AC has the uniform Matkowski property, while Theorem 6.17 does not.

Now, we establish a sufficient condition under which a pair (X, Y) has the uniform weak Matkowski property. To this end, we will use the space $\Phi BV([a, b])$ of functions of bounded Schramm variation introduced in Definition 2.42. Recall that here $\Phi = (\phi_n)_n$ is a sequence of Young functions $\phi_n : [0, \infty) \to [0, \infty)$ such that

$$\sum_{k=1}^{\infty} \phi_k(x) = \infty \quad (x > 0). \tag{6.56}$$

For special choices of $(\phi_n)_n$, the space ΦBV contains many of the spaces of functions of generalized bounded variation we have studied so far, see Proposition 2.43.

Theorem 6.29. *Let X and Y be two function spaces over $[a,b]$. Assume that the space $P_n([a,b])$ of polynomials of degree $\leq n$, equipped with the norm of X, is imbedded into X for each $n \in \mathbb{N}$, and Y is imbedded into some space $\Phi BV([a,b])$ of functions of bounded Schramm variation with norm (2.75). Then (X,Y) has the uniform weak Matkowski property.*

Proof. Given function spaces X and Y with the indicated properties, suppose that the superposition operator (6.1) maps X into Y and satisfies condition (6.49). From our hypotheses $P_n([a,b]) \hookrightarrow X$ and $Y \hookrightarrow \Phi BV([a,b])$ and the monotonicity of γ in (6.49), it follows that

$$\|S_h f - S_h g\|_{\Phi BV} \leq c(Y, \Phi BV)\gamma(\|f - g\|_X) \quad (f, g \in P_n([a,b])), \tag{6.57}$$

where $c(Y, \Phi BV)$ denotes the imbedding constant (0.36). Let $a \leq t < s \leq b$, and let $P_m := \{t_0, t_1, \dots, t_{2m}\} \in \mathcal{P}([t,s])$ be the equidistant partition defined by

$$t_j - t_{j-1} = \frac{s-t}{2m} \quad (j = 1, 2, \dots, 2m). \tag{6.58}$$

Given $u, v \in \mathbb{R}$ with $u \neq v$, let $f : [a,b] \to \mathbb{R}$ be a polynomial satisfying

$$f(t_{2j}) = v \quad (j = 0, 1, \dots, m), \quad f(t_{2j-1}) = \frac{u+v}{2} \quad (j = 1, 2, \dots, m), \tag{6.59}$$

and let $g : [a,b] \to \mathbb{R}$ be the polynomial

$$g(x) := f(x) + \frac{u-v}{2}.$$

Then

$$g(t_{2j}) = \frac{u+v}{2} \quad (j = 0, 1, \dots, m), \quad g(t_{2j-1}) = u \quad (j = 1, 2, \dots, m), \tag{6.60}$$

and the difference $f - g$ trivially satisfies

$$|f(x) - g(x)| \equiv \frac{|u-v|}{2} \quad (a \leq x \leq b).$$

Consequently, substituting these functions f and g into (6.57) yields

$$\|S_h f - S_h g\|_{\Phi BV} \leq c(Y, \Phi BV)\gamma(\|f_1\|_X |u-v|/2),$$

and hence

$$\left\| \frac{S_h f - S_h g}{c(Y, \Phi BV)\gamma(\|f_1\|_X |u-v|/2)} \right\|_{\Phi BV} \leq 1,$$

where again $f_1(x) \equiv 1$. From Proposition 2.44 (b), it follows that

$$\mathrm{Var}_\Phi \left(\frac{S_h f - S_h g}{c(Y, \Phi BV)\gamma(\|f_1\|_X |u-v|/2)} \right) \leq \left\| \frac{S_h f - S_h g}{c(Y, \Phi BV)\gamma(\|f_1\|_X |u-v|/2)} \right\|_{\Phi BV} \leq 1. \tag{6.61}$$

Building on the definition of $\mathrm{Var}_\Phi(f)$ and using (6.59), we therefore get

$$\sum_{j=1}^{m} \phi_j \left(\frac{|h(t_{2j}, f(t_{2j})) - h(t_{2j}, g(t_{2j})) - h(t_{2j-1}, f(t_{2j-1})) + h(t_{2j-1}, g(t_{2j-1}))|}{c(Y, \Phi BV)\gamma(\|f_1\|_X |u-v|/2)} \right)$$
$$= \sum_{j=1}^{m} \phi_j \left(\frac{2\left|h(t_{2j}, v) - h(t_{2j}, \frac{u+v}{2}) - h(t_{2j-1}, \frac{u+v}{2}) + h(t_{2j-1}, u)\right|}{c(Y, \Phi BV)\gamma(\|f_1\|_X |u-v|/2)} \right) \le 1 .$$

Now, since the operator S_h maps the space of all polynomials into the space $\Phi BV([a,b])$, the function $f(\cdot, z)$ has unilateral limits, for all $z \in \mathbb{R}$, in the interval $[a,b]$. Therefore, we may pass to the right-hand limit $s \to t+$ in the last estimate and obtain[8]

$$\sum_{j=1}^{m} \phi_j \left(\frac{\left|h^\#(t, v) - h^\#(t, \frac{u+v}{2}) - h^\#(t, \frac{u+v}{2}) + h^\#(t, u)\right|}{c(Y, \Phi BV)\gamma(\|f_1\|_X |u-v|/2)} \right) \le 1 \tag{6.62}$$

since $s \to t+$ implies $t_j \to t+$ for any j, by (6.58). Now, passing to the limit $m \to \infty$ in (6.62), we obtain

$$\sum_{j=1}^{\infty} \phi_j \left(\frac{\left|h^\#(t, v) - h^\#(t, \frac{u+v}{2}) - h^\#(t, \frac{u+v}{2}) + h^\#(t, u)\right|}{c(Y, \Phi BV)\gamma(\|f_1\|_X |u-v|/2)} \right) \le 1 .$$

However, (6.56) implies that this is possible only if

$$h^\#(t, v) - h^\#\left(t, \frac{u+v}{2}\right) - h^\#\left(t, \frac{u+v}{2}\right) + h^\#(t, u) = 0 .$$

This means that the function $h^\#$ satisfies the Cauchy functional equation

$$2h^\#\left(t, \frac{u+v}{2}\right) = h^\#(t, u) + h^\#(t, v) \quad (a \le t \le b,\ u, v \in \mathbb{R}) .$$

Since $h^\#$ is right-continuous, it follows that (6.46) holds for some functions $\alpha, \beta : [a,b] \to \mathbb{R}$ as claimed. The assertion $\alpha, \beta \in Y$ follows from the fact that $\alpha(x) = h^\#(x, 0)$ and $\beta(x) = h^\#(x, 1) - h^\#(x, 0)$. □

Theorem 6.29 applies to several function spaces which we have discussed in Chapter 1 and Chapter 2. In fact, from Proposition 2.43, we immediately get the following

Corollary 6.30. *The spaces BV, ΛBV, WBV_p, WBV_ϕ, and ΦBV have the uniform weak Matkowski property.*

Corollary 6.30 was proved by a direct construction by Matkowski and Miś for the space BV in [207], see Theorem 6.19, by Merentes and Rivas for the space WBV_p in [224] and for the space WBV_ϕ in [226].

---

8 Letting $s \to t+$ means that we "squeeze" the partition P_m towards the singleton $\{t\}$. Since $h^\#(\cdot, u)$ is right-continuous, in (6.62), we then get the first argument t in $h^\#$ throughout.

Example 6.21 and Corollary 6.30 show that the space BV has the uniform weak Matkowski property, but not the Matkowski property, so neither of the vertical arrows in Table 6.5 may be inverted. We do not know whether or not the horizontal arrows can be inverted either.

Of course, Example 6.21 shows that the spaces mentioned in Corollary 6.30 have the Matkowski property since BV is a special case of the other spaces.

In the following Tables 6.6 and 6.7, we summarize some imbedding conditions on X and Y under which either the global Lipschitz condition (6.30) or the uniform boundedness condition (6.49) for the operator S_h implies a degeneracy for the corresponding function h. Afterwards, we collect some spaces which have the Matkowski property, weak Matkowski property, uniform Matkowski property, or uniform weak Matkowski property in Table 6.8.

Table 6.6. Globally Lipschitz superposition operators.

If X ***satisfies***	***and*** Y ***satisfies***	***then*** (6.30) ***implies***
$\mathbb{R} \hookrightarrow X$	$Y \hookrightarrow B([a,b])$	$h(x,\cdot)$ locally Lipschitz
$P_1([a,b]) \hookrightarrow X$	$Y \hookrightarrow Lip([a,b])$	$h(x,\cdot)$ affine
$P_n([a,b]) \hookrightarrow X$	$Y \hookrightarrow \Phi BV([a,b])$	$h^{\#}(x,\cdot)$ affine

Table 6.7. Uniformly bounded superposition operators.

If X ***satisfies***	***and*** Y ***satisfies***	***then*** (6.49) ***implies***
$\mathbb{R} \hookrightarrow X$	$Y \hookrightarrow B([a,b])$	$h(x,\cdot)$ uniformly bounded
$PL([a,b]) \hookrightarrow X$	$Y \hookrightarrow AC([a,b])$	$h(x,\cdot)$ affine
$P_n([a,b]) \hookrightarrow X$	$Y \hookrightarrow \Phi BV([a,b])$	$h^{\#}(x,\cdot)$ affine

Note that Table 6.8 not only contains results which we have proved before, but also some results from references which are not covered by Corollaries 6.18, 6.28, or 6.30.

Specifically, the uniform Matkowski property of RBV_p was proved in [27], of RBV_ϕ in [1], and for Lip_α in [201]. Historically, C^1, Lip, and Lip_α have been the first spaces for which the Matkowski property in the sense of Definition 5.42 has been discovered. The stronger result that they also have the uniform Matkowski property has been proved more recently in [200, 201].

There are many other spaces with the (uniform or simple) Matkowski property. Let us just remark that in [32] it was shown quite recently that the space κBV of functions of bounded Korenblum variation (Section 2.5) has the Matkowski property. This means that the condition

$$\mathrm{Var}_\kappa(S_h f - S_h g; [0,1]) \le K\,\mathrm{Var}_\kappa(f-g; [0,1]) \quad (f, g \in \kappa BV([0,1])),$$

where $\mathrm{Var}_\kappa(f-g;[0,1])$ denotes the Korenblum variation (2.106), is satisfied only for affine functions (6.32).

Table 6.8. Spaces with the Matkowski property.

function space	*Matkowski property*	*weak M. property*	*uniform M. property*	*uniform weak M. property*
$C([a,b])$	no (Th. 6.14)	–	no (Th. 6.14)	–
$C^1([a,b])$	yes (Cor. 6.18)	–	yes [200]	–
$Lip([a,b])$	yes (Cor. 6.18)	–	yes (Cor. 6.28)	–
$Lip_\alpha([a,b])$	yes [193]	–	yes [201]	–
$AC([a,b])$	yes (Cor. 6.28)	–	yes (Cor. 6.28)	–
$BV([a,b])$	no (Ex. 6.21)	yes (Th. 6.19)	no (Ex. 6.21)	yes (Cor. 6.30)
$WBV_p([a,b])$	no (Ex. 6.21)	yes (Cor. 6.30)	no (Ex. 6.21)	yes (Cor. 6.30)
$RBV_p([a,b])$	yes [27]	–	yes [27]	–
$\Lambda BV([a,b])$	no (Ex. 6.21)	yes (Cor. 6.30)	no (Ex. 6.21)	yes (Cor. 6.30)
$\Phi BV([a,b])$	no (Ex. 6.21)	yes (Cor. 6.30)	no (Ex. 6.21)	yes (Cor. 6.30)
$Lip^1([a,b])$	yes (Cor. 6.18)	–	yes [200]	–
$Lip^1_\alpha([a,b])$	yes (Cor. 6.18)	–	yes	–
$AC^1([a,b])$	yes (Cor. 6.18)	–	yes [200]	–
$BV^1([a,b])$	yes (Cor. 6.18)	–	yes (Cor. 6.28)	–
$WBV^1_p([a,b])$	yes (Cor. 6.18)	–	yes (Cor. 6.28)	–
$RBV^1_p([a,b])$	yes (Cor. 6.18)	–	yes [200]	–

Following [124], we now state two conditions which are equivalent to uniform boundedness.

Proposition 6.31. *Let X and Y be two normed spaces and $A : X \to Y$ be a bounded operator. Then the following three conditions are equivalent:*
(a) *The operator A is uniformly bounded.*
(b) *There exist constants $R, r > 0$ such that $\|Af - Ag\| \le R$ for all $f, g \in X$ satisfying $\|f - g\| = r$.*
(c) *There exist constants $\tilde{R}, \tilde{r} > 0$ such that $\|Af - Ag\| \le \tilde{R}$ for all $f, g \in X$ satisfying $\|f - g\| \le \tilde{r}$.*

Proof. Obviously, (a) implies (b) by taking $R := \gamma(r)$, where γ is the function occurring in (6.49). Suppose that (b) is true, and take $\tilde{r} := 2r$ and $\tilde{R} := 2R$. Given $f, g \in X$ satisfying $\|f - g\| \le \tilde{r}$, we may find an element $h \in X$ such that $\|f - h\| = \|h - g\| = r$ because the two spheres $\{h \in X : \|f - h\| = r\}$ and $\{h \in X : \|g - h\| = r\}$ have a nonempty intersection. By (b), there exists $R > 0$ such that $\|Af - Ah\| \le R$ and $\|Ah - Ag\| \le R$, and hence

$$\|Af - Ag\| \le \|Af - Ah\| + \|Ah - Ag\| \le 2R = \tilde{R},$$

which shows that (c) holds.

The fact that (c) implies (b) is trivial, so it remains to prove that (b) implies (a). Putting, under the hypothesis (b),

$$\tilde{\gamma}(t) := \sup\{\|Af - Ag\| : \|f - g\| = t\} \quad (t \ge 0), \tag{6.63}$$

we see that the function $\tilde{\gamma} : [0,\infty) \to [0,\infty)$ is well-defined and finite. Moreover, $\tilde{\gamma}(0) = 0$ and, given $f, g \in X$ with $\|f - g\| = t$, by (6.63), we have

$$\|Af - Ag\| \le \tilde{\gamma}(t) = \tilde{\gamma}(\|f - g\|). \tag{6.64}$$

However, in contrast to (6.50), the function (6.63) is not necessarily increasing. This flaw may be removed by defining $\gamma : [0,\infty) \to [0,\infty)$ by

$$\gamma(t) := \sup\{\tilde{\gamma}(s) : 0 \le s \le t\} \quad (t \ge 0).$$

Then γ is certainly increasing, and the operator A satisfies (6.49) with this modified function γ. □

Our discussion shows that all three conditions in Proposition 6.31 are equivalent to the uniform boundedness of A in the sense of Definition 6.22. However, another variant of this will be given in Section 6.5.

6.4 Functions of several variables

Given a function $f : [a,b] \to \mathbb{R}$, in (6.43), we have introduced and studied the right regularization $f^{\#}$and, in (6.44), the left regularization $f^{\flat}$ of f. In terms of these regularizations, we could formulate in Theorem 6.19 a necessary conditions for the superposition operator (6.1) to satisfy a global Lipschitz condition in BV and related spaces. We are now going to formulate a parallel result for functions of two variables, which means that we replace the interval $[a,b]$ by the rectangle $[a,b] \times [c,d]$. Following Chistyakov [90], we consider (double) left regularizations here which generalize (6.44).

Definition 6.32. Given $f : [a,b]\times[c,d] \to \mathbb{R}$, we define a function $f^{\flat\flat} : [a,b]\times[c,d] \to \mathbb{R}$ by

$$f^{\flat\flat}(x,y) := \begin{cases} \lim_{(s,t)\to(x-,y-)} f(s,t) & \text{for } a < x \le b \text{ and } c < y \le d, \\ \lim_{(s,t)\to(x-,c+)} f(s,t) & \text{for } a < x \le b \text{ and } y = c, \\ \lim_{(s,t)\to(a+,y-)} f(s,t) & \text{for } x = a \text{ and } c < y \le d, \\ \lim_{(s,t)\to(a+,c+)} f(s,t) & \text{for } x = a \text{ and } y = c. \end{cases} \tag{6.65}$$

In what follows, we call $f^{\flat\flat}$ the *left-left regularization* of f. ■

In Definition 6.32, we distinguish between arguments in the upper right half-open rectangle $(a,b] \times (c,d]$, the half-open bottom side $(a,b] \times \{c\}$, the half-open left side $\{a\} \times (c,d]$, and the left-bottom corner point (a,c). Of course, other distinctions are also possible which lead to other types of regularizations, see Exercise 6.8.

Recall that a function $f : [a,b] \times [c,d] \to \mathbb{R}$ is called *left-left continuous* on $(a,b] \times (c,d]$ if

$$\lim_{(s,t)\to(x-,y-)} f(s,t) = f(x,y) \quad (a < x \le b, c < y \le d). \tag{6.66}$$

By $BV^{bb}([a,b]\times[c,d])$, we denote the set of all functions $f \in BV([a,b]\times[c,d])$ which are left-left continuous on $(a,b]\times(c,d]$.

Proposition 6.33. *From $f \in BV([a,b]\times[c,d])$, it follows that $f^{bb} \in BV^{bb}([a,b]\times[c,d])$. Moreover,*

$$\|f^{bb}\|_{BV} \leq \|f\|_{BV}. \tag{6.67}$$

Proof. A comparison of (6.65) and (6.66) shows that the left-left regularization f^{bb} is always left-left continuous on $(a,b]\times(c,d]$; so, we only have to show that f^{bb} is of bounded variation if f is. Moreover, we modify the proof of (4.64) to prove (6.67).

The finiteness of the one-dimensional variations (1.76) and (1.77) is a consequence of Proposition 4.28. To see that the two-dimensional variation (1.78) is also finite, we fix $\{s_0, s_1, \dots, s_m\} \in \mathcal{P}([a,b])$, $\{t_0, t_1, \dots, t_n\} \in \mathcal{P}([c,d])$, and $\varepsilon > 0$. By definition (6.65) of the left-left regularization, we may find elements $\sigma_i \in (s_{i-1}, s_i)$ $(i = 1, 2, \dots, m)$ and $\tau_j \in (t_{j-1}, t_j)$ $(j = 1, 2, \dots, n)$ as well as $\sigma_0 \in (a, s_1)$ and $\tau_0 \in (c, t_1)$ such that

$$\begin{aligned}&\left|f^{bb}(s_{i-1}, t_{j-1}) + f^{bb}(s_i, t_j) - f^{bb}(s_{i-1}, t_j) + f^{bb}(s_i, t_{j-1})\right| \\ &\qquad \leq \left|f(\sigma_{i-1}, \tau_{j-1}) + f(\sigma_i, \tau_j) - f(\sigma_{i-1}, \tau_j) + f(\sigma_i, \tau_{j-1})\right| + \frac{\varepsilon}{mn}.\end{aligned}$$

It follows that

$$\begin{aligned}&\sum_{i=1}^{m}\sum_{j=1}^{n}\left|f^{bb}(s_{i-1}, t_{j-1}) + f^{bb}(s_i, t_j) - f^{bb}(s_{i-1}, t_j) + f^{bb}(s_i, t_{j-1})\right| \\ &\qquad \leq \sum_{i=1}^{m}\sum_{j=1}^{n}\left|f(\sigma_{i-1}, \tau_{j-1}) + f(\sigma_i, \tau_j) - f(\sigma_{i-1}, \tau_j) + f(\sigma_i, \tau_{j-1})\right| + \frac{\varepsilon}{mn} \\ &\qquad \leq \mathrm{V}_2(f; [a,b]\times[c,d]) + \varepsilon,\end{aligned}$$

where $\mathrm{V}_2(f; [a,b]\times[c,d])$ denotes the variation (1.78). Since $\varepsilon > 0$ was arbitrary, we have shown that $f \in BV([a,b]\times[c,d])$ implies $f^{bb} \in BV([a,b]\times[c,d])$ with (6.67). □

We now state a result for functions of two variables which is parallel to Theorem 6.19. First, we have to define nonautonomous superposition operators acting on functions of two variables.

This is of course straightforward. Given a function $h : [a,b]\times[c,d]\times\mathbb{R} \to \mathbb{R}$ and a space X of functions $f : [a,b]\times[c,d] \to \mathbb{R}$, we define the superposition operator S_h generated by h as

$$S_h f(x,y) := h(x, y, f(x,y)) \quad (a \leq x \leq b,\ c \leq y \leq d). \tag{6.68}$$

Now, we formulate an analogue of Theorem 6.19 for the space $BV([a,b]\times[c,d])$.

Theorem 6.34. *If h has the form*

$$h(x, y, u) = \alpha(x,y) + \beta(x,y)u \quad (a \leq x \leq b, c \leq y \leq d, u \in \mathbb{R}) \tag{6.69}$$

with $\alpha, \beta \in BV([a,b]\times[c,d])$, the corresponding operator (6.68) *satisfies a Lipschitz condition of type* (6.30) *in the space $BV([a,b]\times[c,d])$ with norm* (1.89). *Conversely, suppose that the operator* (6.68) *maps the space $BV([a,b]\times[c,d])$ with norm* (1.89) *into itself and satisfies a Lipschitz condition of type* (6.30). *Then the following is true.*

(a) *The function $h(x,y,\cdot)$ satisfies the Lipschitz condition*

$$|h(x,y,u) - h(x,y,v)| \le 2K|u-v| \quad (a \le x \le b, c \le y \le d, u, v \in \mathbb{R}) .$$

(b) *The left-left regularization* (6.65) *of $h(\cdot,\cdot,u)$ has the form*

$$h^{bb}(x,y,u) = \alpha(x,y) + \beta(x,y)u \quad (a \le x \le b, c \le y \le d,\ u \in \mathbb{R}) \tag{6.70}$$

for some functions $\alpha, \beta \in BV([a,b]\times[c,d])$.

We do not carry out the proof in detail since it is, in large part, parallel to that for functions of one variable. The first statement follows easily from the fact that $BV([a,b] \times [c,d])$ is an algebra (Proposition 1.44). For the proof of (a), one has to consider suitable generalizations of the bridge function (6.37) and to distinguish the four cases arising in the definition (6.65) of the left-left regularization. Finally, statement (b) is proved, with some technical modifications, in the same way as we have done this in Theorem 6.29 for functions of one variable. We drop the details and refer to the paper [90].

The following example which is parallel to Example 6.21 and involves a function h satisfying (6.70), but not (6.69), is also taken from [90].

Example 6.35. Let $\{r_0, r_1, r_2, \ldots\}$ be an enumeration of all rational numbers in $[0,1]$ ($r_0 := 0$), and let $h_0 : \mathbb{R} \to \mathbb{R}$ be any function satisfying $h_0(0) = 0$ and $|h_0(u) - h_0(v)| \le L|u-v|$. We define $h : [0,1]\times[0,1]\times\mathbb{R} \to \mathbb{R}$ by

$$h(x,y,u) := \begin{cases} \frac{h_0(u)}{2^{k+l}} & \text{if } x = r_k \text{ and } y = r_l , \\ 0 & \text{otherwise} . \end{cases}$$

We show that $\mathrm{Var}(S_h f; [0,1]\times[0,1])$ is finite for every function $f \in BV([0,1]\times[0,1])$, where $\mathrm{Var}(f; [a,b]\times[c,d])$ is given by (1.82). For any two partitions $P = \{s_0, s_1, \ldots, s_m\} \in \mathcal{P}([0,1])$ and $Q = \{t_0, t_1, \ldots, t_n\} \in \mathcal{P}([0,1])$, we have

$$\begin{aligned} &\mathrm{V}_2(S_h f, P\times Q; [0,1]\times[0,1]) \\ &\quad = \sum_{i=1}^{m}\sum_{j=1}^{n} |S_h f(s_{i-1}, t_{j-1}) - S_h f(s_i, t_j) - S_h f(s_{i-1}, t_j) + S_h f(s_i, t_{j-1})| \\ &\quad \le 4 \sum_{i=0}^{m}\sum_{j=0}^{n} |h(s_i, t_j, f(s_i, t_j))| \le 4 \sum_{k=1}^{\infty}\sum_{l=1}^{\infty} |h(r_k, r_l, f(r_k, r_l))| \\ &\quad = 4 \sum_{k=1}^{\infty}\sum_{l=1}^{\infty} \frac{|h_0(f(r_k, r_l))|}{2^{k+l}} \le 4L\|f\|_\infty \sum_{k=1}^{\infty}\sum_{l=1}^{\infty} \frac{1}{2^{k+l}} \le 4L\|f\|_\infty , \end{aligned}$$

which shows that $V(S_h f; [0,1]\times[0,1]) \le 4L\|f\|_\infty$.

The one-dimensional variation $\mathrm{Var}(S_h f(\cdot, 0), P; [0, 1])$ occurring in (1.76) may be estimated by

$$\begin{aligned}\mathrm{Var}(S_h f(\cdot, 0), P; [0, 1]) &= \sum_{i=1}^{m} |S_h f(s_i, 0) - S_h f(s_{i-1}, 0)| \\ &\le 2\sum_{i=1}^{m} |h(s_i, 0, f(s_i, 0))| \\ &\le 2\sum_{k=1}^{\infty} |h(r_k, 0, f(r_k, 0))| \le 2L\|f\|_\infty \sum_{k=1}^{\infty} \frac{1}{2^k} = 2L\|f\|_\infty ,\end{aligned}$$

and similarly for the one-dimensional variation $\mathrm{Var}(S_h f(0, \cdot), Q; [0, 1])$ occurring in (1.77). This shows that $\mathrm{Var}(S_h f; [0, 1] \times [0, 1]) \le 8L\|f\|_\infty$, and so the superposition operator (6.68) generated by h maps the space $BV([0, 1] \times [0, 1])$ into itself and is bounded.

Now, we prove that the operator S_h satisfies a global Lipschitz condition (6.30). Fix $f, g \in BV([0, 1] \times [0, 1])$ and $P, Q \in \mathcal{P}([0, 1])$ as above. Then using the shortcut

$$\Delta(x, y) := (S_h f - S_h g)(x, y) = h(x, y, f(x, y)) - h(x, y, g(x, y)),$$

we obtain

$$\begin{aligned}&\mathrm{V}_2(S_h f - S_h g, P \times Q; [0, 1] \times [0, 1]) \\ &= \sum_{i=1}^{m}\sum_{j=1}^{n} |\Delta(s_{i-1}, t_{j-1}) - \Delta(s_i, t_j) - \Delta(s_{i-1}, t_j) + \Delta(s_i, t_{j-1})| \\ &\le 4\sum_{i=0}^{m}\sum_{j=0}^{n} |\Delta(s_i, t_j))| \le 4\sum_{k=1}^{\infty}\sum_{l=1}^{\infty} |\Delta(r_k, r_l))| \\ &= 4\sum_{k=1}^{\infty}\sum_{l=1}^{\infty} \frac{|h_0(f(r_k, r_l)) - h_0(g(r_k, r_l))|}{2^{k+l}} \\ &\le 4L\sum_{k=1}^{\infty}\sum_{l=1}^{\infty} \frac{|(f - g)(r_k, r_l)|}{2^{k+l}} \le 4L\|f - g\|_{BV} ,\end{aligned}$$

which shows that

$$\mathrm{V}_2(S_h f - S_h g; [0, 1] \times [0, 1]) \le 4L\|f - g\|_{BV} .$$

The variation $\mathrm{Var}((S_h f - S_h g)(\cdot, 0), P; [0, 1])$ occurring in (1.76) may be estimated in the form

$$\begin{aligned}\mathrm{Var}((S_h f - S_h g)(\cdot, 0), P; [0, 1]) &= \sum_{i=1}^{m} |\Delta(s_i, 0) - \Delta(s_{i-1}, 0)| \\ &\le 4\sum_{k=1}^{\infty} \frac{|h_0(f(r_k, 0)) - h_0(g(r_k, 0))|}{2^k} \le 4L\|f - g\|_{BV} ,\end{aligned}$$

and similarly for the variation $\mathrm{Var}((S_h f - S_h g)(0,\cdot), Q; [0,1])$ occurring in (1.77). Finally, noting that

$$|(S_h f - S_h g)(0,0)| = |\Delta(0,0)| \le |h_0(f(0,0)) - h_0(g(0,0))| \le L|f(0,0) - g(0,0)|,$$

and combining this with the previous estimates, we conclude that

$$\begin{aligned}
&\|S_h f - S_h g\|_{\mathrm{BV}} \\
&\quad = |\Delta(0,0)| + \mathrm{Var}(\Delta; [0,1] \times [0,1]) \\
&\quad = |\Delta(0,0)| + \mathrm{Var}(\Delta(\cdot,0); [0,1]) + \mathrm{Var}(\Delta(0,\cdot); [0,1]) + \mathrm{V}_2(\Delta; [0,1] \times [0,1]) \\
&\quad \le L|f(0,0) - g(0,0)| + 2L\,\mathrm{Var}(f(\cdot,0) - g(\cdot,0); [0,1]) \\
&\qquad + 2L\,\mathrm{Var}(f(0,\cdot) - g(0,\cdot); [0,1]) + 4L\,\mathrm{V}_2(S_h f - S_h g; [0,1] \times [0,1]) \\
&\quad \le L|(f-g)(0,0)| + 4L\,\mathrm{Var}(f-g; [0,1] \times [0,1]) \le 4L\|f-g\|_{\mathrm{BV}},
\end{aligned}$$

which proves (6.30) with $K = 4L$, although h is not of the form (6.69). ♥

6.5 Comments on Chapter 6

Some of the material treated in Section 6.1 may be found in the monograph [20]. Example 6.3 is taken from [48], see also [46, 47], Theorem 6.4 and Proposition 6.5 from [52], for proofs, see [53]. Theorem 6.7 coincides with [20, Theorem 8.1], where this result is apparently stated for the first time.

The problem to describe sufficient conditions on h (which are possibly close to being necessary) under which the corresponding operator S_h maps BV into itself has a long history. Lyamin [189] has given (without proof) a sufficient condition, but Bugajewska expressed in her paper [66] some doubt about the correctness of this condition. Her objection was fully justified by Maćkowiak's surprising counterexample [190] which is our Example 6.8. Afterwards, Bugajewska gave a correct *sufficient condition*, together with an illuminating discussion, in [66]; this discussion is rephrased in our Theorems 6.11 and 6.13 and in Example 6.12.

The *necessary condition* contained in Theorem 6.10 may be found in an implicit form in the recent beautiful book [107]; we have "extracted" the main idea in the proof of Theorem 6.10.

Of course, there are still many open questions. Among them, we mention the following problem related to BV-functions.

Problem 6.1. *Let $h : [a,b] \times \mathbb{R} \to \mathbb{R}$ be continuous, and suppose that the corresponding operator* (6.1) *maps the space $BV([a,b])$ into itself and is both* bounded and continuous. *Does this imply that $h(x,\cdot) \in Lip_{loc}(\mathbb{R})$ for every*[9] *$x \in [a,b]$?*

9 Observe that the function h in Example 6.9 is discontinuous at each point $(0,u)$ for $u \neq 0$.

Problem 6.2. *For $X \in \{AC, RBV_p\}$, find $h : [a,b] \times \mathbb{R} \to \mathbb{R}$ such that $S_h(X) \subseteq X$, but $h(x,\cdot) \notin Lip_{loc}(\mathbb{R})$ for some $x \in [a,b]$.*

We remark that for $X \in \{BV, WBV_p\}$, we have the counterexample (6.14), while for $X = Lip$ and $X = Lip_\alpha$, we have the counterexample (6.4).

Problem 6.3. *Find $h : [a,b] \times \mathbb{R} \to \mathbb{R}$ such that $S_h(\Lambda BV) \subseteq \Lambda BV$, but $h(x,\cdot) \notin Lip_{loc}(\mathbb{R})$ for some $x \in [a,b]$. What about boundedness and/or continuity of S_h?*

Concerning spaces of differentiable function, the following example which is parallel to Example 6.9 shows that the inclusion $S_h(BV^1) \subseteq BV^1$ does not imply that the function $h(x,\cdot)$ belongs to $Lip^1_{loc}(\mathbb{R})$ for every $x \in [a,b]$.

Example 6.36. Define $h_0 : \mathbb{R} \to \mathbb{R}$ by

$$h_0(u) := \begin{cases} 0 & \text{if } u \le 0\,, \\ \frac{2}{3}u\sqrt{u} & \text{if } 0 < u < 1\,, \\ u - \frac{1}{3} & \text{if } u \ge 1\,, \end{cases}$$

and $h : [0,1] \times \mathbb{R} \to \mathbb{R}$ by (6.14). Then the corresponding operator (6.1) maps the space $BV^1([0,1])$ into itself and is bounded. To see this, fix $f \in BV^1([0,1])$ and observe that the function $g = S_h f$ satisfies

$$g'(0) = h_0'(f(0))f'(0), \quad g'(x) = g(x) \equiv 0 \quad (x > 0)\,.$$

Consequently,

$$\mathrm{Var}(g', P; [0,1]) = |h_0'(f(0))f'(0)| \le |f'(0)| \le \|f\|_{BV^1}$$

for every partition $P = \{t_0, t_1, \dots, t_m\} \in \mathcal{P}([0,1])$. On the other hand, the function $h(0,\cdot) = h_0$ does not belong to $Lip^1_{loc}(\mathbb{R})$ since $h_0'(u) = \sqrt{u}$ is not locally Lipschitz at zero. ♥

For other spaces of differentiable function,[10] the corresponding problem seems to be open:

Problem 6.4. *For $X \in \{Lip, Lip_\alpha, AC, RBV_p\}$, find $h : [a,b] \times \mathbb{R} \to \mathbb{R}$ such that $S_h(X^1) \subseteq X^1$, but $h(x,\cdot) \notin Lip^1_{loc}(\mathbb{R})$ for some $x \in [a,b]$.*

Problem 6.5. *For $X \in \{Lip, Lip_\alpha, AC, BV, WBV_p, RBV_p\}$, find conditions on $h : [a,b] \times \mathbb{R} \to \mathbb{R}$, possibly both necessary and sufficient, under which S_h is bounded and/or continuous in X^1.*

Concerning the Wiener–Young space WBV_ϕ, we also repeat the following open problem which we stated after the proof of Theorem 6.13:

10 In case $X^1 = Lip^1$ and $X^1 = Lip^1_\alpha$, one could try to adapt the proof of Theorem 6.4.

Problem 6.6. *Is Theorem 6.13 true without the requirement $\phi \in \delta_2$?*

In the recent paper [35], it is shown that the operator S_h maps the space $R([a,b])$ of regular functions into itself if $h(x,\cdot)$ is continuous, uniformly with respect to $x \in [a,b]$, and $h(\cdot,u)$ is regular for all $u \in \mathbb{R}$. Moreover, in this case, S_h is bounded and continuous in the norm (0.39). The paper [35] also contains a counterexample which shows that these conditions on h are not necessary for $S_h(R) \subseteq R$. We remark that [310] proves an analogous result, but *assuming* a priori that $h(x,\cdot) \in Lip_{\mathrm{loc}}(\mathbb{R})$.

The Matkowski property and weak Matkowski property is discussed in detail in the book [226]. What we call a uniform (weak) Matkowski property, however, has been analyzed only quite recently in a series of papers. At the risk of being redundant, let us repeat the degeneracy list which we stated at the beginning of Section 5.4, but for the superposition operator (6.1), and in extended form, including the uniform (weak) Matkowski property. It was shown

- in [193] that the space $Lip_\alpha([a,b])$ of Hölder continuous functions of order $\alpha < 1$ with norm (0.71) has the Matkowski property, and in [201], that it even has the uniform Matkowski property:[11]
- in [206] that the space $C^n([a,b])$ of n-times continuously differentiable functions with norm (0.63) has the Matkowski property, and in [200], that it even has the uniform Matkowski property;
- in [194] that the space $AC([a,b])$ of absolutely continuous functions with norm (3.42) has the Matkowski property, and in [124], that it even has the uniform Matkowski property;
- in [162] that the space $Lip^n([a,b])$ of functions with Lipschitz continuous n-th derivative with norm (0.78) has the Matkowski property;
- in [187] that the space $Lip^n_\alpha([a,b])$ of functions with Hölder continuous n-th derivative has the Matkowski property;
- in [289] that the space $AC^n([a,b])$ of functions with absolutely continuous n-th derivative has the Matkowski property;
- in [205] and [224] that the space $RBV_p([a,b]) = W_p^1([a,b])$ of functions of bounded p-variation in Riesz's sense for $1 < p < \infty$ with norm (2.90) has the Matkowski property;
- in [205] that the higher order Sobolev space $W_p^n([a,b])$ has the Matkowski property;
- in [204] that the space $RBV_{2,p}([a,b])$ of functions of bounded $(2,p)$-variation in Riesz's sense with norm (3.77) has the Matkowski property;

11 A generalization to abstract (i.e. Banach space valued) Lipschitz continuous functions may be found in [203].

- in [217] that the space $RBV_\phi([a,b])$ of functions of bounded Riesz–Medvedev variation with norm (2.99) has the Matkowski property, and in [1], that it even has the uniform Matkowski property;[12]
- in [32] that the space $\kappa BV([a,b])$ of functions of bounded Korenblum variation with norm (2.136) has the Matkowski property;
- in [207] that the space $BV([a,b])$ of functions of bounded variation with norm (1.16) has the weak Matkowski property, and in [202], that it even has the uniform weak Matkowski property;
- in [224] that the space $WBV_p([a,b])$ of functions of bounded Wiener p-variation with norm (1.65) has the uniform weak Matkowski property;
- in [135] that the space $WBV_\phi([a,b])$ of functions of bounded Wiener–Young variation with norm (2.11) has the uniform weak Matkowski property;[13]
- in [98] that the space $\Lambda BV([a,b])$ of functions of bounded Waterman variation with norm (2.30) has the weak Matkowski property;
- in [112] that the space $\Phi BV([a,b])$ of functions of bounded Schramm variation with norm (2.75) has the uniform weak Matkowski property.

Recent results connected with the Matkowski or weak Matkowski property may be found in [1, 8, 22, 29–31, 57, 110, 119, 123, 124, 137, 151, 152, 197–199, 208, 233]. In connection with degeneracy phenomena, we state another open question which seems to be difficult and interesting:

Problem 6.7. *Does there exist a pair (X,Y) of Banach spaces X and Y having the Matkowski property, but not the uniform Matkowski property? More specifically, can we construct two function spaces X and Y over $[0,1]$, say, such that Lipschitz continuous superposition operators $S_h : X \to Y$ are generated only by affine functions $h : [0,1] \times \mathbb{R} \to \mathbb{R}$, but there is some nonaffine function $\hat{h} : [0,1] \times \mathbb{R} \to \mathbb{R}$ such that the corresponding operator $S_{\hat{h}} : X \to Y$ is uniformly bounded?*

In [203], the following alternative definition of uniform boundedness is given. A (usually, nonlinear) operator $A : X \to Y$ is called uniformly bounded if there exists a function $\gamma : [0,\infty) \to [0,\infty)$ such that $\gamma(0) = 0$, and for every nonempty set $M \subset X$ from $diam(M) \le t$, it follows that $diam(A(M)) \le \gamma(t)$.

It is not hard to see, however, that this is equivalent to uniform boundedness in the sense of our Definition 6.22. To see this, suppose that $A : X \to Y$ satisfies (6.49), and let $diam(M) \le t$ for some bounded set $M \subset X$. Then $\|f - g\| \le t$ for all $f, g \in M$, and so

$$\|Af - Ag\| \le \gamma(\|f - g\|) \le \gamma(t)$$

12 For the same result in the space $RBV_{\phi,w}([a,b])$ with weighted Riesz variation (2.160), see [27], and for multivalued functions see [28, 29].

13 An analogous result for functions of two variables may be found in [136], for multivalued functions in [36].

by the monotonicity of γ. Since $f, g \in M$ are arbitrary, this precisely means that $diam(A(M)) \leq \gamma(t)$.

Conversely, suppose that $diam(M) \leq t$ implies $diam(A(M)) \leq \gamma(t)$, where $M \subset X$ is bounded. Applying this, in particular, to the set $M := \{f, g\}$, we see that (6.49) is true, and so A is uniformly bounded in the sense of Definition 6.22.

The advantage of the boundedness condition in the above definition consists of its great generality: in fact, this definition carries over to nonlinear operators in arbitrary metric spaces.

Example 6.24 is explained by the fact that the superposition operator S_h is continuous between L_1 and L_∞ only if S_h is *constant*, i.e. (6.32) holds with $\beta(x) \equiv 0$. A proof of this surprising fact can be found in [324] or [20, Theorem 3.17], see also Exercise 6.2.

6.6 Exercises to Chapter 6

We state some exercises on the topics covered in this chapter; exercises marked with an asterisk * are more difficult.

Exercise 6.1. Show that the operator S_h maps $L_p([a, b])$ $(1 \leq p < \infty)$ into $L_\infty([a, b])$ if and only if $|h(x, u)| \leq \alpha(x)$ for some $\alpha \in L_\infty([a, b])$; moreover, in this case, S_h is automatically bounded.

Exercise 6.2. Show that the operator S_h is continuous between $L_p([a, b])$ $(1 \leq p < \infty)$ and $L_\infty([a, b])$ if and only if $h(x, u) = \alpha(x)$ for some $\alpha \in L_\infty([a, b])$, i.e. S_h is constant.

Exercise 6.3. Show that the operator S_h maps $L_\infty([a, b])$ into $L_q([a, b])$ $(1 \leq q < \infty)$ if and only if for each $r > 0$ there exists a function $\alpha_r \in L_q([a, b])$ such that

$$|h(x, u)| \leq \alpha_r(x) \quad (a \leq x \leq b,\ |u| \leq r)\,;$$

moreover, in this case, S_h is automatically bounded and continuous.

Exercise 6.4. Show that the operator S_h maps $L_\infty([a, b])$ into $L_\infty([a, b])$ if and only if for each $r > 0$, there exists a function $\alpha_r \in L_\infty([a, b])$ such that the condition from Exercise 6.3 holds; moreover, in this case, S_h is automatically bounded.

Exercise 6.5. Show that the operator S_h is continuous in $L_\infty([a, b])$ if and only if for each $r > 0$, there exists a continuous function $\beta_r : [0, \infty) \to [0, \infty)$ such that $\beta_r(0) = 0$ and

$$|h(x, u) - h(x, v)| \leq \beta_r(|u - v|) \quad (a \leq x, \leq b,\ |u|, |v| \leq r)\,.$$

Exercise 6.6. Suppose that $h : [a, b] \times \mathbb{R} \to \mathbb{R}$ has the property that the corresponding superposition operator (6.1) maps a function space X into itself and is bounded in the norm of X. In analogy to Exercise 5.11, for $r > 0$, we put

$$\mu_r(h, X) := \sup \{\|S_h f\|_X : \|f\|_X \leq r\}$$

and call this characteristic the *growth function* of h in X. Calculate the functions $\mu_r(h, C)$ and $\mu_r(h, C^1)$ under the hypotheses of Theorem 6.1 and Theorem 6.7, respectively.

Exercise 6.7. Using the notation of Theorem 6.4, show that the growth function $\mu_r(h, Lip_\alpha)$ from Exercise 6.6 satisfies the bilateral estimate

$$\frac{k(r)}{2^{\alpha+1}+1} \leq \mu_r(h, Lip_\alpha) \leq k(r)\,,$$

where $k(r)$ is the local Hölder–Lipschitz constant occurring in (6.5).

Exercise 6.8. In analogy to (6.65), define the right-right regularization $f^{\#\#}$, the right-left regularization $f^{\#\flat}$, and the left-right regularization $f^{\flat\#}$ of $f : [a, b] \times [c, d] \to \mathbb{R}$, and show that Theorem 6.34 also holds for the regularizations $h^{\#\#}(\cdot, \cdot, u)$, $h^{\#\flat}(\cdot, \cdot, u)$, and $h^{\flat\#}(\cdot, \cdot, u)$.

Exercise 6.9. Let $[a_1, b_1], [a_2, b_2], \ldots, [a_n, b_n]$ be real intervals, and denote by

$$I_a^b := [a_1, b_1] \times [a_2, b_2] \times \ldots \times [a_n, b_n] \subset \mathbb{R}^n$$

their Cartesian product. In analogy to (1.82), introduce some kind of variation $\mathrm{Var}(f, P_1 \times P_2 \times \ldots \times P_n; I_a^b)$ for a function $f : I_a^b \to \mathbb{R}$ with respect to partitions $P_k \in \mathcal{P}([a_k, b_k])$ $(k = 1, 2, \ldots, n)$. Given a function of $n + 1$ variables $h : I_a^b \times \mathbb{R} \to \mathbb{R}$, by imitating Theorem 6.11 formulate and prove a sufficient condition under which the superposition operator

$$S_h f(x_1, x_2, \ldots, x_n) := h(x_1, x_2, \ldots, x_n, f(x_1, x_2, \ldots, x_n))$$

generated by this function maps the corresponding space $BV(I_a^b)$ into itself and is bounded.

7 Some applications

This section is concerned with a few applications to problems which lead in a natural way to spaces of functions of bounded (generalized) variation. Historically, convergence criteria for Fourier series have always been one of the main motivations for introducing new concepts of variation. Below, we show that functions from the Waterman space ΛBV are in a certain sense “optimal” for being represented by their Fourier series because they produce sharp convergence results. In the second part of this chapter, we show how our results on composition and superposition operators from the preceding two chapters may be used to prove existence (and even uniqueness) of solutions to nonlinear integral equations with regular or weakly singular kernels.

7.1 Convergence criteria for Fourier series

Almost two centuries ago, Dirichlet [104] proved that every piecewise monotone real function on an interval has a pointwise convergent Fourier series. This result is usually referred to as the *Dirichlet criterion* and may be considered as a first contribution to a rigorous (partial) proof of Fourier’s famous (wrong) conjecture [115], raised in 1807, on the possibility to expand arbitrary real functions into a trigonometric series. According to Szőkefalvi–Nagy [301], the history of Fourier series started with a fruitful controversial discussion in the middle of the nineteenth century between D’Alembert, Euler, and D. Bernoulli regarding the problem of the vibrating string.

An important progress was achieved in 1881 by Jordan [153] who not only introduced functions of bounded variation, but also proved that such functions may be represented as differences of increasing functions, in this way extending the validity of Dirichlet’s result to BV-functions. Subsequently, convergence criteria for Fourier series have been one of the main motivations for introducing and studying new concepts of variation. The purpose of this and the following section is to present some sample results in this spirit.

In this section, we always consider functions f defined on $[a, b] = [0, 2\pi]$ and satisfying $f(0) = f(2\pi)$; thus, they may be extended periodically to the whole real line. Let $f : [0, 2\pi] \to \mathbb{R}$ be continuous of period 2π. The (real) *Fourier series* of f is given by

$$S[f] = S[f](x) = \frac{\alpha_0}{2} + \sum_{n=1}^{\infty} [\alpha_n \cos nx + \beta_n \sin nx] \,, \tag{7.1}$$

where

$$\alpha_n = \alpha_n(f) = \frac{1}{\pi} \int_0^{2\pi} f(t) \cos nt \, dt \,, \quad (n = 0, 1, 2, \ldots) \tag{7.2}$$

and

$$\beta_n = \beta_n(f) = \frac{1}{\pi}\int_0^{2\pi} f(t)\sin nt\,dt \quad (n = 1, 2, 3, \ldots) \tag{7.3}$$

are the (real) *Fourier coefficients* of f. A crucial problem in the theory and applications of Fourier series consists of finding conditions under which the series (7.1) converges pointwise, uniformly, or in some other sense.[1] It is well known that the partial sums

$$s_n(x) = s_n(x; f) = \frac{\alpha_0}{2} + \sum_{k=1}^{n} [\alpha_k \cos kx + \beta_k \sin kx] \tag{7.4}$$

of (7.1) may be written as convolution

$$s_n(x) = (D_n * f)(x) = \int_0^{2\pi} D_n(x-t) f(t)\,dt \tag{7.5}$$

of f and the so-called *Dirichlet kernel*

$$D_n(x) = 1 + 2\sum_{k=1}^{n} \cos kx = \frac{\sin((n+1/2)x)}{\sin(x/2)} \quad (n = 0, 1, 2, \ldots). \tag{7.6}$$

If instead of (7.4) we consider the average of the partial sums, i.e.

$$\sigma_n(x) = \sigma_n(x; f) = \frac{s_0(x) + s_1(x) + \ldots + s_n(x)}{n+1}, \tag{7.7}$$

we may write σ_n as convolution $\sigma_n = K_n * f$ of f and the so-called *Féjèr kernels* K_n. While the Dirichlet kernels D_n are oscillatory, the Féjèr kernels are positive. More precisely, one may show that

$$\int_0^{2\pi} F_n(t)\,dt = \int_0^{2\pi} |F_n(t)|\,dt \equiv 1, \tag{7.8}$$

but

$$\int_0^{2\pi} D_n(t)\,dt \equiv 1, \quad \lim_{n\to\infty} \int_0^{2\pi} |D_n(t)|\,dt = \infty. \tag{7.9}$$

This is one of the reasons why the partial sums s_n present more difficulties[2] than their averages σ_n, which behave quite well. For instance, the sequence $(K_n(t))_n$ converges uniformly to zero on $\mathbb{R} \setminus (-\delta, \delta)$ for every $\delta > 0$, which implies that $(\sigma_n(x; f))_n$ converges uniformly to f, as $n \to \infty$, for each continuous function f.

---

1 The estimates (4.49) and (4.51) in Proposition 4.23 show that the Fourier coefficients of a BV-function converge to zero; of course, this is only necessary for the convergence of (7.1).

2 In fact, one may consider the map $\ell_n(f) := s_n(0; f)$ for each n as a bounded linear functional on the space of 2π-periodic continuous functions. From the functional-analytic *Banach–Steinhaus theorem* and the unboundedness of the sequence $(D_n)_n$ in the L_1-norm, see (7.9), we may then conclude that the sequence $(\ell_n(f))_n$ is also unbounded for some continuous f, and so the Fourier series of f is accordingly divergent at zero.

The existence of continuous functions f whose Fourier series diverge at some point emphasizes the need of imposing additional conditions which guarantee the convergence of $S[f]$. Such conditions are sometimes called *convergence criteria* or *convergence tests*. We recall some of them which will be needed in the sequel. To state one criterion of particular historical interest, we use the conventional notation

$$\varphi(t) = \varphi_x(t) := \frac{f(x+t) + f(x-t) - 2f(x)}{2} \quad (0 \le t \le 2\pi) \tag{7.10}$$

throughout this section, where $x \in [0, 2\pi]$ is fixed.

– **The Lebesgue test.** *Suppose that*

$$\int_0^h |\varphi_x(t)|\,dt = o(h) \quad (h \to 0) \tag{7.11}$$

and

$$\int_{\eta_n}^{2\pi} \frac{|\varphi_x(t) - \varphi_x(t+\eta_n)|}{t}\,dt = o(1) \quad (n \to \infty), \tag{7.12}$$

where $\eta_n := 2\pi/n$. *Then the Fourier series* (7.1) *converges to* $f(x)$. *Moreover, the convergence is uniform over any closed interval of continuity of* f, *where condition* (7.12) *is satisfied uniformly.*

The proof of the Lebesgue test can be found in any book on trigonometric series or harmonic analysis, e.g. [329]. A particularly interesting sufficient condition given in [240] and refined in [49] asserts that one may reach convergence of Fourier series by a suitable homeomorphic change of variables:

– **The Pál–Bohr test.** *Let* $g : [0, 2\pi] \to \mathbb{R}$ *be continuous and* 2π*-periodic. Then there exists a homeomorphism* $\tau : [0, 2\pi] \to [0, 2\pi]$ *such that the Fourier series* (7.1) *of* $f = g \circ \tau$ *converges uniformly.*

The following criteria refer to functions of bounded variation and related classes. They are are due to Jordan [153], Salem [285], Lipschitz and Dini [329], Goffman and Waterman [132], Garcia and Sawyer [116], and Sahney and Waterman [284], respectively. In particular, the Garcia–Sawyer test makes use of the Banach indicatrix I_f of f introduced in (0.106). Recall that f and I_f are related by the equality

$$\mathrm{Var}(f; [0, 2\pi]) = \int_{-\infty}^{\infty} I_f(y)\,dy,$$

and so $f \in BV([0, 2\pi])$ if and only if $I_f \in L_1(\mathbb{R})$, see Proposition 1.27.

– **The Jordan test.** *Let* $f : [0, 2\pi] \to \mathbb{R}$ *be continuous, of bounded variation, and* 2π*-periodic. Then the Fourier series* (7.1) *converges uniformly on* $[0, 2\pi]$ *to* $f(x)$.

- **The Salem test.** *Let $f : [0, 2\pi] \to \mathbb{R}$ be continuous and 2π-periodic. For $n \in \mathbb{N}$, put*

$$\tau_1 := \frac{2\pi}{n},\ \tau_2 := \frac{4\pi}{n},\ \dots,\tau_{n-1} := \frac{2(n-1)\pi}{n},\ \tau_n := 2\pi \tag{7.13}$$

and

$$T_n(x) := \sum_{k=1}^{(n+1)/2} \frac{f(x+\tau_{2k-2}) - f(x+\tau_{2k-1})}{k}.$$

If the sequence $(T_n)_n$ converges uniformly to zero as $n \to \infty$, the Fourier series (7.1) of f converges uniformly.

- **The Lipschitz–Dini test.** *Let $f : [0, 2\pi] \to \mathbb{R}$ be continuous and 2π-periodic. Assume that*

$$\omega_\infty(f;\delta) = o\left(\frac{1}{\log\delta}\right) \quad (\delta \to 0), \tag{7.14}$$

where $\omega_\infty(f;\cdot)$ denotes the modulus of continuity (0.97) of f. Then the Fourier series (7.1) of f converges uniformly.

- **The Goffman–Waterman test.** *Let $\phi : [0,\infty) \to [0,\infty)$ be some Young function, and let ϕ^* denote its conjugate Young function (0.23). Let $f \in C([0,2\pi]) \cap WBV_\phi([0,2\pi])$ be 2π-periodic, where $WBV_\phi([a,b])$ is the Wiener–Young space introduced in Definition 2.2. If*

$$\sum_{n=1}^{\infty} \phi^*\left(\frac{1}{n}\right) < \infty, \tag{7.15}$$

the Fourier series (7.1) of f converges uniformly.

- **The Garsia–Sawyer test.** *Let $f : [0, 2\pi] \to \mathbb{R}$ be continuous and 2π-periodic. If*

$$\int_{-\infty}^{\infty} \log I_f(y)\, dy < \infty, \tag{7.16}$$

the Fourier series (7.1) of f converges uniformly.

- **The Sahney–Waterman test.** *Suppose that the function φ_x defined by (7.10) satisfies (7.11) and*

$$\int_{\eta_n}^{2\pi} \frac{|\varphi_x(t) - \varphi_x(t+\eta_n)|}{t^{\beta+1}}\, dt = o(\eta_n^{-\beta}) \quad (n \to \infty) \tag{7.17}$$

for some $\beta \in (-1, 0)$, where

$$\eta_n := \frac{2\pi}{n + \frac{1}{2}(\beta+1)}. \tag{7.18}$$

Then the Fourier series (7.1) satisfies

$$S[f](x) = o(n^{-\beta}) \tag{7.19}$$

and converges to $f(x)$.

We point out that these criteria are not independent of each other, and some particular cases of them have been known earlier. For example, letting $\phi(t) = t^p$ $(1 < p < \infty)$ in the Goffman–Waterman test, we obtain $\phi^*(t) = t^{p'} = t^{p/(p-1)}$ (up to constants), and so (7.15) reads

$$\sum_{n=1}^{\infty} \frac{1}{n^{p/(p-1)}} < \infty . \tag{7.20}$$

This condition was already known to Young [323]. Interestingly, condition (7.15) also implies that $WBV_\phi([a,b]) \subseteq HBV([a,b])$, the space of functions of bounded harmonic variation, see Definition 2.29. In fact, this follows from Proposition 2.34 for the particular choice $\lambda_n = 1/n$.

Also, the Salem criterion extends the Lipschitz–Dini test. Indeed, if $\omega_\infty(f;\cdot)$ is the modulus of continuity (0.97) of f, it is not hard to see that

$$|T_n(x)| = O(\omega_\infty(f; 2\pi/n) \log n) \quad (n \to \infty) . \tag{7.21}$$

Combining this with condition (7.14) and applying the Salem criterion, we get the uniform convergence of the Fourier series (7.1).

As a second application of the Salem criterion, we obtain the Jordan condition $f \in BV([0,2\pi]) \cap C([0,2\pi])$. For such functions f, we have

$$|T_n(x)| \le \omega_\infty(f; 2\pi/n) \left(1 + \frac{1}{2} + \ldots + \frac{1}{m}\right) + \frac{\mathrm{Var}(f;[0,2\pi])}{m+1} ,$$

where m is any integer smaller than $(n+1)/2$. If we choose m, as we may, so that m goes to infinity, but $\omega_\infty(f; 2\pi/n) \log m$ goes to zero, as $n \to \infty$, we get again the uniform convergence of the Fourier series (7.1). Also, we get the Jordan test from the Garcia–Sawyer test since the integral (7.16) is finite for $f \in BV([0,2\pi]) \cap C([0,2\pi])$.

7.2 Fourier series and Waterman spaces

In this section, we take a closer look at the Fourier series of functions in the Waterman spaces ΛBV which we introduced and discussed in detail in Section 2.2. The first result in this direction [314] is concerned with absolute convergence of the Fourier series (7.1).

Theorem 7.1. *Let $\Lambda = (\lambda_n)_n$ be a Waterman sequence, and let $f \in C([0,2\pi]) \cap \Lambda BV([0,2\pi])$ be 2π-periodic. Suppose that*

$$\sum_{n=1}^{\infty} \frac{\sqrt{\omega_\infty(f; 2\pi/n)}}{n\sqrt{\lambda_n}} < \infty , \tag{7.22}$$

where $\omega_\infty(f;\cdot)$ denotes the modulus of continuity (0.97) of f and the convergence in (7.22) is monotone. Then the Fourier series (7.1) of f converges absolutely.

Proof. Fix $f \in C([0,2\pi]) \cap \Lambda BV([0,2\pi])$, define numbers τ_k as in (7.13), and let $a_k := x + \tau_{k-1}$ and $b_k := x + \tau_k$. For the collection $S := \{[a_1,b_1],\ldots,[a_n,b_n]\} \in \Sigma([0,2\pi])$, we then get $b_k - a_k = 2\pi/n$, and hence

$$|f(b_k) - f(a_k)| \leq \omega_\infty(f; 2\pi/n) \quad (k = 1, 2, \ldots, n),$$

and so

$$\begin{aligned}\sum_{k=1}^{N} |f(b_k) - f(a_k)|^2 &= \sum_{k=1}^{N} \lambda_k |f(b_k) - f(a_k)| \frac{|f(b_k) - f(a_k)|}{\lambda_k} \\ &\leq \mathrm{Var}_\Lambda(f; [0, 2\pi]) \frac{\omega_\infty(f; 2\pi/N)}{\lambda_N}\end{aligned}$$

for $n \in \mathbb{N}$ since the sequence $(\lambda_n)_n$ is decreasing. Consequently,

$$N \int_0^{2\pi} |f(x + \pi/n) - f(x - \pi/n)|^2 \, dx \leq 2\pi \, \mathrm{Var}_\Lambda(f; [0, 2\pi]) \frac{\omega_\infty(f; 2\pi/N)}{\lambda_N}$$

or

$$\sum_{n=1}^{N} (\alpha_n^2 + \beta_n^2) \frac{\sin^2 n\pi}{2N} \leq \mathrm{Var}_\Lambda(f; [0, 2\pi]) \frac{\omega_\infty(f; 2\pi/N)}{2N\lambda_N},$$

where α_n and β_n denote the Fourier coefficients (7.2) and (7.3) of f. Setting now $N = 2^m$, we obtain

$$\sum_{n=2^{m-1}+1}^{2^m} (\alpha_n^2 + \beta_n^2) \leq \mathrm{Var}_\Lambda(f; [0, 2\pi]) \frac{\omega_\infty(f; 2\pi/2^m)}{2^{m+1}\lambda_{2^{m+1}}}.$$

Thus,

$$\begin{aligned}\sum_{n=2^{m-1}+1}^{2^m} (\alpha_n^2 + \beta_n^2)^{1/2} &\leq 2^{m/2} \left\{ \sum_{n=2^{m-1}+1}^{2^m} (\alpha_n^2 + \beta_n^2) \right\}^{1/2} \\ &\leq \sqrt{\mathrm{Var}_\Lambda(f; [0, 2\pi])} \frac{\sqrt{\omega_\infty(f; 2\pi/2^m)}}{\sqrt{2\lambda_{2^{m+1}}}},\end{aligned}$$

and so

$$\sum_{n=2}^{\infty} \sqrt{\alpha_n^2 + \beta_n^2} \leq \sqrt{\frac{\mathrm{Var}_\Lambda(f; [0, 2\pi])}{2}} \sum_{m=1}^{\infty} \frac{\sqrt{\omega_\infty(f; 2\pi/2^m)}}{\sqrt{\lambda_{2^{m+1}}}}. \tag{7.23}$$

However, the convergence of the series on the right-hand side of (7.23) is equivalent to that of the series (7.22) if the terms of this series are decreasing from some point on.[3] This completes the proof. □

For the special Waterman sequence $\Lambda_q = (n^{-q})_n$, see Definition 2.29, we obtain from Theorem 7.1 the following

3 Here, we use the condensation criterion for series with (eventually) decreasing terms.

Corollary 7.2. *For* $0 < q \leq 1$, *let* $f \in C([0, 2\pi]) \cap \Lambda_q BV([0, 2\pi])$ *be* 2π*-periodic. Suppose that*

$$\sum_{n=1}^{\infty} \frac{\sqrt{\omega_\infty(f; 2\pi/n)}}{n^{1-q/2}} < \infty, \tag{7.24}$$

and the convergence in (7.24) *is monotone. Then the Fourier series* (7.1) *of* f *converges absolutely. In particular, this holds for* $f \in C([0, 2\pi]) \cap HBV([0, 2\pi])$ *if*

$$\sum_{n=1}^{\infty} \frac{\sqrt{\omega_\infty(f; 2\pi/n)}}{\sqrt{n}} < \infty. \tag{7.25}$$

Now, we consider some convergence results for functions in the Waterman space ΛBV in more detail. First, we recall a notion from the theory of Fourier series.

Definition 7.3. Let $\alpha, \beta \in \mathbb{R}$. A real series

$$\Gamma := \sum_{n=1}^{\infty} \gamma_n$$

is called (C, α)-*summable* if it converges and $\gamma_n = o(n^{-\alpha})$ as $n \to \infty$. Moreover, the series Γ is called (C, β)-*bounded* if its partial sums are bounded and $\gamma_n = O(n^{-\beta})$ as $n \to \infty$. ■

For example, a classical result [132] asserts that the Fourier series $S[f]$ in (7.1) of a function $f \in BV([0, 2\pi])$ is $(C, -1)$-bounded, convergent everywhere, and even uniformly convergent on each closed interval of continuity. It follows then from a result of Hardy and Littlewood (see, e.g. [140, p. 121]) that $S[f]$ is (C, α)-summable for each $\alpha > -1$, and even uniformly (C, α)-summable on each closed interval of continuity.

Observe that the conclusion (7.19) of the Sahney–Waterman criterion states in this terminology that the Fourier series (7.1) is (C, β)-summable.

Since BV may be considered as a special case of the Waterman spaces $\Lambda_q BV$ (namely, for $q = 0$), it is a tempting idea to try to extend this result to $\Lambda_q BV$, establishing some kind of "interaction" between q, α, and β (if there is any). This has been done by Waterman [315]; his main theorem reads as follows.

Theorem 7.4. *Let* $0 < q < 1$, *and let* $f \in \Lambda_q BV([0, 2\pi])$. *Then the Fourier series* (7.1) *of* f *is everywhere* (C, β)*-bounded for* $\beta = q - 1$, *and* (C, α)*-summable for* $\alpha > q - 1$.

Proof. If we show that $S[f]$ is (C, β)-bounded at a point, it follows from a well-known convexity theorem [140, p. 127] that $S[f]$ is (C, α)-summable at that point for $\alpha > \beta$. So, we need only verify condition (7.17) for $\beta = q - 1$.

For η_n as in (7.18) and $\delta \in (\eta_n, \pi)$, consider

$$I_n(x, \delta) := \int_{\eta_n}^{\delta} \frac{|\varphi_x(t) - \varphi_x(t + \eta_n)|}{t^{\beta+1}}\, dt. \tag{7.26}$$

We estimate this integral in the form

$$I_n(x,\delta) \le \frac{1}{2}[I_n^+(x,\delta) + I_n^-(x,\delta)],$$

where

$$I_n^+(x,\delta) := \int_{\eta_n}^{\delta} \frac{|f(x+t) - f(x+t+\eta_n)|}{t^{\beta+1}}\,dt \tag{7.27}$$

and

$$I_n^-(x,\delta) := \int_{\eta_n}^{\delta} \frac{|f(x-t) - f(x-t-\eta_n)|}{t^{\beta+1}}\,dt\,. \tag{7.28}$$

For $m_n := \mathrm{ent}(\delta/\eta_n)$, the integer part of δ/η_n, and $q = \beta + 1$, we have

$$\begin{aligned}\frac{I_n^+(x,\delta)}{n^\beta} &\le \frac{1}{n^\beta}\sum_{i=1}^{m_n}\int_{i\eta_n}^{(i+1)\eta_n} \frac{|f(x+t) - f(x+t+\eta_n)|}{t^q}\,dt \\ &\le \frac{1}{n^\beta \eta_n^\beta}\sum_{i=1}^{m_n} \frac{\mathrm{osc}(f;[x+i\eta_n, x+\delta+2\eta_n])}{i^q} \\ &\le \frac{2}{\pi^\beta}\,\mathrm{Var}_{\Lambda_q}(f;[x+i\eta_n, x+\delta+2\eta_n])\,,\end{aligned}$$

where $\mathrm{osc}(f;[a,b])$ denotes the oscillation (1.12) of f on $[a,b]$ and $\mathrm{Var}_{\Lambda_q}(f;[a,b])$ the Λ_q-variation of f on $[a,b]$ defined in Definition 2.29. Applying Exercise 2.15, we see that

$$\frac{I_n^+(x,\delta)}{n^\beta} = o(1) \quad (\delta \to 0,\ n \to \infty)\,. \tag{7.29}$$

In the same way, one may show that

$$\frac{I_n^-(x,\delta)}{n^\beta} = o(1) \quad (\delta \to 0,\ n \to \infty)\,,$$

and so

$$\frac{I_n(x,\delta)}{n^\beta} = o(1) \quad (\delta \to 0,\ n \to \infty) \tag{7.30}$$

for each x, and the convergence is uniform at each point of continuity of f.

Suppose now that $f \in \Lambda^c BV([0,2\pi])$, i.e. f is continuous in Λ-variation, see Definition 2.37. Then we may estimate the integral

$$J_n(x,\delta) := \int_{\delta}^{\pi-\eta_n} \frac{|\varphi_x(t) - \varphi_x(t+\eta_n)|}{t^{\beta+1}}\,dt \tag{7.31}$$

in the form

$$J_n(x,\delta) \le \frac{1}{2}[J_n^+(x,\delta) + J_n^-(x,\delta)]$$

as above, where

$$J_n^+(x,\delta) := \int_\delta^{\pi-\eta_n} \frac{|f(x+t) - f(x+t+\eta_n)|}{t^{\beta+1}}\,dt \tag{7.32}$$

and

$$J_n^-(x,\delta) := \int_\delta^{\pi-\eta_n} \frac{|f(x-t) - f(x-t-\eta_n)|}{t^{\beta+1}}\,dt\,. \tag{7.33}$$

For m_n and q, as before, we then have

$$\begin{aligned}\frac{J_n^+(x,\delta)}{n^\beta} &\le \frac{1}{n^\beta}\sum_{i=m_n}^{n-1} \frac{\operatorname{osc}(f;[x+i\eta_n, x+(i+2)\eta_n])}{i^q}\\ &\le \frac{2}{\pi^\beta}\operatorname{Var}_{\Lambda_q^{m_n}}(f;[x+i\eta_n, x+(i+2)\eta_n])\,,\end{aligned}$$

and similarly for $J_n^-(x,\delta)$, where $\operatorname{Var}_{\Lambda_q^m}(f;[a,b])$ denotes the shifted Λ_q-variation of f in the sense of Definition 2.37. Our assumption $f \in \Lambda^c BV([0,2\pi])$ implies that $\operatorname{Var}_{\Lambda_q^m}(f;[0,2\pi])$ is dominated by $\operatorname{Var}_{\Lambda_q}(f;[0,2\pi])$, and so

$$\frac{J_n(x,\delta)}{n^\beta} = O(1) \quad (n\delta \to \infty) \tag{7.34}$$

uniformly in x. Now, choose $(\delta_n)_n$ converging monotonically to zero such that $n\delta \to \infty$ as $n \to \infty$. We may then summarize our results as follows.

If $f \in \Lambda BV([0,2\pi])$, then

$$\frac{I_n(x,\delta_n) + J_n(x,\delta_n)}{n^\beta} = o(1) + O(1) = O(1) \quad (n\delta \to \infty)\,. \tag{7.35}$$

Consequently, $S[f]$ is everywhere (C,β)-bounded, and uniformly (C,β)-bounded on each closed interval of continuity. On the other hand, if $f \in \Lambda^c BV([0,2\pi])$, then even

$$\frac{I_n(x,\delta_n) + J_n(x,\delta_n)}{n^\beta} = o(1) \quad (n\delta \to \infty)\,. \tag{7.36}$$

Consequently, in this case, $S[f]$ converges to $f(x)$ and is everywhere (C,β)-summable. The proof is complete. □

Note that the result for BV-functions mentioned before Theorem 7.4 is formally contained in this theorem if we take $q = 0$. The following theorem, also due to Waterman [315], shows that Theorem 7.4 is, in a certain sense, *sharp* within the class of ΛBV-spaces:

Theorem 7.5. *Let $0 < q < 1$, and let $\Lambda = (\lambda_n)_n$ be some Waterman sequence such that*

$$\Lambda BV([0,2\pi]) \supset \Lambda_q BV([0,2\pi])\,. \tag{7.37}$$

Then there exists a function $f \in C([0,2\pi]) \cap \Lambda BV([0,2\pi])$ whose Fourier series (7.1) is not $(C,q-1)$-bounded at some point.

Proof. By our assumption (7.37), we may find a decreasing positive sequence $(\mu_n)_n$ which converges to zero and satisfies

$$\sum_{n=1}^{\infty} \lambda_n \mu_n < \infty, \quad \sum_{n=1}^{\infty} \frac{\mu_n}{n^q} = \infty . \tag{7.38}$$

For fixed n, let

$$a_{k,n} := \left(2k + \frac{q-1}{2}\right) \frac{\pi}{n + \frac{q}{2}} = \frac{4k + q - 1}{2n + q} \pi$$

and

$$b_{k,n} := \left(2k + 1 + \frac{q-1}{2}\right) \frac{\pi}{n + \frac{q}{2}} = \frac{4k + q + 1}{2n + q} \pi .$$

We define functions $f_n : [0, 2\pi] \to \mathbb{R}$ and $f : [0, 2\pi] \to \mathbb{R}$ by

$$f_n(x) := \sum_{k=1}^{n-1} \mu_k \chi_k(x), \quad f(x) := \sum_{k=1}^{\infty} \mu_k \chi_k(x) \quad (0 \le x \le 2\pi), \tag{7.39}$$

where χ_k denotes the characteristic function of the interval $[a_{k,n}, b_{k,n}]$, and extend f with period 2π to the whole real line. Since the Waterman variations of f_n satisfy

$$\mathrm{Var}_\Lambda(f_n; [0, 2\pi]) \le 2 \sum_{k=1}^{n-1} \lambda_n \mu_n , \tag{7.40}$$

they are uniformly bounded, by (7.38), and so we have $f \in \Lambda BV([0, 2\pi])$. On the other hand, the Fourier series (7.1) of f is not $(C, q-1)$-bounded at zero, which may be seen as follows.

Let $\sigma_n^q(x) = \sigma_n^q(x; f)$ denote the n-th $(C, q-1)$-mean of $S[f]$ at x. Then we have

$$\begin{aligned}\pi \sigma_n^q(0; f) &= \int_0^{2\pi} f_n(t) \frac{\sin\left[\left(n + \frac{q}{2}\right)t - \frac{\pi(q-1)}{2}\right]}{\mu_n^{q-1}(2\sin(t/2))^2}\, dt + 2 \int_0^{2\pi} f_n(t) \frac{\theta(t)(q-1)}{n(\sin(t/2))^2}\, dt \\ &= \sum_{k=1}^{n-1} \int_{a_{k,n}}^{b_{k,n}} \frac{\sin\left[\left(n + \frac{q}{2}\right)t - \frac{\pi(q-1)}{2}\right]}{\mu_n^{q-1}(2\sin(t/2))^2}\, dt + 2 \sum_{k=1}^{n-1} \int_{a_{k,n}}^{b_{k,n}} \frac{\theta(t)(q-1)}{n(\sin(t/2))^2}\, dt ,\end{aligned}$$

where θ is a suitable function satisfying $|\theta(t)| \le 1$ (see, e.g. [329, p. 85]). However,

$$\int_{a_{k,n}}^{b_{k,n}} \frac{\sin\left[\left(n + \frac{q}{2}\right)t - \frac{\pi(q-1)}{2}\right]}{\mu_n^{q-1}(2\sin(t/2))^2}\, dt \ge \frac{1}{\mu_n^{q-1}} \frac{2}{n + q/2} \frac{\mu_k}{b_{k,n}^q} \ge C_1 \frac{\mu_k}{k^q} \tag{7.41}$$

and

$$\int_{a_{k,n}}^{b_{k,n}} \frac{\theta(t)(q-1)}{n(\sin(t/2))^2}\, dt \le \frac{1}{n} \frac{\pi}{n + q/2} \frac{\pi \mu_k}{2a_{k,n}} \le C_2 \frac{\mu_k}{k^2} . \tag{7.42}$$

Taking the sum over $k = 1, 2, \ldots, n-1$ and letting $n \to \infty$, we see that the last expression in (7.41) becomes unbounded, by (7.38), while the last expression in (7.41) remains bounded. We conclude that $\|\sigma_n^q(0;\cdot)\|_\infty \to \infty$ as $n \to \infty$, and so the sequence $(\sigma_n^q(0;f))_n$ is unbounded for the function $f \in \Lambda BV([0,2\pi])$ constructed above.

Moreover, since $C([0,2\pi]) \cap \Lambda BV([0,2\pi])$ is a closed subspace of $\Lambda BV([0,2\pi])$ and the functions f_n in (7.39) can be modified so as to be continuous without any substantial change in the argument, we have the desired result. □

Other convergence criteria which are formulated in terms of moduli of variation and Chanturiya classes (Definition 2.35) will be given in Section 7.4.

7.3 Applications to nonlinear integral equations

In this section, we apply the abstract results of Chapter 5 and Chapter 6 to obtain existence (and sometimes uniqueness) of solutions of certain nonlinear integral equations over the interval $[0,1]$. We start with an existence and uniqueness result for solutions of bounded p-variation in Wiener's or Riesz's sense. To get a unified approach, we use the shortcut $\mathrm{Var}_p(f)$ to denote either the Wiener p-variation $\mathrm{Var}_p^W(f;[0,1])$ defined in (1.61) or the Riesz p-variation $\mathrm{Var}_p^R(f;[0,1])$ defined in (2.88), and BV_p to denote either the space $WBV_p([0,1])$ or the space $RBV_p([0,1])$.

Consider the nonlinear integral equations of Hammerstein type

$$f(s) = \int_0^1 k(s,t)h(t,f(t))\,dt + b(s) \quad (0 \le s \le 1). \tag{7.43}$$

Here, $k : [0,1]\times[0,1] \to \mathbb{R}$ is a given kernel function (whose properties will be made precise below), and $b \in BV_p$ is also given. We are interested in conditions on the nonlinearity $h : [0,1]\times\mathbb{R} \to \mathbb{R}$ under which (7.43) admits a (unique) solution $f \in BV_p$.

Denoting the corresponding linear integral operator in (7.43) by

$$Kg(s) := \int_0^1 k(s,t)g(t)\,dt \quad (0 \le s \le 1), \tag{7.44}$$

we may rewrite (7.43) as operator equation

$$f - KS_h f = b, \tag{7.45}$$

where S_h is the superposition operator (6.1), and try to apply the familiar fixed point principles to (7.45). To this end, we assume that the superposition operator (6.1) induced by h maps the space BV_p into itself and is bounded in the corresponding norm.[4]

4 Sufficient conditions for this in case $p = 1$ have been given in Theorem 6.11.

Now, we impose some hypotheses on the kernel function k in (7.43). Suppose that $k(s,\cdot) \in L_1([0,1])$ for $0 \le s \le 1$, and that the function $v_p : [0,1] \to \mathbb{R}$ defined by

$$v_p(t) := \operatorname{Var}_p(k(\cdot,t))^{1/p}$$

belongs to $L_p([0,1])$. To simplify the notation, we use the shortcut

$$\kappa_p := \|k(0,\cdot)\|_{L_1} + \|v_p\|_{L_p} = \int_0^1 |k(0,t)|\,dt + \left(\int_0^1 \operatorname{Var}_p(k(\cdot,t))\,dt\right)^{1/p}.$$

In the following Lemma 7.6, we give a sufficient condition under which the operator $f \mapsto KS_h f + b$ (whose fixed points coincide with the solutions of (7.45) leaves a closed ball in the space $BV_p([0,1])$ invariant. In case $p = 1$, i.e. in the space $BV([0,1])$, similar conditions of this type have been considered in [67].

Lemma 7.6. *Under the above hypotheses, let $r > 0$ be so large that*

$$\kappa_p \tilde{k}(r) < r, \tag{7.46}$$

where $\tilde{k}(r)$ is given by

$$\tilde{k}(r) := \sup\{|h(t,u)| : 0 \le t \le 1,\ |u| \le r\}. \tag{7.47}$$

Then the operator $f \mapsto KS_h f + b$ maps, for any $b \in BV_p$ satisfying

$$\|b\|_{BV_p} \le r - \kappa_p \tilde{k}(r), \tag{7.48}$$

the closed ball $B_r(BV_p) = \{f \in BV_p : \|f\|_{BV_p} \le r\}$ into itself.

Proof. For the sake of definiteness, let us prove the assertion in the space $WBV_p([0,1])$. Suppose that r satisfies (7.46), and b satisfies (7.48). Given $f \in B_r(WBV_p)$ and a partition $\{s_0, s_1, \dots, s_m\} \in \mathcal{P}([0,1])$, we get[5]

$$\begin{aligned}\sum_{j=1}^m |KS_h f(s_j) - KS_h f(s_{j-1})|^p &= \sum_{j=1}^m \left|\int_0^1 \left[k(s_j,t) - k(s_{j-1},t)\right] h(t,f(t))\,dt\right|^p \\ &\le \sup_{0\le t\le 1} |h(t,f(t))|^p \int_0^1 \sum_{j=1}^m |k(s_j,t) - k(s_{j-1},t)|^p\,dt \\ &\le \tilde{k}(r)^p \int_0^1 v_p(t)^p\,dt,\end{aligned}$$

5 Here, we use the fact that $WBV_p \hookrightarrow B$ with imbedding constant $c(WBV_p, B) = 1$, see (0.36), and hence $\|f\|_\infty \le r$ for $f \in B_r(WBV_p)$.

and so, passing to the supremum over all partitions $\{s_0, s_1, \dots, s_m\} \in \mathcal{P}([0,1])$,

$$\operatorname{Var}_p^W(KS_h f; [0,1]) \le \tilde{k}(r)^p \|v_p\|_{L_p}^p .$$

Consequently, from (7.48), we conclude that

$$\begin{aligned}
\|KS_h f + b\|_{WBV_p} &\le |KS_h f(0)| + \operatorname{Var}_p^W(KS_h f; [0,1])^{1/p} + \|b\|_{WBV_p} \\
&\le \int_0^1 |k(0,t)h(t, f(t))|\, dt + \tilde{k}(r)\|v_p\|_{L_p} + \|b\|_{WBV_p} \\
&\le \tilde{k}(r)\|k(0,\cdot)\|_{L_1} + \tilde{k}(r)\|v_p\|_{L_p} + \|b\|_{WBV_p} \le \tilde{k}(r)\kappa_p + \|b\|_{WBV_p} \le r ,
\end{aligned}$$

which proves the assertion for WBV_p. The proof for $RBV_p([0,1])$ is exactly the same, with obvious modifications in the definition of variations. □

In the following Lemma 7.7, we give a sufficient condition under which the operator $f \mapsto KS_h f + b$ is a contraction in the norm (1.65) (for $WBV_p([0,1])$), respectively (2.90) (for $RBV_p([0,1])$).

Lemma 7.7. *Under the above hypotheses, let* $r > 0$ *be so small that*

$$\kappa_p k(r) < 1 , \tag{7.49}$$

where $k(r)$ *is given by* (6.10). *Then the operator* $f \mapsto KS_h f + b$ *is, for any* $b \in BV_p$, *a contraction on* $B_r(BV_p)$ *with respect to the norm* (1.65), *respectively* (2.90).

Proof. We work again in the space $WBV_p([0,1])$. Suppose that r satisfies (7.49), and let $f, g \in B_r(WBV_p)$ be fixed. We claim that

$$\|KS_h f - KS_h g\|_{WBV_p} \le \kappa_p k(r)\|f - g\|_{WBV_p} ,$$

which, together with (7.49), proves the assertion. First of all, we have

$$\begin{aligned}
|KS_h f(0) - KS_h g(0)| &= \left| \int_0^1 k(0,t)[h(t, f(t)) - h(t, g(t))]\, dt \right| \\
&\le k(r) \int_0^1 |k(0,t)|\, |f(t) - g(t)|\, dt \le k(r)\|k(0,\cdot)\|_{L_1} \|f - g\|_\infty \\
&\le k(r)\|k(0,\cdot)\|_{L_1} \|f - g\|_{WBV_p} ,
\end{aligned} \tag{7.50}$$

where again we have used the fact that $WBV_p([0,1])$ is continuously imbedded into the space of all bounded functions on $[0,1]$ with the supremum norm (0.39).

On the other hand, we show now that also

$$\operatorname{Var}_p^W(KS_h f - KS_h g; [0,1])^{1/p} \le k(r)\|v_p\|_{L_p}\|f - g\|_{WBV_p} .$$

For any partition $\{s_0, s_1, \dots, s_m\} \in \mathcal{P}([0,1])$, we have

$$\begin{aligned}
&\sum_{j=1}^{m} |KS_hf(s_j) - KS_hf(s_{j-1}) - KS_hg(s_j) + KS_hg(s_{j-1})|^p \\
&= \sum_{j=1}^{m} \left| \int_0^1 [k(s_j,t) - k(s_{j-1},t)][S_hf(t) - S_hg(t)]\,dt \right|^p \\
&\le \|S_hf - S_hg\|_\infty^p \sum_{j=1}^{m} \int_0^1 |k(s_j,t) - k(s_{j-1},t)|^p\,dt \\
&\le \|S_hf - S_hg\|_{WBV_p}^p \int_0^1 \sum_{j=1}^{m} |k(s_j,t) - k(s_{j-1},t)|^p\,dt \\
&\le k(r)^p \|f-g\|_{WBV_p}^p \int_0^1 v_p(t)^p\,dt = k(r)^p \|v_p\|_{L_p}^p \|f-g\|_{WBV_p}^p .
\end{aligned}$$

Passing again to the supremum with respect to all partitions $\{s_0, s_1, \dots, s_m\} \in \mathcal{P}([0,1])$, we obtain

$$\mathrm{Var}_p^W(KS_hf - KS_hg; [0,1]) \le k(r)^p \|v_p\|_{L_p}^p \|f-g\|_{WBV_p}^p ,$$

and combining this with (7.50) yields

$$\begin{aligned}
\|KS_hf - KS_hg\|_{WBV_p} &= |KS_hf(0) - KS_hg(0)| + \mathrm{Var}_p^W(KS_hf - KS_hg; [0,1])^{1/p} \\
&\le k(r)\|k(0,\cdot)\|_{L_1}\|f-g\|_{WBV_p} + k(r)\|v_p\|_{L_p}\|f-g\|_{WBV_p} \\
&= k(r)\kappa_p\|f-g\|_{WBV_p} ,
\end{aligned}$$

and establishes the result. The proof for RBV_p is similar. □

Combining Theorem 6.13 (for $\phi(t) = |t|^p$), Lemma 7.6, and Lemma 7.7, and observing that the fixed points of the operator $f \mapsto KS_hf + b$ coincide with the solutions of (7.45), Banach's contraction mapping theorem now implies the following

Theorem 7.8. *Suppose that there exists $r > 0$ such that both estimates* (7.46) *and* (7.49) *hold. Then the nonlinear integral equation* (7.43) *has, for each $b \in BV_p$ satisfying* (7.48), *a unique solution $f \in BV_p$.*

Let us return for a moment to the autonomous case of the operator (5.1) where, as one could expect, our calculations simplify.[6] We suppose, in addition, that the kernel function splits into two functions of one variable, i.e. $k(s,t) = l(s)m(t)$, where $l \in BV_p([0,1])$ and $m \in L_p([0,1])$. In this case, we get

$$v_p(t) = \mathrm{Var}_p(l)^{1/p}|m(t)| ,$$

6 For instance, the characteristic (7.47) takes the simpler form (5.15).

and hence

$$\begin{aligned}\kappa_p &= \int_0^1 |l(0)|\,|m(t)|\,dt + \left(\int_0^1 \operatorname{Var}_p(l)|m(t)|^p\,dt\right)^{1/p}\\ &= |l(0)|\,\|m\|_{L_1} + \operatorname{Var}_p(l)^{1/p}\|m\|_{L_p} \le \|l\|_{BV_p}\|m\|_{L_1}\,.\end{aligned}$$

Therefore, the crucial condition (7.46) in Lemma 7.6 holds if

$$\|l\|_{BV_p}\|m\|_{L_1}\tilde{k}(r) < r\,, \tag{7.51}$$

with $\tilde{k}(r)$ given by (5.15), while the crucial condition (7.49) in Lemma 7.7 holds if

$$\|l\|_{BV_p}\|m\|_{L_1}k(r) < 1 \tag{7.52}$$

with $k(r)$ given by (5.13). We illustrate these conditions by a very elementary example.

Example 7.9. First, let $h(u) = u^\alpha$, where $\alpha \in \mathbb{R}$, $\alpha \ge 2$. The mean value theorem shows that

$$k(r) = \alpha r^{\alpha-1}\,, \quad \tilde{k}(r) = r^\alpha\,, \quad k_1(r) = \alpha(\alpha-1)r^{\alpha-2}\,, \quad \tilde{k}_1(r) = \alpha r^{\alpha-1}\,,$$

where $k_1(r)$ is given by (5.14) and $\tilde{k}_1(r)$ by (5.16). Thus, the estimate (7.52) reads

$$K(r) \le \|l\|_{BV_p}\|m\|_{L_1}\alpha r^{\alpha-1} < 1\,, \tag{7.53}$$

while the general estimate (5.82) from Section 5.5 becomes

$$K(r) \le \|l\|_{BV_p}\|m\|_{L_1}\max\{\alpha(\alpha-1)r^{\alpha-2}, \alpha r^{\alpha-1}\} < 1\,. \tag{7.54}$$

Of course, both estimates may be achieved for r sufficiently small, but (7.53) is better than (7.54) for $r < \alpha - 1$, while (7.54) is better than (7.53) for $r > \alpha - 1$.

Next, let $h(u) = \log(1 + u)$ for $u > 0$ and $h(u) = 0$ for $u \le 0$. Again, the mean value theorem implies that

$$k(r) \equiv 1\,, \quad \tilde{k}(r) = \log(1+r)\,, \quad k_1(r) = \frac{1}{(1+r)^2}\,, \quad \tilde{k}_1(r) = \frac{1}{1+r}\,.$$

So, the estimate (7.52) reads

$$K(r) \le \|l\|_{BV_p}\|m\|_{L_1} < 1\,, \tag{7.55}$$

while the estimate (7.54) becomes

$$K(r) \le \|l\|_{BV_p}\|m\|_{L_1}\max\left\{\frac{r}{(1+r)^?}, \frac{1}{1+r}\right\} = \frac{\|l\|_{BV_p}\|m\|_{L_1}}{1+r} < 1\,. \tag{7.56}$$

The estimate (7.55) does not depend on r, but the estimate (7.56) may be achieved for r sufficiently large; so, in this example, (7.56) is always better than (7.55). ♥

Let us now prove a parallel result in the Waterman space $\Lambda BV([a,b])$ which we introduced in Definition 2.15. Since the space $WBV_p([a,b])$ is a proper subspace of $\Lambda BV([a,b])$ for a suitable Waterman sequence $\Lambda = (\lambda_n)_n$, the results which follow extend those given above.[7]

Consider again (7.43), where now $b \in \Lambda BV([0,1])$ is given, and the hypothetical solution f is searched in the same space $\Lambda BV([0,1])$. As before, we write (7.43) as equivalent operator equation (7.45), where S_h is the superposition operator (6.1) and K is the integral operator (7.44). We impose the following conditions on the functions h and k in (7.43).

Suppose that the superposition operator (6.1) induced by h maps $\Lambda BV([0,1])$ into itself and is bounded. Concerning the kernel function k in (7.43), we assume as before that $k(s,\cdot) \in L_1([0,1])$ for $0 \le s \le 1$, and that the function $v_\Lambda : [0,1] \to \mathbb{R}$ defined by

$$v_\Lambda(t) := \mathrm{Var}_\Lambda(k(\cdot,t))$$

belongs to $L_1([0,1])$. To simplify the notation, we use the shortcut

$$\kappa_\Lambda := \|k(0,\cdot)\|_{L_1} + \|v_\Lambda\|_{L_1} = \int_0^1 \left[|k(0,t)| + \mathrm{Var}_\Lambda(k(\cdot,t))\right] dt\,.$$

The following two lemmas are parallel to Lemma 7.6 and Lemma 7.7.

Lemma 7.10. *Under the above hypotheses, let $r > 0$ be so large that*

$$\kappa_\Lambda \tilde{k}(r) < r\,, \tag{7.57}$$

where $\tilde{k}(r)$ is given by (7.47). Then the operator $f \mapsto KS_h f + b$ maps, for any $b \in \Lambda BV$ satisfying

$$\|b\|_{\Lambda BV} \le r - \kappa_\Lambda \tilde{k}(r)\,, \tag{7.58}$$

the closed ball $B_r(\Lambda BV) = \{f \in \Lambda BV : \|f\|_{\Lambda BV} \le r\}$ into itself.

Proof. Suppose that r satisfies (7.57), and b satisfies (7.58). Given $f \in B_r(\Lambda BV)$ and an infinite collection $S_\infty = \{[a_n,b_n] : n \in \mathbb{N}\} \in \Sigma_\infty([0,1])$, we get

$$\begin{aligned}\sum_{k=1}^\infty \lambda_k |KS_h f(b_k) - KS_h f(a_k)| &= \sum_{k=1}^\infty \lambda_k \left| \int_0^1 [k(b_k,t) - k(a_k,t)]\, h(t,f(t))\, dt \right| \\ &\le \sup_{0\le t\le 1} |h(t,f(t))| \int_0^1 \sum_{k=1}^\infty \lambda_k |k(b_k,t) - k(a_k,t)|\, dt \\ &\le \tilde{k}(r) \int_0^1 v_\Lambda(t)\, dt\,,\end{aligned}$$

7 In fact, as we have shown in Proposition 2.32, the imbedding $WBV_p \hookrightarrow \Lambda_q$ holds for $p'q > 1$.

and so, passing to the supremum over all collections $S_\infty \in \Sigma_\infty([0,1])$,

$$\mathrm{Var}_\Lambda(KS_h f; [0,1]) \le \tilde{k}(r)\|v_\Lambda\|_{L_1}.$$

Consequently, from (7.58), we conclude that

$$\begin{aligned}\|KS_h f + b\|_{\Lambda BV} &\le |KS_h f(0)| + \mathrm{Var}_\Lambda(KS_h f; [0,1]) + \|b\|_{\Lambda BV}\\ &\le \int_0^1 |k(0,t)h(t,f(t))|\,dt + \tilde{k}(r)\|v_\Lambda\|_{L_1} + \|b\|_{\Lambda BV}\\ &\le \tilde{k}(r)\|k(0,\cdot)\|_{L_1} + \tilde{k}(r)\|v_\Lambda\|_{L_1} + \|b\|_{\Lambda BV} \le \tilde{k}(r)\kappa_\Lambda + \|b\|_{\Lambda BV} \le r,\end{aligned}$$

which proves the assertion. □

Lemma 7.11. *Under the above hypotheses, let $r > 0$ be so small that*

$$\kappa_\Lambda \max\{c(\Lambda BV, B), k(r)\} < 1, \tag{7.59}$$

where $c(\Lambda BV, B)$ is the imbedding constant (0.36) *of the imbedding $\Lambda BV([0,1]) \hookrightarrow B([0,1])$, and $k(r)$ is given by* (6.10). *Then the operator $f \mapsto KS_h f + b$ is, for any $b \in \Lambda BV$, a contraction on $B_r(\Lambda BV)$ with respect to the norm* (2.30).

Proof. Suppose that r satisfies (7.59), and let $f, g \in B_r(\Lambda BV)$ be fixed. We claim that

$$\|KS_h f - KS_h g\|_{\Lambda BV} \le \kappa_\Lambda \max\{c(\Lambda BV, B), k(r)\}\|f - g\|_{\Lambda BV},$$

which together with (7.59) proves the assertion. First of all, we have

$$\begin{aligned}|KS_h f(0) - KS_h g(0)| &= \left|\int_0^1 k(0,t)[h(t,f(t)) - h(t,g(t))]\,dt\right|\\ &\le k(r)\int_0^1 |k(0,t)|\,|f(t) - g(t)|\,dt \le k(r)\|k(0,\cdot)\|_{L_1}\|f - g\|_\infty\\ &\le c(\Lambda BV, B)k(r)\|k(0,\cdot)\|_{L_1}\|f - g\|_{\Lambda BV},\end{aligned} \tag{7.60}$$

where we have used the fact that $\Lambda BV([0,1])$ is continuously imbedded[8] into the space of all bounded functions on $[0,1]$ with the supremum norm (0.39).

On the other hand, we now show that also

$$\mathrm{Var}_\Lambda(KS_h f - KS_h g; [0,1]) \le c(\Lambda BV, B)\|v_\Lambda\|_{L_1} k(r)\|f - g\|_{\Lambda BV}.$$

8 Recall that $c(\Lambda BV, B) \le \max\{1/\lambda_1, 1\}$, see Exercise 2.17.

For any collection $S_\infty = \{[a_n, b_n] : n \in \mathbb{N}\} \in \Sigma_\infty([0,1])$, we have

$$\begin{aligned}
&\sum_{k=1}^{\infty} \lambda_k |KS_h f(b_k) - KS_h f(a_k) - KS_h g(b_k) + KS_h g(a_k)| \\
&\quad = \sum_{k=1}^{\infty} \lambda_k \left| \int_0^1 [k(b_k, t) - k(a_k, t)][S_h f(t) - S_h g(t)]\, dt \right| \\
&\quad \le \|S_h f - S_h g\|_\infty \sum_{k=1}^{\infty} \lambda_k \int_0^1 |k(b_k, t) - k(a_k, t)|\, dt \\
&\quad \le c(\Lambda BV, B) \|S_h f - S_h g\|_{\Lambda BV} \int_0^1 \sum_{k=1}^{\infty} \lambda_k |k(b_k, t) - k(a_k, t)|\, dt \\
&\quad \le c(\Lambda BV, B) k(r) \|f - g\|_{\Lambda BV} \int_0^1 v_\Lambda(t)\, dt = c(\Lambda BV, B) k(r) \|v_\Lambda\|_{L_1} \|f - g\|_{\Lambda BV} .
\end{aligned}$$

Passing again to the supremum with respect to all infinite collections $S_\infty \in \Sigma_\infty([0,1])$, we obtain

$$\mathrm{Var}_\Lambda(KS_h f - KS_h g; [0,1]) \le c(\Lambda BV, B) k(r) \|v_\Lambda\|_{L_1} \|f - g\|_{\Lambda BV} ,$$

and combining this with (7.60) yields

$$\begin{aligned}
\|KS_h f - KS_h g\|_{\Lambda BV} &= |KS_h f(0) - KS_h g(0)| + \mathrm{Var}_\Lambda(KS_h f - KS_h g; [0,1]) \\
&\le k(r) \|k(0, \cdot)\|_{L_1} \|f - g\|_{\Lambda BV} + c(\Lambda BV, B) k(r) \|v_\Lambda\|_{L_1} \|f - g\|_{\Lambda BV} \\
&\le \kappa_\Lambda \max\{c(\Lambda BV, B), k(r)\} \|f - g\|_{\Lambda BV} ,
\end{aligned}$$

and establishes the result. □

Combining Lemma 7.10 and Lemma 7.11, and observing that the fixed points of the operator $f \mapsto KS_h f + b$ coincide with the solutions of (7.45), Banach's contraction mapping theorem now implies the following

Theorem 7.12. *Suppose that there exists $r > 0$ such that both estimates* (7.57) *and* (7.59) *hold. Then the nonlinear integral equation* (7.43) *has, for each $b \in \Lambda BV$ satisfying* (7.58), *a unique solution $f \in \Lambda BV$.*

Now, we apply our abstract results to a class of singular integral equations. Consider the nonlinear weakly singular Abel–Volterra equation

$$f(s) - \int_0^s \frac{k(s,t) h(f(t))}{|s-t|^\nu}\, dt = b(s) \quad (0 \le s \le 1), \tag{7.61}$$

where $k : [0,1] \times [0,1] \to \mathbb{R}$ is continuous and $0 < \nu < 1$. We rewrite (7.61) again as operator equation

$$f - K_\nu C_h f = b \quad (f \in X), \tag{7.62}$$

where K_ν is the weakly singular linear integral operator defined by

$$K_\nu g(s) = \int_0^s \frac{k(s,t)g(t)}{|s-t|^\nu}\,dt \quad (0 \le s \le 1)\,, \tag{7.63}$$

and C_h is the composition operator (5.1). A very suitable space for treating such an operator equation is the Hölder space $Lip^o_\alpha([0,1])$ of all functions $f \in Lip_\alpha([0,1])$ satisfying $f(0) = 0$. The so-called second Hardy–Littlewood theorem (see, e.g. [19]) states that the operator (7.63) maps $Lip^o_\alpha([0,1]) \cap L_p([0,1])$ into $Lip^o_\alpha[0,1]$ and is bounded if

$$\frac{1}{1-\nu} < p \le \infty, \quad 0 < \alpha \le 1 - \nu - \frac{1}{p}\,.$$

In order to simplify our computations, let us choose $p = \infty$, and so

$$0 < \alpha \le 1-\nu, \quad Lip^o_\alpha([0,1]) \cap L_\infty([0,1]) = Lip^o_\alpha([0,1])\,.$$

By Theorem 5.51, the local Lipschitz condition (5.14) is necessary and sufficient for the operator (5.1) to satisfy (5.76) in Hölder spaces. Using Banach's fixed point principle in these spaces, we may deduce existence and uniqueness results for equation (7.62), and hence for equation (7.61), in a closed ball of appropriate radius if $\|b\|_{Lip_\alpha}$ is sufficiently small.

To explain this in more detail, we now prove two lemmas which are parallel to Lemmas 7.6 and 7.7. Throughout the remaining part of this section, we suppose for simplicity that $h(0) = 0$ and denote by $\|K_\nu\|_\alpha$ the norm of the operator (7.63) in the space Lip_α. Estimates or even explicit formulas for $\|K_\nu\|_\alpha$ in terms of the kernel k in (7.61) may be found in [19] or [134].

Lemma 7.13. *Under the hypotheses* (5.13) *and* $h(0) = 0$, *the operator* (5.1) *maps the space* $Lip^o_\alpha([0,1])$ *into itself and is bounded. Moreover, if* $r > 0$ *is so large that*

$$\|K_\nu\|_\alpha \mu_r(h, Lip_\alpha) < r\,, \tag{7.64}$$

where $\mu_r(h, Lip_\alpha)$ *is given by Exercise* 5.11, *then the operator* $f \mapsto K_\nu C_h f + b$ *maps, for any* $b \in Lip_\alpha([0,1])$ *satisfying*

$$\|b\|_{Lip_\alpha} \le r - \|K_\nu\|_\alpha \mu_r(h, Lip_\alpha)\,, \tag{7.65}$$

the closed ball $B_r(Lip_\alpha) = \{f \in Lip_\alpha : \|f\|_{Lip_\alpha} \le r\}$ *into itself.*

Proof. The sufficiency of (5.13) for the boundedness of the operator (5.1) in $Lip_\alpha([0,1])$ has already been shown in Theorem 5.24. Moreover, our assumption $h(0) = 0$ guarantees that the operator (5.1) maps $Lip^o_\alpha([0,1])$ into itself.

Fix $f \in Lip^o_\alpha([0,1])$ with $\|f\|_{Lip_\alpha} \le r$, where r satisfies (7.64), and suppose that $b \in Lip^o_\alpha([0,1])$ satisfies (7.65). Then

$$\|K_\nu C_h f + b\| \le \|K_\nu\|_\alpha \mu_r(h, Lip_\alpha) + \|b\|_{Lip_\alpha} \le r\,,$$

by (7.65), which proves the assertion. □

Lemma 7.14. *Suppose that the derivative h' of h exists and satisfies* (5.14). *Moreover, let $r > 0$ be so small that*

$$k_1(r) < \frac{1}{4r\|K_\nu\|_\alpha}, \quad \tilde{k}_1(r) < \frac{1}{\|K_\nu\|_\alpha}, \tag{7.66}$$

where $k_1(r)$ is given by (5.14) *and $\tilde{k}_1(r)$ by* (5.16). *Then the operator $f \mapsto KC_h f + b$ is, for any $b \in Lip_\alpha^o([0,1])$, a contraction on $B_r(Lip_\alpha^o)$ with respect to the norm* (0.71).

Proof. As we have already proved in Theorem 5.51, our hypotheses on h imply that the operator C_h satisfies the Lipschitz condition (5.76) with Lipschitz constant

$$K(r) = \max\{4rk_1(r), \tilde{k}_1(r)\}.$$

Consequently, the assertion follows from our assumption (7.66). □

Combining Lemma 7.13 with Lemma 7.14, and observing that the fixed points of the operator $f \mapsto KC_h f + b$ coincide with the solutions of (7.61), Banach's contraction mapping theorem now implies the following

Theorem 7.15. *Suppose that there exists $r > 0$ such that both estimates* (7.64) *and* (7.66) *hold. Then the nonlinear integral equation* (7.61) *has, for each $b \in Lip_\alpha^o([0,1])$ satisfying* (7.65), *a unique solution $f \in Lip_\alpha^o([0,1])$.*

Analyzing the proof of Theorem 7.15 shows that necessarily $k(r) \to \infty$ as $r \to \infty$ in (7.65), apart from the trivial case of an affine function h, in correspondence with Corollary 5.46. Therefore, we cannot expect *global* existence and uniqueness of solutions of (7.61) in the whole space $Lip_\alpha^o([0,1])$. This again justifies the necessity of replacing the global condition (5.67) by the local condition (5.76).

7.4 Comments on Chapter 7

According to Lakatos [177], functions of bounded variations were discovered by Jordan [153] through a "critical re-examination" of Dirichlet's (flawed) proof that arbitrary functions can be represented by Fourier series. However, as pointed out by Hawkins [143], the key observation that Dirichlet's proof was valid for differences of increasing functions had already been made by DuBois–Reymond [106].

The various convergence tests cited in Section 7.1 may be found in textbooks on harmonic analysis. The Pál–Bohr test asserts that for every $g \in C([0,2\pi])$ of period 2π, there is some homeomorphism τ of $[0,2\pi]$ with itself such that the Fourier series of $g \circ \tau$ converges uniformly. In the paper [131], the authors use the Salem criterion to give a condition on $g \in C([0,2\pi])$, both necessary and sufficient, under which the Fourier series of $g \circ \tau$ converges uniformly for *every* homeomorphism τ of $[0,2\pi]$ with itself. It is clear that this condition should be weaker than bounded p-variation in Wiener's sense because functions $g \in C([0,2\pi]) \cap WBV_p([0,2\pi])$ have uniformly convergent

Fourier series for $p > 1$, and this class of functions is preserved by a homeomorphic change of variables, as Proposition 1.12 shows.[9]

In the paper [78], the authors consider the class $GW([0, 2\pi])$ (respectively the class $UGW([0, 2\pi])$) of all regular functions which have a convergent (respectively uniformly convergent) Fourier series after any change of variable. It is known that

$$HBV([0, 2\pi]) \subseteq GW([0, 2\pi])$$

and

$$HBV([0, 2\pi]) \cap C([0, 2\pi]) \subseteq UGW([0, 2\pi]).$$

In [78], the authors also show that in the terminology of (5.3), we have the equalities

$$COP(GW) = COP(UGW) = Lip_{loc}(\mathbb{R}),$$

which means that $C_h(GW) \subseteq GW$ and $C_h(UGW) \subseteq UGW$ if and only if h is locally Lipschitz; this is perfectly analogous to Theorem 5.10.

As the fundamental Theorems 7.4 and 7.5 show, the Waterman spaces $\Lambda_q BV$ (in particular, the space $HBV = \Lambda_1 BV$) are most suitable for studying the convergence problems of Fourier series. For instance, the Fourier series of functions in HBV converge everywhere pointwise, and converge uniformly on closed intervals of continuity. This result is best possible in the sense that each larger space (within the class of Waterman spaces, see Theorem 7.5 or [314]) contains a continuous function whose Fourier series diverges at some point.

The functions in the Wiener–Young space WBV_ϕ with complementary Young function satisfying (7.15) are contained in HBV, as are the functions with logarithmically integrable Banach indicatrix, see (7.16). However, one may show [314] that all these criteria are contained in the Lebesgue test. Condition (7.15) is also sharp in the sense that if

$$\sum_{n=1}^{\infty} \phi^* \left(\frac{1}{n}\right) = \infty,$$

then there is a continuous function in $WBV_\phi([0, 2\pi])$ whose Fourier series diverges at some point. This was shown by Baernstein [38] and, independently, by Oskolkov [239], and answers a question raised in [132].

Needless to say, the results for Waterman spaces, which are particularly useful and natural, have been extended to more general classes like the space $\Lambda BV_\phi([0, 2\pi])$ of functions of bounded (ϕ, Λ)-variation introduced in Definition 2.84. Recall that given a Young function ϕ and a Waterman sequence $\Lambda = (\lambda_n)_n$, we defined the

---

9 In Proposition 1.12, we have proved this only for $p = 1$, but the proof carries over without changes to the case $p > 1$, see [225].

(ϕ, Λ)-variation of f on $[a, b]$ by

$$\mathrm{Var}_{\phi,\Lambda}(f; [a, b]) = \sup \sum_{n=1}^{\infty} \lambda_n \phi(|f(b_n) - f(a_n)|), \tag{7.67}$$

where the supremum is taken over all collections $\{[a_n, b_n] : n \in \mathbb{N}\} \in \Sigma_\infty([a, b])$. In particular, $f \in \Lambda BV_p([a, b])$ for $1 \le p < \infty$ if

$$\mathrm{Var}_{p,\Lambda}(f; [a, b]) = \sup \sum_{n=1}^{\infty} \lambda_n |f(b_n) - f(a_n)|^p < \infty. \tag{7.68}$$

The following was proved by Shiba [288]:

Theorem 7.16. *Let $1 < q < \infty$, $q' = q/(q-1)$, and $1 \le p < 2q$. Let $f \in \Lambda BV_p([0, 2\pi])$ be 2π-periodic. Suppose that*

$$\sum_{n=1}^{\infty} \frac{\omega_{p+(2-p)q'}(f; 2\pi/n)^{1-p/2q}}{n^{1-1/2q'} \lambda_n^{1/2q}} < \infty, \tag{7.69}$$

where $\omega_p(f; \cdot)$ denotes the p-modulus of continuity (0.98) *of f over $[a, b] = [0, 2\pi]$. Then the Fourier series* (7.1) *of f converges absolutely.*

A somewhat stronger result that was proved in [287] reads as follows:

Theorem 7.17. *Let $1 \le q < \infty$, $q' = q/(q-1)$, and $1 \le p < 2q$. Let $f \in \Lambda BV_p([0, 2\pi])$ be 2π-periodic. Suppose that*

$$\sum_{n=1}^{\infty} \frac{\omega_{p+(2-p)q'}(f; 2\pi/n)^{1-p/2q}}{\sqrt{n}\lambda[1, n]^{1/2q}} < \infty, \tag{7.70}$$

where we have used the shortcut (2.42) *for $\lambda[1, n]$. Then the Fourier series* (7.1) *of f converges absolutely.*

Theorem 7.17 is stronger than Theorem 7.16 because $\lambda[1, n] \ge n\lambda_n$, and hence

$$\sqrt{n}\lambda[1, n]^{1/2q} \ge n^{(q+1)/2q} \lambda_n^{1/2q} = n^{1-1/2q'} \lambda_n^{1/2q}.$$

Observe that in case $q = 1$, i.e. $q' = \infty$, the term $\omega_{p+(2-p)q'}(f; 2\pi/n)$ in (7.70) has to be interpreted as the ordinary modulus of continuity $\omega_\infty(f; 2\pi/n)$ defined in (0.97). In particular, for $p = q = 1$, condition (7.70) becomes

$$\sum_{n=1}^{\infty} \frac{\omega_\infty(f; 2\pi/n)^{1/2}}{\sqrt{n\lambda[1, n]}} < \infty, \tag{7.71}$$

which is the condition used by Wang [313]; compare this with (7.22).

A parallel result for the general space $\Lambda BV_\phi([0, 2\pi])$ of functions of bounded (ϕ, Λ)-variation has been obtained by Schramm and Waterman [287] under the hypothesis that the Young function ϕ satisfies the Δ_2-condition (0.21).

Theorem 7.18. *Let $\Lambda = (\lambda_n)_n$ be a Waterman sequence, $\phi \in \Delta_2$ a Young function, $1 \leq q < \infty$, $q' = q/(q-1)$, and $1 \leq p < 2q$. Let $f \in \Lambda BV_\phi([0, 2\pi])$ be 2π-periodic. Suppose that*

$$\sum_{n=1}^{\infty} \frac{1}{\sqrt{n}} \left[\phi^{-1} \left(\frac{\omega_{p+(2-p)q'}(f; 2\pi/n)^{2q-p}}{\lambda[1,n]} \right) \right]^{1/2q} < \infty, \tag{7.72}$$

where we have again used the shortcut (2.42). Then the Fourier series (7.1) of f converges absolutely.

Putting $\phi(t) = t^p$, i.e. $\Lambda BV_\phi = \Lambda BV_p$ in Theorem 7.18, condition (7.72) becomes

$$\sum_{n=1}^{\infty} \frac{\omega_{p+(2-p)q'}(f; 2\pi/n)^{1/p-1/2q}}{\sqrt{n}\lambda[1,n]^{1/2pq}} < \infty, \tag{7.73}$$

which in case $p > 1$, is different from (7.70) since

$$1 - \frac{p}{2q} = p\left(\frac{1}{p} - \frac{1}{2q}\right) > \frac{1}{p} - \frac{1}{2q}.$$

Therefore, unfortunately Theorem 7.18 does not contain Theorem 7.17 as a special case.

Apart from the moduli of continuity, the modulus of variation (2.52) also turns out to be quite useful in the study of the Fourier series and has been considered by Chanturiya in a series of papers [79–84]. For example, the following result is mentioned in [79]; for a proof, see [303]:

Theorem 7.19. *Let $f \in C([0, 2\pi])$ be 2π-periodic, and suppose that*

$$\sum_{n=1}^{\infty} \frac{\nu(f)_n}{n^2} < \infty, \tag{7.74}$$

where $\nu(f)_n$ is defined by (2.52). Then the Fourier series (7.1) of f converges uniformly.

We point out that Theorem 7.19 implies that if $f \in C([0, 2\pi]) \cap WBV_\phi([0, 2\pi])$ is 2π-periodic and

$$\sum_{n=1}^{\infty} \frac{1}{n}\phi^{-1}\left(\frac{1}{n}\right) < \infty, \tag{7.75}$$

then the Fourier series (7.1) of f converges uniformly. To see this, note that $WBV_\phi([0, 2\pi]) \subseteq V_\nu([0, 2\pi])$ for

$$\nu_n = n\phi^{-1}\left(\frac{1}{n}\right),$$

as Proposition 2.36 (or Table 2.2) shows. Therefore, (7.75) implies (7.74), which establishes the claim. Interestingly, the condition (7.15) (involving the conjugate of ϕ) reduces in the special case $\phi(t) = t^p$ to the condition

$$\sum_{n=1}^{\infty} \frac{1}{n^{p/(p-1)}} < \infty,$$

which is fulfilled for $1 < p < \infty$, while the condition (7.75) (involving the inverse of ϕ) reduces to the condition

$$\sum_{n=1}^{\infty} \frac{1}{n^{(p+1)/p}} < \infty ,$$

which is fulfilled for $1 \le p < \infty$. So, (7.75) also covers the limit case $p = 1$, while (7.15) does not.

If we suppose that the Young function ϕ satisfies condition ∞_1 (Definition 2.11), which excludes the case $p = 1$ for $\phi(t) = t^p$, then one may show that (7.15) and (7.75) are *equivalent*. Moreover, putting

$$\xi_n := (\phi^*)'\left(\frac{1}{n}\right) \quad (n = 1, 2, 3, \ldots) ,$$

Oskolkov [239] has shown that (7.15) is also equivalent to each of the three conditions, namely,

$$\int_0^1 \log\left(\frac{1}{\phi'(t)}\right) dt < \infty , \tag{7.76}$$

$$\lim_{n\to\infty} \xi_n = 0, \quad \sum_{n=1}^{\infty} (\xi_n - \xi_{n+1}) \log n < \infty \tag{7.77}$$

and

$$\sum_{n=1}^{\infty} \frac{\xi_n}{n} < \infty , \tag{7.78}$$

which for $\phi(t) = t^p$, all hold in case $1 < p < \infty$. In fact, the proof of the equivalence of (7.15) and (7.78) relies on the estimate

$$\frac{\xi_{2n}}{2n} \le \phi^*\left(\frac{1}{n}\right) \le \frac{\xi_n}{n}$$

which follows from the convexity of ϕ. Being the derivative (a.e.) of a convex function, ϕ' is monotonically increasing, and so the set

$$M_n := \left\{t : 0 \le t < \infty, \frac{1}{n+1} \le \phi'(t) < \frac{1}{n}\right\}$$

satisfies $(\xi_{n+1}, \xi_n) \subseteq M_n \subseteq [\xi_{n+1}, \xi_n]$, and hence has Lebesgue measure $\lambda(M_n) = \xi_n - \xi_{n+1}$. Consequently,

$$(\xi_n - \xi_{n+1}) \log n \le \int_{M_n} \log\left(\frac{1}{\phi(t)}\right) dt \le (\xi_n - \xi_{n+1}) \log(n+1) ,$$

which establishes the equivalence of (7.76) and (7.77). Finally, the estimate

$$\frac{t}{2}\phi'\left(\frac{t}{2}\right) \le \phi(t) \le t\phi'(t) ,$$

together with the convexity of ϕ and the condition $\phi(0) = 0$, implies the equivalence of (7.15) and (7.76).

In Proposition 2.85, we have shown that $\Lambda BV_\phi([0, 2\pi]) \subseteq V_\nu([0, 2\pi])$ for

$$\nu_n = n\phi^{-1}\left(\frac{1}{\lambda[1, n]}\right).$$

So, as a consequence of Theorem 7.19, we obtain the following

Corollary 7.20. *Let $f \in C([0, 2\pi])$ be 2π-periodic, and suppose that*

$$\sum_{n=1}^{\infty} \frac{1}{n}\phi^{-1}\left(\frac{1}{\lambda[1, n]}\right) < \infty, \tag{7.79}$$

where ϕ is some Young function, $\Lambda = (\lambda_n)_n$ is a Waterman sequence, and $\lambda[1, n]$ is defined by (2.42). *Then the Fourier series* (7.1) *of f converges uniformly.*

In connection with the convergence criteria we just discussed, the following interesting result involving the Banach indicatrix (0.106) of a continuous function is proved in [239, Theorem 3]:

Proposition 7.21. *Let $\tau : [0, \infty) \to [0, \infty)$ be an increasing function such that $\tau(0) = 0$, $\tau(t) > 0$ for $t > 0$, and*

$$\int_0^1 \log\left(\frac{1}{\tau(t)}\right) dt = \infty. \tag{7.80}$$

Then there exists a 2π-periodic function $f \in C([0, 2\pi])$ whose Banach indicatrix (0.106) *satisfies*

$$I_f(x) \le \frac{1}{\tau(x)} \quad (x > 0)$$

and whose Fourier series (7.1) *diverges at some point.*

Proposition 7.21 shows that the Goffman–Waterman criterion (7.15) is, in a certain sense, *sharp*. To see this, observe that under the hypotheses of Proposition 7.21, we have $f \in WBV_\phi([0, 2\pi])$, where[10]

$$\phi(t) := \int_0^t \tau(s)\, ds \quad (t \ge 0).$$

Condition (7.80) then means that (7.76) *fails*, and thus also (7.15).

Let us mention another result of Chanturiya [79] which is formulated in terms of a combination of both the modulus of continuity (0.97) and the modulus of variation (2.52).

10 The convexity of ϕ follows from the fact that τ is increasing.

Theorem 7.22. *Suppose that there exists some $\alpha \in (0, 1]$ such that either*

$$\nu(f)_n = O\left(\frac{n}{(\log n)(\log\log n)^\alpha}\right)$$

and

$$\omega_\infty(f;\delta) = O\left(\frac{1}{\log(1/\delta)}\right)\exp(o(\log\log(1/\delta))^\alpha),$$

or

$$\nu(f)_n = O\left(\frac{n}{(\log n)(\log\log n)(\log\log\log n)^\alpha}\right)$$

and

$$\omega_\infty(f;\delta) = O\left(\log(1/\delta)^{-\exp(o(\log\log\log(1/\delta))^\alpha)}\right).$$

Then the Fourier series (7.1) *of f converges uniformly.*

Apart from convergence criteria for the Fourier series (7.1), summability results for the Fourier coefficients (7.2) and (7.3) are also of interest. For example, in [317], the author shows that for $f \in \Lambda BV$, the Fourier coefficients of f satisfy

$$\alpha_n(f), \beta_n(f) = O\left(\frac{1}{n\lambda_n}\right) \quad (n \to \infty).$$

Finally, let us briefly discuss two results on the rate of convergence of the partial Fourier sums (7.4), the first in the Wiener–Young space $WBV_\phi([0, 2\pi])$, the second in the special Waterman space $\Lambda_q BV([0, 2\pi])$.

Proposition 7.23. *Let ϕ be a Young function and $f \in C([0, 2\pi]) \cap WBV_\phi([0, 2\pi])$. Then*

$$\|f - s_n(\cdot; f)\|_{WBV_\phi} \leq c \int_0^{\omega_\infty(f;\pi/n)} \log \frac{\mathrm{Var}_\phi^W(f;[0, 2\pi])}{\phi(t)}\,dt,$$

where $\omega(f;\delta)$ denotes the modulus of continuity (0.97) *of f, and $\mathrm{Var}_\phi^W(f;[a, b])$ the Wiener–Young variation* (2.2) *of f.*

Proposition 7.23 has been proved by Oskolkov [239]; in the special case $\phi(t) = t$ (i.e. $WBV_\phi = BV$), the result goes back to Stechkin [298]. Convergence results involving the average sums (7.7) may be found in [5, 23].

The following proposition provides a *pointwise* estimate for the rate of convergence of (7.4) to $f(x)$.

Proposition 7.24. *Let $0 < q < 1$, and let $f \in C([0, 2\pi]) \cap \Lambda_q BV([0, 2\pi])$. Then*

$$|f(x) - s_n(x; f)| \leq \frac{(2 - q)(1 + 2/\pi)}{(n + 1)^{1-q}} \sum_{k=1}^{n} \frac{\mathrm{Var}_{\Lambda_q}(\varphi_x;[0, \pi/k])}{k^q},$$

where φ_x denotes the auxiliary function (7.10), *and $\mathrm{Var}_{\Lambda_q}(f;[a, b])$ the Waterman variation of f given in Definition 2.29.*

Proposition 7.24 is due to Bojanić and Waterman [51], in the special case $q = 0$ (i.e. $\Lambda_q BV = BV$) this is an earlier result of Bojanić [50]. For a more general result (classes of type ΛBV which can be closer to HBV), see Waterman [318].

A completely different approach to the convergence of Fourier series based on the Korenblum variation can be found in [164]. As Exercise 2.40 shows, a function $f \in \kappa BV([a,b]) \cap C([a,b])$ is not necessarily of vanishing κ-variation. In [164, Theorem 3], it is shown that if $f \in \kappa BV([0,2\pi])$, where $\kappa(t) = 1/(1 - \frac{1}{2}\log t)$, and the function φ_x defined in (7.10) has vanishing κ-variation at t_0, then $s_n(t_0; f) \to f(t_0)$ as $n \to \infty$, with $s_n(\cdot; f)$ defined by (7.4).

To conclude, let us make some comments on Section 7.3. Solutions of nonlinear integral equations in spaces of type WBV_p or ΛBV have been studied by Bugajewska, Bugajewski, and others [67–72]. One could ask why we did not give more explicit conditions on the functions h which guarantee that the superposition operator S_h in (7.45) maps the space BV_p or the space ΛBV into itself. The reason is simply that conditions which are both necessary and sufficient are not known, and conditions which are only necessary (like Theorem 6.10) or only sufficient (like Theorem 6.11) are rather technical. The "natural" requirement that $h(x,\cdot)$ is locally Lipschitz, uniformly in x, and $h(\cdot,u)$ belongs to BV_p, respectively ΛBV, uniformly in u, is not sufficient, as the surprising Example 6.8 shows.

In the last part of Section 7.3, we have studied the singular nonlinear equation (7.61) in Hölder spaces. Such equations are considered in so-called generalized Hölder spaces by Babaev [37] and Mukhtarov [234], see also Chapter 5 of the book [138]. Existence and uniqueness results for solutions of (7.61) are closely related to existence and uniqueness results for solutions of initial or boundary value problems for ordinary differential equations. In view of similar applications to partial differential equations, it is useful to formulate analogous results for functions which are defined on a domain $\Omega \subseteq \mathbb{R}^n$ (not necessarily bounded). To this end, one has to work in the space $Lip_\alpha(\Omega)$ of all bounded continuous functions $f : \Omega \to \mathbb{R}$ with norm

$$\|f\|_{Lip_\alpha} := \|f\|_\infty + \sup_{x \neq y} \frac{|f(x) - f(y)|}{|x-y|^\alpha}, \tag{7.81}$$

where $\|f\|_\infty$ now denotes the supremum of $|f(x)|$ on Ω. Here, we have the following analogue to Theorem 5.51:

Theorem 7.25. *Suppose that the function $h : \mathbb{R} \to \mathbb{R}$ is differentiable on $\mathbb{R}$ and satisfies condition* (5.14). *Then the composition operator $C_h f = h \circ f$ maps the space $X = Lip_\alpha(\Omega)$ into itself and satisfies the local Lipschitz condition* (5.76) *in the norm* (7.81).

The proof consists of an evident modification of the arguments used in the proof of Theorem 5.51 in Chapter 5.

References

[1] A. Acosta, W. Aziz, J. Matkowski and N. Merentes: Uniformly continuous composition operator in the space of φ-variation functions in the sense of Riesz, *Fasc. Math.* **43** (2010), 5–11.

[2] C. R. Adams: The space of functions of bounded variation and certain general spaces, *Trans. Amer. Math. Soc.* **40** (1936), 421–438.

[3] C. R. Adams and J. A. Clarkson: On convergence in variation, *Bull. Amer. Math. Soc.* **40** (1934), 413–417.

[4] C. R. Adams and J. A. Clarkson: Properties of functions $f(x, y)$ of bounded variation, *Trans. Amer. Math. Soc.* **36** (1934), 711–730.

[5] T. I. Akhobadze: *Convergence of Cesaro means of negative orders of functions of bounded generalized second variation* (Russian), Matem. Zametki **20** (1976), 631–644; Engl. transl.: Math. Notes **20** (1976), 914–922.

[6] R. R. Akhmerov, M. I. Kamenskij, A. S. Potapov, A. E. Rodkina and B. N. Sadovskij: *Measures of Noncompactness and Condensing Operators* (Russian), Nauka, Novosibirsk 1986; Engl. transl.: Birkhäuser, Basel 1992.

[7] L. Ambrosio, N. Fusco and D. Pallara: *Functions of Bounded Variation and Free Discontinuity Problems*, Oxford Univ. Press, New York 2000.

[8] L. Anzola, N. Merentes and J. L. Sánchez: *Uniformly continuous composition operators in spaces of λ-bounded variation functions in the sense of Waterman*, J. Math. Control Sci. Appl. (to appear).

[9] J. Appell: On the solvability of nonlinear noncompact problems in function spaces with applications to integral and differential equations, *Boll. Unione Mat. Ital.* **6** (1982), 1161–1177.

[10] J. Appell: Measures of noncompactness, condensing operators and fixed points: An application-oriented survey, *Fixed Point Theory (Cluj)* **6** (2005), 157–229.

[11] J. Appell: Some counterexamples for your calculus course, *Analysis* **31** (2010), 1001–1012.

[12] J. Appell: *Analysis in Beispielen und Gegenbeispielen*, Springer, Heidelberg 2010.

[13] J. Appell, J. Banaś and N. Merentes: Some quantities related to monotonicity and bounded variation of functions, *J. Math. Anal. Appl.* **367** (2010), 476–485.

[14] J. Appell, N. Guanda, N. Merentes and J. L. Sánchez: Boundedness and continuity properties of nonlinear composition operators: A survey, *Comm. Applied Anal.* **15** (2011), 153–182.

[15] J. Appell, N. Guanda and M. Väth: Function spaces with the Matkowski property and degeneracy phenomena for nonlinear composition operators, *Fixed Point Theory (Cluj)* **12**, Number 2 (2011), 265–284.

[16] J. Appell, Z. Jesús and O. Mejía: Some remarks on nonlinear composition operators in spaces of differentiable functions, *Boll. Unione Mat. Ital.* **9-4** (2011), 321–326.

[17] J. Appell and N. Merentes: Composing functions of bounded Korenblum variation, *Dynam. Systems Appl.* **22** (2013), 197–206.

[18] J. Appell, N. Merentes and J. L. Sánchez: Locally Lipschitz composition operators in spaces of functions of bounded variation, *Annali Mat. Pura Appl.* **190** Number 1 (2011), 33–43.

[19] J. Appell and M. Väth: *Elemente der Funktionalanalysis*, Vieweg, Wiesbaden 2005.

[20] J. Appell and P. P. Zabrejko: *Nonlinear Superposition Operators*, Cambridge Univ. Press, Cambridge 1990; Paperback Ed.: Cambridge 2008.

[21] J. Appell and P. P. Zabrejko: Remarks on the composition operator problem in various function spaces, *Complex Var. Elliptic Equ.* **55** Number 8 (2010), 727–737.

[22] F. Armao, D. Głazowska, S. Rivas and J. Rojas: Uniformly bounded composition operators in the Banach space of bounded (p, k)-variation functions in the sense of Riesz and Popoviciu, *Central Europ. J. Math.* **11** (2013), 357–367.

[23] V. D. Asatiani: *On the uniform convergence of Cesaro means of negative order of Fourier series* (Russian), Soobshch. Akad. Nauk Gruz. SSR **92** (1978), 549–552.

[24] M. Avdipahić: On the classes ΛBV and $V[\nu]$, *Proc. Amer. Math. Soc.* **95** (1985), 230–234.

[25] M. Avdipahić: Concepts of generalized bounded variation and the theory of Fourier series, *Intern. J. Math. Math. Sci.* **9** (1986), 223–244.

[26] J. M. Ayerbe Toledano, T. Domínguez Benavides, G. López Acedo: *Measures of Noncompactness in Metric Fixed Point Theory*, Birkhäuser, Basel 1997.

[27] W. Aziz, A. Azocar, J. Guerrero and N. Merentes: Uniformly continuous composition operators in the space of functions of ϕ-variation with weight in the sense of Riesz, *Nonlin. Anal. TMA* **74** Number 2 (2011), 573–576.

[28] W. Aziz, J. Giménez, N. Merentes and J. L. Sánchez: Uniformly continuous set-valued composition operators in the spaces of total φ-bidimensional variation in the sense of Riesz, *Opusc. Math.* **30** Number 3 (2010), 241–248.

[29] W. Aziz, J. A. Guerrero and N. Merentes: Uniformly continuous set-valued composition operators in the spaces of functions of bounded variation in the sense of Riesz, *Bull. Pol. Acad. Sci. Math.* **58** Number 1 (2010), 39–45.

[30] W. Aziz, J. A. Guerrero and N. Merentes: Nemytskii operators in the space of set-valued functions of bounded p-variation, *J. Math. Control Sci. Appl.* **4** Number 1 (2011), 85–94.

[31] W. Aziz, J. A. Guerrero and N. Merentes: On Nemytskii operators in the space of set-valued functions of bounded p-variation in the sense of Riesz with respect to weight functions, *Fasciculi Math.* **50** (2013), 85–94.

[32] W. Aziz, J. A. Guerrero, J. L. Sánchez and M. Sanoja: Lipschitzian composition operators in the space $\kappa BV[a,b]$, *J. Math. Control Sci. Appl.* **4** Number 1 (2011), 67–73.

[33] W. Aziz, H. Leiva, N. Merentes and B. Rzepka: A representation theorem for ϕ-bounded variation of functions in the sense of Riesz, *Ann. Soc. Math. Pol. Comment. Math.* **50** Number 2 (2010), 109–120.

[34] W. Aziz, H. Leiva, N. Merentes and J. L. Sánchez: Functions of two variables with bounded φ-variation in the sense of Riesz, *J. Math. Appl. (Rzeszów)* **32** (2010), 5–23.

[35] W. Aziz, N. Merentes and J. L. Sánchez: *A note on the composition of regular functions*, Zeitschr. Anal. Anw. (to appear).

[36] A. Azócar, J. A. Guerrero, J. Matkowski and N. Merentes: Uniformly continuous set-valued composition operators in the space of functions of bounded variation in the sense of Wiener, *Opusc. Math.* **30** Number 1 (2010), 53–60.

[37] A. A. Babaev: *On the structure of a certain nonlinear operator and its applications* (Russian), Uchen. Zap. Azerbajdzh. Gos. Univ. **4** (1961), 13–16.

[38] A. Baernstein: On the Fourier series of functions of bounded φ-variation, *Studia Math.* **42** (1972), 243–248.

[39] A. N. Bakhvalov, M. I. Dyachenko, K. S. Kazaryan, P. Sifuentes and P. L. Ul'yanov: *Real Analysis in Exercises* (Russian), Fizmatlit, Moscow 2005.

[40] S. Banach: Sur les lignes rectifiables et les surfaces dont l'aire est finie, *Fund. Math.* **7** (1925), 225–236.

[41] J. Banaś and K. Goebel: Measures of Noncompactness in Banach Spaces, *Lect. Notes Pure Appl. Math.* **60**, M. Dekker, New York 1980.

[42] J. Banaś and L. Olszowy: Measures of noncompactness related to monotonicity, *Comm. Math.* **41** (2001), 13–23.

[43] J. Banaś and T. Zając: A new approach to the theory of functional integral equations of fractional order, *J. Math. Anal. Appl.* **375** (2011), 375–387.

[44] N. K. Bary: Mémoire sur la répresentation finie des fonctions continues I, *Math. Ann.* **103** (1930), 185–248.

[45] N. K. Bary: Mémoire sur la répresentation finie des fonctions continues II, *Math. Ann.* **103** (1930), 598–653.

[46] M. Z. Berkolajko: *On a nonlinear operator acting in generalized Hölder spaces* (Russian), Voronezh. Gos. Univ. Sem. Funk. Anal. **12** (1969), 96–104.

[47] M. Z. Berkolajko: *On the continuity of the superposition operator acting in generalized, Hölder spaces* (Russian), Voronezh. Gos. Univ. Sbornik Trudov Aspir. Mat. Fak. **1** (1971), 16–24.

[48] M. Z. Berkolajko and Ya. B. Rutitskij: *On operators in Hölder spaces* (Russian), Doklady Akad. Nauk. SSSR **192** Number 6 (1970), 1199–1201; Engl. transl.: Soviet Math. Doklady **11** Number 3 (1970), 787–789.

[49] H. Bohr: Über einen Satz von J. Pál, *Acta Szeged* **7** (1935), 129–135.

[50] R. Bojanić: An estimate of the rate of convergence for Fourier series of functions of bounded variation, *Publ. Inst. Math. Belgrade* **26** (1979), 57–60.

[51] R. Bojanić and D. Waterman: On the rate of convergence of Fourier series of functions of generalized bounded variation, *Odjeljenje Prir. Mat. Nauka* **22** (1983), 5–11.

[52] V. A. Bondarenko and P. P. Zabrejko: *The superposition operator in Hölder functions spaces* (Russian), Doklady Akad. Nauk. SSSR **222** Number 6 (1975), 1265–1268; Engl. transl.: Soviet Math. Doklady **16** Number 3 (1975), 739–743.

[53] V. A. Bondarenko and P. P. Zabrejko: *The superposition operator in Hölder spaces* (Russian), Yaroslav. Gos. Univ. Kach. Pribl. Metody Issled. Oper. Urav. nauk. SSSR **3** (1978), 17–29.

[54] M. Boni and P. Brandi: Variazione classica e generalizzata con peso, *Atti Sem. Mat. Fis. Univ. Modena* **23** (1974), 286–307.

[55] G. Bourdaud, M. Lanza de Cristoforis and W. Sickel: Superposition operators and functions of bounded variation II, *Nonlin. Anal. TMA* **62** (2005), 483–517.

[56] G. Bourdaud, M. Lanza de Cristoforis and W. Sickel: Superposition operators and functions of bounded variation, *Rev. Mat. Iberoamer.* **22** Number 2 (2006), 455–487.

[57] M. Bracamonte, T. Ereú, J. Giménez and N. Merentes: *Uniformly continuous superposition operators in the space of functions of bounded n-dimensional Φ-variation*, Demonstr. Math. (to appear).

[58] M. Bracamonte, J. Giménez and N. Merentes: On generalized second Φ-variation of normed space valued maps, *J. Math. Control Sci. Appl.* **4** Number 1 (2011), 75–82.

[59] M. Bracamonte, J. Giménez and N. Merentes: On first and second Φ-variation in the sense of Schramm-Riesz, *Publ. Ciencias Tech.* **7** Number 1 (2013), 1–5.

[60] M. Bracamonte, J. Giménez, N. Merentes and J. L. Sánchez: On second Riesz Φ-variation of normed space valued maps, *Adv. Pure Math.* **2** Number 1 (2012), 45–58.

[61] H. Brézis: *Analyse Fonctionelle – Théorie et Applications*, Masson, Paris 1983.

[62] G. Brown: *Continuous functions of bounded n-th variation*, Proc. Edinburgh Math. Soc. **16** (1969), 205–214.

[63] A. M. Bruckner and C. Goffman: Differentiability through change of variables, *Proc. Amer. Math. Soc.* **61** (1976), 235–241.

[64] A. M. Bruckner and J. L. Leonhard: On differentiable functions having an everywhere dense set of intervals of constancy, *Canad. Math. Bull.* **8** (1965), 73–76.

[65] M. Bruneau: Variation Totale d'une Fonction, *Lect. Notes Math.* **413**, Springer, Berlin 1974.

[66] D. Bugajewska: On the superposition operator in the space of functions of bounded variation, revisited, *Math. Comp. Modelling* **52** (2010), 791–796.

[67] D. Bugajewska and D. Bugajewski: On nonlinear integral equations and nonabsolutely convergent integrals, *J. Dynam. Systems Appl.* **14** (2005), 135–148.

[68] D. Bugajewska, D. Bugajewski and H. Hudzik: BV_ϕ-solutions of nonlinear integral equations, *J. Math. Anal. Appl.* **287** (2003), 265–278.

[69] D. Bugajewska, D. Bugajewski and G. Lewicki: On nonlinear integral equations in spaces of functions of bounded generalized φ-variation, *J. Integr. Equ. Appl.* **21** Number 1 (2009), 1–20.
[70] D. Bugajewska and D. O'Regan: On nonlinear integral equations and λ-bounded variation, *Acta Math. Hung.* **107** Number 4 (2005), 295–306.
[71] D. Bugajewski: On BV-solutions of some nonlinear integral equations, *Integr. Equ. Oper. Th.* **46** (2003), 387–398.
[72] D. Bugajewski and D. O'Regan: Existence results for BV-solutions of nonlinear integral equations, *J. Integr. Equ. Appl.* **15** Number 4 (2003), 343–357.
[73] P. S. Bullen: An inequality for variations, *Amer. Math. Monthly* **90** (1983), 560.
[74] V. I. Burenkov: *On the superposition of absolutely continuous functions and on the superposition of functions with bounded variation* (in Russian), Doklady Akad. Nauk SSSR **189** Number 2 (1969), 234–236; Engl. transl.: Soviet Math. Doklady **10** Number 6 (1969), 1379–1381,
[75] V. I. Burenkov: *On integration by parts and a problem on composition of absolutely continuous functions which arises in this connection* (in Russian), Trudy Mat. Inst. Steklova **134** (1975), 38–46.
[76] N. L. Carothers: *Real Analysis*, Cambridge Univ. Press, Cambridge 2000.
[77] P. Cerone, S. S. Dragomir and C. E. M. Pearce: A generalized trapezoid inequality for functions of bounded variation, *Turk. J. Math.* **24** (2000), 147–163.
[78] M. Chaika and D. Waterman: On the invariance of certain classes of functions under composition, *Proc. Amer. Math. Soc.* **43** Number 2 (1974), 345–348.
[79] Z. A. Chanturiya: *The modulus of variation and its application in the theory of Fourier series* (Russian), Doklady Akad. Nauk SSSR **214** (1974), 63–66; Engl. transl.: Soviet Math. Doklady **15** (1974), 67–71.
[80] Z. A. Chanturiya: *Absolute convergence of Fourier series* (Russian), Matem. Zametki **18** (1975), 185–192; Engl. transl.: Math. Notes **18** (1975), 695–700.
[81] Z. A. Chanturiya: *On uniform convergence of Fourier series* (Russian), Matem. Sborn. **100** (1976), 534–554; Engl. transl.: Math. USSR Sborn. **29** (1976), 475–495.
[82] Z. A. Chanturiya: On the absolute convergence of classes $V[n^\alpha]$, *Coll. Math. Soc. Janos Bolyai* **19** (1978). 219–240.
[83] Z. A. Chanturiya: On the absolute convergence of Fourier series of the classes $H^\omega \cap V[\nu]$, *Pacific J. Math.* **103** (1981), 37–61; Errata: ibid. **103** (1982), 611.
[84] Z. A. Chanturiya: *The modulus of variation of a function and continuity* (Russian), Soobshch. Akad. Nauk Gruz. SSR **80** (1975), 281–283.
[85] V. V. Chistyakov: On mappings of bounded variation, *J. Dynam. Control Syst.* **3** Number 2 (1997), 261–289.
[86] V. V. Chistyakov: Mappings of bounded Φ-variation with arbitrary functions Φ, *J. Dynam. Control Syst.* **4** Number 2 (1998), 217–247.
[87] V. V. Chistyakov: Lipschitzian superposition operators between spaces of bounded generalized variation with weight, *J. Appl. Anal.* **6** Number 2 (2000), 173–186.
[88] V. V. Chistyakov: Generalized variation of mappings with applications to composition operators and multifunctions, *Positivity* **5** Number 4 (2001), 323–358.
[89] V. V. Chistyakov: Mappings of generalized variation and composition operators, *J. Math. Sci. (New York)* **110** Number 2 (2002), 2455–2466.
[90] V. V. Chistyakov: Superposition operators in the algebra of functions of two variables with finite total variation, *Monatsh. Math.* **137** Number 2 (2002), 99–114.
[91] V. V. Chistyakov: *Lipschitzian Nemytskii operators in the cones of mappings of bounded Wiener φ-variation*, Folia Math. (Łódź) **11** Number 1 (2004), 15–39.
[92] V. V. Chistyakov: A Banach algebra of functions of several variables of finite total variation and Lipschitzian superposition operators I, *Nonlin. Anal. TMA* **62** Number 3 (2005), 559–578.

[93] V. V. Chistyakov: A Banach algebra of functions of several variables of finite total variation and Lipschitzian superposition operators II, *Nonlin. Anal. TMA* **63** Number 1 (2005), 1–22.

[94] V. V. Chistyakov: *Abstract superposition operators on mappings of bounded variation of two real variables I* (Russian), Sibir. Mat. Zhurn. **46** Number 3 (2005), 698–717; Engl. transl.: Siber. Math. J. **46** Number 3 (2005), 555–571.

[95] V. V. Chistyakov: *Abstract superposition operators on mappings of bounded variation of two real variables II* (Russian), Sibir. Mat. Zhurn. **46** Number 4 (2005), 942–957; Engl. transl.: Siber. Math. J. **46** Number 4 (2005), 751–764.

[96] V. V. Chistyakov and O. E. Galkin: On maps of bounded p-variation with $p > 1$, *Positivity* **2** (1998), 19–45.

[97] V. V. Chistyakov and O. E. Galkin: Mappings of bounded Φ-variation with arbitrary function Φ, *J. Dynam. Control Syst.* **4** Number 2 (1998), 217–247.

[98] V. V. Chistyakov and O. M. Solycheva: Lipschitzian operators of substitution in the algebra λBV, *J. Difference Equ. Appl.* **9, 3/4** (2003), 407–416.

[99] J. Ciemnoczołowski and W. Orlicz: Composing functions of bounded φ-variation, *Proc. Amer. Math. Soc.* **96** Number 3 (1986), 431–436.

[100] J. A. Clarkson and R. Adams: On definitions of bounded variation for functions of two variables, *Trans. Amer. Math. Soc.* **35** (1933), 824–854.

[101] Z. Cybertowicz and W. Matuszewska: Functions of bounded generalized variation, *Ann. Soc. Math. Pol.* **20** (1977), 29–52.

[102] D. S. Cyphert and J. A. Kelingos: The decomposition of functions of bounded κ-variation into differences of κ-decreasing functions, *Studia Math.* **81** (1985), 185–195.

[103] Ch. J. de la Vallée-Poussin: Sur la convergence des formules entre ordenées équidistantes, *Bull. Acad. Sci. Belg.* **8** (1908), 314–410.

[104] P. L. Dirichlet: Sur la convergence des séries trigonométriques que servent á représenter une fonction arbitraire entre des limites donnés, *J. Reine Angew. Math.* **4** (1829), 157–159.

[105] S. S. Dragomir: On the trapezoid quadrature formula and applications, *Kragujevac J. Math.* **23** (2001), 25–36.

[106] P. DuBois-Reymond: *Zur Geschichte der trigonometrischen Reihen: Eine Entgegnung*, H. Laupp, Tübingen 1880.

[107] R. M. Dudley and R. Norvaiša: *Concrete Functional Calculus*, Springer, Berlin 2011.

[108] N. Dunford and J. T. Schwartz: *Linear Operators I*, Interscience, New York 1958.

[109] T. Ereú, J. Giménez and N. Merentes: On bi-dimensional second variation, *Ann. Soc. Math. Pol. Comment. Math.* **52** Number 1 (2012), 39–59.

[110] T. Ereú, N. Merentes, B. Rzepka and J. L. Sánchez: On composition operator in the algebra of functions of two variables with bounded total Φ-variation in Schramm sense, *J. Math. Appl. (Rzeszów)* **33** (2010), 35–50.

[111] T. Ereú, N. Merentes and J. L. Sánchez: Some remarks on the algebra of functions of two variables with bounded total Φ-variation in Schramm's sense, *Ann. Soc. Math. Pol. Comment. Math.* **50** Number 1 (2010), 23–33.

[112] T. Ereú, N. Merentes, J. L. Sánchez and M. Wróbel: Uniformly continuous composition operators in the space of bounded Φ-variation functions in the Schramm sense, *Opusc. Math.* **32** Number 2 (2012), 239–247.

[113] H. Federer: *Geometric Measure Theory*, Springer, Berlin 1969.

[114] G. M. Fichtengol'ts: *On absolutely continuous functions* (in Russian), Matem. Sbornik **31** (1924), 286–296.

[115] J. Fourier: *Théorie analytique de la chaleur*, F. Didot, Paris 1881; Engl. transl.: Dover Publications, New York 1995.

[116] A. M. Garsia and S. Sawyer: On some classes of continuous functions with convergent Fourier series, *J. Math. Mech.* **13** (1964), 589–601.
[117] F. W. Gehring: A study of α-variation, *Trans. Amer. Math. Soc.* **76** (1954), 420–443.
[118] B. R. Gelbaum and J. M. H. Olmstedt: *Counterexamples in Analysis*, Holden Day, San Francisco 1964.
[119] J. Giménez, L. López and N. Merentes: *The Nemitskij operator on $Lip^{(k)}$-type and $BV^{(k)}$-type spaces*, Demonstr. Math. **46** Number 3 (2013), 28–37.
[120] J. Giménez, L. López, N. Merentes and J. L. Sánchez: On bounded second variation, *Adv. Pure Math.* **2** Number 1 (2012), 22–26.
[121] J. Giménez, N. Merentes and S. Rivas: Integral representation of functions of bounded second Φ-variation in the sense of Schramm, *Opusc. Math.* **32** Number 1 (2012), 137–151.
[122] E. Giusti: *Minimal Surfaces and Functions of Bounded Variation*, Birkhäuser, Basel 1984.
[123] D. Głazowska, J. A. Guerrero, J. Matkowski and N. Merentes: Uniformly bounded composition operators on a Banach space of functions of bounded Wiener-Young variation functions, *Bull. Korean Math. Soc.* **50** Number 2 (2013), 675–683.
[124] D. Głazowska, J. Matkowski, N. Merentes and J. L. Sánchez: Uniformly bounded composition operators in the Banach space of absolutely continuous functions, *Nonlin. Anal. TMA* **75** Number 13 (2012), 4995–5001.
[125] M. Goebel and F. Sachweh: On the autonomous Nemytskij operator in Hölder spaces, *Zeitschr. Anal. Anw.* **18** Number 2 (1999), 205–229.
[126] C. Goffman: Bounded variation and absolute continuity in the theory of surface area, *Real Anal. Exchange* **4** (1979), 9–47.
[127] C. Goffman: A geometric characterization of bounded variation and absolute continuity, *Bull. Inst. Math. Acad. Sinica* **9** Number 3 (1981), 395–398.
[128] C. Goffman and F. C. Liu: Lusin type theorems for functions of bounded variation, *Real Anal. Exchange* **5** (1979), 261–266.
[129] C. Goffman and F. C. Liu: A remark on absolutely continuous functions, *Real Anal. Exchange* **15** Number 1 (1989), 384–385.
[130] C. Goffman, T. Nishiura and D. Waterman: *Homeomorphisms in Analysis*, Amer. Math. Soc., Providence 1997.
[131] C. Goffman and D. Waterman: Functions whose Fourier series converge for every change of variable, *Proc. Amer. Math. Soc.* **19** (1968), 80–86.
[132] C. Goffman and D. Waterman: Some aspects of Fourier series, *Amer. Math. Monthly* **77** (1970), 119–133.
[133] U. Goginava: *On the imbedding of the Waterman class in the class H_p^ω* (Russian), Ukrain. Mat. Zh. **57** Number 11 (2005), 1557–1562; Engl. transl.: Ukrain. Math. J. **57** Number 11 (2005), 1818–1824.
[134] R. Gorenflo and S. Vessella: Abel Integral Equations, *Lect. Notes Math.* **1461**, Springer, Berlin 1991.
[135] J. A. Guerrero, H. Leiva, J. Matkowski and N. Merentes: Uniformly continuous composition operators in the space of bounded φ-variation functions, *Nonlin. Anal. TMA* **72** Number 6 (2010), 3119–3123.
[136] J. A. Guerrero, J. Matkowski and N. Merentes: Uniformly continuous composition operators in the space of functions of two variables with bounded total φ-variation in the sense of Wiener, *Ann. Soc. Math. Pol. Comment. Math.* **50** Number 1 (2010), 41–48.
[137] J. A. Guerrero, J. Matkowski, N. Merentes and J. L. Sánchez: *Uniformly continuous set-valued composition operators in the space of functions of the Wiener bounded p-variation*, J. Math. Appl. (Rzeszów) (to appear).

[138] A. I. Gusejnov and Kh. Sh. Mukhtarov: *Introduction to the Theory of Nonlinear Singular Integral Equations* (Russian), Nauka, Moscow 1980.
[139] W. Hahn: *Theorie der Reellen Funktionen*, Springer, Berlin 1921.
[140] G. H. Hardy: *Divergent Series*, Clarendon Press, Oxford 1949.
[141] G. H. Hardy and J. E. Littlewood: Some properties of fractional integrals, *Math. Zeitschr.* **27** (1928), 565–606.
[142] G. H. Hardy and J. E. Littlewood: A convergence criterion for Fourier series, *Math. Zeitschr.* **28** (1928), 612–634.
[143] T. Hawkins: *Lebesgue's Theory of Integration: Its Origins and Developments*, Chelsea Publ., New York 1975.
[144] E. Helly: Über lineare Funktionaloperationen, *Sitzungsber. Kaiserl. Akad. Wiss.* **121** Number AI (1912), 265–297.
[145] H. Herda: Modular spaces of generalized variation, *Studia Math.* **30** (1968), 21–42.
[146] E. Hewitt and K. Stromberg: *Real and Abstract Analysis*, Springer, Heidelberg 1978.
[147] T. H. Hildebrandt: *Introduction to the Theory of Integration*, Acad. Press, New York 1963.
[148] E. W. Hobson: *Theory of Functions of a Real Variable*, Cambridge Univ. Press, Cambridge 1927.
[149] F. N. Huggins: Some interesting properties of the variation function, *Amer. Math. Monthly* **83** (1976), 538–546.
[150] R. L. Jeffrey: Generalized integrals with respect to bounded variation, *Canad. J. Math.* **10** (1958), 617–628.
[151] Z. Jesús, O. Mejía and N. Merentes: *The composition operator between the Lipschitz space* $Lip[a,b]$ *and the Schramm space* $\Phi BV[a,b]$, submitted.
[152] Z. Jesús, O. Mejía, N. Merentes and S. Rivas: *The composition operator and the space of functions of bounded variation in the sense of Schramm-Korenblum*, J. Funct. Spaces Appl. (to appear).
[153] C. Jordan: *Sur la série de Fourier*, C. R. Acad. Sci. Paris **2** (1881), 228–230.
[154] C. Jordan: *Cours d'Analyse*, Gauthier-Villars, Paris 1893.
[155] M. Josephy: Composing functions of bounded variation, *Proc. Amer. Math. Soc.* **83** Number 2 (1981), 354–356.
[156] R. Kannan and C. K. Krueger: *Advanced Analysis on the Real Line*, Springer, Berlin 1996.
[157] Y. Katznelson and K. Stromberg: Everywhere differentiable, nowhere monotone, functions, *Amer. Math. Monthly* **81** (1974), 349–354.
[158] H. J. Kim: *Some properties of functions of generalized bounded variation*, Kangweon-Kyungki Math. J. **12** Number 2 (2004), 117–125.
[159] H. J. Kim: Absolute continuity of functions of $\phi\Lambda BV$, *J. Appl. Math. Comp.* **22** (2006), 557–562.
[160] S. K. Kim and J. Kim: Functions of $\kappa\Phi$-bounded variation, *Bull. Korean Math. Soc.* **23** Number 2 (1986), 171–175.
[161] S. K. Kim and J. Yoon: Riemann-Stieltjes integral of functions of κ-bounded variation, *Comm. Korean Math. Soc.* **5** Number 2 (1990), 65–73.
[162] J. Knop: On globally Lipschitzian Nemytskii operator in special Banach space of functions, *Fasciculi Math.* **21** (1990), 79–85.
[163] B. Korenblum: An extension of the Nevanlinna theory, *Acta Math.* **135** (1975), 187–219.
[164] B. Korenblum: A generalization of two classical convergence tests for Fourier series, and some new Banach spaces of functions, *Bull. Amer. Math. Soc.* **9** (1983), 215–218.
[165] M. A. Krasnosel'skij: *Continuity conditions for some nonlinear operators* (Russian), Ukrain. Mat. Zhurn. **2** Number 3 (1950), 70–86.
[166] M. A. Krasnosel'skij: *On the continuity of the operator* $Fu(x) = f(x, u(x))$ (Russian), Doklady Akad. Nauk. SSSR **77** Number 2 (1951), 185–188.

[167] M. A. Krasnosel'skij: *Topological Methods in the Theory of Nonlinear Integral Equations* (Russian), Gostekhizdat, Moskva 1956; Engl. transl.: Macmillan, New York 1964.

[168] M. A. Krasnosel'skij: *On quasilinear operator equations* (Russian), Doklady Akad. Nauk. SSSR **214** Number 2 (1974), 761–764; Engl. transl.: Soviet Math. Doklady **15** Number 1 (1974), 237–241.

[169] M. A. Krasnosel'skij and Ya. B. Rutitskij: *Convex Functions and Orlicz Spaces* (Russian), Fizmatgiz., Moscow 1958; Engl. transl.: Noordhoff, Groningen 1961.

[170] M. A. Krasnosel'skij, P. P. Zabrejko, E. I. Pustyl'nik and P. E. Sobolevskij: *Integral Operators in Spaces of Summable Functions* (in Russian), Nauka, Moscow 1966; Engl. transl.: Noordhoff, Leyden 1976

[171] E. Kreyszig: *Introductory Functional Analysis with Applications*, Wiley & Sons, New York 1978.

[172] A. Kufner, O. John and S. Fučik: *Function Spaces*, Noordhoff, Leyden 1977.

[173] R. G. Kuller: *Topics in Modern Analysis*, Prentice-Hall, Englewood Cliffs 1969.

[174] W. Küstermann: Funktionen von beschränkter Schwankung in zwei reellen Veränderlichen, *Math. Ann.* **77** (1916), 474–481.

[175] W. Küstermann: Functions of bounded variation, *Bull. Amer. Math. Soc.* **22** (1916), 288.

[176] H. Leiva, N. Merentes, S. Rivas and J. L. Sánchez: *On functions of bounded (φ, k)-variation*, submitted.

[177] I. Lakatos: *Proofs and Refutations*, Cambridge Univ. Press, Cambridge 1976.

[178] A. S. Leonov: *On the total variation for functions of several variables and a multidimensional analog of Helly's selection principle* (Russian), Mat. Zametki **63** (1998), 69–80; Engl. Transl.: Math. Notes **63** (1998), 61–71.

[179] D. Lesnic: *Characterizations of the functions with bounded variation*, Proc. Int. Conf. Theory Appl. Math. Inform. Alba Iulia (2003), 47–54.

[180] R. Leśniewicz and W. Orlicz: *On generalized variations* (II), Studia Math. **45** (1973), 71–109.

[181] E. L. Lima: *Curso de Análise V.1*, Instituto de Matemática Pura e Aplicada. CNPq (Projeto Euclides), Brasil, 1976.

[182] S. Łojasiewicz: *An Introduction to the Theory of Real Functions*, J. Wiley & Sons, Chichester 1988.

[183] E. R. Love: A generalization of absolute continuity, *J. London Math. Soc.* **26** (1951), 1-13.

[184] E. R. Love and L. C. Young: Sur une class des fonctionelles linéaires, *Fund. Math.* **28** (1937), 243–257.

[185] E. R. Love and L. C. Young: On fractional integration by parts, *Proc. London Math. Soc.* **44** (1938), 1–35.

[186] S. M. Lozinskij: *On the Banach indicatrix* (Russian), Vestnik Leningr. Gos. Univ. **7** (1958), 70–87.

[187] M. Lupa: *Form of Lipschitzian operator of substitution in some class of functions*, Zeszyty Nauk. Politech. Łódz **21** (1989), 87–96.

[188] N. N. Luzin: *Theory of Functions of a Real Variable* (Russian), Uchpedgiz, Moskva 1940.

[189] A. G. Lyamin: *On the acting problem for the Nemytskij operator in the space of functions of bounded variation* (Russian), in: 11th School Theory Oper. Function Spaces, Chel'yabinsk (1986), 63–64.

[190] P. Maćkowiak: *A counterexample to Lyamin's theorem*, Proc. Amer. Math. Soc. (to appear).

[191] L. Maligranda, V. V. Mykhaylyuk and A. Plichko: On a problem of Eidelheit from the Scottish Book concerning absolutely continuous functions, *J. Math. Anal. Appl.* **375** (2011), 401–411.

[192] L. Maligranda and W. Orlicz: On some properties of functions of generalized variation, *Monatsh. Math.* **104** (1987), 53–65.

[193] A. Matkowska: *On characterization of Lipschitzian operator of substitution in the class of Hölder's functions*, Zeszyty Nauk. Politech. Łódz. Mat. **17** (1984), 81–85.

[194] A. Matkowska, J. Matkowski and N. Merentes: Remark on globally Lipschitzian composition operators, *Demonstratio Math.* **28** Number 1 (1995), 171–175.

[195] J. Matkowski: Functional equations and Nemytskij operators, *Funkc. Ekvacioj Ser. Int.* **25** (1982), 127–132.

[196] J. Matkowski: *Form of Lipschitz operators of substitution in Banach spaces of differentiable functions*, Zeszyty Nauk. Politech. Łódz. Mat. **17** (1984), 5–10.

[197] J. Matkowski: On Nemytskii Lipschitzian operator, *Acta Univ. Carolinae* **28** Number 2 (1987), 79–82.

[198] J. Matkowski: On Nemytskij operators, *Math. Japonica* **33** Number 1 (1988), 81–86.

[199] J. Matkowski: Lipschitzian composition operators in some function spaces, *Nonlin. Anal. TMA* **30** Number 2 (1997), 719–726.

[200] J. Matkowski: Uniformly continuous composition operators in the spaces of differentiable functions and absolutely continuous functions, *Intern. Ser. Numer. Math.* **157** (2008), 155–166.

[201] J. Matkowski: Uniformly continuous superposition operators in spaces of Hölder functions, *J. Math. Anal. Appl.* **359** (2009), 56–61.

[202] J. Matkowski: Uniformly continuous superposition operators in space of bounded variation functions, *Math. Nachr.* **283** (2010), 1060–1064.

[203] J. Matkowski: Uniformly bounded composition operators between general Lipschitz function normed spaces, *Topol. Methods Nonlin. Anal.* **38** (2011), 395–405.

[204] J. Matkowski and N. Merentes: Characterization of globally Lipschitzian composition operators in the Banach space $BV_p^2[a,b]$, *Archivum Math. (Brno)* **28** (1992), 181–186.

[205] J. Matkowski and N. Merentes: *Characterization of globally Lipschitzian composition operators in the Sobolev space $W_p^n[a,b]$*, Zeszyty Nauk. Politech. Łódz. Mat. **24** (1993), 91–99.

[206] J. Matkowski and N. Merentes: *Remarks on the characterization of globally Lipschitzian composition operators in C^n space*, unpublished manuscript.

[207] J. Matkowski and J. Miś: On a characterization of Lipschitzian operators of substitution in the space $BV\langle a,b\rangle$, *Math. Nachr.* **117** (1984), 155–159.

[208] J. Matkowski and M. Wróbel: Uniformly bounded set-valued Nemytskij operators acting between generalized Hölder function spaces, *Central Europ. J. Math.* **10** (2010), 609–618.

[209] W. Matuszewska and W. Orlicz: On some properties of functions of bounded φ-variation in sense of Riesz, *Ann. Soc. Math. Polon.* **32** (1992), 91–102.

[210] E. J. McShane: Extension of range of functions, *Bull. Amer. Math. Soc.* **40** (1934), 837–842.

[211] Yu. T. Medvedev: *A generalization of a certain theorem of Riesz* (Russian), Uspekhi Mat. Nauk. **6** (1953), 115–118.

[212] M. V. Medvedeva: *On embeddings of the classes H^ω* (Russian), Mat. Zametki **64** Number 5 (1998), 713–719; Engl. transl.: Math. Notes **64** Number 5 (1998), 616–621.

[213] M. V. Medvedeva: *Embeddings of classes of continuous functions into classes of functions of generalized bounded variation* (Russian), Matem. Sbornik **193** Number 7 (2002), 109–130; Engl. transl.: Math. Sbornik **193** Number 7 (2002), 1049–1070.

[214] M. V. Medvedeva: *On embeddings of the function classes H^ω into the function class of generalized bounded variation* (Russian), Anal. Math. **29** Number 1 (2003), 29–57.

[215] N. Merentes: A new characterization of the Riesz class A_p, *Ann. Univ. Sci. Budapest* **32** (1989), 91–95.

[216] N. Merentes: *Composition of functions of bounded $\vec{p}$-variation*, Pure Math. Appl. Ser. B **1** (1990), 39–45.

[217] N. Merentes: On a characterization of Lipschitzian operators of substitution in the space of bounded Riesz φ-variation, *Ann. Univ. Sci. Budapest* **34** (1991), 139–144.

[218] N. Merentes: Functions of bounded $(\varphi, 2)$-variation, *Ann. Univ. Sci. Budapest* **34** (1991), 145–154.

[219] N. Merentes: On the composition operator in $AC[a, b]$, *Collect. Math.* **42** Number 3 (1991), 237–243.

[220] N. Merentes: On functions of bounded $(p, 2)$-variation, *Collect. Math.* **43** Number 2 (1992), 117–123.

[221] N. Merentes: On the composition operator in $RV_\varphi[a, b]$, *Collect. Math.* **46** Number 3 (1995), 231–238.

[222] N. Merentes and K. Nikodem: On Nemytskij operator and set-valued functions of bounded p-variation, *Radovi Math.* **8** Number 1 (1992), 139–145.

[223] N. Merentes, K. Nikodem and S. Rivas: Continuity of the superposition of set-valued functions, *J. Appl. Anal.* **3** Number 1 (1997), 43–48.

[224] N. Merentes and S. Rivas: *On characterization of the Lipschitzian composition operator between spaces of functions of bounded p-variation*, Czechosl. Math. J. **45** Number 4 (1995), 627–637.

[225] N. Merentes and S. Rivas: On Nemytskii operator in the space of set-valued functions of bounded p-variation in the sense of Riesz, *Publ. Math. Debrecen* **47** Number 1-2 (1995), 15–27.

[226] N. Merentes and S. Rivas: *El operador de composición en espacios de funciones con algún tipo de variación acotada*, Novena Escuela Venez. Mat., Mérida (Venezuela) 1996.

[227] N. Merentes and S. Rivas: On the composition operators between $RV_p[a, b]$ and $BV[a, b]$, *Sci. Math.* **1** Number 3 (1998), 287–292.

[228] N. Merentes and S. Rivas: *A characterization of the functions of (φ, k)-variation*, submitted.

[229] N. Merentes, S. Rivas and J. L. Sánchez: *On functions of bounded (p, k)-variation*, J. Function Spaces Appl. (to appear).

[230] N. Merentes, S. Rivas and J. L. Sánchez: Locally Lipschitz composition operators in the space $\Phi BV[a, b]$, *Nonlin. Anal. TMA* **75** Number 4 (2012), 1751–1757.

[231] N. Merentes and J. L. Sánchez: Characterization of globally Lipschitz Nemytskii operators between spaces of set-valued functions of bounded φ-variation in the sense of Riesz, *Bull. Pol. Acad. Sci.* **52** Number 4 (2004), 417–430.

[232] N. Merentes and J. L. Sánchez: BV_p-functions and change of variable, *Real Anal. Exchange* **37** Number 1 (2012), 177–188.

[233] J. Miś: Lipschitzian operators of substitution in the space of vector functions of bounded variation, *Ann. Polon. Math.* **50** (1989), 71–80.

[234] Kh. Sh. Mukhtarov: *On the properties of the operator $Fu = f(u(x))$ in the space H_φ* (Russian), Sbornik Nauchn. Rabot Mat. Kaf. Dagestan Univ. (1967), 145–150.

[235] J. Musielak and W. Orlicz: *On generalized variations* (I), Studia Math. **18** (1959), 11–41.

[236] J. Musielak and W. Orlicz: On modular spaces, *Studia Math.* **18** (1959), 49–65.

[237] A. D. Myshkis: *Linear Differential Equations with Retarded Arguments* (Russian), Nauka, Moscow 1972.

[238] I. P. Natanson: *Theory of Function of a Real Variable* (Russian), Fizmatgiz, Moskva-Leningrad 1950; Engl. transl.: Ungar Publ., New York 1967.

[239] K. I. Oskolkov: *Generalized variation, the Banach indicatrix, and the uniform convergence of Fourier series* (Russian), Matem. Zametki **12** (1972), 313–324; Engl. transl.: Math. Notes **12** (1972), 619–625.

[240] J. Pál: *Sur des transformations de fonctions qui font converger leurs séries de Fourier*, C. R. Acad. Sci. Paris **158** (1914), 101–103.

[241] J. Park: Note on the generalized bounded variations, *J. Appl. Math. Inform.* **3** (1996), 285–294.

[242] J. Park: Riemann-Stieltjes integration of functions of bounded $\kappa\phi$-variations, *J. Appl. Math. Inform.* **4** (1997), 545–553.

[243] J. Park: On the functions of bounded $\kappa\phi$-variations, *J. Appl. Math. Inform.* **28** (2010), 487–498.

[244] J. Park and S. H. Cho: Functions of $\kappa G\phi$-bounded variations, *J. Appl. Math. Comp.* **13** (2003), 447–455.

[245] J. Park and S. K. Kim: Remark on the functions of $\kappa G\phi$-bounded variations, *Proc. Coll. Natur. Sci. Seoul* **13** Number 2 (1988), 1–6.

[246] S. Perlman: Functions of generalized variation, *Fund. Math.* **105** (1980), 199–211.

[247] S. Perlman and D. Waterman: Some remark on functions of Λ-bounded variation, *Proc. Amer. Math. Soc.* **74** Number 1 (1979), 113–118.

[248] W. F. Pfeffer: The generalized Riemann-Stieltjes integral, *Real Anal. Exchange* **21** Number 2 (1996), 521–547.

[249] L. Pick, A. Kufner, O. John and S. Fučik: *Function Spaces, Vol. 1*, De Gruyter, Berlin 2012.

[250] P. B. Pierce and D. J. Velleman: Some generalizations of the notion of bounded variation, *Amer. Math. Monthly* **113** (2006), 897–904.

[251] P. B. Pierce and D. Waterman: On the invariance of classes ΦBV and ΛBV under composition, *Proc. Amer. Math. Soc.* **132** Number 3 (2003), 755–760.

[252] M. T. Popoviciu: Sur quelques propriétés des fonctions d'une variable réelle convexes d'ordre superieur, *Mathematica (Cluj)* **8** (1934), 1–85.

[253] M. T. Popoviciu: *Les Fonctions Convexes*, Hermann, Paris 1944.

[254] F. Prus-Wiśniowski: On superposition of functions of bounded φ-variation, *Proc. Amer. Math. Soc.* **107** (1989), 361–366.

[255] F. Prus-Wiśniowski: Some remarks on functions of bounded φ-variation, *Ann. Soc. Math. Pol. Comment. Math.* **30** (1990), 147–166.

[256] F. Prus-Wiśniowski: Continuous functions of bounded φ-variation, *Ann. Soc. Math. Pol. Comment. Math.* **31** (1991), 127–146.

[257] F. Prus-Wiśniowski: λ-variation and Hausdorff distance, *Math. Nachr.* **158** (1992), 283–297.

[258] F. Prus-Wiśniowski: A continuous function of bounded φ-variation that is not φ-absolutely continuous, *Ann. Soc. Math. Pol. Comment. Math.* **34** (1994), 165–172.

[259] F. Prus-Wiśniowski: A decomposition of Wiener ϕ-variation, *Ann. Soc. Math. Pol. Comment. Math.* **43** (2003), 113–128.

[260] F. Prus-Wiśniowski: Λ-absolute continuity, *Rocky Mt. J. Math.* **39** Number 5 (2009), 1613–1656.

[261] F. Prus-Wiśniowski and G. Szkibiel: Wiener ϕ-variation, *Real Anal. Exch.* **25** Number 1 (2000), 171–173.

[262] F. Prus-Wiśniowski and D. Waterman: Smoothing Λ sequences, *Real Anal. Exch.* **20** Number 2 (1995), 647–650.

[263] M. M. Rao and Z. D. Ren: *Theory of Orlicz spaces*, M. Dekker, New York 1991.

[264] F. Riesz: Untersuchungen über Systeme integrierbarer Funktionen, *Math. Annalen* **69** (1910), 449–497.

[265] F. Riesz: Sur certains systèmes singuliers d'équations intégrales, *Ann. Sci. Ecole Norm. Sup. Paris* **28** (1911), 33-68.

[266] F. Riesz and B. Szőkefalvi-Nagy: *Functional Analysis*, Ungar Publ., New York 1995.

[267] A. W. Roberts and D. E. Varberg: *Convex Functions*, Academic Press, New York 1973.

[268] A. C. M. van Rooij and W. H. Schikhof: *A Second Course on Real Functions*, Cambridge Univ. Press, Cambridge 1982.

[269] W. Rudin: *Principles of Mathematical Analysis*, McGraw Hill, New York 1964.

[270] W. Rudin: *Functional Analysis*, McGraw Hill, New York 1973.

[271] T. Runst and W. Sickel: *Sobolev Spaces of Fractional Order, Nemytskij Operators, and Nonlinear Partial Differential Equations*, De Gruyter, Berlin 1996.

[272] A. M. Russell: Functions of bounded second variation and Stieltjes-type integrals, *J. London Math. Soc.* **2** Number 2 (1970), 193–208.

[273] A. M. Russell: On functions of bounded k-th variation, *J. London Math. Soc.* **2** Number 3 (1971), 742–746.

[274] A. M. Russell: Functions of bounded k-th variation, *Proc. London Math. Soc.* **26** Number 3 (1973), 547–563.

[275] A. M. Russell: Functions of bounded k-th variation and Stieltjes-type integrals, *Bull. Austral. Math. Soc.* **10** (1974), 475–476.

[276] A. M. Russell: An integral representation for a generalised variation of a function, *Bull. Austral. Math. Soc.* **11** (1974), 225–229.

[277] A. M. Russell: Stieltjes-type integrals, *J. Austral. Math. Soc.* **20** (1975), 431–448.

[278] A. M. Russell: A Banach space of functions of generalized variation, *Bull. Austral. Math. Soc.* **15** (1976), 431–438.

[279] A. M. Russell: Further results on an integral representation of functions of generalised variation, *Bull. Austral. Math. Soc.* **18** (1978), 407–420.

[280] A. M. Russell: A commutative Banach algebra of functions of bounded variation, *Pacific J. Math.* **84** Number 2 (1979), 455–463.

[281] A. M. Russell: Further comments on the variation function, *Amer. Math. Month.* **86** (1979), 480–482.

[282] A. M. Russell: A commutative Banach algebra of functions of bounded variation, *Amer. Math. Month.* **87** (1980), 39–40.

[283] A. M. Russell and C. J. F. Upton: A generalization of a theorem by F. Riesz, *Analysis Math.* **9** (1983), 69–77.

[284] B. N. Sahney and D. Waterman: On the summability of Fourier series, *Rev. Roumaine Math. Pures Appl.* **26** (1981), 237–330.

[285] R. Salem: Essais sur les séries trigonométriques, *Act. Sci. Industr. Paris* **862** (1940), 1–88.

[286] M. Schramm: Functions of Φ-bounded variation and Riemann-Stieltjes integration, *Trans. Amer. Math. Soc.* **287** Number 1 (1985), 49–63.

[287] M. Schramm and D. Waterman: Absolute convergence of Fourier series of functions of ΛBV_p and $\phi\Lambda BV$, *Acta Math. Acad. Sci. Hung.* **40** Number 3-4 (1982), 273–276.

[288] M. Shiba: *On absolute convergence of Fourier series of functions of class* $\Lambda BV^{(p)}$, Sci. Rep. Fukushima Univ. **30** (1980), 7–10.

[289] A. Sieczko: *Characterization of globally Lipschitzian Nemytskii operators in the Banach space* AC_{r-1}, Math. Nachr. **141** (1989), 7–11.

[290] W. Sierpiński: Sur une propriété des fonctions qui n'ont que des discontinuités de première espèce, *Bull. Acad. Sci. Roumaine* **16** (1933), 1–4.

[291] R. Sikorski: *Funkcje Rzeczywiste*, PańStv. Wyd. Naukowe, Warsaw 1958.

[292] A. Smajdor and W. Smajdor: Jensen equation and Nemytskij operator for set-valued functions, *Radovi Mat.* **5** (1989), 311–319.

[293] E. P. Sobolevskij: *The superposition operator in Hölder spaces* (Russian), VINITI 3765-84, Voronezh 1984.

[294] E. P. Sobolevskij: *On a superposition operator* (Russian), VINITI 8803-V, Voronezh 1985.

[295] E. P. Sobolevskij: *The superposition operator in Hölder spaces* (Russian), VINITI 8802-V, Voronezh 1985.

[296] Y. U. Sok and J. K. Park: A study on the functions of $\kappa\phi$-bounded variations, *J. Chungcheong Math. Soc.* **2** (1989), 55–64.

[297] J. Šremr: A note on absolutely continuous functions of two variables in sense of Carathéodory, *Inst. Math. Acad. Sci. Czech Republic* **144** (2008), 1–12.

[298] F. Stechkin: *On the approximation of continuous functions by Fourier series* (Russian), Uspekhi Mat. Nauk **7** Number 4 (1952), 139–141.
[299] F. Szigeti: On Niemitzki operators in Sobol'ev spaces, *Zeitschr. Angew. Math. Mech.* **63** Number 5 (1983), T 332.
[300] F. Szigeti: Multivariable composition of Sobol'ev functions, *Acta Sci Mat.* **48** (1985), 469–476.
[301] B. Szőkefalvi-Nagy: *Introduction to Real Functions and Orthogonal Expansions*, Académiai Kiadó, Budapest 1964.
[302] A. E. Taylor: *Introduction to Functional Analysis*, John Wiley & Sons, New York 1958.
[303] V. I. Tevzadze: *On the convergence of Fourier series at a point* (Russian), Soobshch. Akad. Nauk Gruz. SSR **84** (1976), 289–292.
[304] L. Tonelli: Sulle funzioni a variazione limitata, *Annali Scuola Norm. Sup.* **8** (1938), 1–13.
[305] P. L. Ul'yanov: *Absolute convergence of Fourier-Haar series for the superposition of functions* (Russian), Anal. Math. **4** (1978), 225–236.
[306] P. L. Ul'yanov: *Some results on series in the Haar system* (Russian), Doklady Akad. Nauk SSSR **262** Number 3 (1982), 542–545; Engl. transl.: Soviet Math. Doklady **25** Number 1 (1982), 87–90.
[307] P. L. Ul'yanov: *On an algebra of functions and Fourier coefficients* (Russian), Doklady Akad. Nauk SSSR **269** Number 5 (1983), 1054–1056; Engl. transl.: Soviet Math. Doklady **27** Number 2 (1982), 462–464.
[308] P. L. Ul'yanov: *Composition of functions and Fourier coefficients* (Russian), Trudy Mat. Inst. Steklova **172** (1985), 338–348; Engl. transl.: Proc. Steklov Inst. Math. **3** (1987), 367–378.
[309] D. E. Varberg: On absolutely continuous functions, *Amer. Math. Monthly* **72** (1965), 831–841.
[310] N. Viloria: El operador de Nemytskij en el espacio de funciones regladas, *Divulg. Mat.* **12** (2004), 149–153.
[311] B. Z. Vulikh: *A Brief Course in the Theory of Functions of a Real Variable* (Russian), Nauka, Moskva 1973.
[312] H. Wang: Embedding of Lipschitz classes into classes of functions of Λ-bounded variation, *J. Math. Anal. Appl.* **354** Number 2 (2009), 698–703.
[313] S. Wang: Some properties of the functions of Λ-bounded variation, *Sci. Sinica* **A-25** (1982), 149–160.
[314] D. Waterman: On convergence of Fourier series of functions of generalized bounded variation, *Studia Math.* **44** (1972), 107–117; Errata: ibid. **44** (1972), 651.
[315] D. Waterman: On the summability of Fourier series of functions of Λ-bounded variation, *Studia Math.* **55** (1976), 87–95.
[316] D. Waterman: On Λ-bounded variation, *Studia Math.* **57** (1976), 33–45.
[317] D. Waterman: Fourier series of functions of Λ-bounded variation, *Proc. Amer. Math. Soc.* **74** Number 1 (1979), 119–123.
[318] D. Waterman: Estimating functions by partial sums of their Fourier series, *J. Math. Anal. Appl.* **87** (1982), 51–57.
[319] D. Werner: *Funktionalanalysis*, Springer, Heidelberg 2000.
[320] D. Werner: *Einführung in die höhere Analysis*, Springer, Heidelberg 2006.
[321] N. Wiener: The quadratic variation of a function and its Fourier coefficients, *J. Math. Phys. MIT* **3** (1924), 73–94.
[322] L. C. Young: *Sur une généralisation de la notion de variation de puissance pième bornée au sens de M. Wiener, et sur la convergence des séries de Fourier*, C. R. Acad. Sci. Paris **204** (1937), 470–472.
[323] L. C. Young: General inequalities for Stieltjes integrals and the convergence of Fourier series, *Math. Ann.* **115** (1938), 581–612.

[324] P. P. Zabrejko, A. I. Koshelev, M. A. Krasnosel'skij, S. G. Mikhlin, L. S. Rakovshchik, V. Ya. Stetsenko: *Integral Equations - A Reference Text* (Russian), Nauka, Moskva 1968; Engl. transl.: Noordhoff, Leyden 1975.
[325] Z. Zahorski: *Über die Menge der Punkte in welchen die Ableitung unendlich ist*, Tôhoku Math. J. **48** (1941), 321–330.
[326] Z. Zahorski: Sur la première dérivée, *Trans. Amer. Math. Soc.* **69** (1950), 1–54.
[327] G. Zawadzka: On Lipschitzian operators of substitution in the space of set-valued functions of bounded variation, *Radovi Math.* **6** (1990), 279–293.
[328] T. Zerekidze: Functions of classes V and the Banach indicatrix (Russian), in: *Conf. Young Scient. Postgrad. Students*, Tbilissi (1974), 63–64.
[329] A. Zygmund: *Trigonometric Series*, Cambridge Univ. Press, Cambridge 1959.

List of functions

In many places within the first chapters of this book, we have shown that the inclusion $X \subset Y$ between two function spaces X and Y is strict by constructing a function $f \in Y \backslash X$. For the reader's ease, we have provided some of these functions here and indicate where to find them.

- For $1 \leq p < \infty$, a function $f \in L_p([0,1]) \setminus \left(\cup_{q>p} L_q([0,1])\right)$: Example 0.11.
- For $1 < p \leq \infty$, a function $f \in \left(\cap_{q<p} L_q([0,1])\right) \setminus L_p([0,1])$: Example 0.12.
- For $1 < p \leq \infty$, a function $f \in L_p([1,\infty)) \setminus \left(\cup_{q<p} L_q([1,\infty))\right)$: Example 0.13.
- For $1 \leq p \leq \infty$, a function $f \in L_p([0,\infty)) \setminus \left(\cup_{q\neq p} L_q([0,\infty))\right)$: Example 0.14.
- For $1 \leq p < q \leq \infty$, a function $f \in \left(\cap_{r\in[p,q]} L_r([0,1])\right) \setminus \left(\cup_{s\notin[p,q]} L_s([0,\infty))\right)$: Example 0.15.
- A function $f \in C([a,b]) \setminus (\cup_{\alpha<1} Lip_\alpha([a,b]))$: Example 0.41.
- A function $f \in C([a,b]) \setminus BV([a,b])$: Example 1.8.
- A function $f \in BV([a,b]) \setminus CBV([a,b])$: Example 1.19.
- A function $f \in CBV([a,b]) \setminus \{C([a,b]) \cap BV([a,b])\}$: Example 1.19.
- For $0 < \beta < \alpha \leq 1$, a function $f \in Lip_\beta([a,b]) \setminus Lip_\alpha([a,b])$: Exercise 0.1.
- For $0 < \beta < 1$, a function $f \in Lip_\beta([a,b]) \setminus \left(\cup_{\alpha>\beta} Lip_\alpha([a,b])\right)$: Exercise 0.1.
- For $0 < \alpha < 1$ fixed, a function $f \in Lip_\alpha([a,b]) \setminus BV([a,b])$: Example 1.23.
- A function $f \in (\cap_{\alpha<1} Lip_\alpha([a,b])) \setminus BV([a,b])$: Example 1.24.
- A function $f \in \{C([a,b]) \cap BV([a,b])\} \setminus (\cup_{\alpha<1} Lip_\alpha([a,b]))$: Example 1.25.
- A function $f \in AC([a,b]) \setminus (\cup_{\alpha<1} Lip_\alpha([a,b]))$: Example 3.5.
- For $1 \leq p < q$, a function $f \in WBV_q([a,b]) \setminus WBV_p([a,b])$: Example 1.39.
- For $p \geq 1$, a function $f \in \left(\cap_{q>p} WBV_q([a,b])\right) \setminus WBV_p([a,b])$: Example 1.39.
- For $0 < p < q \leq 1$, a function $f \in \Lambda_q BV([a,b]) \setminus \Lambda_p BV([a,b])$: Example 2.30.
- For $0 < p < 1$, a function $f \in \left(\cap_{q>p} \Lambda_q BV([a,b])\right) \setminus \Lambda_p([a,b])$: Exercise 2.30.
- For $p > 1$ and $p - 1 < pq$, a function $f \in \Lambda_q BV([a,b]) \setminus WBV_p([a,b])$: Proposition 2.32.
- For $p > 1$, a function $f \in \left(\cap_{p-1<pq} \Lambda_q BV([a,b])\right) \setminus WBV_p([a,b])$: Exercise 2.6.
- For $p > 1$ and $p - 1 \geq pq$, a function $f \in WBV_p([a,b]) \setminus \Lambda_q BV([a,b])$: Exercise 2.5.
- For $p > 1$ and $p - 1 \geq pq$, a function $f \in WBV_p([a,b]) \setminus \left(\cup_{p-1\geq pq} \Lambda_q BV([a,b])\right)$: Exercise 2.49.
- A function $f \in HBV([a,b]) \setminus BV([a,b])$: Example 2.30.
- A function $f \in HBV([a,b]) \setminus \left(\cup_{p>1} WBV_p([a,b])\right)$: Exercise 2.7.
- For $1 \leq p < q$, a function $f \in RBV_p([a,b]) \setminus RBV_q([a,b])$: Exercise 2.1.
- For $0 < q < 1$, a function $f \in Lip_{1-q}([a,b]) \setminus \Lambda_q BV([a,b])$: Exercise 2.83.
- For $0 < q < 1$, a function $f \in V_{\nu^q}([a,b]) \setminus WBV_{1/(1-q)}([a,b])$: Example 2.40.
- For $0 < q < r < 1$, a function $f \in \Lambda_r^c BV([a,b]) \setminus V_{\nu^q}([a,b])$: Example 2.41.
- A function $f \in (\cap_{\alpha<1} Lip_\alpha([a,b])) \setminus \left(\cap_{p\geq 1} RBV([a,b])\right)$: Example 2.53.

List of symbols

Norms

Other symbols

Index

De Gruyter Studies in Nonlinear Analysis and Applications

Volume 21
Moshe Marcus, Laurent Véron
Nonlinear Second Order Elliptic Equations Involving Measures, 2013
ISBN 978-3-11-030515-9, e-ISBN 978-3-11-030531-9, Set-ISBN 978-3-11-030532-6

Volume 20
John R. Graef, Johnny Henderson, Abdelghani Ouahab
Impulsive Differential Inclusions: A Fixed Point Approach, 2013
ISBN 978-3-11-029361-6, e-ISBN 978-3-11-029531-3, Set-ISBN 978-3-11-029532-0

Volume 19
Petr Hájek, Michal Johanis
Higher Smoothness in Banach Spaces, 2013
ISBN 978-3-11-025898-1, e-ISBN 978-3-11-025899-8, Set-ISBN 978-3-11-220385-9

Volume 18
Smaïl Djebali, Lech Górniewicz, Abdelghani Ouahab
Solution Sets for Differential Equations and Inclusions, 2012
ISBN 978-3-11-029344-9, e-ISBN 978-3-11-029356-2, Set-ISBN 978-3-11-029357-9

Volume 17
Jürgen Appell, Józef Banaś, Nelson Merentes
Bounded Variation and Around, 2014
ISBN 978-3-11-026507-1, e-ISBN 978-3-11-026511-8, Set-ISBN 978-3-11-026624-5

Volume 16
Martin Väth
Topological Analysis: From the Basics to the Triple Degree for Nonlinear Fredholm Inclusions, 2012
ISBN 978-3-11-027722-7, e-ISBN 978-3-11-027733-3, Set-ISBN 978-3-11-027734-0